Sven Sebastian Uhlmann • Clara Ulrich
Steven J. Kennelly
Editors

The European Landing Obligation

Reducing Discards in Complex, Multi-Species and Multi-Jurisdictional Fisheries

Editors
Sven Sebastian Uhlmann
Flanders Research Institute
for Agriculture, Fisheries and Food
Oostende, Belgium

Clara Ulrich
Technical University of Denmark
Kgs. Lyngby, Denmark

Steven J. Kennelly
IC Independent Consulting
Cronulla, NSW, Australia

https://doi.org/10.1007/978-3-030-03308-8

For all who are interested in sustainable fisheries

Foreword

In 2011, European policymakers initiated a major reform of the Common Fisheries Policy (CFP) to improve the Union's fisheries management system, a challenge that proved necessary given the state of fish stocks in European waters. Balancing environmental and socio-economic objectives in fisheries management was imperative. The main objective of the reform was therefore to ensure the preservation of marine resources while increasing the competitiveness of European fishing fleets. These two aspects are not antagonistic but complementary, because the preservation of fisheries resources constitutes a prerequisite for a successful fishing sector. This principle has been assimilated by fishers, who also understood that an overhaul was necessary to ensure the sustainability of their activities.

There is no need to remind the reader that the task incumbent on the European Parliament and the Council was far-reaching. After months of intense institutional negotiations, the revised CFP was introduced in 2013, articulated as two flagship measures: (i) reach the objective of maximum sustainable yield (MSY) in 2020 and (ii) introduce the Landing Obligation. Whether or not we consider the Landing Obligation as being adapted to the overall objectives of the CFP will not be the subject of my remarks. Whilst it is crucial that the Union's fisheries management system should improve selectivity of catches and so contribute to the sustainability of fisheries, I have personally been a long-time opponent to the obligation to land. But only time will tell whether this measure will prove successful. If the means of action might be controversial and open to debate, the overall sustainability objective is undeniable.

For the past 5 years, we have gradually introduced the measures necessary for the implementation of the CFP. In 2015, I was acting as the rapporteur for the implementation of the Landing Obligation (omnibus regulation) where my objective was to ensure a smooth transition for the fisheries sector while respecting the principle of sustainability and preservation of marine resources. The CFP imposes a step-by-step organisation of the Landing Obligation with full implementation on 1 January 2019. This represents a huge challenge for both the fisheries sector, the legislators and the scientific community, whose support are essential to ensure the implementation of the Landing Obligation.

The scientific community is a crucial ally in this process and plays a central role in helping fishers and the sector to adapt to the Landing Obligation through innovative means. The Landing Obligation has been the subject of a lot of scientific research, regarding both its relevance and feasibility as well as the means to achieve it. The Committee on Fisheries that I chair at the European Parliament was regularly informed of the latest studies available. The main observation about the Landing Obligation remains the same: it will be hard to implement and will require a lot of mobilisation from the sector. The Landing Obligation may lead to clear socio-economic consequences for the fisheries sector. A major challenge is the so-called choke effect in mixed fisheries. This matter has real socio-economic consequences for many fisheries. Therefore, significant efforts must be made in research and development to ease the implementation of the Landing Obligation.

For the Landing Obligation to prove successful, another crucial aspect must be taken into account – an appropriate and effective control regulation. Without proper control and enforcement, the Landing Obligation will fail in its objectives and will not fulfil its expectations. Scientific research and new technologies will also be very helpful in that area.

Once again, I would like to underline the importance of science in designing a sustainable fisheries management system. Reliable scientific data is a key component of the CFP. In that sense, scientists work in a remarkable way to help create a better understanding of maritime resources and ecosystems. The Landing Obligation is an example of the close interaction of the scientific community with the CFP, as shown by this book.

European Parliament, Brussels, Belgium
August 2018

Alain Cadec

Preface

The Common Fisheries Policy of the European Union was reformed in 2013 to improve the conservation of marine biological resources and the viability of the fishing sector and reduce unsustainable fishing practices (European Union 2013). One of the cornerstones of the reform is Article 15 (termed the Landing Obligation, LO), stipulating the obligation to bring to land all catches of quota- or size-regulated species with the overall aim to gradually eliminate discards.

The United Nations Food and Agricultural Organisation's (FAO) Fisheries Glossary (FAO 2018) describes discards as

> the proportion of the total organic material of animal origin in the catch, which is thrown away or dumped at sea, for whatever reason. It does not include plant material and post-harvest waste such as offal

and bycatch as

> the part of a catch of a fishing unit taken incidentally in addition to the target species towards which fishing effort is directed. Some or all of it may be returned to the sea as discards, usually dead or dying.

Bycatch and discards may be dead or alive, depending on the severity of any sustained injury and stress suffered by being caught and discarded (Davis 2002).

Globally, it has been estimated that between 7 and 10 million tonnes of commercial fisheries catches are discarded annually (Kelleher 2005; Zeller et al. 2018). In Europe, the North-East Atlantic and North Sea have been identified as "discard hotspots" with a number of discard-intensive fisheries operating in the area (Guillen et al. 2018). The levels of discards vary across regions, species and fisheries (Uhlmann et al. 2013; Catchpole et al. 2017), and the reasons for discarding usually fall within four categories: (i) fish are too small (minimum size restrictions), (ii) quota restrictions (no right to land), (iii) low market value and/or (iv) fish are damaged. Discarding possibly contributes to European fish stocks being fished at levels above those delivering maximum sustainable yields (MSY), and that total removals are, at least for some fisheries, higher than reported due to unreported landings and discards (IUU; Zeller et al. 2018).

To reduce unwanted catch, European institutions have developed and introduced the Landing Obligation (LO) policy whereby catches of regulated species in European waters, or by Union vessels in international waters, must be brought back to shore and are deducted from applicable quotas, unless exempted.

Paradoxically, under the LO, fishers are asked to potentially increase mortality of unwanted catches by bringing them to shore instead of discarding them at sea (Borges 2015). The reasoning behind this is that bringing in unwanted catches of very low market value will incur additional costs, and this should incentivise fishers to avoid catching them in the first place (European Commission 2012; Condie et al. 2013). But until this happens, there remains a risk that fishing mortality will increase under the LO instead of decreasing.

Due to the LO, for the first time in its history, the Common Fisheries Policy is shifting its focus from landed catches to all catches, including discards. Deducting unwanted catches from quota shares increases variability in fishing opportunities, especially in mixed, multispecies fisheries. Fishing opportunities are traditionally distributed based on constant schemes (one particular scheme known as "relative stability", Sobrino and Sobrido 2017), both across Member States and within Member States (with different allocation schemes between fishers). When discarding is allowed, the potential mismatches between what is caught and what can be legally landed can be dealt with by discarding surplus catches. But when all catches must be landed, "choke" situations may arise, where a quota for a regulated species may become exhausted before another target species' quota, leading to the possibility of early closures of some fisheries. New mechanisms must therefore be established under the LO to reduce unwanted catches without jeopardising the viability of the fishing sector (Salomon et al. 2014). The LO represents a major paradigm shift in the history of EU fisheries management with potentially far-reaching consequences. Considerable efforts have been dedicated in recent years to understand these consequences and address the many questions surrounding the implementation of this new policy. This book aims to assemble these efforts into one place and so provide a comprehensive overview of the Landing Obligation.

The book consists of five major parts. The first part provides global and European perspectives on current discard ban policies in the world and reviews the new knowledge needs which they require. Chapter 1 by Karp et al. profiles various discard ban legislations that have been introduced as a top-down management measure throughout the world to reduce unwanted catches and the waste of natural resources. Together with the full or partial prohibition to discard unwanted catch at sea, country-specific combinations of input control measures such as gear modifications to increase selectivity, effort restrictions on where to fish (e.g. spatial discard avoidance, real-time closures) and improved utilisation of catches from quota trading and catch valorisation are described. Effective catch documentation may also provide the option to fully account for all removals to evaluate the impact on fish stocks and ecosystems. Next, Chap. 2 by Borges and Penas Lado describes the evolution of the European discard policy from 1992 onwards and identifies the internal and external drivers in European institutions that shaped the Landing Obligation with its inclusion of the de minimis, high survival exemptions and inter-stock quota flexibility. Finally, Chap. 3 by Rihan et al. describes the new needs for scientific evidence and documentation triggered by the Landing Obligation. It summarises the process by which exemptions are applied for by Member States, evaluated by the EU Scientific, Technical and Economic Committee for Fisheries (STECF) and enacted by the European Commission. The implications of changes on how unwanted catches are registered and incorporated in the advice process by the International Council for the Exploration of the Sea (ICES) are also detailed.

The second part of this book broadly sketches some consequences that the introduction of the LO may bring to the livelihoods of commercial fishers throughout Europe. A perspective piece sets the scene and fuels the debate with arguments about why fishers throughout Europe show very little acceptance for this policy, foreseeing more bad than good from its implementation. Chapter 4 by Fitzpatrick et al. summarises fishers' arguments based on a range of interviews. Chapter 5 by Villasante et al. collected more specifically concerns by small-scale fishers in the Mediterranean Sea and compared their reasoning for discarding. More concrete and quantitative projections are carried out in Chap. 6 by Hoff et al. using bioeconomic models to foresee undesirable economic effects of the policy for fishing fleets and their communities and evaluate alternative policy scenarios. Finally, the potential impacts of the LO on an ecosystem scale were investigated in Chap. 7 by Depestele et al., reviewing evidence of scavengers' reliance on marine carrion and summarising ecosystem-scale modelling results of food-web interdependencies. This chapter deviates slightly from the other chapters in this book in the sense that it is less directly linked to the LO, but it represents a very comprehensive and novel overview on the resilience of scavengers in the sea and on their potential ability to switch to other food sources if fisheries discards are reduced.

Following on from these potential impacts, the third part explores the cultural, institutional and multi-jurisdictional challenges that need to be overcome to achieve a successful implementation of the LO and minimise undesirable and deleterious impacts. However, as Stockhausen in Chap. 8 warns, through his eyes, the Landing Obligation was weakened after its adoption in 2013 by including exemptions and delaying the control and enforcement of any infringements, risking that discarding may continue at undesirable levels. In the Netherlands, fishers' protests against the Landing Obligation were eminent and, in Chap. 9 by van Hoff et al., partial blame is levelled at a multi-governance structure, where regionalisation left stakeholders in limbo on how to reach common ground about the policy goals. In another case study focusing on demersal cod fisheries in the Baltic Sea, Chap. 10 by Valentinsson et al. demonstrates that the introduction of the Landing Obligation has been unsuccessful and has so far failed to deliver any of the expected benefits. Finally, using an alternative approach, Kraak and Hart in Chap. 11 hypothesise that ingrained behavioural mechanisms may hamper an understanding of, uptake of, and compliance with the objectives of the Landing Obligation.

The fourth part of the book features scientific and technical results from recent research efforts (mainly, but not exclusively, from the EU Horizon 2020 DiscardLess and MINOUW and the Life-iSEAS projects) to suggest options for either avoiding the capture of unwanted catches in the first place or, if unavoidably caught, evaluating how to use and valorise them to reduce any wastage. Chapter 12 by Bellido et al. introduces a Marine Spatial Planning approach to avoid discards by guiding fishers with maps to areas where undersized fish are less likely to occur. Chapter 13 by Reid et al. summarises several studies conducted in different regions of Europe, including, among others, some "challenge trials" in Denmark, Ireland and France, where fishers were asked to reduce their discards by whatever legal means they would prefer. Chapter 14 by O'Neill et al. summarises knowledge about gear selectivity and describes recent innovations for a range of gear types, while Chap. 15 by Marçalo et al. specifically focuses on solutions to mitigate slipping mortality in purse seine fisheries. Chapter 16 by Viðarsson et al. discusses which technology and vessel modifications are possible to handle and stow unwanted catches onboard. Finally, Chap. 17 by Iñarra et al. reviews a great number of uses of unwanted catches other than for human consumption and provides a decision support tool to help prioritise the most suitable ones.

The final part of this book tackles the key elements of control, monitoring and surveillance, without which the LO cannot be enforced and successfully implemented. To facilitate an accurate documentation of fishing activity, effort and catches, Chap. 18 by James et al. compares various sensor-based on-board technologies to monitor fishing operations. Chapter 19 by Nuevo et al. describes how the EU Member States in collaboration with the European Fisheries Control Agency are streamlining their control programmes in the so-called "Last Haul" inspection programme. Finally, Chap. 20 by Jacobsen et al. looks to the future, gauging the potential of genetic techniques in monitoring and reducing bycatches.

We believe that the 20 chapters of this book provide a comprehensive examination of the European Landing Obligation from all relevant perspectives. The book was timely published exactly at the time where the Landing Obligation was expected to be fully implemented. The book should hopefully be of significant interest not only to all stakeholders involved but also to the general public of Europe and other jurisdictions throughout the world that are also searching for ways to deal with bycatch and discard issues.

Oostende, Belgium — Sven Sebastian Uhlmann
Kgs. Lyngby, Denmark — Clara Ulrich
Cronulla, Australia — Steven J. Kennelly

References

Borges, L. (2015). The evolution of a discard policy in Europe. Fish and Fisheries, 16(3), 534–540. https://doi.org/10.1111/faf.12062.

Catchpole, T.L., Ribeiro-Santos, A., Mangi, S.C., Hedley, C., Gray, T.S. (2017). The challenges of the landing obligation in EU fisheries. Marine Policy, 82, 76–86.

Condie, H.M., Grant, A., Catchpole, T.L. (2013). Does banning discards in an otter trawler fishery create incentives for more selective fishing? Fisheries Research, 148(2013), 137–146. https://doi.org/10.1016/j.fishres.2013.09.011.

Davis, M.W. (2002). Key principles for understanding fish bycatch discard mortality. Canadian Journal of Fisheries and Aquatic Sciences, 59, 1834–1843. https://doi.org/10.1139/f02-139.

European Commission (2012). Proposal for a regulation of the European Parliament and of the Council on the Common Fisheries Policy – general approach. 11322/12.

European Union (2013). Regulation (EU) No 1380/201308 of the European Parliament and of the Council of 11 December 2013 on the Common Fisheries Policy, amending Council Regulations (EC) No 1954/2003 and (EC) No 1224/2009 and repealing Council Regulations (EC) No 2371/2002 and (EC) No 639/2004 and Council Decision 2004/585/EC. Official Journal of the European Union, L 354: 22–61. Brussels, Belgium.

FAO (2018). Fisheries Glossary. http://www.fao.org/faoterm/collection/fisheries/en. Accessed 30 August 2018.

Guillen, J., Holmes, J.S., Carvalho, N., Casey, J., Dörner, H., Gibin, M., et al. (2018). A review of the European Union landing obligation focusing on its implications for fisheries and the environment. Sustainability. https://doi.org/10.3390/su10040900.

Gilman, E., Suuronen, P., Hall, M., Kennelly, S. (2013). Causes and methods to estimate cryptic sources of fishing mortality. Journal of Fish Biology, 83, 766–803.

Kelleher, K. (2005). Discards in the world's marine fisheries. An update. FAO Fisheries Technical Paper (Vol. No. 470. R). 22 p.

Sobrino, J.M., & Sobrido, M. (2017). The common fisheries policy: A difficult compromise between relative stability and the discard ban. In G. Andreone (Ed.), The future of the law of the sea: Bridging gaps between national, individual and common interests (pp. 23–43). Cham: Springer. https://doi.org/10.1007/978-3-319-51274-7_2.

Salomon, M., Markus, T., Dross, M. (2014). Masterstroke or paper tiger – The reform of the EU′s Common Fisheries Policy. Marine Policy, 47, 76–84. https://doi.org/10.1016/J.MARPOL.2014.02.001.

Uhlmann, S.S., van Helmond, A.T.M., Stefánsdóttir, E.K., Sigurðardóttir, S., Haralabous, J., et al. (2014). Discarded fish in European waters: general patterns and contrasts. ICES Journal of Marine Science, 71, 1235–1245

Zeller, D., Cashion, T., Palomares, M., Pauly, D. (2018). Global marine fisheries discards: A synthesis of reconstructed data. Fish and Fisheries, 19(1), 30–39. https://doi.org/10.1111/faf.12233.

Acknowledgements

We are grateful for the financial support received from the Horizon 2020 Research Projects DiscardLess (Grant Agreement 633680) and MINOUW (Grant Agreement 634495), which made the publication as an open access book possible. Our thanks are extended to Alexandrine Cheronet and Judith Terpos from Springer for the excellent assistance throughout the manuscript preparation and publishing phase, Miriam Levenson for language editing and Kristian Schreiber Plet-Hansen for checking references.

Sven Sebastian Uhlmann would like to thank his colleagues, his family, Tuia and his lady luck (at the time) for the (moral) support in this undertaking.

Clara Ulrich would like to thank all the participants from the DiscardLess project for the collective support to this book and for the many valuable contributions to it. She is also grateful to her ever-supportive husband and children.

Steven J. Kennelly acknowledges the support of his company, IC Independent Consulting (www.icic.net.au), in funding his role in this book as well as the support of his wife, Peta.

Collectively, we would like to express our sincere gratitude for the invaluable comments received from the following reviewers:

Nick Bailey (Independent Expert, Scotland)
Lee Benaka (National Oceanic and Atmospheric Administration, USA)
Tom Catchpole (Centre for Environment, Fisheries and Aquaculture Science, UK)
Barrie Deas (The National Federation of Fishermen's Organisations, UK)
Steven Degraer (Royal Belgian Institute of Natural Sciences, Belgium)
Elisabeth Druel (ClientEarth, Belgium)
Steve Dunn (Alliance Legal Services, Australia)
Eric Gilman (Pelagic Ecosystems Research Group, USA)
Jordi Guillen (European Commission – Joint Research Centre, Italy)
Mark Hager (Gulf of Maine Research Institute, USA)
Katell Hamon (Wageningen Economic Research, The Netherlands)
Dan Holland (National Oceanic and Atmospheric Administration, USA)

Claudia Junge (Institute of Marine Research, Norway)
Alexander Kempf (Thünen Institut, Germany)
Arne Kinds (Université de Bretagne Occidentale, France)
Leyla Knittweis (University of Malta, Malta)
Josep Lloret (University of Girona, Spain)
Colm Lordan (Marine Institute, Ireland & International Council for the Exploration of the Sea, Denmark)
Tara Marshall (University of Aberdeen, Scotland)
Francesc Maynou (Institut de Ciències del Mar, Spain)
Howard McElderry (Archipelago Marine Research Ltd., Canada)
Christopher McGuire (Nature Conservancy, USA)
Matthew McHugh (Bord Iascaigh Mhara, Ireland)
Michael Morrissey (Oregon State University, USA)
Kåre Nolde Nielsen (The Arctic University of Norway, Norway)
Catherine O'Keefe (Massachusetts Division of Marine Fisheries, USA)
Martin Pastoors (Pelagic Freezer Trawler Association, The Netherlands)
Kathleen J. Pitz (Monterey Bay Aquarium Research Institute, USA)
Mike Pol (Commonwealth of Massachusetts, USA)
Hans Polet (Flanders Research Institute for Agriculture, Fisheries and Food, Belgium)
Graham Raby (University of Windsor, Canada)
Antonello Sala (Italian National Research Council & Institute of Marine Sciences, Italy)
Milena Schreiber (University of Gothenburg, Sweden)
Sevaly Sen (Oceanomics Pty Ltd, Sydney, Australia)
Petri Suuronen (Luke, Finland)
Ian Tuck (National Institute of Water & Atmospheric research, New Zealand)
Sara Vandamme (North Western Waters Advisory Council, Ireland)
Gert Van Hoey (Flanders Research Institute for Agriculture, Fisheries and Food, Belgium)
Ingrid Van Putten (CSIRO & University of Tasmania, Australia)
Mike van 't Land (Flanders Research Institute for Agriculture, Fisheries and Food, Belgium)

Contents

Part I Global and European Perspectives on Discard Policies

1 Strategies Used Throughout the World to Manage Fisheries Discards – Lessons for Implementation of the EU Landing Obligation 3
William A. Karp, Mike Breen, Lisa Borges, Mike Fitzpatrick, Steven J. Kennelly, Jeppe Kolding, Kåre Nolde Nielsen, Jónas R. Viðarsson, Luis Cocas, and Duncan Leadbitter

2 Discards in the Common Fisheries Policy: The Evolution of the Policy 27
Lisa Borges and Ernesto Penas Lado

3 Requirements for Documentation, Data Collection and Scientific Evaluations 49
Dominic Rihan, Sven S. Uhlmann, Clara Ulrich, Mike Breen, and Tom Catchpole

Part II Potential Social, Economic and Ecological Impacts of the Landing Obligation

4 Fishing Industry Perspectives on the EU Landing Obligation 71
Mike Fitzpatrick, Katia Frangoudes, Laurence Fauconnet, and Antoni Quetglas

5 The Implementation of the Landing Obligation in Small-Scale Fisheries of Southern European Union Countries 89
Sebastian Villasante, Manel Antelo, Maria Christou, Laurence Fauconnet, Katia Frangoudes, Francesc Maynou, Telmo Morato, Cristina Pita, Pablo Pita, Konstantinos I. Stergiou, Celia Teixeira, George Tserpes, and Vassiliki Vassilopoulou

6 **Potential Economic Consequences of the Landing Obligation** 109
Ayoe Hoff, Hans Frost, Peder Andersen, Raul Prellezo, Lucía Rueda, George Triantaphyllidis, Ioanna Argyrou, Athanassios Tsikliras, Arina Motova, Sigrid Lehuta, Hazel Curtis, Gonzalo Rodríguez-Rodríguez, Hugo M. Ballesteros, Julio Valeiras, and José María Bellido

7 **The Impact of Fisheries Discards on Scavengers in the Sea** 129
Jochen Depestele, Jordan Feekings, David G. Reid, Robin Cook, Didier Gascuel, Raphael Girardin, Michael Heath, Pierre-Yves Hernvann, Telmo Morato, Ambre Soszynski, and Marie Savina-Rolland

Part III Cultural, Institutional and Multi-Jurisdictional Challenges

8 **How the Implementation of the Landing Obligation Was Weakened** 165
Björn Stockhausen

9 **Muddying the Waters of the Landing Obligation: How Multi-level Governance Structures Can Obscure Policy Implementation** 179
Lucvan Hoof, Marloes Kraan, Noor M. Visser, Emma Avoyan, Jurgen Batsleer, and Brita Trapman

10 **The Baltic Cod Trawl Fishery: The Perfect Fishery for a Successful Implementation of the Landing Obligation?** 197
Daniel Valentinsson, Katja Ringdahl, Marie Storr-Paulsen, and Niels Madsen

11 **Creating a Breeding Ground for Compliance and Honest Reporting Under the Landing Obligation: Insights from Behavioural Science** 219
Sarah B. M. Kraak and Paul J. B. Hart

Part IV Tactical and Technological Options for Reducing Unwanted Catches

12 **A Marine Spatial Planning Approach to Minimize Discards: Challenges and Opportunities of the Landing Obligation in European Waters** 239
José M. Bellido, Iosu Paradinas, Raúl Vilela, Guillermo Bas, and Maria Grazia Pennino

13 The Best Way to Reduce Discards Is by Not Catching Them! 257
David G. Reid, Julia Calderwood, Pedro Afonso, Pierre Bourdaud, Laurence Fauconnet, José Manuel González-Irusta, Lars O. Mortensen, Francesc Ordines, Sigrid Lehuta, Lionel Pawlowski, Kristian S. Plet-Hansen, Zachary Radford, Marianne Robert, Marie-Joelle Rochet, Lucía Rueda, Clara Ulrich, and Youen Vermard

14 Discard Avoidance by Improving Fishing Gear Selectivity: Helping the Fishing Industry Help Itself 279
Finbarr G. O'Neill, Jordan Feekings, Robert J. Fryer, Laurence Fauconnet, and Pedro Afonso

15 Mitigating Slipping-Related Mortality from Purse Seine Fisheries for Small Pelagic Fish: Case Studies from European Atlantic Waters 297
Ana Marçalo, Mike Breen, Maria Tenningen, Iñigo Onandia, Luis Arregi, and Jorge M. S. Gonçalves

16 Onboard and Vessel Layout Modifications 319
Jónas R. Viðarsson, Marvin I. Einarsson, Erling P. Larsen, Julio Valeiras, and Sigurður Örn Ragnarsson

17 What to Do with Unwanted Catches: Valorisation Options and Selection Strategies 333
Bruno Iñarra, Carlos Bald, Marta Cebrián, Luis T. Antelo, Amaya Franco-Uría, José Antonio Vázquez, Ricardo I. Pérez-Martín, and Jaime Zufía

Part V Control, Monitoring and Surveillance

18 Tools and Technologies for the Monitoring, Control and Surveillance of Unwanted Catches 363
Kelly M. James, Neill Campbell, Jónas R. Viðarsson, Carlos Vilas, Kristian S. Plet-Hansen, Lisa Borges, Óscar González, Aloysius T. M. van Helmond, Ricardo I. Pérez-Martín, Luis Taboada Antelo, Jorge Pérez-Bouzada, and Clara Ulrich

19 Monitoring the Implementation of the Landing Obligation: The Last Haul Programme 383
Miguel Nuevo, Cristina Morgado, and Antonello Sala

20 Possible Uses of Genetic Methods in Fisheries Under the EU Landing Obligation 407
Magnus Wulff Jacobsen, Brian Klitgaard Hansen, and Einar Eg Nielsen

Index 429

Part I
Global and European Perspectives on Discard Policies

Chapter 1
Strategies Used Throughout the World to Manage Fisheries Discards – Lessons for Implementation of the EU Landing Obligation

William A. Karp, Mike Breen, Lisa Borges, Mike Fitzpatrick, Steven J. Kennelly, Jeppe Kolding, Kåre Nolde Nielsen, Jónas R. Viðarsson, Luis Cocas, and Duncan Leadbitter

Abstract In many countries, policies regarding reduction of unwanted catch and discards are crafted in response to concerns regarding accountability, conservation, and waste as well as scientific needs to fully account for all sources of fishing mortality. It is important to note, however, that unwanted catch is minimal and most, or all, of the catch has value in some fisheries. Utilisation rates are very high, and discarding is generally not of concern in such fisheries which occur primarily, but not entirely, in developing countries. Where unwanted catch and discards are a

W. A. Karp (✉)
School of Aquatic and Fishery Sciences, University of Washington, Seattle, WA, USA
e-mail: bkarp@uw.edu

M. Breen
Institute of Marine Research (IMR), Bergen, Norway

L. Borges
FishFix, Brussels, Belgium

M. Fitzpatrick
Marine Natural Resource Governance, Cork, Ireland

S. J. Kennelly
ICIC Independent Consulting, Cronulla, Australia

J. Kolding
Department of Biological Sciences, University of Bergen, Bergen, Norway

K. N. Nielsen
The Arctic University of Norway, Tromsø, Norway

J. R. Viðarsson
Matís ohf. Icelandic food and biotech R&D, Reykjavík, Iceland

L. Cocas
Subsecretaría de Pesca y Acuicultura, Gobierno de Chile, Valparaíso, Chile

D. Leadbitter
Fish Matter, Wollongong, Australia

S. S. Uhlmann et al. (eds.), *The European Landing Obligation*,
https://doi.org/10.1007/978-3-030-03308-8_1

concern, legislation may be prescriptive, as can be seen in the EU Landing Obligation (LO), and programmes established in e.g. Norway, Iceland, Argentina, Chile and New Zealand. Elsewhere, legislative language is intended to minimize unwanted catch but allows for some flexibility in developing strategies and solutions, as in the USA. The effectiveness of these approaches depends on many factors and all require effective cross-sectoral collaboration. Also essential is a comprehensive monitoring and control system which insures regulatory compliance and collection of adequate data to address scientific and management information needs. In this chapter, we evaluate the effectiveness of discard and unwanted catch reduction approaches under diverse legislative systems in different parts of the world, with reference to emerging practices under the LO. We consider the importance of finding the balance between top-down and bottom-up processes and look carefully at different governance/regulatory frameworks (e.g. input controls, output controls, quota management and transferability, cooperative/collaborative management), factors which encourage or discourage innovation and collaborative problem solving, monitoring and accountability. This is accomplished through case studies from selected fisheries around the world.

Keywords Avoidance · Discards · Full retention · Selectivity · Unwanted catch · Utilisation

1.1 Introduction

Waste is associated with many contemporary food production processes, especially in developed countries. In general, producers seek to reduce waste by improving utilisation of raw materials and avoiding unwanted materials during harvesting. In wild capture fisheries, many strategies are available for reducing unwanted catch (UWC) during harvesting. However, these strategies are rarely completely effective with the result that parts of the catch are discarded at sea. Although avoidance is preferable, discarding might be perceived as an acceptable practice under certain circumstances, e.g. when discard-related mortality is low and, especially for rare, endangered or protected species caught incidentally, when animals can be released uninjured. However, for UWC where chances of survival may be small, discarding at sea is generally considered wasteful and undesirable.

Conflicting regulatory and economic drivers often create perverse incentives for fishers to discard fish. This can occur when unmarketable or undersized fish are taken, when a vessel operator is not permitted to retain marketable catch or when catch of lower value is discarded so that fishing for more valuable catch can continue (high-grading; FAO 2016). Discarding is minimal or entirely absent in some fisheries where most, if not all, of the catch has a value and is fully utilised. This typically occurs in small-scale/artisanal/traditional fisheries (Kolding et al. 2014; Damalas

2015) but examples can be found elsewhere such as in industrial fisheries which produce fish meal and oil.

The 2013 reform of the Common Fisheries Policy (CFP) included an "obligation to land all catches", referred to as the Landing Obligation (LO). The objective of the LO is to

> "gradually eliminate discards, on a case-by-case basis, taking into account the best available scientific advice, by avoiding and reducing, as far as possible, unwanted catches, and by gradually ensuring that catches are landed". An additional objective is to "make the best use of unwanted catches, without creating a market for such of those catches that are below the minimum conservation reference size" (European Union 2013).

The LO is not a fully comprehensive discard ban as it only applies to TAC (Total Allowable Catch) regulated species. In addition, it applies to species in effort-regulated fisheries in the Mediterranean for which a minimum size has been defined. Exemptions for the LO apply to species and fisheries for which high survival rates can be demonstrated for discarded fish. Furthermore, up to 5% of the total catch of species may be discarded in cases where selectivity increases are difficult to achieve or where handling of unwanted catches creates disproportionate costs. The LO requires that fish under a Minimum Conservation Reference Size (MCRS) are landed but prohibits their use for direct human consumption.

In principle, the LO involves a shift from landing quota to catch quota management as all catches must be recorded and accounted for against quotas (Mortensen et al. 2017). The LO has raised serious concerns in the fishing industry, especially where choke species issues arise in mixed fisheries regulated by TACs (Schrope 2010). Other concerns include the perception that there is insufficient knowledge to allow implementation of the LO, and that the period during which the LO is to be phased in (2015–2019) is insufficient to allow necessary preparation by those that are involved and affected (Fitzpatrick et al. 2017).

In this chapter, we compare worldwide examples of strategies for reducing discards in order to inform discussions on questions such as: "what makes a successful discard mitigation strategy?"; and "what lessons can be learned with regards to the successful implementation of the EU Landing Obligation (LO)?". We do so by briefly reviewing approaches in several countries. The cases were selected to offer insights regarding (a) scale and drivers of discarding issues – where they exist, (b) motivation, objectives and legal status of a discard mitigation policy – if applicable, (c) technical and management approaches employed to mitigate discarding, and (d) perceived outcomes of the discard policy – or lack thereof.

1.2 Case Studies

1.2.1 Norway

Norwegian fisheries are managed through a complex system of regulations which aims to control both input (i.e., fishing licences) and output (i.e. quotas), as well as the exploitation pattern, through a multi-facetted collection of regulations and technical measures referred to as the "Discard Ban Package" (DBP: Johnsen and Eliasen 2011; Gullestad et al. 2015).

The DBP is an integrated suite of regulatory and technical measures to minimise unwanted catch including: the regulatory discard ban; gear selectivity technical measures; closed areas; and monitoring and control measures. The Norwegian Discard Ban evolved over ~30 years in response to specific fisheries problems, starting in 1987 with an *ad hoc* strategy to save the 1983 NE Arctic cod cohort from "high-grading" (Gullestad et al. 2015). Between 1987 and 2008, regulations and technical measures progressively extended this discard ban to include a further 17 commercially important species, to address bycatch issues in several fisheries, including demersal trawling for shrimp (e.g., Isaksen et al. 1992) and gadoid trawling (Larsen and Isaksen 1993). In 2009, the "Marine Resources Act" (section §15: Duty to Land Catches), in principle, extended the DBP to encompass all living marine resources with the phrase: "All catches of fish shall be landed . . .". However, in practice, the Norwegian Seawater Fisheries Regulations (2014) (section §48: Prohibition against discarding fish), which are enforced by the Fisheries Directorate and the Coastguard, effectively limit the ban to 55 commercially important species. In addition, the regulation introduces several "pragmatic exemptions" (Gullestad et al. 2015), e.g., that surviving fish may be released, and that damaged catch (unfit for human consumption) can be discarded, "in small quantities".

Key features of the discard package include:

- Improved Selectivity – development and regulated application of bycatch reduction devices, and other technologies and technical measures to address fishery-specific UWC issues.
- Real-Time Closures – triggered by the UWC limits being exceeded in an area and dimensions are defined by the distribution of the UWC. They are perceived by most stakeholders as a fair and effective way of addressing unpredictable distributions of UWC, although there has been discontent amongst fishers due to extended delays in re-opening some areas, primarily due to a lack of resources to survey them.
- Monitoring – using a "Reference Fleet" and targeted programmes to identify emerging UWC issues in different fisheries. However, there is no systematic at-sea monitoring by observers.
- Decriminalisation – of catching UWC for fishers who demonstrate responsible behaviour and conduct in avoiding further UWC.
- Pragmatic application of regulations – by the authorities to engage the stakeholders and ensure a "level playing field".

- Improved dialogue – between management, fishers and scientists – to ensure the individual concerns of each have been recognised and that all stakeholders are working towards a common goal.

The greatest beneficial impact from the Norwegian Discard Ban is generally considered to have been its role in catalysing a cultural shift among all stakeholders, towards the recognition that needless waste of a living resource is no longer acceptable (Gullestad et al. 2014). The burden of proof on fishers to demonstrate responsible behaviour and conduct is arguably resulting in increased professional responsibility (Johnsen and Eliasen 2011). This change in mindset has driven an ambition to minimise UWC and contributed to the development of a "common code of conduct" in relation to resources and compliance (Gezelius 2006). Recently, this has stimulated the development of strategies and technologies to avoid issues related to excessive catches in some fisheries, due to localised high densities of fish, e.g. high-grading of cod in the Barents Sea (Grimaldo et al. 2014; Underwood et al. 2014) and slipping of mackerel from purse seines (Breen et al. 2012). It has also promoted greater utilisation of the retained catch (Richardsen et al. 2017).

1.2.2 Iceland

A ban on discarding six primary commercial species was introduced in the Icelandic fishery in 1977 (European Commission 2007). Requirements evolved as management strategies progressed from effort- to quota restrictions (Johnsen and Eliasen 2011). The ban has gradually expanded, and now applies to all species, including those with no market value. Catches that marginally exceed quotas can be landed legally since the law allows 5% of the quota to be transferred between years. Fishers can also land up to 5% of catches without deduction from quota but must then forfeit most of the value of the surplus catch (Fiskistofa 2018a). Fishers are also allowed to land catches under minimum size limits and only 50% of the weight of this portion of the catch is deducted from their quotas. This creates an incentive to land undersized catches (Fiskistofa 2018b). Larger overruns and non-target catch can be covered through the purchase or leasing of additional quota. Failure to cover excess catch with allowed overages or purchased quota can result in fines and/or revocation of licenses (Fiskistofa 2018c).

Capture of juvenile fish is discouraged through real-time area closures if catches below minimum sizes exceed prescribed limits (Johnsen and Eliasen 2011). Catches are monitored by on-board inspectors from the Directorate of Fisheries, dockside surveillance and use of electronic logbooks. The coastguard also inspects vessels to verify catch reporting. Coastguard surveillance and on-board inspector coverage levels are quite limited; inspectors from the Directorate of Fisheries are at sea for around 1200 days a year in total and the coastguard only patrols Icelandic waters for around 300 days a year using four patrol vessels, one fixed-wing aircraft and two helicopters (Fiskistofa 2018d; Landhelgisgæslan 2018).

Catch records and size composition information are matched and compared to identify possible discarding (Fiskistofa 2017). Despite a mandatory landings policy, discarding still occurs, but the level has gradually declined since the early 1990s. According to estimates from the Icelandic Marine Research Institute, haddock discard rates have, for example, fallen from 22% in 1997 to 0.12% in 2013, and cod discard rates have not exceeded 2% since 2001 and were estimated at 0.60% in 2013 (Pálsson et al. 2015; Pálsson 2003). These estimates are based on sampling by the Directorate of Fisheries and the Marine Research Institute. Some stakeholders question the validity of these estimates and believe the discard numbers are substantially underestimated while others consider them to be reliable. Limited coverage by onboard inspectors and patrol vessels is the main reason for distrust of the official discard estimates. Therefore, the Ministry of Fisheries is currently preparing a regulation which will require all commercial fishing vessels to be equipped with Electronic Monitoring (EM) equipment including video cameras to remotely and electronically monitor potential discarding (Ministry of Fisheries 2018). The regulation will also require all official dockside weighing of catches to be electronically monitored. If the regulation is passed in its current form, dockside monitoring will come into effect immediately and onboard monitoring will be in full effect by January 1, 2020.

Improved selectivity is important in reducing discards in Iceland. Selectivity increases are primarily due to advances in gear technology, regulations on gear selectivity devices, and widespread use of voluntary move-on solutions based on real-time information shared between fishers (Margeirsson et al. 2008).

The extensive consolidation that has occurred in the Icelandic fishing fleet over the past 20–30 years has contributed markedly to the overall success in reducing discards. This consolidation is one of the side-effects of the individual transferable quota (ITQ) system, as smaller and less profitable businesses merge with (or are acquired by) larger entities that benefit from economies of scale and better access to capital. Small companies operating with limited quotas have almost disappeared and large vertically integrated seafood companies that include catching, processing and marketing are now in possession of the majority of the quota (Íslandsbanki 2017). Consequently, the numbers of vessels and fishermen have been reduced significantly, such that capacity and quota allocations are now better matched, and vessels generally have enough quota to operate at full capacity throughout the year. This has eliminated many of the incentives to discard. Nevertheless, allegations of illegal discarding on larger vessels persist, despite allocation of adequate quota (RÚV 2018).

Key features to the success of the discard ban include:

- Improved selectivity.
- Real-time closures.
- Monitoring, Control and Surveillance (MCS) mechanisms that include discard monitoring (although of limited in scope).
- Regulatory incentives to land undersized catch with partial or zero deduction from quota.

- Voluntary move-on measures are used to collect and share information on where and when to avoid unwanted catch.

The discard ban is now accepted by all stakeholders and by the public such that it is generally considered unacceptable to discard catch in Icelandic fisheries. The focal point of the "discard discussion" in Iceland has shifted towards the utilisation of by-products, as full utilisation of the entire catch has been encouraged with a high degree of success (Vigfusson et al. 2013).

1.2.3 USA

Science-based management has been the hallmark of US national policy since enactment of the Magnuson-Stevens Fishery Conservation and Management Act (MSA) in 1976. Under the current version of this act

> Conservation and management measures shall, to the extent practicable, (a) minimize bycatch and (b) to the extent bycatch cannot be avoided, minimize the mortality of such bycatch"; bycatch is defined as fish which are harvested in a fishery, but which are not sold or kept for personal use [...],

and includes economic and regulatory discards. Regionalization underpins MSA and the eight Regional Fishery Management Councils are allowed flexibility in developing Fishery Management Plans (FMPs) that comply with the MSA. Therefore, many approaches to reducing bycatch can be found in US fisheries regulations, but discarding is not banned in any US fishery.

Fisheries for walleye pollock (*Gadus chalcogrammus*), the nation's largest single-species fishery, take place throughout Alaska but primarily in the Eastern Bering Sea and Aleutian Islands (BSAI). Since the late 1970s, catches in this region have ranged between 0.8 and 1.5 million t (Ianelli and Stram 2015). The regulations have been amended many times to reduce bycatch and discard in the pollock fishery (NPFMC 2016). The American Fisheries Act was passed in 1998 and mandated significant changes in management of the fishery. Almost the entire quota (which is reviewed annually) was permanently divided among three sectors: inshore catcher vessels delivering to shore-side plants, catcher/processors, and motherships (catcher vessels delivering to floating processors). This eliminated competition among and within sectors. Provisions for binding agreements within or among co-operatives were also established.

Bycatch challenges impacting the pollock fleet include incidental catches of Pacific salmon and other ground-fish, and undersized pollock. Through a combination of regulatory action, internal contractual arrangements, and cross-sectoral collaboration, innovative changes in fishing practice have been developed and implemented to address these challenges.

Regulations were amended in 1998 to require all ground-fish vessels to retain all pollock and Pacific cod (*Gadus macrocephalus*) and achieve defined utilisation standards (NPFMC 2016). Between 1997 and 1998, overall cod discard was reduced from 8.6% to 2.2% of cod harvest and overall pollock discard was reduced from 8.2% to 1.6% of pollock harvest (Witherell et al. 2000). Subsequently discards have been further reduced. Pollock and cod constitute a very high proportion of the overall ground-fish harvest in this region (85.7% in 2016, 80.1% in 2017; NPFMC 2017), so overall discard rates were greatly reduced.

Bycatch of Pacific salmon cannot be retained in the walleye pollock fishery and the fishery can be penalized if salmon bycatch exceeds proscribed limits. Abundance of limiting salmon species is highly variable and influenced by multiple factors including spatial and temporal co-occurrence, and variability in the abundance of salmon year classes. Concern about salmon bycatch is exacerbated by its cultural and economic importance and long-term trends of reduced abundance of many stocks (Stram and Ianelli 2015). NPFMC has taken many actions to reduce salmon bycatch. These include time and area closures, hard caps, gear modifications, and incentive-based measures (NPFMC 2016). The most recent action, approved in 2016, constitutes a comprehensive approach including rolling hotspots (real-time, temporary closures), required use of gear modifications and contractual arrangements which obligate vessels to avoid high bycatch areas based on observer data (NPFMC 2016; Karp et al. 2005; Stram and Ianelli 2015).

The effectiveness of the increased retention requirement can be evaluated through catch and discard data obtained by observers. However, it is not possible to measure the effectiveness of the salmon bycatch reduction programme directly because bycatch rates are influenced by many factors including highly-variable patterns of salmon distribution and abundance (Stram and Ianelli 2015); furthermore, once vessels relocate, bycatch rates that would have occurred subsequently are unknown. The programme has been effective in the sense that time and area closures and hard caps have not constrained overall pollock harvests. Also, requirements for vessels to relocate to avoid high bycatch areas have been enforced effectively.

For both the increased retention and the salmon bycatch avoidance measures, elimination of the race for fish through permanent allocation of quota to cooperatives has enabled collaboration and innovation in reducing bycatch. Also, for both programmes, several best practices in bycatch management are satisfied; these include observer coverage and data collection, explicit performance standards, and adequate surveillance (Gilman 2011).

1.2.4 Chile

Chile has historically been among the 10 major fishing nations in the world, with annual landings (including aquaculture) exceeding 2.8 million t in 2016 (Sernapesca 2016). However, important fishing stocks have shown signs of overexploitation and consequently catches decreased from the maximum historical level of 7 million t in the mid-1990s (Sernapesca 2017). Consequently, management changes have been made to assure fisheries and ecosystem sustainability.

In this context the term "discard" first appeared in the Fisheries Law amendment of 2001, which also introduced an ITQ system. Before 2001, Chile had a "race for fish", i.e. vessels would fish until the total allowable catch (TAC) was reached and the fishery closed. The 2001 amendment prohibited discards without distinguishing among species and sizes, and established penalties that would apply to offenders, such as ITQ deductions of 30% (MEFT 2011). Without a concurrent extensive enforcement system or changes in selectivity, these penalties discouraged fishers from complying with the requirements and also from reporting discards. As a result, the magnitude of discards in Chilean fisheries was unknown and the practice of discarding continued (Borges et al. 2016).

The discard problem was revisited in the 2012 revision of the Fisheries Law. The "discard ban" was re-defined and the term "incidental catch" was applied to seabirds, marine mammals and sea turtles caught during fishing operations. This revision included provisions for compulsory industry-funded electronic monitoring (EM) on-board fishing vessels (including video sensors); and strengthened the penalties for discard offences. However exemptions from the discard ban and its penalties are permitted to identify and quantify the causes of discards and incidental catch. Through this process mitigation measures could be later tailored for each fishery and binding mitigation plans put in place as of 2018. Exemptions are conditional on a minimum 2-year at-sea monitoring and research programme with observers on-board commercial vessels. Suspending penalties under these exemptions largely eliminates incentives for fishers to alter their fishing behaviour and therefore reduces bias in the information collected. Once these penalty-exempt research programmes are completed, mitigation plans are put in place where further exemptions can be applied for as long as several requirements are met, such as on-board monitoring by observers continues and a global catch quota, which accounts for discards, has been set for the target species.

By February 2018, eight binding mitigation plans had been agreed upon, covering 11 fisheries within the artisanal and industrial fleets (Subpesca 2018). Even though draft mitigation plans were proposed by the Under-Secretariat for Fisheries to the respective Management Committees, they were further developed through a participatory, bottom-up process that included skippers, vessel owners, fishers and NGOs, in addition to members of the Management Committees. Mitigation plans and compliance monitoring by EM is required for all vessels in the industrial fleets by 2018 and by 2020 for artisanal vessels over 15 m (MEFT 2017).

The impact of the Chilean discard ban is not fully understood because it is still being implemented. In some cases, these penalty-exempt programmes have enabled fishery-specific changes in regulations including changes in minimum landing size (MLS) and authorization to catch and land species which were previously prohibited for some gears or fleets. In other cases, data collected has led to increases in target species TAC, since discard mortality estimates are now included in the stock assessments. However, restrictions and uncertainty related to choke species remain in multispecies fisheries. Quota transferability under the ITQ system (a potential approach for reducing the impact of choke species) is limited. When the discard ban is fully operational, and EM is in place, discarding choke species will no longer be possible and this may constrain fisheries for which selectivity improvements are not possible.

1.2.5 Argentina

In Argentina, fisheries resources are regulated and managed under federal law which provides a framework for the management of discards. Furthermore, each of the five provinces with maritime coasts has its own administration and applicable fishing legislation. Under article 21 of the Federal Fisheries Law "throwing discards and wastes into the sea, which is contrary to responsible fishing practices" is prohibited, while the General Environmental Law obliges the state "to promote the rational and sustainable use of natural resources".

In 1996, regulations established minimum total length or length at sexual maturity for several demersal and coastal species, prohibiting the capture of specimens below those sizes. Further amendments restricted fishing vessels to capturing no more than 10% of sizes smaller than those fixed under this resolution. After this threshold is reached, captured specimens must be returned to the sea immediately. This regulation also details provisions for the establishment of move-on rules when catches of undersized fish are predominant, and for area closures when the presence of undersized fish in an area is predominant. These minimum size requirements were abolished in 2006 since vessels were discarding catches above the 10% threshold at sea to avoid penalties.

Control and enforcement of fishing activities in Argentina is weak, and while discarding is illegal, it occurs frequently. A recent audit of the Under-secretariat of Fisheries and Aquaculture[1] has identified significant weaknesses in all their areas of operation, from inspections to data reporting and recording.

Argentina has had a National On-Board Observer Programme (PNOB) for the last 30 years to monitor activities of the main fishing fleets, and to obtain high quality information to be used exclusively for scientific purposes. The data collect by the observers is confidential and cannot be used for control and enforcements purposes.

Argentinean trawl fisheries are estimated to discard between 25% and 30% of their total catch. These discards comprise ~85 fish and invertebrate species in the bottom trawl fleet, and ~60 and ~32 species, respectively, in the factory trawl and shrimp trawl fleets. In addition, there is bycatch of more than 20 species of sharks and rays, many of them considered to be highly vulnerable to exploitation[2].

---

[1] https://www.agn.gov.ar/informes-resumidos/subsecretaria-de-pesca-y-acuicultura-gestion-de-la-tecnologia-informatica

[2] https://www.researchgate.net/profile/Alicia_Boraso/publication/309672657_La_zona_costera_patagonica_Vol_III_Pesca_y_Conservacion/links/581c829808aea429b291bdaa/La-zona-costera-patagonica-Vol-III-Pesca-y-Conservacion.pdf

In summary, even though there is a legislative framework to reduce UWC and prohibit catches of undersized fish, discarding continues to occur frequently in Argentinean fisheries, especially when undersized individuals and/or non-commercial species are caught.

1.2.6 Australia

Australia's commercial fisheries are managed by eight jurisdictions that use the full range of contemporary fisheries management methods (most in combination). However, discarding is not banned in Australian commercial fisheries. Instead, the approach generally used to manage commercial discards involves: (i) targeted observer programmes to identify problematic discarding issues; (ii) research into, and the implementation of, modifications to fishing gears that reduce problematic discards – using a variety of Bycatch Reduction Devices (BRDs) such as grids, panels and escape vents; and (iii) if no such modifications can be developed, spatial and/or temporal closures may be used.

For example, in the oceanic and estuarine prawn trawl fisheries of New South Wales (NSW) focused observer data collection in 1989–1991 identified problematic discards of juvenile finfish of species that were targeted by other commercial and recreational fisheries (Kennelly and Broadhurst 2002). Experiments were conducted to develop grids and panels to reduce UWC that resulted in discarding. These were used by fishermen voluntarily for many years before they were required by regulation. It has been estimated that these BRDs are now saving over 27 million juvenile fish per year in this fishery (Kennelly 2017; Kennelly and Broadhurst 2002).

1.2.7 New Zealand

Borges et al. (2016) and Telesetsky (2016) reviewed New Zealand's management of fisheries discards and a summary is provided below. As in Australia, New Zealand's fisheries are managed through use of the full range of contemporary fisheries management methods. However, unlike Australia, discarding/dumping by commercial fishers of fish, aquatic life, or seaweed covered by the New Zealand Quota Management System is prohibited –with a few exceptions. Specifically, a commercial fisher is not permitted to "return to or abandon in the sea or any other waters any fish, aquatic life, or seaweed of legal size. . .that is subject to the quota management system". What this means in practice is that New Zealand operates a "discard ban", with a prohibition on the discarding of catches of quota species over their minimum legal size (i.e. to prevent high-grading) and all catches for quota species for which a minimum has not been established. A catch does not need to be brought onboard for it to be considered "abandoned". Where quota fish are caught but left dead at sea, a harvester has a duty to prevent dumping and "make reasonable efforts to retrieve" such fish or be subject to an offence. The NZ Fisheries Act is silent on whether it is legal or not to discard non-quota species but there is some obligation to report discards of such species.

Any quota species that is less than the minimum legal size must be "immediately" returned "whether alive or dead to the sea or waters", where the fish or aquatic life were harvested. Generally, sub-MLS catches do not need to be reported except for certain specified stocks. However, only 11 finfish, rock lobsters, scallops, oysters, and paua (abalone) have minimum legal sizes.

New Zealand deploys an instrument to mitigate discarding of over quota catches, referred to as the "deemed value" system, by which the government puts a tax on the landing of over quota catches. The level of the tax is it set through a complex set of principles to discourage discarding, without creating incentives for targeting species not covered by quota (Telesetsky 2016; Sanchirico et al. 2006).

Under Schedule 6 of the NZ Fisheries Act, a fishing operator is authorised to legally discard thirty-two stocks in New Zealand if they comply with area and practice requirements. Three scallop stocks must be returned if they are collected during a closed scallop fishery season or in an area that has been closed to scallop fishing. Other scallops and dredge oysters may be returned as long as the shellfish is likely to survive. Most of the fish and shellfish listed in Schedule 6 "may" be returned to "the waters from which it was taken" if (1) the species is "likely to survive on return" and (2) "the return takes places as soon as practicable" after the species has been taken. This introduces a degree of discretion for the skipper and his crew for the return of such species without any explicit duty spelled out in the Statute to report such decisions. In practice, the Ministry requires that all Schedule 6 species returned to the sea be reported on both catch and landing returns even though Schedule 6 returns will not be counted against TACs.

In general Schedule 6 species are those understood to have high survivability when discarded. Certain species such as cockles, scallops, oysters, mussels, lobsters, and clams are likely to have high survival, depending on how they have been caught. For finfish, even if a fish is alive when it is brought on the deck, those with swim bladders that inflate after capture because of barotrauma are less likely to survive capture and discarding.

Exemptions may protect fishers who discard fish from prosecution. For example, as long as fishers report a discard as part of his/her returns, the fisher may return any fish, aquatic life, and seaweed to the sea where a fisheries officer or observer was present, the officer or observer authorized its return, and the fisher returned it under supervision of the officer or observer. Fishers may also legally discard parts of fish if the fish were lawfully processed and the parts of the fish that are retained on board allow for the accurate calculation of greenweight (weight prior to processing). Fishers may also return catch where there are concerns for the safety of the vessel or crew but there is no explicit requirement to report quota species that were returned for safety reasons even though the Ministry has an expectation that these species will be reported under the Fisheries (Reporting) Regulations on Catch Effort Returns.

While discards below MLS are not reported, discarding increased for legally sized fish following the introduction of the ITQ system in New Zealand (MRAG

2007). This occurred despite the provisions in the regulations that enabled fishers to match their catch to fishing opportunities by leasing, borrowing, banking or purchasing quota and the deemed value system (Lock and Leslie 2007). By 2007, discarding was considered to be widespread and an increasing problem in New Zealand's fisheries (MRAG 2007). According to Condie et al. (2014), at the same time as the introduction of the discard ban, there was a small increase in selectivity, but it is difficult to assess the impact of the discarding prohibition because accurate statistics on discards are unavailable. An analysis has suggested that total catches have been about the double of reported catches since the introduction of the quota management system in 1986, indicating that discarding remains a significant problem in New Zealand's fisheries (Hersoug 2018 and references therein). In line with this perception, the New Zealand government recently commenced on an agenda of mandatory EM of fishing operations including on-board cameras[3]).

1.2.8 Asia

1.2.8.1 Southeast Asia

In developing countries, including those in South-East Asia, small-scale fisheries generally harvest more fish than large-scale fisheries (Mills et al. 2011) and usually have low discard because utilisation is high (World Bank et al. 2012). Trawling makes a major contribution to seafood production in Asia with about 50% of seafood produced by an estimated 100,000 trawlers, ranging in size from those towing a net behind a 5 m boat with a 2 hp engine up to vessels of 80 m length or more. In Southeast Asia, large areas of trawlable continental shelf, fed by river inflows make for some of the most productive fishing grounds in the world. There is a wide range of sizes and shapes of fishes such that it is impossible to ensure that all species are caught at sizes above their size at first maturity (especially using trawls but also for other gears).

Even though trawls have been used in the region since the early twentieth century, after World War II there was a major increase in trawl fleets as developing countries sought to increase the supply of seafood for domestic consumption, export revenue and job creation. Aid funding and technical advice from developed countries led to a rapid increase in capacity which created large discarding issues and the depletion of stocks. Concerns about the wastage of useable protein, at a time when governments were seeking to increase protein supplies supply to the rural poor generated a debate about solutions, at a time when the developed countries were also beginning to address issues concerning bycatch, discards and wastage (e.g., FAO 1996).

[3]https://www.mpi.govt.nz/protection-and-response/sustainable-fisheries/strengthening-fisheries-management/future-of-our-fisheries/digital-monitoring-of-commercial-fishing/

Thailand's discard problem was resolved in the late 1960s and early 1970s with the development of a fish meal industry which supported poultry and shrimp farming industries. In Vietnam, previously discarded fish were used as food for pigs. Concerns about feeding animals encouraged a shift of interest towards products for human consumption. As a result, the surimi industry developed in the mid-1980s and has since grown throughout the region. In the Gulf of Thailand, the amount of fish used for fish meal has declined considerably due (in part) to overfishing but also the redirection of fish to higher value products. The same has occurred in Vietnam and to a lesser degree in China. In Myanmar there is still a big focus on fish meal. Throughout South-east Asia, large quantities of small fish are also used for fish sauces and pastes or are dried. And there is increasing evidence that these 'low-value' products have disproportional high value for alleviating prevalent micro-nutrient deficiencies (Roos et al. 2003; Kawarazuka and Béné 2011).

1.2.8.2 India

One of the first examples of the implementation of a discard ban in legislation occurred in India. Pramod (2010) describes Regulation 5, of the Maritime Zones of India Rules 1982: "Crews may not discard substantial surplus catch, catch exceeding authorized quantities shall be retained onboard, recorded, and surrendered as required by authorized officers". But Pramod goes on to note that mechanized vessels have never abided by this law and in any case, India keeps no formal records on quantities of discards. Nevertheless, over the last three decades, fisheries discard in India has been assumed to be low because, as in the other Asian countries mentioned above, most catches are fully utilized given the demand for protein-rich foods.

1.2.9 Africa

The following information summarizes observations from three African countries as reported by Kennelly (2014).

Discarding is not prohibited in Nigeria, and its shrimp trawl fisheries make extensive use of bycatch reduction technologies. This is mostly driven by European Union requirements for shrimp imports. Virtually all trawlers in Nigeria use Turtle Exclusion Devices (TEDs) and quite well-designed square mesh panels that effectively reduce the bycatch of large quantities of juvenile fish.

However, finfish bycatch from the shrimp fishery has a well-established market where it is retained and sold from trawlers to small-scale canoe operators. These operators sell the bycatch to onshore buyers (mostly women) who, in turn, dry and smoke the fish for sale at local and regional markets. This multi-layered sector (termed the "buyemsellum" sector) provides significant seafood protein to a large number of people who are often unable to find sufficient protein to meet their dietary needs. The use of BRD in to these fisheries is effectively reducing the bycatch available for local consumption. The current fisheries challenge in Nigeria, therefore, is to resolve this issue. Options for identifying alternative sources of seafood and new employment possibilities are now being investigated.

Madagascar currently has a very well managed shrimp trawl fishery, with no buyemsellum sector and significant use of bycatch reduction technologies, under drivers that include the shrimp import requirements of the EU and a desire to achieve Marine Stewardship Council certification. The priorities for Madagascar's trawlers are to improve the performance of the BRDs currently used so that they release more discards while increasing the retention of shrimp. Quite straightforward modifications (that have been developed elsewhere) to the gears currently used are being examined to assist with these priorities. It should also be noted that there is strong growth in the aquaculture industry in Madagascar, which brings with it the need for fish meal. As for Asian countries, increasing the utilisation of bycatch for aquaculture feed will increase its value and reduce discards.

The Cameroon trawl sector is characterized by little formal fisheries management, no implementation of sustainable fishing practices, and no pressing drive to improve fishing methods due to export requirements (most of the targeted shrimp is not exported to Europe). There is also a significant buyemsellum sector that, as in Nigeria, complicates the perceived need to reduce bycatches with the need to provide fish for undernourished people. The current challenges for this fishery therefore encompass most aspects of fisheries management, especially implementation of effective MCS, a programme to quantify and then mitigate bycatch issues, and identifying ways to manage the buyemsellum sector.

In Africa is an overarching need for an "awareness and enlightenment" campaign to educate the public and key stakeholders about the need for sustainable fisheries management, conservation of resources and effective use of fishing practices. The negative consequences of reducing access to important sources of protein and micronutrients for the local people must be recognized, however and the ecosystem consequences of reducing bycatch (or lack thereof) is also an important consideration (Garcia et al. 2012; Kolding and van Zwieten 2011).

1.3 Discussion: What Makes a Discard Mitigation Policy Work?

The variety of the described cases highlights that there is no single or simple answer to the question of what makes a discard policy work. Thus, what works in one fishery or region may not be suitable elsewhere. For example, a policy to reduce unwanted catch through selectivity improvements in countries such as Nigeria or Cameroon may conflict with local food security needs. This situation is very different to northern European fisheries, for example, with high discard rates of undersized fish of low market value. Accordingly, we have split the discussion of the cases we have described into two categories. The first category covers fisheries where discard rates are, or were, high and where there has been an emphasis on the reduction of UWC. The second category includes fisheries where discard rates are low, because most catches have value.

1.3.1 Fisheries with a Focus on Reducing Unwanted Catches

In this category we have included Norway, Iceland, the US, Argentina, Chile, Australia and New Zealand. Most of the discard policies in this category are highly complex. Each of these examples includes measures to reduce UWC, including selectivity improvements, various spatial measures, quota related measures and economic incentives. In addition, most the policies include monitoring and control measures and improved utilisation. A similar range of measures is being used to implement these discard policies regardless of whether the policy is defined as a discard ban (Norway, Iceland, Argentina, Chile and New Zealand) or as a policy aimed at achieving discard reduction (US and Australia).

The term "discard ban" suggests elimination of all discards. However, none of the discard bans currently in place (or under implementation) prohibit all discarding (Borges et al. 2016). Instead, they focus on specific components of the catch and exemptions apply (e.g. those with high chance of survival if released, those designated as endangered or threatened species). Moreover, there are no examples of simple discard bans, with Chile's first and unsuccessful discard policy representing a possible exception.

In some of the cases there has been a gradual evolution from a specific discard-related issue, such as high-grading of cod in Norway, towards a more comprehensive discard policy. In Iceland and Norway, for instance, the achievement of discard rate reductions and mindset changes regarding the wastefulness of discarding required periods of 30–40 years to achieve. The US walleye pollock fishery is an exception in that regard as discard rates were reduced rapidly, perhaps due to the initial requirement for full observer coverage. This case is possibly the most demonstrably successful discard policy described here. Furthermore, some countries (e.g., Norway, Iceland, USA and Chile) stress stakeholder involvement as an essential aspect of their discard policies. This may include aspects such as: consultative processes with the regulatory authorities, modification of regulations based on stakeholder feedback and initiatives for implementing selective technical measures (i.e., increased quota/access to closed areas).

An important component of many discard reduction strategies is the use of more selective fishing gears which may also be associated with temporal and spatial closures to reduce UWC (see Bellido et al., this volume; O'Neill et al., this volume; Reid et al., this volume).

TAC management, often together with individual quota allocations (or allocations to groups or cooperatives) predominates in the fisheries in this category. Accordingly, policy initiatives to address discarding often feature changes in the arrangements for implementing and allocating quotas. Quota related measures are, in most cases, a fundamental element of the overall discard policy and are generally considered to at least partially remove incentives for discarding. This issue is complex because inflexible quota management systems can cause or exacerbate choke species problems, while quota management programmes, which allocate quota to individuals and or groups of fishers and include mechanisms for transfer or leasing, can help reduce choke species effects and related discarding (Poos 2015). However, assessing the contribution of quota-based measures to the overall effectiveness of discard policies in isolation from the other policy elements is difficult. In New Zealand, the move to ITQs created an increase in discarding of legally sized fish and whether that situation has improved is difficult to ascertain due to a lack of accurate discarding data.

For a number of the cases examined, increased utilisation has occurred as a means of reducing UWC when it cannot be avoided completely. In both Iceland and Norway, the utilisation initiatives include viscera and other processing waste generated at sea (e.g., Richardsen et al. 2017).

Within fisheries focused on discard reduction, perhaps the greatest divergence in approach concerns accountability and compliance monitoring. Accountability in the context of discard restrictions or prohibitions is problematic. While it is relatively straightforward to estimate and verify the retained or landed catch, estimation and verification of discards is difficult (See Chap. 13). At-sea observers provide the data necessary for discard estimation, but observer programmes are expensive, and fishers may behave differently when observed, which can result in biased discard estimates (Benoît and Allard 2009). Technological approaches such as EM can be effective, especially for compliance purposes when discards are prohibited (and may also be an effective means of directly estimating catch and discard for certain gear types). Self-reporting of discards by vessel operators is generally considered unreliable (Kraan et al. 2013) unless verified by observers, EM or some other method (Plet-Hansen et al. 2017).

The US walleye pollock fishery stands out from the other cases described here due to its comprehensive accountability measures. This fishery requires a sufficient observer coverage to ensure that all hauls are monitored (one or two observers per vessel), which not only provides confidence that vessels are compliant, but also ensures data quality that may be lacking in other countries and fisheries (Stanley et al. 2011). The other cases have less stringent observer coverage or other forms of monitoring; although some, such as Norway and Iceland, feature significant penalties for non-compliance. It remains to be seen how the new discard regime in Chile, which requires EM on board all vessels over 15 m by 2020, will affect compliance levels and data quality. A similar question applies to the planned implementation of EM in New Zealand fisheries.

One of the striking findings of this comparative analysis is that discard data quality appears to be problematic in the majority of cases (Borges et al. 2016). This is despite the essential importance of a discard policy in many of the management frameworks described and the data collection resources available. Several issues arise from this, not least that it makes the task of determining whether a discard policy works or not significantly more difficult. At least one of the cases, the first iteration of Chile's discard policy, provides evidence that an overly rigid and punitive discard ban can have a negative impact on data provision and compliance.

1.3.2 High Utilisation Fisheries

In many African and Asian fisheries, a local demand exists for non-targeted fish, which serves an important role by contributing to human nutrition and animal feed. In such cases, discards are low due to high utilisation levels. Accordingly, discarding becomes less of an issue than whether the fishery is sustainable. Improvements in selectivity in such fisheries would have a negative impact on important local supplies of fish protein. Of the five cases described in this category, only India has a regulatory discard ban which, however, has not been effectively implemented.

Although there is not a general management interest in improving selectivity in these fisheries, there remains concern over the capture and mortality of threatened, endangered or protected species. These may be encountered frequently in the tropical waters of Africa and Asia where, like elsewhere, retention is usually illegal. The use of bycatch reduction devices (BRDs) in these fisheries is uncommon and generally occurs where the target species is destined for export (Kennelly 2014).

1.4 Lessons Learned

Reducing or eliminating discards is a public policy priority in some regions although this is not always the case. For example, very high utilisation rates can be seen in many small-scale and artisanal fisheries and some industrial fisheries, such as those for fish meal and oil. Furthermore, there is much debate in the literature regarding the factors which lead to high levels of discarding, and the costs and benefits (in economic and ecosystem contexts) of improving selectivity and avoidance compared with increasing utilisation.

In fisheries with a focus on reducing UWC, policy makers seek to implement changes through a mixture of measures which include improved utilisation as well as avoidance and improved selectivity. To some extent, we can draw from these examples and experiences to better understand and mitigate problems associated with implementation of the LO.

The LO constitutes a far-reaching change in EU fisheries policy which impacts, or will impact, fishers and stakeholders throughout Europe (Condie et al. 2014). Yet the goals and implementation process are poorly understood, and legitimacy of this policy has been questioned by many fishers, which tends to impede the implementation process (de Vos et al. 2016). Success in implementing policies to reduce discards is generally easier to achieve if fishers and other stakeholders are involved in the formulation and implementation of policies. From our case studies, we can see examples of good and poor communication of goals, and the extent to which formulation and implementation is guided by a participatory process and how these have impeded or facilitated success. For example, under the initial Chilean discard ban a top-down regulatory approach was employed with no accountability or enforcement; this resulted in an ineffective policy. During reformulation of the Chilean discard ban, fishers and stakeholders played a substantial role in policy development and in implementation through research and drafting of discard mitigation plans together with the high mandatory monitoring levels. The new programme is better understood and accepted, and initial indications suggest a strong likelihood of success. The US programmes for increased utilisation and avoidance of salmon bycatch in the walleye pollock fishery involved a great deal of participation by fishers and stakeholders that resulted in policies which were well-understood and implemented successfully.

Effective Monitoring, Control and Surveillance (MCS) are difficult to achieve and poor MCS may render regulations ineffective. This has certainly been the case in New Zealand and Argentina and under the first iteration of the Chilean programme. MCS in Iceland and Norway has also been limited and this makes verification of programme effectiveness difficult, if not impossible. It should be noted, however, that moving towards industry and stakeholder recognition of the importance of reducing discards and developing a culture of compliance is integral to the long-term strategies employed by Iceland and Norway. In addition, poor or inadequate monitoring can make it difficult or impossible to evaluate the effectiveness of a discard reduction (or elimination) programme. In some parts of the world, MCS is entirely independent of observer coverage because observers focus entirely on scientific data collection. MCS is effective in the US walleye pollock fishery so that compliance is high, and accountability is fully documented (note that industry pays for observer coverage in this fishery). In this fishery much of the compliance monitoring is carried out by observers. In the US and elsewhere EM is beginning to be used as a compliance monitoring tool and has been shown to be effective in some instances. Regardless, an effective discard policy must have adequate MCS as well at-sea monitoring through observers and/or EM. This is necessary to ensure the accuracy of catch and discard data.

Requirements for reducing or eliminating discards work best as part of a broader fisheries management policy that integrates regulatory measures to address an overarching goal. This can be seen in the Norwegian DBP which is comprised of a suite of regulatory and technical measures to minimise unwanted catch, including: the regulatory discard ban; gear selectivity technical measures; closed areas; and MCS measures. Elements of a more comprehensive and integrated approach can be found elsewhere such as the broad guidance to minimise discards in the US and in the Chilean and Icelandic approaches. However, the EU LO was intended to be an all-inclusive measure to eliminate unwanted catch and this will be more difficult to implement.

Integration of discard reduction policies within a broader strategy also requires a measured, adaptive approach such has occurred in Iceland and Norway, and to some extent in Chile. This facilitates alignment of the policy with industry incentives and fostering a culture of compliance and allows for incremental improvements in the overall programme over time. The EU LO was originally conceived to be fully effective after a relatively brief phase-in period (2014–2019), which therefore could have undermined its effectiveness. However, its implementation through multi-annual plans and discard plans could enable the adaptive, incremental and case-specific approaches seen as favourable characteristics in some of the case-studies reviewed here.

Also related to the need for a broader and more comprehensive, strategic approach is the issue of governance as it relates to quota management. Quota management programmes that provide for individual or group (co-operative) ownership or leasing of quota, such as those that occur in US, Iceland and New Zealand, allow fishers to manage and trade quotas so as to avoid or minimise choke species problems which often result in discarding. In Europe, national quota management regimes can differ but how they may be adapted to cope with the challenges of the landing obligation is beyond the scope of this chapter. They may also allow for improved efficiencies in fishing operations which can facilitate change in operations and uptake of innovative technologies. Furthermore, flexibilities in these types of quota management programmes can make it easier for fishers to carry-over small amounts of quota across fishing years, or deduct overages from the next year‘s allocation. These types of mechanisms greatly enhance the ability of the fleet and individual fishers to adapt to regulatory change such as discard restrictions. Such opportunities are not currently available to all EU fishers working under the CFP although flexibility is allowed in some countries such as Denmark and the Netherlands.

A successful discard policy requires careful balancing of top-down factors such as legislative requirements and effective controls with bottom-up factors including participation, stakeholder acceptance and cultural attitudes towards compliance. The cases we have described have balanced these aspects differently, with some taking a long-term approach emphasising a gradual development of a low discard mindset and less restrictive and rigorous controls. Others have focussed on strong controls, while attempting to retain stakeholder support, in an effort to achieve positive results over a shorter timeframe. With the exception of India, the high utilisation fisheries described have not established discard policies and discard levels are low because utilisation, especially for human consumption, is high. Balance among the factors described above will differ within Europe due to the diversity and complexity of European fisheries and will most likely include aspects of all the approaches described in this chapter.

Acknowledgments We would like to thank two anonymous reviewers and the editorial team for their constructive comments and suggestions.

References

Bellido, J.M., Paradinas, I., Vilela, R., Bas, G., Grazia Pennino, M. (this volume). A marine spatial planning approach to minimize discards: Challenges and opportunities of the Landing Obligation in European waters. In S.S. Uhlmann, C. Ulrich, S.J. Kennelly (Eds.), *The European Landing Obligation – Reducing discards in complex, multi-species and multi-jurisdictional fisheries*. Cham: Springer.

Benoit H.P., & Allard J. (2009). Can the data from at-sea observer surveys be used to make general inferences about catch composition and discards? *Canadian Journal of Fisheries and Aquatic Sciences, 66*, 2025–2039. https://doi.org/10.1139/f09-116.

Borges, L., Cocas, L., Nielsen, K. (2016). Discard ban and balanced harvest: A contradiction? *ICES Journal of Marine Science, 73*, 1632–1639.

Breen, M., Isaksen, B., Ona, E., Pedersen, A.O., Pedersen, G., Saltskår, J., et al. (2012). A review of possible mitigation measures for reducing mortality caused by slipping from purse-seine fisheries. ICES CM 2012/C:12.

Condie, H.M., Grant, A., Catchpole, T.L. (2014). Incentivising selective fishing under a policy to ban discards; lessons from European and global fisheries. *Marine Policy, 45*, 287–292.

Damalas, D. (2015). Mission impossible: Discard management plans for the EU Mediterranean fisheries under the reformed Common Fisheries Policy. *Fisheries Research, 165*, 96–99.

de Vos, B.I., Döring, R., Aranda, M., Buisman, F.C., Frangoudes, K., Goti, L., et al. (2016). New modes of fisheries governance: Implementation of the landing obligation in four European countries. *Marine Policy, 64,* 1–8.

European Commission. (2007). Impact assessment of discard policy for specific fisheries. European Commission, Brussels. https://ec.europa.eu/fisheries/sites/fisheries/files/docs/publications/impact_assessment_discard_policy_2007_en.pdf

European Union. (2013). Regulation (EU) No 1380/2013 of the European Parliament and of the Council of 11 December 2013 on the Common Fisheries Policy, amending Council Regulations (EC) No 1954/2003 and (EC) No 1224/2009 and repealing Council Regulations (EC) No 2371/2002 and (EC) No 639/2004 and Council Decision 2004/585/EC. *Official Journal of the European Union*, L354, 22–61.

FAO. (1996). Report of the technical consultation on reduction of wastage in fisheries. Tokyo, Japan, 28 October – 1 November, 1996. FAO Fisheries Report No. 547, FAO, Rome, p. 27.

FAO. (2016). The state of world fisheries and aquaculture 2016, p. 200.

Fiskistofa. (2017). Ársskýrsla 2016. Fiskistofa, Hafnarfjörður, Iceland.

Fiskistofa. (2018a). Yfirlit yfir VS-afla eftir tegundum. Retrieved from www.fiskistofa.is: http://www.fiskistofa.is/veidar/aflastada/vs-afli/

Fiskistofa. (2018b). Undirmál. Retrieved from www.fiskistofa.is: http://www.fiskistofa.is/veidar/aflaupplysingar/undirmal/undirmal_fiskveidiar.jsp

Fiskistofa. (2018c). Um fiskveiðistjórnun. Retrieved from www.fiskistofa.is: http://www.fiskistofa.is/fiskveidistjorn/stjornfiskveida/

Fiskistofa. (2018d). Eftirlit á sjó. Retrieved from www.Fiskistofa.is: http://www.fiskistofa.is/umfiskistofu/arsskyrsla-2013/eftirlit-a-sjo/

Fitzpatrick, M., Quetglas, T., Frangoudes, K., Triantaphyllidis, G., Nielsen, K.N. (2017). DiscardLess Policy Brief No 2: Year 2 of the Landing Obligation: Key Issues in Mediterranean Fisheries. https://doi.org/10.5281/zenodo.573666.

Garcia, S.E., Kolding, J., Rice, J., Rochet, M.-J., Zhou, S., Arimoto, T., et al. (2012). Reconsidering the consequences of selective fisheries. *Science, 335*, 1045–1047.

Gezelius, S.S. (2006). Monitoring fishing mortality: Compliance in Norwegian offshore fisheries. *Marine Policy, 30*, 462–469.

Gilman, E. (2011). Bycatch governance and best practice mitigation technology in global tuna fisheries. *Marine Policy, 35*, 590–609.

Grimaldo, E., Sistiaga, M., Larsen, R.B. (2014). Development of catch control devices in the Barents Sea cod fishery. *Fisheries Research, 155*, 122–126.

Gullestad, P., Aglen, A., Bjordal, Å., Blom, G., Johansen, S., Krog, J., et al. (2014). Changing attitudes 1970–2012: Evolution of the Norwegian management framework to prevent overfishing and to secure long-term sustainability. *ICES Journal of Marine Science, 71* (2), 173–182.

Gullestad, P., Blom, G., Bakke, G., Bogstad, B. (2015). The "Discard Ban Package": Experiences in efforts to improve the exploitation patterns in Norwegian fisheries. *Marine Policy, 54,* 1–9. https://doi.org/10.1016/j.marpol.2014.09.025.

Ianelli, J.M., & Stram, D.L. (2015). Estimating impacts of the pollock fishery bycatch on western Alaska Chinook salmon. *ICES Journal of Marine Science, 72*, 1159–1172.

Isaksen, B., Valdemarsen, J.W., Larsen, R., Karlsen, K. (1992). Reduction of fish by-catch in shrimp trawl using a rigid separator grid in the aft belly. *Fisheries Research, 13*, 335–352.

Íslandsbanki. (2017). Íslenskur Sjávarútvegur. (2017). Reykjavík: Íslandsbanki. http://www.hafro.is/Bokasafn/Greinar/fish_res_59-437.pdf

Johnsen, P.J., & Eliasen, S. (2011). Solving complex fisheries management problems: What the EU can learn from the Nordic experiences of reduction of discards. *Marine Policy, 35*, 130–139.

Karp, W.A., Haflinger, K., Karp, J.G. (2005). Intended and unintended consequences: Fisher responses to bycatch reduction requirements in the Alaska Groundfish Fisheries. CM 2005/V:03. Available online at http://www.ices.dk/sites/pub/CM%20Doccuments/2005/V/V0305.pdf

Kawarazuka, N., & Béné, C. (2011). The potential role of small fish species in improving micronutrient deficiencies in developing countries: Building evidence. *Public Health and Nutrition, 14*, 1927–1938.

Kennelly, S.J. (2014). Solving by-catch problems: Successes in developed countries and challenges for protein-poor countries. *Journal of the Marine Biological Association of India, 56*, 19–27.

Kennelly, S.J. (2017). Developing a National Bycatch Reporting System. Final FRDC Report, ISBN 978-0-9924930-5-9, October, 2017.

Kennelly, S.J., & Broadhurst, M.K. (2002). Bycatch begone: Changes in the philosophy of fishing technology. *Fish and Fisheries, 3*, 340–355.

Kolding, J., & van Zwieten, P.A.M. (2011). The tragedy of our legacy: How do global management discourses affect small-scale fisheries in the South? *Forum for Development Studies, 38*, 267–297.

Kolding, J., Béné, C., Bavinck, M. (2014). Chapter 22: Small-scale fisheries – Importance, vulnerability, and deficient knowledge. In S. Garcia, J. Rice, A. Charles (Eds.), *Governance for marine fisheries and biodiversity conservation* (Interaction and coevolution, pp. 317–331). Wiley-Blackwell. https://doi.org/10.1002/9781118392607.

Kraan, M., Uhlmann, S., Steenbergen, J., Van Helmond, A.T.M., Van Hoof, L. (2013). The optimal process of self-sampling in fisheries: Lessons learned in the Netherlands. *Journal of Fish Biology, 83*, 963–973. https://doi.org/10.1111/jfb.12192.

Landhelgisgæslan. (2018). Arsyfirlit 2013. Retrieved from www.lhg.is: http://www.lhg.is/media/arsskyrslur/LHG_Arsyfirlit_2013_LWR2.pdf

Larsen, R.B., & Isaksen, B. (1993). Size selectivity of rigid sorting grids in bottom trawls for Atlantic cod (*Gadus morhua*) and haddock (*Melanogrammus aeglefinus*). *ICES Marine Science Symposia, 196*, 178–182.

Lock, K., & Leslie, S. (2007). *NewZealand's quota management system: A history of the first 20 years* (Motu Working Paper 07–02). Motu Economic and Public Policy Research, p. 75

Margeirsson, S., Jónsson, G.R., Hrafnkelsson, B., Jensson, P., Arason, S. (2008). Processing forecast of cod. Decision making in the cod industry based on recording and analysis of value chain data. University of Iceland. Reykjavík, Iceland. http://www.matis.is/media/utgafa/krokur/PhDThesisSveinnMarg.pdf

MEFT. (2011). Ley 19713/2001 de18 de enero de 2001 del Ministerio de Economía, Fomento y Reconstrucción; Subsecretaría de Pesca. Establece como medida de administración el límite máximo de captura por armador a las principales pesquerías industriales nacionales y la regularización del registro pesquero artesanal. Diario Oficial de la República de Chile el25 de enero de2001.16 p. http://www.leychile.cl/Navegar?idNorma=180659

MEFT. (2017). Decreto Supremo N°76 de 2015. Aprueba Reglamento del Dispositivo de Registro de Imágenes para Detectar y Registrar el Descarte. https://www.leychile.cl/Navegar?idNorma=1100026

Mills, D.J., Westlund, L., de Graaf, G., Kura, Y., Willmann, R., Kelleher, K. (2011). Under-reported and undervalued: Small-scale fisheries in the developing world. In N.L. Andrew, & R. Pomeroy (Eds.), *Small-scale fisheries management: Frameworks and approaches for the developing world* (pp. 1–15). Wallingford: CABI.

Ministry of Fisheries. (2018). Frumvarp til laga um breyting á lögum lögum nr. 57/1996, um umgengni um nytjastofna sjávar og lögum nr. 116/2006, um stjórn fiskveiða (rafræn vöktun, eftirlit með vigtun, fjarstýrð loftför o.fl.) og um breyting á lögum nr. 36/1992 um Fiskistofu. Retrieved from https://www.stjornarradid.is/lisalib/getfile.aspx?itemid=2a0680ae-8056-11e8-942c-005056bc530c

Mortensen, L.O., Ulrich, C., Eliasen, S., Olesen, H.J. (2017). Reducing discards without reducing profit: Free gear choice in a Danish result-based management trial. *ICES Journal of Marine Science, 74*,1469–1479.

MRAG. (2007). Impact assessment of discard policy for specific fisheries. Studies and Pilot Projects for Carrying Out the Common Fisheries Policy No FISH/2006/17.

Norwegian Seawater Fisheries Regulations. (2014). Regulations relating to sea-water fisheries. Norwegian Ministry of Fisheries and Coastal Affairs. Amended 7th April 2014. https://www.fiskeridir.no/.../20140407-regulations-relating-to-sea-water-fisheries.pdf

NPFMC. (2016). Bering Sea/Aleutian Islands groundfish fishery management plan amendment action summaries. North Pacific Fishery Management Council. Anchorage, Alaska, p. 23.

NPFMC. (2017). Stock assessment and fishery evaluation report for the groundfish resources of the Bering Sea/Aleutian Islands regions. North Pacific Fishery Management Council. Anchorage, Alaska.

O'Neill, F.G., Feekings, J., Fryer, R.J., Fauconnet, L., Afonso, P. (this volume). Discard avoidance by improving fishing gear selectivity: Helping the fishing industry help itself. In S.S. Uhlmann, C. Ulrich, S.J. Kennelly (Eds.), *The European Landing Obligation – Reducing discards in complex, multi-species and multi-jurisdictional fisheries*. Cham: Springer.

Pálsson, Ó.K. (2003). A length-based analysis of haddock discards in Icelandic fisheries. *Fisheries Research, 59* 437–446. http://www.hafro.is/Bokasafn/Greinar/fish_res_59-437.pdf.

Pálsson, O.K., Bjornsson, H., Gudmundsson, S., Ottesen, Þ. (2015). Mælingar á brottkasti þorsks og ýsu. (2013). Reykjavik: Marine Research Institute. http://www.hafro.is/Bokasafn/Timarit/fjolrit-183.pdf

Plet-Hansen, K.S., Eliasen, S.Q., Mortensen, L.O., Bergsson, H., Olesen, H.J., Ulrich, C. (2017). Remote electronic monitoring and the landing obligation – Some insights into fishers' and fishery inspectors' opinions. *Marine Policy, 76*, 98–106.

Poos, J.J. (2015). Update of "Delineating catch quotas for Dutch demersal fisheries: A theoretical pilot study." IMARES Wageningen Research Report number C139/15. Ijmuiden, The Netherlands, p. 19.

Pramod, G. (2010). Illegal, unreported and unregulated marine fish catches in the Indian exclusive economic zone. Field report, policy and ecosystem restoration in fisheries, Fisheries Centre, University of British Columbia, BC, Vancouver, Canada, p. 30.

Reid, D.G., Calderwood, J., Afonso, P., Fauconnet, L., Pawlowski, L., Plet-Hansen, K.S., et al. (this volume). The best way to reduce discards is by not catching them! In S.S. Uhlmann, C. Ulrich, S.J. Kennelly (Eds.), *The European Landing Obligation – Reducing discards in complex, multi-species and multi-jurisdictional fisheries*. Cham: Springer.

Richardsen, R., Nystøyl, R., Strandheim, G., Marthinussen, A. (2017). Analyse marint restråstoff, 2016. SINTEF Report OC2017A-095. ISBN 978-82-7174-286-7. https://brage.bibsys.no/xmlui/handle/11250/2446152

Roos, N., Islam, M.M., Thilsted, S.H. (2003). Small indigenous fish species in Bangladesh: Contribution to vitamin A, calcium and iron intakes. *Journal of Nutrition, 133*, 4021S–4026S.

RÚV. (2018). Brottkast, ís-svindl og uppgjöf Fiskistofu. Retrieved from http://www.ruv.is/frett/brottkast-is-svindl-og-uppgjof-fiskistofu

Sanchirico, J.N., Holland, D., Quigley, K., Finad, M. (2006). Catch-quota balancing in multispecies individual fishing quotas. *Marine Policy, 30*, 767–785.

Schrope, M. (2010). What's the catch? *Nature, 465*, 540–542.

Sernapesca. (2016). Anuarios Estadísticos del Servicio Nacional de Pesca y Acuicultura. http://sernapesca.cl/index.php?option=com_remository&Itemid=246&func=startdown&id=261

Sernapesca. (2017). Anuarios Estadísticos del Servicio Nacional de Pesca y Acuicultura. http://sernapesca.cl/index.php?option=com_remository&Itemid=246&func=select&id=2

Stanley, R.D., McElderry, H., Mawani, T., Koolman, J. (2011). The advantages of an audit over a census approach to the review of video imagery in fishery monitoring. *ICES Journal of Marine Science, 68*, 1621–1627. https://doi.org/10.1093/icesjms/fsr058.

Stram, D.L., & Ianelli, J.N. (2015). Evaluating the efficacy of salmon bycatch measures using fishery-dependent data. *ICES Journal of Marine Science, 72*, 1173–1180.

Subpesca. (2018). Planes de Reducción del descarte y la pesca incidental. http://www.subpesca.cl/portal/615/w3-propertyvalue-62973.html

Telesetsky, A. (2016). *Fishing for the future: Addressing fisheries discards and increasing export value for New Zealand's sustainable fisheries*. Fulbright New Zealand, ISBN 978-0-473-36903-3, p. 128.

Underwood, M., Engås, A., Rosen, S., Aasen, A. (2014). Excess Fish Exclusion Device (ExFED): How to passively release fish at depth during trawling. In Escape slots in front of a demersal seine codend combined with a new "Fish lock". In ICES. 2014. First Interim Report of the ICES-FAO Working Group on Fishing Technology and Fish Behaviour (WGFTFB), 5–9 May 2014, New Bedford, USA. ICES CM 2014/SSGESST:08.

Vigfusson, B., Sandholt, G., Gestsson, H.M., Sigfusson, Þ. (2013). Tveir fyrir einn. Reykjavik: Íslenski sjávarklasinn. https://issuu.com/sjavarklasinn/docs/tveir_fyrir_einn_fullvinnsla_aukaaf

Witherell, D., Pautzke, C., Fluharty, D. (2000). An ecosystem-based approach for Alaska groundfish fisheries. *ICES Journal of Marine Science, 57*, 771–777.

World Bank, FAO & World Fish Center. (2012). Hidden harvest: The global contribution of capture fisheries. Report No. 66469-GLB. Washington, DC: World Bank.

Chapter 2
Discards in the Common Fisheries Policy: The Evolution of the Policy

Lisa Borges and Ernesto Penas Lado

Abstract This chapter deals with the development of the European Union (EU) discard policy over time. It describes the process from 1992, when the issue of discards was first recognised in the Common Fisheries Policy (CFP) reform process, to the Landing Obligation (LO) adopted in 2013. It analyses the context to which policy choices were made that shaped the present format of the EU LO, how it is being implemented and the impact it is having on associated fisheries management measures. Finally, future possible policy developments are examined.

Keywords *de minimis* and high survival exemptions · Discards · European Commission · Landing Obligation

2.1 Introduction: Historical Background

2.1.1 Discards in the Common Fishery Policy

Discarding of fish has always taken place in European fisheries and has continued since the inception of the Common Fisheries Policy (CFP). Although data are incomplete, the levels of discards in some fisheries have been well known for several decades (CEC 1992).

Traditionally, discards are driven by economic reasons: for many species and sizes there is a limited market acceptability, they have no or low commercial value,

---

Disclaimer: The views and opinions expressed in this chapter are those of the authors and do not reflect the official position of any employer

L. Borges (✉)
FishFix, Brussels, Belgium
e-mail: info@fishfix.eu

E. Penas Lado
Fisheries Biologist/Manager, Brussels, Belgium

S. S. Uhlmann et al. (eds.), *The European Landing Obligation*,
https://doi.org/10.1007/978-3-030-03308-8_2

but also, in a system of limited landings by stock, fishers have traditionally discarded the smaller, lower-value fish to maximise the value of the fish landed and counted against quotas, a practice known as 'highgrading'. However, some rules of the CFP have also traditionally contributed to discarding: certain provisions of the CFP made discarding compulsory, namely for (a) catches over quota, (b) catches of undersized fish and (c) catches not corresponding to the expected catch composition of the legal mesh size (Council Regulation (EC) No 850/98; EC 1998).

The consideration of discarding as a problem for the Common Fisheries Policy was recognised many years ago. As early as 1992, at the time of the first reform of the CFP, the European Commission raised the need to address and resolve this problem in its communication on reform (CEC 1992). However, the question was not seen as a significant problem by the Council of Ministers, and no specific action followed.

In the 2002 reform, discards were also raised as an issue (CEC 2001). However, no specific proposals were made in that regard because the debate following the 2001 communication revealed that discards were not seen as a significant problem, and policy makers focused on other priorities, such as stakeholder consultation, long-term management plans and vessel construction subsidies. Discards were seen, in that context, as an inevitable result of the economic activity and partially a result of the management system, which few seriously questioned at the time.

2.1.2 The Communications of 2007 and 2011

The perceptions toward discarding started to change in the mid-2000s. A report from the EU Scientific, Technical and Economic Committee for Fisheries (STECF) in 2006 illustrated the magnitude of the discard problem in EU fisheries (STECF 2006). Following this report in 2007, the European Commission made a first attempt to address the issue systematically in policy terms, through a communication (CEC 2007). But this had little impact: the only result was the introduction of a ban on 'highgrading' in all areas in 2010. This ban was poorly enforced and its effect on the real level of discards was never evaluated.

By 2010 the problem of discarding was being increasingly referred to in the European media and this influenced European public opinion which started to put the EU under pressure to tackle the issue (Borges 2015). Pressure was particularly important in the United Kingdom, where the high levels of discards were used to criticise the CFP and, by extension, the European Union. The compulsory collection of discard data from 2003 as part of the EU Data Collection Framework (DCF), and the publication of some of these data in the mid-2000s contributed to raise awareness by showing the high level of discarding in many EU fisheries. The question of discarding started to be a significant source of embarrassment for the fisheries policy and, by extension, for the European Union.

In 2011 the European Commission published a second communication proposing to set up a process to resolve the problem gradually through cooperation with the fishing industry. However, this communication did not bring a real change in policy. It has been suggested that the main reason for this was a lack of incentive for the main players (the industry and Member State administrations) to engage. This, in turn, was considered to be due to the fishing industry believing that there would be no advantage in reducing discards (Penas Lado 2016), coupled with a general perception by most that it was too technically difficult to implement (Fitzpatrick et al. 2011). Finally, it was believed that the absence of a legally-binding common approach would result in a piecemeal approach that would not provide a level playing field for different fleets.

However, this experience was very important to shape the future policy. In a policy characterised by top-down prescriptive approaches, only very few took seriously an alternative, bottom-up, cooperative approach. This was also the lesson learnt by non-EU countries that had previously implemented a non-discard policy, such as Norway (Karp et al., this volume). This eventually led to the consideration by the European Commission that a non-discard policy could only be established through top-down legislation.

2.2 The Landing Obligation and the CFP Reform of 2013

2.2.1 Why a Ban on Discards? The Proposal by the European Commission

The question of discarding was one of the most prominent issues in the reform of the CFP in 2013. Why? It is interesting to note that the Commission's 'Green paper' on CFP reform, published in 2009 (CEC 2009) did not even consider discarding as one of the five structural deficiencies of the policy. The Commission viewed discarding as an undesirable, but largely inevitable consequence of a management system based on individual stocks, allocated through fixed shares for each Member State. Together with other provisions, such as catch composition rules under the technical measures regulation (EC 1998), discarding was made compulsory under certain circumstances. Since these policy elements were not being questioned, a ban on discarding was initially considered as not appropriate.

In the public debate following the publication of the Green Paper, several contributions touched upon the issue of discards. But these did not come from the industry which, by and large, did not consider discards as a problem. But for a number of environmental NGOs this was an example of the CFP's inability to resolve its many shortcomings, particularly with regards to ecological impacts. In addition, some British media presented the discarding of good, large fish under the CFP as the ultimate demonstration of the absurdity of the policy, and abundantly used the images of fish being discarded as an argument to discredit the CFP (Borges 2015; Gambarato and Medvedev 2016).

In a context where fisheries issues were no longer confined to fisheries constituencies, the problem of discarding was becoming a major iconic issue with political consequences that reached beyond fisheries policy itself. This was the fundamental reason why an issue largely ignored in the preparation of the fishery policy reform, went to the top of the agenda almost overnight (Joa 2015).

The Commission understood that the high level of discards seriously undermined the credibility of the policy, over and beyond the objective reasons and difficulties behind the phenomenon. Furthermore, the failure to conceptualize the bottom-up approach proposed in the Communication of 2007 illustrated the need for a new approach to address the problem. Last but not least, the experience from abroad (i.e., Norway), having applied a non-discard policy for years, clearly indicated that for the discard ban to work, it needed to be imposed top-down in law, as the only way to overcome the initial resistance from the industry.

These considerations led the Commission to include in its proposal for the reformed CFP (CEC 2011a) the idea of a phased-in ban on discarding by species. This came about relatively late in the process of preparing the reform proposal, to the extent that the initial Impact Assessment of the reform (CEC 2011b) did not evaluate the effects of such a measure, which had to be incorporated at a later stage.

The Commission was fully aware that imposing a Landing Obligation as a top-down measure in a policy that hitherto had made discarding compulsory under a number of circumstances would present many difficulties and some contradictions. However, it was felt that these difficulties could and should be resolved if all players involved, from administrations to the industry, were jointly forced to make it happen. The failure of the bottom-up approach of 2007 was believed to demonstrate that the practical difficulties could be resolved only if the players were all under legal pressure to deliver the results.

2.2.2 *The Debate with the Council of the EU and the European Parliament*

The EU co-legislators reacted to the proposal with mixed feelings: while all agreed that discards in EU fisheries had to be addressed, several Member States and Members of the European Parliament (MEPs) were critical with the specific proposal. Nobody opposed the principle of eliminating or at least reducing discards, but many disagreed on the level of ambition and on the specific details proposed.

The main issue in the negotiation concerned the level of flexibility to apply to the Landing Obligation. This divided the opinions between Member States in Council and MEPs in Parliament. The positions largely reflected the relative weight of industry versus the society at large in establishing the political positions (see also Fitzpatrick et al., this volume; van Hoof et al., this volume). While those more influenced by industry's positions were in favour of a high level of flexibility in the objectives and specific provisions, those more influenced by other societal interests favoured more limited flexibility. In any case, the need for at least some mechanisms for flexibility was recognised by all, notably due to the fact that some discard practices were actually the result of the CFP rules themselves. These mechanisms were included to facilitate implementation in practice.

While it was recognised that discard levels were highly variable between different fisheries and areas, the flexibility mechanisms were the same for all cases. This was the result of the emphasis by the co-legislators to ensure the level playing field, and the difficulty to demonstrate beyond doubt that the different discard levels of different fisheries were inevitable.

2.2.3 The Flexibility Mechanisms

The result of the negotiations was the adoption of a series of mechanisms of flexibility as follows: species for which fishing is prohibited, species that have high survival rates after being discarded and catches which fall under the *de minimis* exemption.

2.2.3.1 The *de minimis* Allowance

The *de minimis* allowance was the clearest recognition by the co-legislators, that achieving zero discards in some fisheries would be almost impossible, notably on account of quota allocations and the cost of handling unwanted fish. So certain allowances should be established for well-defined cases.

The most significant point of friction was the maximum level of this allowance to continue discarding at sea, and whether such allowance should be consistent in all fisheries or modulated according to different discard levels and different difficulty levels to reduce discarding.

Regarding the former, the decision taken (a maximum of 5% after a transitional period starting at 7%) corresponded to the political wish to avoid a bad image of the policy: if the allowance was too high, then the policy would 'look' as if current discard levels could in many cases be maintained, and the co-legislators wanted to send a clear political message that current fishing practices should change.

More difficult was the second question. The evidence that different European fisheries have very different levels of discarding (for example in the North Sea discard levels vary between practically zero in pelagic fisheries to nearly 50% in demersal ones, ICES 2017a) was compounded by the desire of the co-legislators to establish measures that would not look discriminatory. And in the absence of accurate data on different discard levels, providing for different allowance levels for different fisheries could be seen as discriminatory. The CFP's traditional priority to establish a 'level playing field' translated into a *de minimis* provision with the same ceiling for all fisheries. This was introduced under the expectation that discard plans in certain areas would not need to make full use of the maximum allowance and that this maximum allowance would be used only in the most difficult cases. In any case, it was believed by the co-legislators that there should not be, under any circumstance, discard levels higher than the 5% of the *de minimis* rule (although a phasing-in starting at 7% was agreed). This was not based on science, but on the political drive to look ambitious: some co-legislators indicated that a tolerance level of two digits would look like nothing would change.

2.2.3.2 The High Survival Exemption

This exemption was an obvious one: there is no point in landing and eventually killing individuals which otherwise would have survived being discarded. Therefore a high survival clause was included to allow discarding of individuals and species that were scientifically demonstrated to be resilient towards capture-and-discarding processes. However, a definition of what threshold level may be considered as "high" was not given.

Some species are known to survive outside seawater for a time, such as molluscs, skates and rays. This may also be true for some crustaceans. It becomes more complex for fish, whose survival after discarding is not well known in many cases, nor how survival is influenced by technical, biological and environmental factors. This complexity resulted in very loose provisions which have ample room for interpretation.

2.2.3.3 The Inter-Stock Flexibility

This flexibility mechanism (up to 9% of the TAC of a certain stock could be counted against another stock) was established in recognition that under relative stability, the shares of quota allocated to Member States could, in some cases, make it extremely difficult to comply with the Landing Obligation. Co-legislators, however, were very conscious of the risks of this mechanism, and they established two associated conditions: (a) that the stocks from which the catches are attributed to other stocks would be inside safe biological limits – to prevent the mechanism to result in higher catches of weak stocks and (b) that the real catches by species would be recorded, to avoid catch data being misleading.

Catches that are caught in excess of quotas or for which the Member State has no quota, may be deducted from the quota of the target species provided that they: (i) do not exceed 9% of the quota of the target species and (ii) the stock of the non-target species is within safe biological limits (i.e. $F < F_{lim}$ and $SSB > B_{lim}$).

The inter-species quota flexibility (and *de minimis*) can provide flexibility in the system to better adjust catch compositions to resemble fishing opportunities and increase both ecological and economic sustainability. However, a STECF report (STECF 2013) warns that these provisions could be used to legally increase catches well above desired or intended levels, and that they will require careful consideration if negative and unintended consequences are to be avoided.

2.2.4 *What to Do with the Unwanted Fish?*

This was also a hotly debated question concerning whether the undersized fish could be sold for direct human consumption.

Some Member States (particularly from the Baltic basin) considered that, once the undersized fish are caught and counted against quota, it was better to give them some value. They expected this value to be low, so that consuming a significant proportion of the quota with low value fish would be sufficient deterrent not to catch too many undersized fish in the first place.

Other Member States (notably from the Mediterranean basin) considered that this possibility would actually 'legalise' the catch of undersized fish and could lead to an increase in the level of their catches, considering that, unlike the Baltic, small fish can fetch very good prices in that region.

To resolve this dilemma, the co-legislators adopted an intermediate solution: undersized fish inevitably caught should be landed and (where relevant) counted against quota, but then sold only for non-human consumption. The idea behind this solution was twofold:

– if the undersized fish was devoid of any sales value, there would be little incentive for fishermen to land it in the first place, and any amount landed would create a disposal problem for port authorities if nobody could buy it. In addition, the image of –even undersized- fish being shoveled away and dumped with benefit to no-one was not considered a good solution in terms of addressing the credibility problem referred to above.
– If the undersized fish were given some value, but not the full value of the market for direct human consumption, there would be a partial incentive to land the fish inevitably caught, but there would still be an incentive to use the quotas to catch fish that could fetch the full price.

These considerations were based on general knowledge, not on any specific study. This compromise is still largely to be tested in practice (see Iñarra et al., this volume).

2.3 Implementation

The Landing Obligation introduced in the 2013 CFP in its article 15 (EU 2013) includes four specific exemptions: species for which fishing is prohibited, species that have high survival rates after being discarded, catches which fall under the *de minimis* exemption, and catches damaged by predators. It should be noted that the interpretation of these exemptions is not always straightforward. This must be taken into account when evaluating the implementation.

2.3.1 The Flexibility Mechanisms

Species with scientific evidence of high survival rates after being discarded can have an exemption from the obligation to be landed. However, the definition of what is 'high' survival is unclear, while STECF (2013) has concluded that defining a single value cannot be scientifically rationalised and therefore assessing proposed exemptions on the basis of high survival need to be considered on a case-by-case basis, taking account of the species and fisheries involved (Rihan et al., this volume).

This provision has been adopted in several fisheries since 2015. In pelagic fisheries, the application of the exemptions for survival (and the *de minimis*) in some fisheries have provided the flexibility that allowed the industry to formally comply with the Landing Obligation without any significant change in their operations (PELAC 2015; MEDAC 2017).

As for the inter-species quota flexibility, this instrument had not yet been used by April 2017 (Veits 2017).

2.3.2 Predator-Damaged Fish

A further exemption to the LO was introduced with the Omnibus regulation (European Union 2015a), where caught fish damaged by predators should be returned to sea. The reason detailed in the regulation was that such catches "can constitute a risk to humans, to pets and to other fish by virtue of pathogens and bacteria which might be transmitted by such predators".

Nevertheless, this exemption was seen as particularly important for Baltic Sea fisheries targeting salmon, due to the increase in the predatory behaviour of seals consuming salmons caught mainly in longlines (Fitzpatrick and Nielsen 2016). This exemption has indeed been applied to salmon fisheries in the Baltic Sea. But there are no safeguards to limit this mechanism to any particular species or area.

2.3.3 Discard Plans and Minimum Sizes

2.3.3.1 Discard Plans

With the delay in the agreement of the multispecies, multiannual plans foreseen in the 2013 CFP between European institutions, and under the objective of introducing the LO progressively, several so-called discard plans were, in accordance with the CFP, adopted by the EC between 2014 and 2017 through delegated acts. The discard plans identify the specific fisheries and species entering the LO and applicable exemptions by sea area for a period of 3 years, based on joint recommendations by regional Member States groups in consultation with the relevant Advisory Councils.

Although the discard plans were originally planned as an intermediate legislative measure to be substituted gradually by the agreed multiannual management plans in each sea basin, these are now well-established legislative procedures that continue to be adopted and amended regardless of the adoption of a corresponding multiannual plan.

2.3.3.2 Reduction of Minimum Sizes

The 2013 CFP reform introduced specific provisions which allow changing minimum landings/conservation sizes under discard plans and multiannual plans, still under the prevailing aim of ensuring the protection of juveniles of marine organisms. Catches below minimum conservation reference sizes (MCRS, comparable, but not equivalent, to the previously known MLS) have limited use and cannot be sold for human consumption to avoid creating markets for undersized fish.

In the Baltic Sea, the size at which cod can be sold for human consumption was reduced in 2015, i.e. the MLS of 38 cm changed to a MCRS of 35 cm (EU 2014a). As expected, there was an increase in cod landings between 35 and 38 cm, which in turn caused an increase in national quota consumption, because the average fish size of Baltic cod stocks is small (MRAG 2016). At the same time, the industry reported difficulty in increasing gear selectivity due to the restrictions in the trawl gear allowed in the Baltic Sea (BSAC 2016; Valentinsson et al., this volume). This resulted in decreased selectivity by incentivizing commercialization of smaller size eastern cod, while there was no apparent reduction in discard rates (ICES 2017a). This ran counter to the idea of the legislators about the need for increasing selectivity in order to facilitate the implementation of the Landing Obligation.

Furthermore, in south western Atlantic waters, anchovy caught in CECAF area 34.1.2 and in ICES subarea 9 also had a reduction in minimum size with the introduction of the LO in 2015, from a MLS of 12 cm to a MCRS of 9 cm (EU 2014b). In the Skagerrak and Kattegat in 2016, *Nephrops* human consumption size was also reduced from 130 mm total length and 40 mm carapace length to 105 mm and 32 mm respectively (EU 2015b); while clam (Venus spp.) size limits in the Adriatic Sea went from 25 mm to 22 mm in 2017 (EU 2016a).

2.3.4 Additional Regulatory Mechanism: TACs and Prohibited Species

With the phased introduction of the LO between 2015 and 2019, several other regulatory mechanisms have since been used to deal with the LO.

2.3.4.1 TAC Footnotes

Historically, the TAC and quota regulations have included footnotes for some pelagic stocks TACs (e.g. horse mackerel) detailing specific percentages (2% or 5%) of catches of non-target species (e.g. boarfish, haddock, whiting and mackerel) that can be taken as bycatch in pursuit of that target TAC, without being accounted for in the respective non-target stock but on the target pelagic TAC. However, only in 2018 have the footnotes included the LO provisions on interspecies flexibility and its 9% maximum combined catch, and more importantly that the non-target stocks be within safe biological limits.

As the catches of non-target stocks are not necessarily accounted for in their respective TACs, there is a risk of overexploitation of those non-target stocks. STECF (2013, 2017b) highlighted that there is the potential to significantly increase the mortality on non-targeted bycatch species to levels inconsistent with achieving F_{MSY} to the extent that stock biomass could be reduced below safe biological limits. This underlines the importance of ensuring that the use of flexibility mechanisms needs to be consistent with the achievement of the Maximum Sustainable Yield (MSY) objective of the CFP.

2.3.4.2 TAC Increases

Since 2015, catches by fisheries subject to the phased introduction of the LO (with some exemptions) should have been brought to shore and landed. To accommodate the predicted increase in landed catch from such fisheries, the relevant 2015, 2016 and 2017 TACs were increased in accordance with the estimated catch that formerly would have been discarded (Borges 2018). According to the European Commission (2017a, b), TAC adjustments are part of the overall package of measures to implement the LO but they should nevertheless not jeopardise the F_{MSY} objective or increase fishing mortality.

According to Borges (2018), of the 40, 64 and 88 TACs under the LO between 2015 and 2017, respectively, around 30% (the majority of which being TACs for demersal stocks) were increased in 2016–2017 to account for the LO, and of these, 10 TACs were set already above landings advice before any adjustments were made. Therefore, the author concludes that the LO is likely to have contributed to TAC increases above maximum advised catch in 2016 and particularly in 2017 to accommodate the predicted increase in landed catch, and will continue to do so until 2019 when all EU TAC regulated stocks and fisheries in the Atlantic come under the LO. In addition, the fact that these TAC increases have been allocated according to relative stability limits considerably compromises their ability to resolve choke species situations for Member States having zero quota or very low quotas of the stocks concerned.

2.3.4.3 TACs Suppression

Removing TACs from annual TAC regulations so that associated stocks are removed from the LO has been put forward by several stakeholders as a way to deal with problematic stocks, i.e. where discarding is high due to low commercial value and/or where quotas are insufficient to cover catches.

In 2017, following a request from the EC, ICES assessed the sustainability risk to the stock of dab and flounder of having no catch limits as low, as long as dab and flounder remains largely bycatch species (ICES 2017b). With this advice, the EC proposed, and Council agreed, to delete the combined TAC for dab and flounder in the North Sea (EC 2017a). With the suppression of the TAC, these two stocks were removed from the LO and no longer constitute a risk for premature closure of the target fisheries for plaice and sole where they are bycaught. However, they continue to be discarded in high numbers, likely to have between 10% and 30% survival after discarding and low commercial value, but continue to be caught in fisheries that no longer have the incentive to improve selectivity.

In 2018a, discussion is going on regarding several additional TAC removals (EC 2017b).

2.3.4.4 Zero TACs

Zero TACs have been used for a number of years in cases where specific stocks are considered extremely overexploited and in need of complete protection from fishing. In the TAC 2017 regulation (EU 2017), picked dogfish (spurdog, *Squalus acanthias*) were listed as a prohibited species. Specimens should therefore not be harmed and if caught should be released immediately, with the exception for vessels operating in a specific area where landings of up to 270 tonnes of dead picked dogfish are allowed, as long as vessels are engaged in a 'bycatch avoidance programme'. Furthermore, any vessel engaged in such a programme may land not more than 2 tonnes per month of picked dogfish that is dead at the moment when the fishing gear is hauled on board.

The listing of spurdog on the prohibited species list, but with a TAC, has initiated a discussion on how to deal with zero TAC species under the LO: in terms of which species should be listed as prohibited, how species are chosen to be on the list, their level of protection and the level of enforcement. NGOs' position is that designating stocks with zero TAC advice as prohibited species will not protect them from overfishing, and provides little incentive for fishers to improve the selectivity of their fishing practices.

Listing a species on the prohibited species list means discarding can continue, and without a post-release high survival, this measure adds little to the sustainability of the stock.

2.3.5 Prohibited Species List

Both the annual TAC and the technical measures regulations include a list of species for which deliberate catching, retention on board, transshipment or landing is prohibited. Furthermore, when caught, specimens should not be harmed and should be promptly released back into the sea. Species listed in the Convention on International Trade in Endangered Species of Wild Fauna and Flora (CITES) Appendix I are included on the prohibited list. However, except for when listed in CITES, no other specific criteria are known for granting inclusion. As stated in the previous section about zero TACs, several species and stocks have been recently added and removed from the list when they pose a specific issue with the implementation of the LO.

According to STECF (2017c), the prohibited species list should ideally only be used for species which are biologically sensitive to any exploitation. Without additional management measures to improve survival, listing a species will not necessarily lead to a decrease in mortality. Furthermore, the decision to include, or remove, any species on or off the list should be made according to transparent criteria developed in a participatory process.

2.3.6 Technical Measures

Technical measures at EU level are specified through several regulations dating back to 1998, when the original technical measures regulation was adopted (Council Regulation (EC) No 850/98). This regulation specifies areas and seasons where fishing is prohibited, prohibited fishing methods, minimum landing sizes, minimum mesh sizes, among many other measures to minimise the impact of fishing on the marine environment.

In light of the reformed CFP, to simplify rules, and to allow for the introduction of new mechanisms to increase selectivity to facilitate the Landing Obligation, the EC has proposed a new framework for technical conservation measures (EC 2016a). At the time of writing, this proposal was still being discussed by the EU co-legislators.

2.3.7 Multiannual Management Plans

There have also been indirect effects of the LO in other fisheries management measures. The argument that the reality of mixed fisheries associated with the LO is incompatible with reaching MSY for all stocks simultaneously has been gaining momentum in Europe and has been addressed inter alia by Sissenwine et al. (2014) and in the EU research project MYFISH (Rindorf et al. 2017). The discussion concerns different maximum allowed MSY catch opportunities, associated with premature closure of mixed fisheries under the LO, and under- or over-utilization of some of those catch opportunities.

2.3.7.1 F_{MSY} Upper Range

The management plans in the reformed CFP have no explicit harvest control rules (HCRs) but include F_{MSY} ranges between which fishing opportunities can be set when pre-determined conditions are met (EC 2016b; EU 2016b).

The use of the F_{MSY} upper range has been justified to allow for mixed fisheries to adapt to the LO. Managers argued that extra flexibility is needed to cope with the LO and avoid premature fisheries closures, while NGOs defend that the objective of "above MSY levels" enshrined in the CFP is not in line with any F value above F_{MSY}.

Actually, the case for a flexible consideration of F_{MSY} values is based on various arguments: from the need to address choke species problems to the multi-objective nature of the legal basis of the policy both in the EU Treaty and in international fisheries law.

This question has been resolved in the framework of the two first management plans adopted after the reform: the plans for the Baltic (EC 2016b) and the plan for demersal stocks in the North Sea (EU 2018). In both cases, the implementation of F_{MSY} as a range was consolidated, albeit under certain conditions: that the upper limit be precautionary and that the upper part be used only when SSB is above certain thresholds.

It is too soon to evaluate the results of these conditions, whether they will provide the necessary flexibility to address choke species issues (at the level of the TAC) or whether the conditions associated will limit this flexibility to a very low number of cases.

2.3.7.2 Target and Bycatch Species

The new multiannual management plan for the North Sea basin (EU 2018) considers species: (a) that should be managed through MSY (F_{MSY} by 2020) (b) species that may be managed by precautionary approach if MSY advice is not available and (c) other species not subject to catch limits to be managed according to the precautionary approach.

Although the Landing Obligation must apply to all stocks under TAC management, regardless of whether they are target or bycatch stocks, this classification has implications for the implementation of the policy: the classification of the stocks concerned as 'target' or 'bycatch' stocks seems to respond, at least to some extent, to their consideration as 'choke' species, so this would seem to be a case where the implementation of the Landing Obligation has implications for the approach towards MSY objectives.

2.3.8 *Monitoring, Control and Enforcement*

Monitoring and enforcement are essential parts of the implementation of the Landing Obligation. Not only the credibility of the policy depends on effective enforcement, but also its expected positive effects will only be realised if adequate monitoring takes place. For example, the improvement of data (due to reporting on everything that is removed from the sea) is an expected side benefit of the Landing Obligation, but only if effective monitoring takes place. On the other hand, the flexibility mechanisms available will only be used to the full extent if the industry feels the pressure of enforcement; otherwise there would be little incentive to use such mechanisms, which would in turn give the impression that the whole policy is inapplicable.

The move to managing catches instead of landings requires new forms of monitoring (James et al., this volume; Nuevo et al., this volume) and it is partly through this monitoring that the incentive for compliance with the LO and the motivation to avoid unwanted catches can be generated (see also Kraak and Hart, this volume).

2.3.8.1 Postponement of Serious Infringement

Failing to comply with the Landing Obligation is categorized as a serious infringement under the control regulation (Regulation (EC) No. 1224/2009; EC 2009), but a 2 year delay was agreed in the so-called Omnibus regulation in 2015. Sanctions only took effect from the January 1st, 2017 (EU 2015a); from this date MSs had to start applying the points system for illegal discarding.

Although the omnibus regulation did not delay the enforcement and control of the LO, MSs have taken a soft approach in these two first years of the introduction of the LO and have focused on information sharing and training activities rather than on LO compliance (STECF 2017a).

2.3.8.2 Reporting on the Implementation of the LO

The Omnibus regulation (EU 2015a) introduced the obligation to the EC to submit an annual report, starting in 2016 until 2020, on the implementation of the Landing Obligation. This report should be based on information given by the Member States, the Advisory Councils and other relevant sources.

The 2015, 2016 and 2017 MSs reports were reviewed and summarised by STECF (2016, 2017a, 2018) and show a qualitative analysis of the efforts made by MSs on the different areas of the LO implementation (Rihan et al., this volume). Nevertheless, in 2017 STECF noticed that, "*overall, Member States indicate that difficulties encountered so far have been minimal but several reports have highlighted that the most significant issue they face is the industries' reluctance to implement the*

Landing Obligation, despite considerable efforts to disseminate information to them. They also report that fishers seem slow to change fishing practices; and in many areas, a "business as usual" mentality seems to prevail".

2.3.8.3 Revision of the Control Regulation

EC is proposing a revision of the Control Regulation (EC) No. 1224/2009 (EC 2009) with a draft proposal adopted in 2018 (EC 2018b). The EC has started a discussion on the elements of this review, and have stated their intention to include the use of Remote Electronic Monitoring (REM or EM, video and sensors system; James et al., this volume) to monitor the implementation of the LO on selected fisheries on a voluntary basis (EC 2017b).

In this context, and in a letter inviting inputs from Advisory Councils on its proposal for establishing Specific Control and Inspection Programmes, the EC stated that "*independent research, audits of the MS control systems conducted by the Commission, and the 'last-haul'* (Nuevo et al., this volume) *and other project initiatives driven by the European Fisheries Control Agency (EFCA) alongside the MS authorities, all indicate a general lack of compliance with the LO and that illegal and unrecorded discarding is widespread*".

2.4 Future Perspectives

At this point in time it is difficult to foresee future developments. The effects of the Landing Obligation as established cannot be assessed ex ante because these effects depend largely on whether Member States will use existing flexibility mechanisms to the full. It also depends on the level of enforcement. However, the ongoing experience in the search for practical solutions for implementation allows certain initial conclusions to be drawn.

2.4.1 Facilitating Implementation

The effective implementation of this policy as of 2019 will require progress in a number of areas of the policy. Possible ideas are, inter alia:

- Strengthening of at-sea Monitoring, Control and Surveillance (MCS) systems to allow for a level playing field between fisheries and Member States. Already new technologies are being considered and made possible in the revised Control regulation to allow for this.
- Using the flexibility mechanisms to the full. This is not yet the case, in particular with the inter-species cross-reporting mechanism, barely used so far;

- Consolidating fishing mortality ranges within new multiannual plans. These ranges were first introduced in the Baltic management plan (EU 2016b) largely due to avoiding choke species effects, but have been seriously questioned by environmental NGOs and subject to strict conditions by policy makers;
- Enhancing quota swaps through increasing transparency, providing the European industry with a better knowledge of the swapping opportunities available in other member States;
- Introducing bycatch quotas for stocks where some Member States have zero quotas and where they cannot manage to obtain small quotas through swaps. Although this option may be seen by many as a breach of relative stability, the annual fishing opportunities regulation already contain a number of such bycatch quotas;
- Supporting the establishment of real-time closures, at least in certain fisheries with high proportions of juvenile fish. A legal basis for this already exists in the control regulation (EC 2009);
- Facilitating (through funding of the European Maritime and Fisheries Fund EMFF; EU 2014c) the investment in more selective gear, and in land-based equipment to store and process unwanted catches. At present, very little investment of this kind is taking place.

These ideas may be necessary in different combinations in different areas/fisheries, and the EC has taken the initiative to discuss them with Advisory Councils and Member States' regional groupings to identify the most serious cases and the mechanisms to resolve choke species issues.

2.4.2 Possible Legislative Changes?

The European Commission has no plan to propose an amendment to the existing legislation. This indicates that the European Commission believes that it is necessary to try to implement the 2013 policy, and has started a proactive campaign to discuss and promote a variety of ideas to allow for its efficient implementation. The reason is clear: before thinking about a legislative change, it is necessary to use all the elements of flexibility and effective MCS systems to allow for effective implementation.

For the future, the Commission must submit a report on the application of the policy by 31 December 2022. Although it is premature to prejudge what the Commission will decide to report on the Landing Obligation, one can, already at this stage, suggest some ideas to consider regarding a future policy:

- The modulation of objectives. The current legislation provides for exactly the same levels of discard allowances in all areas and all fisheries, thus ignoring very different current levels of discarding and very different levels of difficulty to achieve to those target levels. In the North Sea alone, discard levels in different fisheries range from 1% to 2% to around 50% (ICES 2017a). This harmonisation may need to be revised to fix more achievable objectives for different fisheries;

- Establishing clearer objectives for the improvement of selectivity in the context of technical measures. While the 5% above mentioned may be too simplistic and poorly adapted to different circumstances, the legislation could at least establish as a principle the improvement of exploitation patterns to levels that would maximize MSY yields and minimize the catch of juvenile fish. These principles would then translate into different objectives in different fisheries and areas.
- Advancing towards multispecies approaches would also help reduce choke species effects. The scientific basis for this is being developed, at least for certain areas such as the North Sea (ICES 2013). This could allow a new policy focus where greater flexibility for individual stocks would be compensated with more emphasis on the total removals from the fishery (and thus a more ecosystem-based management). Allocation-related chokes would also benefit from a new approach to implement relative stability: more focused on the total bonus that Member States would obtain from the fishery, and less on individual stock allocations; this could be done based on an overall evaluation of the value of fishing rights by Member States and a redistribution of individual quotas within that overall value.
- A new approach to the implementation of relative stability would help address the problem of allocation-related chokes. Indeed, relative stability was established in 1983 as a principle, with the specific allocation keys being open to adaptive revision (EEC 1983). The possibility to include bycatch quotas (already existing in some TACs) would be a solution in some cases. A more ambitious approach would be to recast relative stability as a total percentage of the value that each Member State would obtain from the EU, and then allow for a reallocation of national allocations within these overall percentages, to better match fishing rights with real catches.

In any case, there seems to be no way back from the non-discard policy. The evolution of fisheries policy around the world is going in that direction (Karp et al., this volume), and the practice of discarding is increasingly considered unacceptable from different points of view, including considerations about food security. Any possible future change in the policy is likely to be about adjustments to improve practical implementation and mitigate possible negative results, but a CFP with high levels of discarding will never return.

Acknowledgements Part of this work has received funding from the Horizon 2020 Programme under grant agreement DiscardLess number 633680. This support is gratefully acknowledged.

References

Borges, L. (2015). The evolution of a discard policy in Europe, *Fish and Fisheries, 16* (3), 534–540. https://doi.org/10.1111/faf.12062.

Borges, L. (2016, February 26). One year on: The landing obligation in Europe. *ICES Newsletter*.

Borges, L. (2018). Setting of total allowable catches in the 2013 EU common fisheries policy reform: Possible impacts. *Marine Policy, 91*, 97–103.

Borges, L., Cocas, L., Nielsen, K.N. (2016). Discard ban and balanced harvest: A contradiction? *ICES Journal of Marine Science, 73* (6), 1632–1639. https://doi.org/10.1093/icesjms/fsw065.

BSAC. (2016). Joint Working Group (Demersal + Pelagic) to continue the discussions on technical measures for the Baltic and draft amendments to the technical measures regulation for the Baltic 2187/2005. 26–27 January 2016. Gdynia, Poland, p. 12.

CEC. (1992). Report from the Commission to the Council on the discarding of fish in Community fisheries: Causes, impact, solutions. SEC (92) 423 final, 12 March 1992. Brussels, p. 54.

CEC. (2001). Green paper on the future of the common fisheries policy. COM (2001) 135 final, 20 March 2001. Brussels, p. 40.

CEC. (2007). Communication from the Commission to the Council and the European Parliament: A policy to reduce unwanted by-catches and eliminate discards in European fisheries. COM(2007) 136 final, p. 8.

CEC. (2009). Commission of the European Communities Green Paper – Reform of the Common Fisheries Policy. COM (2009)163 final, p. 28.

CEC. (2011a). Communication from the Commission to the European Parliament, the Council, the European Economic and Social Committee and the Committee of the Regions. Reform of the Common Fisheries Policy. COM(2011) 417 final, 13 July 2011. Brussels, p. 12.

CEC. (2011b). Commission Staff Working Paper: Impact Assessment Accompanying the document "Commission proposal for a Regulation of the European Parliament and of the Council on the Common Fisheries Policy". SEC(2011) 891 final.

EC. (1998). Council Regulation (EC) No 850/98 of 30 March 1998 for the conservation of fishery resources through technical measures for the protection of juveniles of marine organisms. *Official Journal of the European Union*, L125, 1–55.

EC. (2009). Council Regulation (EC) No 1224/2009 of 20 November 2009 establishing a Community control system for ensuring compliance with the rules of the common fisheries policy, amending Regulations (EC) No 847/96, (EC) No 2371/2002, (EC) No 811/2004, (EC) No 768/2005, (EC) No 2115/2005, (EC) No 2166/2005, (EC) No 388/2006, (EC) No 509/2007, (EC) No 676/2007, (EC) No 1098/2007, (EC) No 1300/2008, (EC) No 1342/2008 and repealing Regulations (EEC) No 2847/93, (EC) No 1627/94 and (EC) No 1966/2006. *Official Journal of the European Union*, L343, 1–50.

EC. (2016a). Proposal for a Regulation of the European Parliament and of the Council on the conservation of fishery resources and the protection of marine ecosystems through technical measures, amending Council Regulations (EC) No 1967/2006, (EC) No 1098/2007, (EC) No 1224/2009 and Regulations (EU) No 1343/2011 and (EU) No 1380/2013 of the European Parliament and of the Council, and repealing Council Regulations (EC) No 894/97, (EC) No 850/98, (EC) No 2549/2000, (EC) No 254/2002, (EC) No 812/2004 and (EC) No 2187/2005. COM (2016) 134 final, 11 March 2016. Brussels, p. 44.

EC. (2016b). Proposal for a Regulation of the European Parliament and of the Council on establishing a multi-annual plan for demersal stocks in the North Sea and the fisheries exploiting those stocks and repealing Council Regulation (EC) 676/2007 and Council Regulation (EC) 1342/2008. COM(2016) 492 final, p. 23.

EC. (2017a). Communication from the Commission on the State of Play of the Common Fisheries Policy and Consultation on the Fishing Opportunities for 2018. COM (2017) 368 final. http://eur-lex.europa.eu/legal-content/EN/TXT/PDF/?uri=CELEX:52017DC0368&from=EN. Accessed 29 Aug 2018.

EC. (2017b). Landing Obligation seminar November 2017 – Summary. https://ec.europa.eu/fisheries/sites/fisheries/files/docs/pages/landing-obligation-seminar-november-2017-summary_en.pdf. Accessed 29 Aug 2018

EC. (2018a). Discarding and the Landing Obligation. https://ec.europa.eu/fisheries/cfp/fishing_rules/discards_en. Accessed 29 Aug 2018.

EC. (2018b). Proposal for a Regulation of the European Parliament and of the Council amending Council Regulation (EC) No 1224/2009, and amending Council Regulations (EC) No 768/2005, (EC) No 1967/2006, (EC) No 1005/2008, and Regulation (EU) No 2016/1139 of the European Parliament and of the Council as regards fisheries control. COM(2018) 368 final, 30 May 2018. Brussels, p. 91.

EEC. (1983). Council Regulation (EEC) No 170/83 of 25 January 1983 establishing a Community system for the conservation and management of fishery resources. *Official Journal of the European Communities,* L24, 1–13.

EU. (2013). Regulation (EU) No 1380/2013 of the European Parliament and of the Council of 11 December 2013 on the Common Fisheries Policy. Brussels, Belgium. http://eur-lex.europa.eu/LexUriServ/LexUriServ.do?uri=OJ:L:2013:354:0022:0061:EN:PDF. Accessed 29 Aug 2018.

EU. (2014a). Commission Delegated Regulation (EU) No 1396/2014 of 20 October 2014 establishing a discard plan in the Baltic Sea. *Official Journal of the European Union*, L370, 40–41.

EU. (2014b). Commission Delegated Regulation (EU) No 1394/2014 of 20 October 2014 establishing a discard plan for certain pelagic fisheries in south-western waters. *Official Journal of the European Union,* L370, 31–34.

EU. (2014c). Regulation (EU) No 508/2014 of the European Parliament and of the Council of 15 May 2014 on the European Maritime and Fisheries Fund and repealing Council Regulations (EC) No 2328/2003, (EC) No 861/2006, (EC) No 1198/2006 and (EC) No 791/2007 and Regulation (EU) No 1255/2011 of the European Parliament and of the Council. *Official Journal of the European Union,* L149, 1–66.

EU. (2015a). Regulation (EU) 2015/812 of the European Parliament and of the Council of 20 May 2015 amending Council Regulations (EC) No 850/98, (EC) No 2187/2005, (EC) No 1967/2006, (EC) No 1098/2007, (EC) No 254/2002, (EC) No 2347/2002 and (EC) No 1224/2009, and Regulations (EU) No 1379/2013 and (EU) No 1380/2013 of the European Parliament and of the Council, as regards the Landing Obligation, and repealing Council Regulation (EC) No 1434/98. *Official Journal of the European Communities*, L133, 1–20.

EU. (2015b). Commission Delegated Regulation (EU) No 2015/2440 of 22 October 2015 establishing a discard plan for certain demersal fisheries in the North Sea and in Union waters of ICES Division IIa. *Official Journal of the European Union*, L336, 42–48.

EU. (2016a). Commission Delegated Regulation (EU) No 2016/2376 of 13 October 2016 establishing a discard plan for mollusc bivalve Venus spp. in the Italian territorial waters. *Official Journal of the European Union,* L352, 48–49.

EU. (2016b). Regulation (EU) No 2016/1139 of the European Parliament and of the Council of 6 July 2016 establishing a multiannual plan for the stocks of cod, herring and sprat in the Baltic Sea and the fisheries exploiting those stocks, amending Council Regulation (EC) No 2187/2005 and repealing Council Regulation (EC) No 1098/2007. *Official Journal of the European Union,* L191, 1–15.

EU. (2017). Regulation (EU) 2017/127 of the Council of 20 January 2017 fixing for 2017 the fishing opportunities for certain fish stocks and groups of fish stocks, applicable in Union waters and, for Union vessels, in certain non-Union waters. *Official Journal of the European Union,* L24, 1–172.

EU. (2018). Regulation (EU) No 2018/973 of the European Parliament and of the Council of 4 July 2018 establishing a multiannual plan for demersal stocks in the North Sea and the fisheries exploiting those stocks, specifying details of the implementation of the Landing obligation in the North Sea and repealing Council Regulations (EC) No 676/2007 and (EC) No 1342/2008. *Official Journal of the European Union,* L179, 1–13.

Fitzpatrick, M., Graham, N., Rihan, D.J., Reid, D.G. (2011). The burden of proof in co-management and results-based management: The elephant on the deck! *ICES Journal of Marine Science, 68*(8), 1656–1662.

Fitzpatrick, M., & Nielsen, K.N. (2016). DiscardLess Policy Brief No1: Year 1 of the Landing Obligation, key issues from the Baltic and Pelagic fisheries, 30 September 2016. https://doi.org/10.5281/zenodo.215155.

Fitzpatrick, M., Frangoudes, K., Fauconnet, L., Quetglas, A. (this volume). Fishing industry perspectives on the EU Landing Obligation. In S.S. Uhlmann, C. Ulrich, S.J. Kennelly (Eds.), *The European Landing Obligation – Reducing discards in complex, multi-species and multi-jurisdictional fisheries*. Cham: Springer.

Gambarato, R.P., & Medvedev, S.A. (2016). Transmedia storytelling impact on government policy change. In *Politics, protest, and empowerment in digital spaces* (pp. 31–51). Hershey: IGI Global. https://doi.org/10.4018/978-1-5225-1862-4.ch003.

ICES. (2013). North Sea. Multispecies considerations for North Sea stocks. ICES Advice 2013, 6.3.1, p. 9. http://www.ices.dk/sites/pub/Publication%20Reports/Advice/2013/2013/mult-NS.pdf. Accessed 29 Aug 2018.

ICES. (2017a). Greater North Sea Ecoregion – Fisheries overview. Published 4 July 2017. https://doi.org/10.17895/ICES/pub.3116.

ICES. (2017b). EU request on a combined dab and flounder TAC and potential management measures besides catch limits. ICES Advice 2017. ICES Special Request Advice, p. 8. http://ices.dk/sites/pub/Publication%20Reports/Advice/2017/Special_requests/eu.2017.04.pdf. Accessed 29 Aug 2018.

Iñarra, B., Bald, C., Cebrián, M., Antelo, L.T., Franco-Uría, A., Vázquez, J.A., Pérez-Martín, R., Zufía, J. (this volume). What to do with unwanted catches: Valorisation options and selection strategies. In S.S. Uhlmann, C. Ulrich, S.J. Kennelly (Eds.), *The European Landing Obligation – Reducing discards in complex, multi-species multi-juridictional fisheries*. Cham: Springer.

James, K.M., Campbell, N., Viðarsson, J.R., Vilas, C., Plet-Hansen, K.S., Borges, L., et al. (this volume). Tools and technologies for the monitoring, control and surveillance of unwanted catches. In S.S. Uhlmann, C. Ulrich, S.J. Kennelly (Eds.), *The European Landing Obligation – Reducing discards in complex, multi-species and multi-jurisdictional fisheries*. Cham: Springer.

Joa, A. (2015). *Fish fight in Europe: A process-analysis of the campaign for a discard ban provision in the EU's CFP Reform*. BSc thesis. University of Twente, The Netherlands, p. 54.

Karp, W.A., Breen, M., Borges, L., Fitzpatrick, M., Kennelly, S.J., Kolding, J., et al. (this volume). Strategies used throughout the world to manage fisheries discards – Lessons for implementation of the EU Landing Obligation. In S.S. Uhlmann, C. Ulrich, S.J. Kennelly (Eds.), *The European Landing Obligation – Reducing discards in complex, multi-species and multi-jurisdictional fisheries*. Cham: Springer.

Kraak, S.B.M., & Hart, P.J.B. (this volume). Creating a breeding ground for compliance and honest reporting under the Landing Obligation: Insights from behavioural science. In S.S. Uhlmann, C. Ulrich, S.J. Kennelly (Eds.), *The European Landing Obligation – Reducing discards in complex, multi-species and multi-jurisdictional fisheries*. Cham: Springer.

MRAG. (2016). 4th Surveillance report, DFPO Denmark Eastern Baltic Cod Fishery (p. 28). https://www.msc.org/track-a-fishery/fisheries-in-the-program/certified/north-east-atlantic/Denmark-Eastern-Baltic-cod/assessment-downloads-1/20160128_SR_COD140.pdf

MEDAC. (2017). Annual report on the implementation of the LO for small pelagics. MEDAC Contribution. Ref.: 38/2017, p. 7.

Nuevo, M., Morgado, C., Sala, A. (this volume). Monitoring the implementation of the Landing Obligation: Last Haul programme. In S.S. Uhlmann, C. Ulrich, S.J. Kennelly (Eds.), *The European Landing Obligation – Reducing discards in complex, multi-species and multi-jurisdictional fisheries*. Cham: Springer.

PELAC. (2015). *Annex II: Experiences with the Landing Obligation in pelagic fisheries* (pp. 9–12). http://www.pelagicac.org/media/pdf/1516PAC12%20Recommendations%20on%20control%20of%20LO.pdf

Penas Lado, E. (2016). *The common fisheries policy. The quest for sustainability* (p. 395). Ames: Wiley Blackwell.

Rihan, D., Uhlmann, S.S., Ulrich, C., Breen, M., Catchpole, T. (this volume). Requirements for documentation, data collection and scientific evaluations. In In S.S. Uhlmann, C. Ulrich, S.J. Kennelly (Eds.), *The European Landing Obligation – Reducing discards in complex, multi-species and multi-jurisdictional fisheries*. Cham: Springer.

Rindorf, A., Dichmont, C.M., Thorson, J., Charles, A., Clausen, L.W., Degnbol, P., et al. (2017). Inclusion of ecological, economic, social, and institutional considerations when setting targets and limits for multispecies fisheries. *ICES Journal of Marine Science, 74*(2), 453–463.

Sissenwine, M.M., Mace, P., Lassen, H.J. (2014). Preventing overfishing: Evolving approaches and emerging challenges. *ICES Journal of Marine Science, 71*(2), 153–156.

STECF. (2006). Discards from community vessels. STECF Plenary meeting, Ispra, 6–10 November 2006. Publications Office of the European Union, Luxembourg. https://stecf.jrc.ec.europa.eu/documents/43805/99464/2006-11_23rd+report+of+the+STECF.pdf. Accessed 29 Aug 2018.

STECF. (2013). Landing Obligation in EU fisheries (STECF-13-23). Publications Office of the European Union, Luxembourg, EUR 26330 EN, JRC 86112, p. 115.

STECF. (2016). Methodology and data requirements for reporting on the Landing Obligation (STECF-16-13). Publications Office of the European Union, Luxembourg, EUR 27758 EN. https://doi.org/10.2788/984496.

STECF. (2017a). 54th Plenary meeting report (PLEN-17-01). Publications Office of the European Union, Luxembourg; EUR 28569 EN. https://doi.org/10.2760/33472.

STECF. (2017b). 55th Plenary meeting report (PLEN-17-02). Publications Office of the European Union, Luxembourg. EUR 28359 EN. https://doi.org/10.2760/53335.

STECF. (2017c). Long-term management of skates and rays (STECF-17-16). Publications Office of the European Union, Luxembourg.

STECF. (2018). 57th Plenary meeting report (PLEN-18-01). Publications Office of the European Union, Luxembourg. ISBN 978-92-79-85804-8. https://doi.org/10.2760/088784.

Valentinsson, D., Ringdahl, K., Storr-Paulsen, M., Madsen, N. (this volume). The Baltic cod trawl fishery: The perfect fishery for a successful implementation of the Landing Obligation? In S.S. Uhlmann, C. Ulrich, S.J. Kennelly (Eds.), *The European Landing Obligation – Reducing discards in complex, multi-species and multi-jurisdictional fisheries*. Cham: Springer.

van Hoof, L., Kraan, M., Visser, N. M., Avoyan, E., Batsleer, J., Trapman, B. (this volume). Muddying the waters of the Landing Obligation: How multi-level governance structures can obscure policy implementation. In S.S. Uhlmann, C. Ulrich, S.J. Kennelly (Eds.), *The European Landing Obligation – Reducing discards in complex, multi-species and multi-jurisdictional fisheries*. Cham: Springer.

Veits, V. (2017). Presentation given at the European Parliament Public Hearing on the state of play of the implementation of Landing Obligation and allocation of quotas (24 April 2017). http://www.europarl.europa.eu/cmsdata/117542/Veronika%20Veits_DG%20Mare.pdf. Accessed 29 Aug 2018.

Chapter 3
Requirements for Documentation, Data Collection and Scientific Evaluations

Dominic Rihan, Sven S. Uhlmann, Clara Ulrich, Mike Breen, and Tom Catchpole

Abstract This chapter describes the process and evaluation behind several of the flexibility elements introduced into the Basic Regulation. Firstly, it describes how a fishery may be granted a "high survival exemption" (HSE) or "*de minimis*" to the Landing Obligation, for a particular regulated species. It details the process of generating any exemption supporting evidence by Member States and the evaluation process. The impact of the Landing Obligation on the scientific advice for setting fishing opportunities is also described as well as the annual reporting process. The final section deals with the challenges faced by the main players in this whole process.

Keywords *De minimis* · Delegated act · European Commission · Flexibility · High survival · ICES · Joint recommendations · Landing Obligation · STECF · TACs and quotas

---

Electronic Supplementary Material The online version of this chapter (https://doi.org/10.1007/978-3-030-03308-8_3) contains supplementary material, which is available to authorized users.

D. Rihan (✉)
Bord Iascaigh Mhara, Dun Laoghaire, Co. Dublin, Ireland
e-mail: Dominic.Rihan@bim.ie

S. S. Uhlmann
Flanders Research Institute for Agriculture, Fisheries and Food, Oostende, Belgium

C. Ulrich
DTU Aqua, Technical University of Denmark, Kgs Lyngby, Denmark

M. Breen
Institute of Marine Research (IMR), Bergen, Norway

T. Catchpole
Centre for Environment, Fisheries & Aquaculture Science (Cefas), Lowestoft, UK

S. S. Uhlmann et al. (eds.), *The European Landing Obligation*,
https://doi.org/10.1007/978-3-030-03308-8_3

3.1 Introduction

According to the Basic Regulation (EU) 1380/2013, the Common Fisheries Policy (CFP) recognises that to ensure good governance, decision-making should, among others, be based on the principles of best available scientific advice. Scientific advice comprises the provision of catch advice for setting fishing opportunities, and evaluation and assessment of specific elements of the CFP. This chapter describes the new challenges that the Landing Obligation (Article 15) has posed to the two main advisory bodies: ICES (the International Council for the exploration of the Sea) and STECF (the EU Scientific, Economic and Technical Committee for Fisheries).

The main role of ICES has been to change the focus of the advice provided to the EU Commission from landings to catches to take account of the new management approach under the Landing Obligation. In this context, ICES has considered how best to provide information corresponding to wanted (i.e. catches above minimum conservation reference sizes) and unwanted catches (i.e. catches below minimum conservation reference sizes) and also to estimate catches discarded under exemptions to the Landing Obligation. ICES has also developed advice and methodologies to support experimental work in relation to high survivability exemptions, building on initial work by STECF. Additionally, ICES has received specific requests from the Commission to consider the impacts on the removal of TACs (ICES 2017) and amending specific Regulations (e.g., the sprat box in the North Sea).

STECF performs scientific work directly requested by the EU Commission. The main body of work carried out by STECF with regard to the Landing Obligation has been in assessing the main provisions and flexibilities provided for in Article 15 and the likely impacts of applying such mechanisms. To date STECF has held nine Expert Working Groups to advise on various aspects of the Landing Obligation (STECF 2013, 2014a, b, c, d, 2015a, b, 2016a). STECF (2015c) has also developed a methodology for the calculation of increases in TACs to take account of catches previously discarded (i.e., so-called TAC uplifts). Finally, STECF has reviewed the mandatory annual reports provided by Member States and the Advisory Councils on the implementation of the Landing Obligation (STECF 2016a).

3.2 High Survival Exemption

Article 15, paragraph 2(b), of the CFP describes an exemption from the Landing Obligation for *"species for which scientific evidence demonstrates high survival rates, taking into account the characteristics of the gear, of the fishing practices and of the ecosystem"*. Clear, defensible, scientific evidence is required to support any proposal for such an exemption for selected species or fisheries. The inclusion of this provision within the CFP has sparked a sudden and uniquely synchronized interest in discard survival assessments and mobilised Member States and fishing industry representative organisations to apply for research projects to generate supporting evidence.

In the 1990s and early 2000s, the importance of quantifying, and mitigating sources of unaccounted fishing mortality (e.g., discards, escapees, ghost fishing, illegal and unreported fishing) in addition to landed catches was recognised at ICES advisory level. This led to the establishment of several expert groups on unaccounted fishing mortality which also considered the survivability of escaping fish and discards (e.g., ICES 1995, 1997, 2005). However, these groups ceased to exist beyond 2009 due to a lack of activity in this research area, with funding prioritised elsewhere. With the inclusion of High Survival Exemptions (HSE) in Article 15, interest in researching discard survivability has grown considerably. This section describes the process on (a) how HSE evidence is generated, (b) how it is submitted, and (c) which fisheries (species/métier) have been subsequently granted a HSE to the Landing Obligation after review by the European Commission and evaluation by STECF.

3.2.1 Description of the Evidence

Following STECF (2013) advice, ICES established a workshop on Methods for Estimating Discard Survival (WKMEDS) in January 2014, to provide guidance on how to quantify discard survival robustly. WKMEDS published its first draft guidance in April 2014 (ICES 2014), ahead of the evaluation of the first Joint Recommendations (JR) submitted by the regional groups of Member States. WKMEDS recommended: (i) assessments should be representative of discarded catch and practices, ideally at a métier scale; (ii) methods should avoid biasing results through observation induced mortality, and wherever possible demonstrated with appropriate controls; and (iii) the monitoring period should be sufficiently long to observe any delayed mortality attributable to the catch-and-discarding process.

To quantify (sub-) lethal stress and discard survival, three methodologies were identified: captive observation; tagging/biotelemetry; and vitality/reflex assessments; (ICES 2014; Breen and Catchpole in press). In captive observation, sub-samples of animals are selected from the discarded catch and monitored to provide estimates of survival rates. Tagging/biotelemetry use tagging technologies to monitor post-release mortality of (tagged) organisms. Vitality assessments quantify the health of the organism at the point of discarding. By combining vitality assessments with one or both of the other two techniques, the at-capture condition may be correlated with an individual's likelihood of post-release survival (Davis 2010). Depending on the strength of such a correlation, a vitality index may be used as a proxy for survival (e.g., Barkley and Cadrin 2012; Morfin et al. 2017).

Over a series of five subsequent meetings, WKMEDS provided an open forum for researchers and stakeholders actively involved in survival assessments to discuss and develop their methods (ICES 2015a, b, 2016a, b, c). Results from this group were disseminated to other interested ICES expert groups and were used by scientists to guide research efforts in this field (e.g., Uhlmann et al. 2016; Methling et al. 2017).

The WKMEDS group also developed protocols for systematically reviewing survival assessments and meta-analysing survival data. Key questions were established (ICES 2015a, b; Catchpole et al. in submission) which should be considered by researchers and which may be used by external reviewers (i.e., STECF) to gauge the quality and robustness of any collected evidence.

3.2.2 Evidence Collected So Far

Spanning from the Mediterranean to the Baltic, more than 20 studies have been commissioned, in most cases with public funding, in at least 11 Member States, and Turkey between 2013 and 2018. An overview of the studied species, their survival rates and corresponding references has been included in an online supplement (Table 3.3; in Electronic supplementary material). A critical review of the survival rates of flatfish and *Nephrops* is in preparation (Catchpole et al. unpubl. data).

At least 20 species have been assessed for vitality status at the point of discarding and monitored for delayed mortality through captivity experiments. Species of such interest included flatfish such as Common sole (*Solea solea*), European plaice (*Pleuronectes platessa*), turbot (*Scophthalmus maximus*); pelagic species such as sardines (*Sardina pilchardus*), Atlantic salmon (*Salmo salar*), European seabass (*Dicentrarchus labrax*) and crustaceans such as Norway lobster (*Nephrops norvegicus*) and common spiny lobsters (*Palinurus elephas*). The prioritisation of species and fisheries to investigate was influenced by the potential for high survival, based on existing knowledge, and the likely timing for introduction under the Landing Obligation during the phasing-in period. The emphasis for research has recently turned to commercial skates and ray species, such as thornback ray (*Raja clavata*). This is, because it has been recognised that, despite indications that the HSE was originally intended for such relatively robust elasmobranch species in the first place (see Borges and Penas Lado, this volume), there are gaps in the evidence to support this.

Species discarded from both active and passive gear types have been studied including beam and otter trawls, Danish and purse seines, trammel-nets, gill nets, pots, creels and fish traps and some recreational hook-and-line fisheries.

It is fair to say that generating robust evidence on discard survival estimates that is representative of a fishery is challenging. Most studies have faced logistical difficulties in planning field trips and accommodating equipment onboard commercial or recreational charter vessels. Organisms monitored in captivity need to be housed in conditions sympathetic to their biological needs and should be monitored long enough for any fishing-related mortality to occur (Uhlmann et al. 2016; van der Reijden et al. 2017). Similarly, tagged organisms need to be monitored/detected/recaptured in sufficient numbers to produce meaningful survival estimates based on detection/recapture rates. Some of these limitations mean that the generated evidence is not purely empirical, but also model-based. For example, extension models were developed to estimate survival in cases where monitoring periods could not be extended (Catchpole et al. 2015) or where it was not possible to house fish holding facilities on-board (Morfin et al. 2017).

3.2.3 Review of the Evidence

The evidence submitted by Member States to support requests for a HSE has been evaluated by STECF (STECF 2013, 2014a). These evaluations considered whether the survival assessment methods are appropriate, and whether the limitations of the results have been fully explored. The ICES (2014) guidance on survival assessment protocols has been used to promote best practice and harmonisation of recently conducted assessments (ICES 2014; 2015a). Furthermore, the guidance has provided a reference for critical review which is used by STECF to apply a systematic and consistent evaluation process. Assessing the robustness of evidence, and its representativeness across entire fisheries, has proved challenging for STECF, and many studies have been criticized for lacking sufficient data to adequately describe the potential variability in survival at the fishery scale (STECF 2014b, 2015a, 2016b). STECF has repeatedly emphasized, "*that before considering the implementation of a high survival exemption, it should be remembered that avoidance of unwanted catch, through improved selectivity or other means, is the primary objective of the Landing Obligation*" (STECF 2016b).

Defining what "high survival" means has also been challenging. STECF has highlighted that this is a subjective term that involves trade-offs between different management and societal objectives, driven by the management priority for that fishery at that particular time (e.g., improving stock sustainability; improving financial viability; or avoiding waste). However, survivability should be considered in the context of the discard rates in the fishery. If only a relatively small proportion of discarded fish are likely to survive then granting such an exemption would seem counter intuitive (STECF 2016b). This could be interpreted by some as opening a loophole in the legislation that could be used to circumvent the Landing Obligation (Stockhausen, this volume). But rigorous stock assessments accounting for discard survival are needed to evaluate what the consequences may be for a stock.

Using the evaluation by STECF on the discard survival evidence provided by the Member States, the EU Commission decides whether to grant exemptions. To date, eight of the three-year Discard Plans, contained in the 13 Commission Delegated Regulations enacted, have included survival exemptions, which have been approved by the EU Commission. These apply to five regions (Baltic Sea, Black Sea, North Western Waters, South Western Waters and the North Sea) and provide survival exemptions for a total of 11 species such as anchovy (*Engraulis encrasicolus*), herring (*Clupea harengus*), mackerel (*Scomber scombrus*), horse mackerel (*Trachurus trachurus*), jack mackerel (*Trachurus picturatus*), Atlantic cod (*Gadus morhua*), Atlantic salmon (*Salmo salar)*, Norway lobster (*Nephrops norvegicus*), common sole (*Solea solea*), turbot (*Scophthalmus maximus*), and scallop (*Pecten jacobeus*) caught in specific areas and fishing gears. The empirical estimates of discard survival derived from experimentation for these species caught in specific fisheries and areas range from 46–90% survival. Several high survivability exemptions are also in place for highly migratory species, such as tunas and billfish. These exemptions implement recommendations emanating from relevant Regional Fisheries Management Organisations (RFMOs), such as ICCAT (STECF 2017a).

3.3 De Minimis

Article 15, paragraph 2(c) of the CFP allows for *de minimis* exemptions of up to 5% of total annual catches of all species subject to the Landing Obligation that can be discarded. Such catches are not counted against quotas, but must be recorded in fisher's logbooks, such as Electronic Reporting Systems (ERS). According to Recitals 29 & 31 of the CFP, the application of *de minimis* should be considered as a "last resort" mechanism and only after other technical or tactical approaches to avoid capture of unwanted catch have been fully explored. *De minimis* exemptions in the CFP are subject to two conditions, "*where scientific evidence indicates that increases in selectivity are very difficult to achieve*" or "*to avoid disproportionate costs of handling unwanted catches, for those fishing gears where unwanted catches per fishing gear do not represent more than a certain percentage, to be established in a plan, of total annual catch of that gear*".

The first of the STECF meetings in 2013 took this provision to mean that the spirit and general purpose of the *de minimis* provision ('a small discard proportion'), was to provide a 'safety valve' allowing for some discarding in the most difficult circumstances (STECF 2013). Accordingly, STECF outlined the types of information that would be needed to justify such exemptions, based on the conditions contained in the Regulation. While the *de minimis* exemption allows for minimal levels of discarding to be authorised, it does not provide a solution to quota restrictions that could result in the cessation of fishing (a so called 'choke' event). The *de minimis* provision does not enable more catches to be taken, but rather provides flexibility on how catches are handled onboard and the destination of those catches.

3.3.1 Description of the Evidence

At an early stage, STECF identified that the concept of *de minimis* exemptions was largely an economic argument (STECF 2013, 2014a, d). It was based on the logic that at a certain level of improvement in selectivity, losses of marketable catches would render the fishery in question uneconomic or that handling of unwanted catches on board would significantly increase the costs to the vessel owner, similarly threatening the viability of the business. On this basis STECF advised that, Member States should provide evidence to support these arguments (STECF 2013, 2014a).

For the first condition, in respect of improvements in selectivity, STECF (2013) put forward an approach using a 'break-even indicator' as a tool for evaluating potential *de minimis* cases. This was designed to show that the larger the change in selectivity towards unwanted fish, the more significant the decrease in revenue could be, due to losses of marketable catches. This was described as the ratio of current revenue to break even revenue (CR/BER) and shows how close the current revenue of a vessel or fleet is to the revenue required for it to break even. If the ratio is negative, variable costs exceed current revenue, indicating that the more revenue is generated, the greater the losses.

For the other condition, STECF interpreted that, based on the wording in the Regulation, there was no need to identify and justify what disproportionate costs would be. However, providing studies that demonstrate the scale of additional costs of handling and sorting unwanted catches on board, coupled with the increased labour costs associated with these operations on board, would be useful. The wording in the article suggests that disproportionate costs of handling unwanted catch are assumed when the unwanted catch of a specific fishing gear is below a certain percentage of the total catch of that gear, and that the percentage threshold would be established in a discard plan. The key question to STECF appeared to relate to 'the percentage unwanted'. On this basis STECF did not explore this issue any further at that stage.

In 2016, STECF re-visited the *de minimis* issue and provided further guidance in the form of an analytical framework to assist in the submission of economic cases for *de minimis* exemptions. The framework proposed was based on an option appraisal methodology, applying a multi-criteria performance matrix to structure the analysis and present the results (STECF 2016b). The objective was to create a methodological framework to improve consistency in the economic analysis provided in support of *de minimis* submissions.

3.3.2 Evidence Collected So Far

Despite the provision of these detailed guidelines, in reality, unlike the proposals on the basis of high survivability, requests for *de minimis* exemptions submitted by the regional groups have varied considerably in terms of the quality and type of the evidence supplied. It is evident from the Joint Recommendations submitted, Member States have struggled to put together compelling cases to support *de minimis* exemptions, while STECF has found it difficult to evaluate the proposals, particularly those justified under the disproportionate costs conditionality. The CR/BER approach and multi-criteria performance matrix have not been used by Member States and instead the approach taken has been *ad hoc* and based on rather generic and simplistic arguments.

To justify requests under the first condition, Member States have focused on the results from one or more selectivity studies. The results of such studies generally show losses of marketable catch associated with the use of selective gears. Using simple analyses of the economic impacts of such losses on the whole fleet, Member States have attempted to demonstrate that improvements in selectivity are difficult to achieve. An example of this is in the beam trawl fishery for sole in the North Sea, where there is considerable evidence to suggest that increasing selectivity by increasing mesh size will result in high losses of sole with only marginal gains in terms of reduction in unwanted catches (STECF 2015a; Bayse and Polet 2015).

This argument has been used in most cases for *de minimis* exemptions. However, there are examples where Member States have attempted to demonstrate the impacts of handling and storing unwanted catches to justify these as "disproportionate costs".

Examples include the French "EODE" study (Balazuc et al. 2016) and a Dutch impact assessment carried out by Buisman et al. 2013. In several other cases the argument for the *de minimis* exemption is on the basis that the costs for handling and disposing of unwanted catches ashore are disproportionate given there are limited outlets for such catches. Such a justification has been used in the cases of *de minimis* exemptions in the Mediterranean and in this case has been based on information supplied by the Mediterranean Advisory Council (STECF 2014b, 2017b).

3.3.3 Review of the Evidence

STECF has continued to struggle to review the evidence provided to support *de minimis* exemptions, although over time improvements in the quality of cases put forward by Member States have been noted by STECF (STECF 2016b, 2017b). Increasingly, particularly in the North Sea and North Western Waters, the case for an exemption has been conditional on vessels using a selective gear, the justification being that it is difficult for selectivity to be improved any further and the *de minimis* covers the residual unwanted catches. For instance, an exemption in the beam-trawl fisheries for sole in the North Sea and NWW (North Western Waters) is linked to the use of an escape panel to reduce unwanted catches of sole and plaice (STECF 2015a). Such cases seem much more in the spirit of the Regulation than cases where the *de minimis* seems to be requested to support a *status quo* of discarding practices in specific fisheries.

It has now been more or less accepted that the original break-even indicator approach and option appraisal framework are too complex; requiring data that are not available in many cases. In response, STECF has refined its advice and developed a simple template for the regional groups to provide an overview of the fleets involved and the catch data used to calculate the *de minimis* volume (STECF 2016b). STECF has then evaluated whether the supporting information supplied sufficiently demonstrates significant economic losses in the absence of a *de minimis* exemption. Regional groups have tended to follow this approach and the evaluation process has become more consistent; accepting that in many cases there is still a lack of relevant supporting information that demonstrates economic losses rendering fisheries potentially uneconomic. In these cases, STECF have tended to give "neutral" advice which neither endorses nor rejects out of hand any exemption. In all cases, the ultimate decision to grant an exemption lies with the Commission.

To date, all the discard plans have contained *de minimis* exemptions, which have been approved by the Commission, following the STECF evaluation. In total 42 exemptions are in place. These apply across a wide range of fisheries and apply to single stocks. Table 3.1 summarizes the extent of *de minimis* across the different sea basins and shows the progression since implementation of the Landing Obligation began in 2014. Once a *de minimis* provision is awarded, it is deducted from the TAC before the remainder is allocated to the relevant Member States, based on the understanding that this amount will be caught and discarded. So far, the *de minimis*

Table 3.1 Number and progression of *de minimis* exemptions across sea basins since the implementation of the Landing Obligation in 2014

Region	Fishery	Years							
		2014		2015		2016		2017	
		New	Existing	New	Existing	New	Existing	New	Existing
NWW	Pelagic	4	–	–	4	–	4	–	3
	Demersal	–	–	7	–	–	7	–	7
SWW	Pelagic	4	–	–	4	–	4	–	4
	Demersal	–	–	3	–	–	3	–	3
North Sea	Pelagic	1	–	–	1	–	1	–	1
	Demersal	–	–	5	–	7	1	–	7
Baltic	All	–	–	–	–	–	–	–	–
Mediterranean	Pelagic	7	–	–	7	–	7	–	7
	Demersal	–	–	–	–	–	–	10	–
Black Sea	All	–	–	–	–	–	–	–	–

exemptions have been based on single stock, however, Member States have also requested evaluations on using *de minimis* for combinations of stocks. The regulation refers to the '*total annual catches of all species subject to an obligation to land to which de minimis can apply*' and therefore does not preclude a *de minimis* being applied to a combination of stocks. STECF has noted that, while a combined *de minimis* may provide some flexibility where some unwanted species may be more difficult to handle than others, they may result in higher deductions from the TACs than for the single stock approach, and therefore reduced fishing opportunities (STECF 2018).

3.4 Impact of the Landing Obligation on the Scientific Advice on Fishing Opportunities

3.4.1 ICES Advice

Scientific advice for fishing opportunities for North East Atlantic stocks is provided by ICES. The advice involves two processes. First, stock assessments estimate the current status of the stock and its historical development. Second, short-term forecasts estimate the catches in the following year corresponding to a given management target. The target is usually expressed in terms of a fishing mortality which reflects a broad management objective (management plans, MSY or Precautionary Approach). Stock assessments for the main commercial stocks are based on catch-at-age estimates. Initially, catches comprised only landings, but discard estimates by fishing activity ("*metier*") have been increasingly available through improved sampling programs financed by the EU Data Collection Framework (Council Regulation (EC) No 199/2008). There is also a requirement for vessel operators to record discard amounts for commercial species, however, limited compliance has meant that the log book data provide a gross underestimate of discard levels and reliance on these data is inadvisable (STECF 2013). Because of the limited coverage of scientific discard sampling, estimates are not available for all stocks and *métiers* in all Member States. Informed procedures have thus been developed over time to raise estimates of discard ratios and allocate age structure to the unsampled strata, usually through a dedicated ICES Database called InterCatch.[1] In 2015, the scientific advice therefore included a complete provision for both landings and discards by *métiers* for most commercial stocks.

The Landing Obligation poses a significant conceptual and technical challenge, and has rendered the provision of scientific advice more complex to perform and quality-check, and more difficult to formulate. Since 2015, two main changes have been triggered in the ICES process, involving the way catch data are collected and the way forecasts are performed and presented.

[1] http://ices.dk/marine-data/data-portals/Pages/InterCatch.aspx

Changes in catch data have emerged from the need to quantify and raise new categories. It is still considered that even if landed, a part of the catch is still not targeted by fishermen, but they cannot be called discards anymore and are referred to as "unwanted catch", while landings are now referred to as "wanted catch". To keep track of these categories, the InterCatch database was expanded in November 2015, and now includes five categories:

- **L = Landings**. Landings above minimum size;
- **B = BMS Landings**. Landings below minimum conservation reference size, BMS. Relevant for stocks under the landing obligation. The BMS landing will consist of BMS landings and predator damaged fish.
- **D = Discards**. The part of the catch which is thrown overboard into the sea and not registered in the logbook. This is still based on fishery observer estimates and applying to all stocks, both under and outside the Landing Obligation.
- **R = Logbook Registered Discard**. Relevant for stocks under the Landing Obligation. Logbook registered discard are discards, which are registered in the logbook and are under the exemption rules (e.g. *de minimis*). Damaged fish can be included under this Logbook registered discard.
- **C = Catch** can be used for a few species, for which there is no separation in the information of landings or discards

The sum of (B + D + R) is the "unwanted catch" and corresponds to what was previously recorded as D alone. Thus, there are now many more strata to fill and raise in the stock assessment process. Even until now, the R and B categories have remained negligible, and conceptual decisions must be made on how to sample, monitor and include these catch components in the stock assessment. The categories B and R are usually not sampled directly, but being considered "unwanted", they are allocated an age distribution from the category D.

Additionally, the logbook systems in place prior to the Landing Obligation were not able to deal with the new categories, and software updates have been required in all Member States. Depending on when and how updates took place, and on whether B and R information are transmitted to the scientific institutes, differences in reporting of all these categories to ICES have been observed across Member States.

The second major change in the process relates to the advice itself; i.e. to the maximum tonnage advised by ICES to be in line with the management objective. The target fishing mortality (e.g., F_{msy}) corresponds to the quantity of dead fish to be removed from the population, regardless of whether they are landed or discarded. Previously, the advice was expressed in terms of landings, assuming a given share of discards based on previous years' observations. In theory, the Landing Obligation would ensure that all catches would be landed, and a single catch advice would suffice. In practice, this poses a number of quantitative challenges, linked to the facts that: (i) discarding still takes place and cannot be ignored; and (ii) legal provisions (e.g., high survivability, *de minimis* and predator damage fish) in article 15 mean that the Landing Obligation is only partially applicable, particularly during the transition period 2015–2018 where not all fleets catching the same stocks have been phased-in at the same time, leading to partial uplifts (see below). It has been impossible to formulate single catch advice that would encompass this management complexity. ICES has therefore chosen to issue advice as a single maximum catch value split between wanted and unwanted shares, leaving it to the Commission to decide the actual level of the TAC. Combined together, the scientific information has become more complex to collect, to use and to quality-check, and to explain to clients in a simple and transparent manner.

3.4.2 TAC Uplifts

An added complexity in setting fishing opportunities is the inclusion of a provision within the CFP that allows for TAC adjustments to be made for those stocks under the Landing Obligation. This takes account of fish that otherwise would have been discarded but are now to be landed. These adjustments are to be made based on the contribution by the fleets under the Landing Obligation to total catches and discards of the concerned stocks. This is contained in Article 16(2) of the CFP and, at the time of the CFP negotiations, this was a key condition for Member States in agreeing to the Landing Obligation.

This would seem a relatively straightforward exercise under full implementation of the Landing Obligation, as the start point for the TAC would be the full catch advice provided by ICES less any deduction for fish discarded under exemptions. However, the methodology used for calculating TAC adjustments when setting the fishing opportunities has been the subject of extensive discussion; particularly in cases where available discard data is incomplete, or Member States chose to use catch thresholds based on historic landings to determine whether a vessel was subject or not to the Landing Obligation. This has been the case in the NWW, SWW and the North Sea where, in agreement with the Commission, Member States chose a phased implementation approach with partial coverage of certain stocks (e.g. In the NWW – "*Where total landings per vessel of all species in 2013 and 2014 consist of more than 10 % of the following gadoids: cod, haddock, whiting and saithe combined, the landing obligation shall apply to haddock*"). This is illustrated in Table 3.2.

STECF provided advice to the Commission on how to calculate TAC adjustments and have developed a methodology to facilitate these calculations (STECF 2015c, 2017c):

- The ICES forecasts of catch and discards should be used as a starting point for calculating TAC adjustments consistent with FMSY principles. Fleet specific discard rates derived from the STECF database should then be used to determine fleet specific TAC adjustments.
- For stocks where the entire fishery or clearly defined fleet segments, are subject to the landing obligation, and for which specific discard rates are available, then calculation of TAC adjustments can simply be based on the ICES catch forecast.

Table 3.2 An extract showing an example of the contribution (%) of each fleet segment identified under the Member States' joint recommendations to total catches and discards of the stocks/TACs for relevant stock in the North Western Waters region used to calculate the TAC adjustment

Fisheries	Gear	Mesh Size	LO	ICES area	STECF/ Annex/ Area	Gear	2013			2014			Average		
							Landings	Discards	Catch	Landings	Discards	Catch	Landings	Discards	Catch
Gadoids	Trawls & seines: OTB, SSC, OTT, PTB, SDN, SPR, TBN, TB, TB, SX, SV, OT, PT, TX	All	Where total landings per vessel of all species in 2013 and 2014 consist of more than 10% of the following gadoids: Cod, haddock, whiting and saithe combined, the LO shall apply to **haddock**	Vb and VIa	IIa, 3D	TR1	97%	18%	82%	96%	24%	83%	96%	21%	83%
						TR2	3%	82%	18%	2%	76%	15%	3%	79%	17%
					Overall trawls		**100%**	**100%**	**100%**	**98%**	**100%**	**99%**	**99%**	**100%**	**99%**

Source: Ribeiro et al. (2015)

- In situations where only a proportion of vessels (within a specific fleet segment) and that also meet a historic landing composition threshold are subject to the Landing Obligation, then the average ICES discard rate should be applied to reflect the proportion of vessels and landings covered under the landing obligation.

This methodology has largely been followed up to and including the fishing opportunities set for 2018. The exception has been in the Baltic Sea, where the Landing Obligation has been fully implemented since 2016, with all stocks managed under TACs and the relevant fisheries being subject to the Landing Obligation. In these cases, the ICES catch advice has been applied. From 2019 onwards, in the other sea basins TAC adjustments will no longer be relevant and, as in the Baltic, the catch advice will be applied.

3.5 Annual Reporting

The CFP did not include a reporting requirement on the Landing Obligation. However, during the negotiation of the so-called "Omnibus Regulation" (Regulation (EU) No 2015/812), which aimed to remove incompatibilities of existing technical and control measures with the Landing Obligation by amending or repealing such measures, a mandatory reporting requirement was introduced. It covers seven elements and is based on information from, among others, the Member States and the Advisory Councils concerned. The elements are as follows:

- steps taken by Member States and producer organisations to comply with the landing obligation;
- steps taken by Member States regarding control of compliance with the landing obligation;
- information on the socioeconomic impact of the landing obligation;
- information on the effect of the landing obligation on safety on board fishing vessels;
- information on the use and outlets of catches below the minimum conservation reference size of a species subject to the landing obligation;
- information on port infrastructures and of vessels' fitting with regard to the landing obligation; for each fishery concerned; and
- information on the difficulties encountered in the implementation of the landing obligation and recommendations to address them.

After submission of the first reports in 2016, covering implementation in 2015, the Commission requested STECF review the reports submitted and also to provide guidance on the structure of future reports to improve their utility (STECF 2016a). STECF was also requested to identify additional elements that could be usefully reported on.

Following the assessment of the 2015 submissions from Member States and Advisory Councils, STECF concluded that, while they provided some insights into the operation of the Landing Obligation, they were generally lacking in structure and quantitative information (e.g., catch documentation, levels of funding, number of vessels impacted) making it difficult to undertake any substantive evaluation. For this reason, STECF developed a questionnaire to allow for ease of reporting in a structured way. At the same time STECF discussed other metrics that might improve the monitoring of the Landing Obligation given that the reporting requirements contained in Article 15(14) focus on certain aspects of the Landing Obligation and its potential impacts mostly ashore at port level. STECF identified that there was a lack of emphasis relating to the monitoring of effects and impacts of the Landing Obligation in terms of what happens at sea and in the environment. In particular, impacts on catch and catch profiles, compliance, selectivity, spatial and temporal changes in fishing operations, longer term socio-economic and environmental effects are not covered.

Subsequently, STECF has carried out reviews of the annual reports in 2017 (based on information from 2016; STECF 2017a) and in 2018 (reporting on 2017; STECF 2018). In 2017, 21 Member States and three Advisory Councils submitted reports for 2016, while in 2018 STECF received reports from 15 Member States and two Advisory Councils, despite the reporting being a mandatory requirement. The information supplied has remained qualitative rather than quantitative, but as most Member States have followed the questionnaire, the information has at least been supplied in a semi-structured way.

Based on the most recent assessment by STECF it is interesting to note that implementation measures reported by Member States in response to the questionnaire do not necessarily imply successful implementation of the Landing Obligation or that it is achieving its aims. STECF base this on the fact that there is little or no evidence that there has been significant relevant change in fishing practices or adequate monitoring and control of fishing operations to implement the Landing Obligation (STECF 2018). This supports claims by Member States and the fishing industry that the transitional phase of implementation has been challenging. Many of the concerns are anticipated for the future and not yet necessarily observed and STECF concluded that the reports of limited impact in some regions such as the Mediterranean and Black Seas may also be related to non-implementation, rather than because the Landing Obligation does not pose any issue.

It remains to be seen whether the quality of the information supplied will improve and allow for a meaningful assessment of implementation of the Landing Obligation and its associated impacts. However, at the present time it provides the best opportunity for monitoring progress in the short to medium term.

3.6 Conclusion

It is well documented that the Landing Obligation is challenging for Member States to implement and for the fishing industry to comply with, but it has also created several challenges for science. STECF has found it difficult to provide conclusive advice on some aspects of the flexibilities and exemptions in the legislative text. The subjective nature of the conditions – "high survival", "very difficult to achieve" or "disproportionate costs" – and limited information in many cases means that there is a large element of expert judgement required rather than empirically assessing the quality of the science. In saying this, the quality of submissions to support the exemptions has improved since the first JR's were submitted in 2014. In particular, progress has been made in the execution of survival experiments, which in most cases closely follow the recommendations made by STECF and also ICES (T. Catchpole, unpubl., data).

STECF has also highlighted the limitations in the information presented, in the methodologies used or where there are inconsistences in the information provided. On many occasions, information is presented that shows, for example, increasing selectivity results in losses of marketable fish. But whether this constitutes a technical difficulty or whether these losses result in a reduction in revenue significant enough to render the fishery uneconomic are not questions that can be readily answered without further data and more detailed analysis. STECF has consistently acknowledged that providing detailed information for individual fisheries is challenging. It therefore has only been able to consider the validity of the supporting information underpinning proposed exemptions. The lack of economic data in many cases makes it impossible to do any meaningful analysis of the economic impacts.

With regard to exemptions based on high survivability, the wide range of factors that can affect survival lead to considerable variability in the survival estimates observed. However, identifying and quantifying these is difficult due to the relatively limited species-specific information and differences between assessment protocols. Judging the representativeness of individual or limited studies as an indicator of discard survival across an entire fishery is therefore difficult, given the range of factors that can influence survival and how they may vary in time even within a fishery. The choice of acceptable survival thresholds, in the context of article 15.2 (b), will depend on which objective in the CFP (e.g., avoidance of waste; improve stock sustainability; improve financial viability) is set as a priority.

Finally, the introduction of the Landing Obligation has increased the potential for uncertainty in the science supporting fisheries management in the EU. An increased emphasis on the use of selective technical measures to reduce unwanted catches (O'Neill et al., this volume; Reid et al., this volume) is likely to increase uncertainty in our understanding of harvesting/exploitation patterns (e.g., Breen and Cook 2002; Breen et al. 2016). That is, it cannot be assumed that all of the escaping/released animals will survive (e.g., Breen et al. 2007; Ingolfsson et al. 2007). Therefore, the actual reduction in fishing mortality, in real terms, will only be a proportion of what is perceived from the reduction in discards caught. Moreover, the magnitude and pattern of the escape/release related mortality is far more challenging to quantify and monitor than discard related mortality. Furthermore, both ICES and STECF have consistently highlighted the lack of reporting by vessel operators of fish discarded under exemptions, discards of fish currently not subject to the Landing Obligation and catches of fish below MCRS, which compromises the quality of stock assessments and the accuracy of catch forecast. The quality of science supporting the implementation of the Landing Obligation would strongly benefit from provisions that strengthen data collection in this respect (e.g. through innovative monitoring measures such as Remote Electronic Monitoring, see Chap. 18). STECF concluded that if the data situation does not improve and the true quantities being caught as reported do not reflect the actual removals, then this will have a significant impact on the quality of scientific advice and, without a precautionary approach in setting fishing opportunities, a key objective of the CFP in achieving MSY may be compromised.

Acknowledgments We are grateful to the contributions by and discussions with members of the ICES expert working group on methods for estimating discard survival (WGMEDS).

References

Balazuc, A., Goffier, E., Soulet, E., Rochet, M.J., Leleu K. (2016). *EODE – Expérimentation de l'Obligation de Debarquement à bord de chalutiers de fond artisans de Manche Est et mer du Nord, et essais de valorisation des captures non désirées sous quotas communautaires*, 136 + 53 pp. Version Février 2016. https://www.comitedespeches-hautsdefrance.fr/wp-content/uploads/2016/03/Rapport-final-EODE-Exp%C3%A9rimentation-de-lObligation-de-DEbarquement-CRPMEM-NPdCP-Version-f%C3%A9vrier-2016.pdf. Accessed 18 June 2018.

Barkley, A.S., & Cadrin, S.X. (2012). Discard mortality estimation of yellowtail flounder using reflex action mortality predictors. *Transactions of the American Fisheries Society, 141*, 638–644. https://doi.org/10.1080/00028487.2012.683477.

Bayse, S., & Polet, H. (2015). *Evaluation of a large mesh extension in a Belgian beam trawl to reduce the capture of sole (Solea solea).* ILVO Instituut voor landbouwen visserijonderzoek report, February 2015. 8 pp. https://www.vissersbond.nl/wp-content/uploads/2015/11/Evaluation-of-a-Large-Mesh-Extension-in-a-Belgian-Beam-Trawl.pdf. Accessed 3 July 2018.

Borges, L., & Penas Lado, E. (this volume). Discards in the common fisheries policy: The evolution of the policy. In S.S. Uhlmann, C. Ulrich, S.J. Kennelly (Eds.), *The European Landing Obligation - Reducing discards in complex, multi-species and multi-jurisdictional fisheries*. Cham: Springer.

Breen, M. & Cook, R. (2002). *Inclusion of discard and escape mortality estimates in stock assessment models and its likely impact on fisheries management* (ICES CM 2002/V: 27, p. 15).

Breen, M., Huse, I., Ingolfsson, O.A., Madsen, N., Soldal, A.V. (2007). *SURVIVAL: An assessment of mortality in fish escaping from trawl codends and its use in fisheries management.* EU Contract Q5RS-2002-01603. Final Report.

Breen, M., Graham, N., Pol, M., He, P., Reid, D., Suuronen, P. (2016). Selective fishing and balanced harvesting. *Fisheries Research, 184*, 2–8. https://doi.org/10.1016/j.fishres.2016.03.014.

Buisman, E., Van Oostenbrugge, H., Beukers, R. (2013). *Economische effecten van een aanlandplicht voor de Nederlandse visserij.* LEI-rapport 2013-062. ISBN/EAN: 978-90-8615-657-3. 48 pp. https://library.wur.nl/WebQuery/wurpubs/fulltext/283011. Accessed 3 July 2018.

Catchpole, T., Randall, P., Forster, R., Smith, S., Ribeiro Santos, A., Armstrong, F., et al. (2015). Estimating the discard survival rates of selected commercial fish species (plaice – *Pleuronectes platessa*) in four English fisheries (MF1234). *Cefas report* (p.108).

Catchpole, T., Breen, M., Depestele, J., Kopp, D., Méhault, S., Madsen, N., et al. (submitted) A critical review of European discard survival assessments. *Fish and Fisheries*.

Davis, M.W. (2010, March). Fish stress and mortality can be predicted using reflex impairment. *Fish and Fisheries 11*:1–11. https://doi.org/10.1111/j.1467-2979.2009.00331.x.

ICES. (1995, April). *Report from the study group on unaccounted mortality in fisheries* (ICES CM 1995/B:1). Aberdeen, Scotland, UK.

ICES. (1997). *Report from the Study Group on Unaccounted Mortality in Fisheries* (ICES CM 1997). Hamburg Germany, April 1997.

ICES. (2005). *Joint report of the study group on unaccounted fishing mortality (SGUFM) and the workshop on unaccounted fishing mortality (WKUFM)* (ICES CM 2005/B:08. 68 pp). 25–27 September 2005, Aberdeen, UK.

ICES. (2014). *Report of the workshop on methods for estimating discard survival (WKMEDS)* (ICES CM 2014/ACOM:51, p. 114). 17–21 February 2014, ICES HQ, Copenhagen, Denmark.

ICES. (2015a). *Report of the workshop on methods for estimating discard survival 2 (WKMEDS 2)* (ICES CM 2014\ACOM:66, p. 35). 24–28 November 2014, ICES HQ.

ICES. (2015b, April 20–24). *Report of the workshop on methods for estimating discard survival 3 (WKMEDS 3)* (ICES CM 2015\ACOM:39, p. 47). London, UK.

ICES. (2016a). *Report of the workshop on methods for estimating discard survival 4 (WKMEDS 4)* (ICES CM 2015\ACOM:39, p. 57). 30 November–4 December 2015, Ghent, Belgium.

ICES. (2016b, May 23–27). *Report of the workshop on methods for estimating discard survival 5 (WKMEDS 5)* (ICES CM 2016\ACOM: 56, p. 51). Lorient, France.

ICES. (2016c, December 12–16). *Report of the workshop on methods for estimating discard survival 6 (WKMEDS 6)* (ICES CM 2016/ACOM:56, p. 49). Copenhagen, Denmark.

Ingolfsson, O.A., Soldal, A.V., Huse, I., Breen, M. (2007). Escape mortality of cod, saithe, and haddock in a Barents Sea trawl fishery. *ICES Journal of Marine Science, 64*(9), 1836–1844. https://doi.org/10.1093/icesjms/fsm150.

Methling, C., Skov, P.V., Madsen, N. (2017). Reflex impairment, physiological stress, and discard mortality of European plaice *Pleuronectes platessa* in an otter trawl fishery. *ICES Journal of Marine Science, 74*(6), 1660–1667. doi: https://doi.org/10.1093/icesjms/fsx004.

Morfin, M., Kopp, D., Benoît, H.P., Méhault, S., Randall, P., Foster, R., et al. (2017). Survival of European plaice discarded from coastal otter trawl fisheries in the English Channel. *Journal of Environmental Management, 204*(1), 404–412. https://doi.org/10.1016/j.jenvman.2017.08.046.

O'Neill, F. G., Feekings, J., Fryer, R.J., Fauconnet, L., Afonso, P. (this volume). Discard avoidance by improving fishing gear selectivity: Helping the industry help themselves. In S.S. Uhlmann, C. Ulrich, S.J. Kennelly (Eds.), *The European Landing Obligation – Reducing discards in complex, multi-species and multi-jurisdictional fisheries*. Cham: Springer.

Reid, D.G., Calderwood, J., Afonso, P., Fauconnet, L., Pawlowski, L., Plet-Hansen, K.S., et al. (this volume). The best way to reduce discards is never to catch them in the first place! In S.S. Uhlmann, C. Ulrich, S.J. Kennelly (Eds.), *The European Landing Obligation – Reducing discards in complex, multi-species and multi-jurisdictional fisheries*. Cham: Springer.

Ribeiro Santos, A., Dolder, P., Reeves, S., Catchpole, T. (2015). *Information to support decisions on TAC adjustments for stocks subject to the landing obligation in 2016*. Request for services – Ad hoc contract on TAC adjustments, 5th November 2014, (p. 41). https://www.researchgate.net/publication/306001036_SCIENTIFIC_TECHNICAL_AND_ECONOMIC_COMMITTEE_FOR_FISHERIES_-_TAC_adjustments_for_stocks_subject_to_the_landing_obligation_STECF-15-17. Accessed 13 July 2018.

Scientific, Technical and Economic Committee for Fisheries (STECF) – Landing obligation in EU fisheries (STECF-13-23). (2013). Publications Office of the European Union, Luxembourg, EUR 26330 EN, JRC 86112, (p. 115).

Scientific, Technical and Economic Committee for Fisheries (STECF) – Landing Obligation in EU Fisheries - part II (STECF-14-01). (2014a). Publications Office of the European Union, Luxembourg, EUR 26551 EN, JRC 88869, (p. 67).

Scientific, Technical and Economic Committee for Fisheries (STECF) – 46th Plenary Meeting Report (PLEN-14-02). (2014b). Publications Office of the European Union, Luxembourg, EUR 26810 EN, JRC 91540, (p. 117).

Scientific, Technical and Economic Committee for Fisheries (STECF) – Landing Obligations in EU Fisheries - part 3 (STECF-14-06). (2014c). Publications Office of the European Union, Luxembourg, EUR 26610 EN, JRC 89785, (p. 56).

Scientific, Technical and Economic Committee for Fisheries (STECF) – Landing Obligations in EU Fisheries - part 4 (STECF-14-19). (2014d). Publications Office of the European Union, Luxembourg, EUR 26943 EN, JRC 93045, (p. 96).

Scientific, Technical and Economic Committee for Fisheries (STECF) – Landing Obligation - Part 5 (demersal species for NWW, SWW and North Sea) (STECF-15-10). (2015a). Publications Office of the European Union, Luxembourg, EUR 27407 EN, JRC 96949, (p. 62).

Scientific, Technical and Economic Committee for Fisheries (STECF) – Landing Obligation - Part 6 (Fisheries targeting demersal species in the Mediterranean Sea) (STECF-15-19). (2015b). Publications Office of the European Union, Luxembourg, EUR 27600 EN, JRC 98678, (p. 268).

Scientific, Technical and Economic Committee for Fisheries (STECF) – TAC adjustments for stocks subject to the landing obligation (STECF-15-17). (2015c). Publications Office of the European Union, Luxembourg, EUR 27547 EN, JRC 98384, (p. 16).

Scientific, Technical and Economic Committee for Fisheries (STECF) – Methodology and data requirements for reporting on the Landing Obligation (STECF-16-13). (2016a). Publications Office of the European Union, Luxembourg, EUR 27758 EN, https://doi.org/10.2788/984496.

Scientific, Technical and Economic Committee for Fisheries (STECF) – Evaluation of the landing obligation joint recommendations (STECF-16-10). (2016b). Publications Office of the European Union, Luxembourg; EUR 27758 EN; https://doi.org/10.2788/59074.

Scientific, Technical and Economic Committee for Fisheries (STECF) – 54th Plenary Meeting Report (PLEN-17-01). (2017a). Publications Office of the European Union, Luxembourg; EUR 28569 EN; https://doi.org/10.2760/33472.

Scientific, Technical and Economic Committee for Fisheries (STECF) – Evaluation of the landing obligation joint recommendations (STECF-17-08). (2017b). Publications Office of the European Union, Luxembourg, 2017, ISBN 978-92-79-67480-8, https://doi.org/10.2760/149272, JRC107574.

Scientific, Technical and Economic Committee for Fisheries (STECF) – Data and information requested by the Commission to support the preparation of proposals for fishing opportunities in 2018 (STECF-17-13). (2017c). Publications Office of the European Union, Luxembourg, 2017, ISBN 978-92-79-67485-3, https://doi.org/10.2760/628725, JRC108053.

Scientific, Technical and Economic Committee for Fisheries (STECF) – 57th Plenary Meeting Report (PLEN-18-01). (2018). Publications Office of the European Union, Luxembourg, ISBN 978-92-79-85804-8. https://doi.org/10.2760/088784. JRC111800.

Stockhausen, B. (this volume). How the implementation of the Landing Obligation was weakened. In S.S. Uhlmann, C. Ulrich, S.J. Kennelly (Eds.), *The European Landing Obligation – Reducing discards in complex, multi-species and multi-jurisdictional fisheries*. Cham: Springer.

Uhlmann, S. S., Theunynck, R., Ampe, B., Desender, M., Soetaert, M., Depestele, J. (2016). Injury, reflex impairment, and survival of beam-trawled flatfish. *ICES Journal of Marine Science, 73* (4), 1244–1254. https://doi.org/10.1093/icesjms/fsv252.

van der Reijden, K.J., Molenaar, P., Chen, C., Uhlmann, S. S., Goudswaard, P. C., van Marlen, B. (2017). Survival of undersized plaice (*Pleuronectes platessa*), sole (*Solea solea*), and dab (*Limanda limanda*) in North Sea pulse-trawl fisheries. *ICES Journal of Marine Science, 74*, 1672–1680. https://doi.org/10.1093/icesjms/fsx019.

Part II
Potential Social, Economic and Ecological Impacts of the Landing Obligation

Chapter 4
Fishing Industry Perspectives on the EU Landing Obligation

Mike Fitzpatrick, Katia Frangoudes, Laurence Fauconnet, and Antoni Quetglas

Abstract The Landing Obligation (LO) represents a fundamental change in European Union fisheries policy and it has a particularly significant bearing on the activities of Europe's fishing industry. This chapter provides an account of European fishing industry engagement with the discard issue prior to the LO and industry attitudes towards the LO. A discussion about discard management in Europe follows. The fishing industry had a consistent approach to discard management in the run-up to the LO enactment: they favoured fishery-specific discard reduction plans and were unanimously opposed to an outright 'discard ban'. Canvassing fishers' opinions from the North Sea (Denmark, France), Eastern and Western Mediterranean (Greece, Spain and France), the Celtic Sea (France, the UK and Ireland), Western English Channel (France) and the Azores between 2015 and 2018 reveals a consistent negative attitude towards the LO. We found that choke species are the main concern outside the Mediterranean Sea while in the Mediterranean region, the cost of disposal and the creation of a black market for juvenile fish are seen as the main negatives. Fishers recognise the necessity of reducing discards although zero discard fisheries are not seen as attainable. They favour a combination of selectivity improvements and spatial management as the best discard reduction measures. New measures to deal with intractable choke species problems are being sought by industry and Member State groups but the European Commission want existing measures to be utilised first. We discuss some potential consequences of negative stakeholders' attitudes towards this key element of EU fisheries management policy.

M. Fitzpatrick (✉)
Marine Natural Resource Governance, Cork, Ireland
e-mail: mike@irishobservernet.com

K. Frangoudes
University of Brest, Ifremer, CNRS, UMR 6308, AMURE, IUEM, Plouzané, France

L. Fauconnet
Departamento de Oceanografia e Pescas – IMAR Institute of Marine Research, University of Azores, Horta, Portugal

A. Quetglas
IEO, Instituto Español de Oceanografía, Centre Oceanogràfic de les Balears, Palma de Mallorca, Spain

S. S. Uhlmann et al. (eds.), *The European Landing Obligation*,
https://doi.org/10.1007/978-3-030-03308-8_4

These include control and compliance challenges, associated business reputation problems for the industry, a longer LO implementation timescale, and deterioration in the quality of scientific data about discards.

Keywords EU landing obligation · Fisheries control · Fisheries governance · Industry-science collaboration · Stakeholder engagement · Top-down policy

4.1 Introduction

Since 1984 when the annual Total Allowable Catch (TAC) regulation restricted the carrying of catches onboard vessels beyond those allowed by quotas (EC 1983) EU fishers were frequently obliged to discard fish, for reasons ranging from lack of quota to minimum size restrictions. With the introduction of the LO in 2015 the situation has been reversed, and fishers are now required to land catches of all fish subject to TACs or minimum sizes. This basic shift is expected to have implications for the industry at all levels from operational aspects (like sorting and storing of fish that were previously discarded) to management and governance issues such as quota management regimes.

This chapter first provides an account of how the European fishing industry addressed the issue of fisheries discards prior to the implementation of the LO. Second it presents a synthesis of recent research on industry attitudes to the LO and opinions from two individual EU fishers on the LO. Finally, it discusses the implications of these industry attitudes for discard management in Europe. Source material was drawn from interviews and surveys with fishers conducted as part of the DiscardLess research project (http://www.discardless.eu), the lead authors' own experiences working in the industry, literature sources such as European Commission publications on discards, relevant EU legislation, published books and articles, Advisory Councils' advice and reports and fishing industry press articles.

We have included the views of Advisory Councils (ACs) in some sections although we recognise that they are not solely industry bodies as their membership also comprises other groups such as environmental NGOs. However, the majority of Advisory Council members are from the fishing industry and as such their views and positions are generally reflective of the industry. For opinions from NGOs on the LO see e.g. Borges et al. (2018) and Stockhausen (this volume).We have noted cases where there is a non-consensus or majority position that is unsupported by at least some of the other interest groups within an Advisory Council. The chapter on small scale fisheries (Villasante et al., this volume) also contains some more detailed information on the views of small scale sectors on the LO which may differ from those reported here where we attempt to represent a broad cross section of fishing sectors.

4.2 European Fishing Industry Engagement with the Discard Issue up to the LO

In the run up to the 2012 CFP reform, public interest in the issue of fisheries discards was raised due to the UK TV programme and online campaign, "Hugh's Fish Fight", which gathered over 800,000 signatures for a petition to ban discarding (Borges 2015). Within EU fisheries policy circles, however, discards had been acknowledged as an important issue at least two decades previously (Borges and Penas Lado this volume). In 1991, the first 10 year review of the CFP noted the fact that the TAC system and relative stability would lead to "inevitable discards at sea" (CEC 1991). Also at that time, a discard ban was considered but rejected as unenforceable (CEC 1992). In 2001, as part of the next CFP reform a Commission Green Paper (CEC 2001) proposed pilot discard bans but the 2002 CFP (EC 2002) did not specifically address the discard issue.

In 2007, the European Commission took a position on the discards issue in an official communication, stating that the EC aimed to "*reduce unwanted by-catches and progressively eliminate discards in European fisheries*" (CEC 2007). In response, the European Association of Fish Producer Organisations (EAPO) stated that while the industry recognized discarding as a significant problem, there were complex drivers behind it, including the TAC system and catch composition rules. This, they argued, precluded a simple solution such as an absolute discard ban for all species. The socio-economic implications of a discard ban were also poorly understood (EAPO 2007).

The action plan on tackling discards proposed in the 2007 Commission communication never materialized. This has been interpreted by a member of the Commission as evidence that addressing discards was either not a priority for the industry and Member States or that the Commission "*failed to find the right incentives for that to happen*" (Penas Lado 2016). An industry representative, in an interview conducted for this chapter, points to an 8-page long response produced by the North Sea Advisory Council (NSAC) as evidence that the industry took the Commission's discards paper seriously (NSAC 2007). The NSAC response welcomed the Commission's approach while highlighting that a discard ban would necessitate fundamental changes across a range of fisheries regulations and require huge increases in enforcement efforts. The NSAC document proposed fishery specific, long-term management plans as a more favourable and practical means of achieving discard reductions. It also mentioned the use of overages or the ability for vessels "*to acquire quota post-landing for over quota species caught*" as a useful mechanism employed in the operation of the Norwegian and Icelandic discard bans. The industry representative stated that the Commission did not respond to their paper and that momentum on the issue was lost.

The next significant milestone in the EU fisheries discards debate was the publication of the 2009 Green Paper on reform of the CFP (CEC 2009) that contained multiple references to the elimination of discards. Simultaneously, public pressure on the discards issue was considerably ramped up through the aforementioned Hugh's Fish Fight campaign. Industry responses to the Green paper highlighted their opposition to discarding as "*it is a nuisance to their business, and it is detrimental to rebuilding of stocks*" (Danish Fishermen's Association 2009). Industry also supported the reduction of "*landings of unwanted fish to lowest possible levels*" (Federation of Irish Fishermen 2009).

But they rejected an outright ban which they considered overly simplistic and instead looked for a gradual reduction of discards through fishery and gear-specific discard plans (EAPO 2009). Industry organisations also stressed their viewpoint that greater industry participation in the development of discard plans would "*reduce discards significantly*" (Danish Fishermen's Association 2009).

In 2011, a Commission proposal for a new CFP introduced the term Landing Obligation (LO) to refer to a requirement to land all catches of species regulated by TAC or Minimum Landing Size (MLS) (CEC 2011). Article 15 of the proposal set out the conditions and timeline by which the LO would be gradually implemented. In response, fishing industry organisations came together to issue an alternative proposal (Europêche, EAPO, Cogeca 2012). The industry suggestion was for a landing obligation only in fisheries where Spawning Stock Biomass (SSB) was below Blim for 3 consecutive years. All other fisheries (regulated by TAC or MLS), would be obliged to reduce discards "*to the lowest possible level*" by 2019 through fishery specific management plans. Where reduction targets specified in the management plans were not achieved by 2019, landing obligations should be introduced.

The LO was finally agreed in May 2013 by EU fisheries ministers. Despite the multiple consultation processes that were a feature of the 2012 CFP reform (there were over 400 submissions on the 2009 Green Paper), the final iteration of the discard policy was essentially a political creation. It featured a 4 year implementation timeline, which could be regarded as short for such a fundamental policy change. A member of the commission staff described the situation as one where the experience of other countries on gradual approaches had to be ignored as "*the political pressure was there for quick solutions*" (Penas Lado 2016).

4.3 Fishing Industry Stakeholders' Opinions of the LO

The DiscardLess project gathered industry stakeholders' opinions towards the LO. This section synthesises the results from this research. Opinions from different groups of stakeholders are collected every year throughout the project using different methodologies (interviews, focus groups, opinion survey). Industry participants included fishers, fisheries organisations (Producers' Organisations (POs) or associations) and the seafood processing industry. One of the principal tasks is aimed at monitoring changes in economic and social factors during and after the actual implementation of the Landing Obligation policy.

The selected case studies represent different regional seas and different types of fleets from the North Sea (Denmark, France), Eastern and Western Mediterranean (Greece, Spain and France), the Celtic Sea (France, the UK and Ireland), the Western English Channel (France) and the Azores. The results are based on a combination of 70 individual interviews (see e.g. Reid et al., this volume) and 200 responses to a postal survey (in France and Greece) that were conducted between 2015 and 2018.

A very broad range of issues and views were raised during these interviews and surveys and in order to present them coherently we have organised them into the following broad categories: (i) knowledge of the LO, the implementation process and participation in it; (ii) likely impact of the LO; (iii) adaptation or mitigation strategies.

4.3.1 Knowledge of the Landing Obligation and Participation in the Implementation Process

There are diverse views among fishers regarding their knowledge and awareness of the LO. French fishers in the Eastern English Channel, who fish from a mix of larger and smaller scale vessels mainly using towed gears, felt they have a good knowledge of the LO. This is due to an industry-led project initiated by local PO's and the Regional Fisheries Committee which simulated and tested in real time the operational and economic impacts of the LO (Balazuc et al. 2016). The project also included some gear selectivity trials but these were restricted due to fishers' perceptions about the negative economic impacts of new trawl gears. Some aspects of the LO remain poorly understood by fishers, such as the implementation timeline and issues such as how unwanted catches should be disposed of.

A significant number of fishers, in particular but not exclusively within the small scale sectors, claim that they are aware of the implementation of the LO but have very little detailed knowledge about it. Small scale fishers from the Azores raised some fundamental questions about its application such as whether they would be subject to it due to their current low discard rates. Many Azorean fishers felt that the policy was artificial, disconnected from the reality of fisheries in the region and would benefit neither themselves nor the resource. Greek trawl fishers had very low awareness of the LO, perhaps due to the absence of a representative organisation, but in discussions they felt it went against the work done recently in eliminating catches and sales of smaller fish.

According to a French fishers' representative, fishers need to understand the rationale of the LO regulation as this would provide some incentive to comply. This problem is summarised by the following quote from a French Eastern Channel fisher "*There is no need to increase our work and our costs, spend our quotas and not get a good price for our fish*". This sentiment appears to be mirrored by at least some of the EU Commission staff. One of them described the policy's lack of clarity in the following terms: "*A non discard policy would imply changes that may affect direct fishermen's revenues without a clear perspective of possible tradeoffs.*" (Penas Lado 2016).

Almost all fishers' general attitudes towards the LO were negative and this perception has not changed significantly over the past 3 years. For Mediterranean fishers in particular, the LO is perceived as being designed for quota managed fisheries in the Atlantic as it doesn't take into account the specific context of Mediterranean fisheries. All fishers met in Boulogne-sur-Mer (France) expressed their opposition to the LO and its implementation in EU waters.

Regarding participation of fishers in the LO implementation process, Danish fishers consider that their voice was partially, but insufficiently, heard. In France, fishers consider that the national fisheries administration did represent their views in LO negotiations. Catalan, Greek and Azorean fishers felt that they had not participated in any negotiations related to the LO.

Some fishers have expressed concern that the LO implementation time frame is too short for such a radical change and that it will create economic viability problems for them.

Among representatives of fishers' organisations, there is a much better knowledge about the in's and out's of the LO and its implementation process. However, they feel that their views were not sufficiently taken into account during the design of the policy. They also argue that their role in the implementation process, specifically in drafting Joint Recommendations for regional discard plans, should be clearer. Fishers' representatives feel that the LO is taking up a lot of time which detracts from their ability to deal with other important fisheries management issues. They are still uncertain about aspects of LO implementation despite their sound knowledge about the subject and participation in meetings. This uncertainty extends to the handling of choke situations, conflicts between the LO and technical and control regulations and how discard plans will be integrated into Multi-Annual Plans.

4.3.2 Impacts of the Landing Obligation

The impacts of the LO were highlighted by fishers in particular with regard to working conditions, safety, economic viability and ecosystems.

Fishers across all case studies are concerned that the LO will increase the time to sort out catches and to store additional quantities of fish. These issues are also viewed as a safety problem onboard arising from an increased workload related to sorting the catch, as well as the transfer at sea of unwanted catches.

Fishers think that the LO will negatively impact on their economic viability as operational costs (fuel, ice, disposal costs of unwanted catches) will increase and new investments will be required (e.g., purchasing of more selective gears and increasing the storage capacity of vessels). Many fishers mentioned that they cannot afford such investments and would require support from the European Maritime and Fisheries Fund (EMFF). Small-scale fishers, in particular small trawlers, stated that their fuel consumption will increase as storage capacity is limited and they will have to return to harbour more often. The lack of utilisation options for unwanted catch at a reasonable cost, linked to the absence of a processing industry capable of dealing with discards, was highlighted in the majority of regions.

The most pressing and significant issue for fishers (outside the Mediterranean) is undoubtedly the problem of dealing with choke situations and the potential negative effects on fishing fleets. A choke species is "*a species for which the available quota is exhausted (long) before the quotas are exhausted of (some of) the other species that are caught together in a (mixed) fishery*" (Zimmermann et al. 2015). All of the relevant fishery Advisory Councils have been conducting some form of risk analysis of fisheries likely to cause choke problems in their region. Some have taken a stock-specific approach (NWWAC 2017), while others have looked at which mitigation measures may work in a general sense (NSAC 2017). All ACs agreed that some residual choke problems will persist even when all available measures are applied and are looking to both Member States and the Commission for guidance on how these can be resolved. A NSAC Demersal Working Group meeting in February 2018 stated that due to Relative Stability problems, Member States were unwilling to use quotas to address residual chokes (NSAC 2018a). This uncertainty was also summarised at a 2018 LO seminar by the Scottish White Fish Producers Association when the LO challenge for industry was described as needing to satisfy legal, societal and market demands without going out of business but also without knowing how (Pew 2018).

Fishers felt that the LO will also have a negative impact on ecosystem health, referring to birds and other organisms that have fed on discards until now.

4.3.3 Adaptation and Mitigation Strategies

The most commonly proposed mitigation strategies by fishers across case studies are selectivity improvements and exemptions. Mediterranean fishers said that selectivity of the bottom trawl fleet has already improved with the introduction of 40 mm square mesh cod-ends but that this measure would need to be applied to the whole Mediterranean (including in non-EU countries). Fishers in a number of cases mentioned selectivity trials which have been conducted, as well as their desire to use EMFF funding to support adoption of more selective gears (East and West Mediterranean, Celtic Sea, North Sea).

Fishers across all case studies also mention spatial and temporal closures as possible mitigation strategies (see also Reid et al., this volume). They stress that such closures should be scientifically based and that mapping programmes of zones with concentrations of juvenile fish are required. The need to integrate fishers' local ecological knowledge into discard plans was mentioned as a strategy to address choke problems.

The fishing industry has also pointed out that there is a trade-off between selectivity improvements and economic losses in mixed fisheries which limits the extent to which selectivity can resolve discard problems (NSAC 2018a).

4.3.4 Control and Monitoring

Some LO specific issues have been raised by stakeholders on the revision of the EU Fisheries Control System proposed by the European Commission in 2017 (EC 2017).

The North Western Waters Advisory Council (NWWAC) response agreed that there is a need for full control of high-risk vessels and that dedicated programmes to measure compliance with the Landing Obligation should be implemented (NWWAC 2018a). However they point out that the use of Electronic Monitoring (EM) with video on vessels is a controversial tool for some fishers and that "*good communication will be needed to ensure buy-in on the use of this technology by the industry*".

The NSAC response to an earlier version of the proposal supported a risk-based monitoring approach but pointed out that the majority of fishers do not see the LO as fair or rational and thus there is an associated compliance problem (NSAC 2016). They also drew attention to the controversial nature of EM with video on vessels at a NSAC meeting in April 2018 (NSAC 2018a). French fishers consider that fishing vessels are private spaces and vessel owners are concerned that crew members will protest against videos on this basis.

4.3.5 Industry-Science Collaboration

A concern often mentioned by fishers is that discard data provided by them under the LO could negatively impact their fishing opportunities in the long term. This concern is manifested in declining observer coverage in some Member States and regions (Our Fish 2017), which in turn could have a negative impact on the quality of scientific data about discards. Fishers are concerned that in the context of uncertainty around how choke situations will be dealt with, there could be a negative impact on data provision and industry-science collaboration which many fishers felt was improving prior to the LO.

In contrast to this position, there have been a number of strategic collaborations between fishers and scientists arising directly from the LO. These have arisen mainly as initiatives to examine survival of discarded fish as such information is required in order to justify exemptions from the LO on the basis of high survival (see also Rihan et al., this volume). An example of such collaborations is research on survival of flatfish and rays in pulse trawls (Schram and Molenaar 2018). This research was conducted by Wageningen Marine Research and commissioned by a Dutch fishers organisation VISNED.

4.3.6 Opposition to the LO

Some sectors within the industry are taking a more oppositional approach to the LO. The South Western Waters Advisory Council (SWWAC) communicated some very clear statements regarding their difficulties with the LO in a recent opinion document (SWWAC 2017) submitted to the Commission. (This position was not supported by all of the non-industry groups in the SWWAC).

Their proposals included:

- Compensation of crews for losses associated with unwanted catches.
- Greater flexibility in granting exemptions as all of the requests for information sought by STECF are not "humanly or financially possible" to provide.
- Simplification of the exemption process by, for example, granting high-survival exemptions for all hook or pot fisheries.
- Application of fixed multi-year TACs.
- Deferral of any further extension and implementation of the LO beyond 2018 until agreement can be reached on the points above.

A Fisher's Thoughts on the LO:

Joan J. Vaquero, bottom trawl fisherman, Mallorca-Balearic Islands, (Spain).

"Our fisheries are highly multispecific with more than a hundred commercial species. We can improve the selectivity using larger mesh sizes but we're going to lose a lot of small-sized species (but adult individuals), which will endanger the economic viability of the fishery. Adult fish here in the Mediterranean do not reach the sizes they reach in the Atlantic; this is only due to differences in conditions, the Mediterranean is a poor sea compared to the NE Atlantic.

As it stands, the LO will produce a lot of problems on board. We use relatively small bottom trawlers compared to the Atlantic fisheries. Our vessels are not prepared to store large quantities of fish because we work on a daily basis, returning to our homeport every afternoon (we work from 5:00 am to 17:00 pm), so we can only store the commercial catch of the day. Taking large volumes of catch on board (landings plus discards) could be also dangerous under bad weather conditions. We would also have a problem in case we have to sort out the discards by species because nowadays the crew is reduced to the minimum to allow the economic viability of the fishery; we cannot afford to contract an additional person on board. Finally, but more importantly, if we land the discards, what do we do with them? We do not have fish processing industries so what are the alternatives for discard disposal? It seems the only alternative here in Mallorca would be bringing them to the incineration plant. Imagine what society would say about this practice,

(continued)

because at the end we fishers would be the focus of the criticism. And the bottom trawl fishery is already seen as the bad boy of the fishing family. . .

The future of fishing in the Mediterranean was bleak even before the LO. The number of vessels has been reduced a lot during the last 20 years, mainly because of socioeconomic aspects. We live in a highly touristic area so young people prefer working on in the tourist sector rather than at sea. Apart from the hard work at sea, fishing has been burdened with a lot of administrative commitments making it even harder to maintain the activity. The LO is a new, imposed load to the sector. Consumer preferences have also changed: consumers now buy processed, frozen fish. At the end, a very reduced number of vessels will persist and I think they will have no problem maintaining their activity as with a reduced fishing effort stocks will be healthier. Also, lower supplies will mean higher prices for fish. I cannot imagine the future under the LO. Are we going to build fish processing industries here? I do not think so.

As I said before, increasing mesh size could be a problem for the viability of the fishing sector in the Mediterranean. In my case, if I see that I take large volumes of undersized fish in some areas I reduce the trawling time to reduce the unwanted catch. We also avoid working in areas and periods of the year where and when we take a lot of undersized fish. But in most cases these choices are done on a daily basis, based on day-to-day experience, because it's not possible to foresee the areas and times with large discard volumes. We try to avoid large fish shoals, not only to reduce discards but also because we will saturate the market if all of us land such large volumes. For some species we have adopted voluntary daily quotas to avoid affecting market prices (e.g., for picarel, 150 kg per vessel per day) which also has the effect of reducing discards.

In the Balearic Islands the largest discard volumes are taken on the continental shelf. In our port, for instance, we reduce the time fishing on the shelf and focus on slope grounds. In summer, we do not work on the shelf at all and leave these fishing grounds to small-scale fishers targeting red lobster, mainly to avoid towing in areas where they set their trammel nets. As such, we reduce both the effort and the discards on the shelf grounds.

Maybe the only positive aspect of the LO would be to raise awareness among fishers about the discard problem and the need to reduce unwanted catches; in essence, adopting practices in line with sustainable exploitation. But I do not see any other positive aspect. On the contrary, I see a lot of negative ones. I'm not a biologist so I don't know the best fate for undersized fish. But I see the sea as a harvested field: discards are like manure to feed fish and the entire food chain, I do not think it is a good idea to remove all this biomass from the sea."

(continued)

A Fisher's Thoughts on the LO:

John Lynch, Fisherman, Howth, Co. Dublin, Ireland.

"When the idea of a landing obligation, or discard ban as it was then, was first proposed by the Commission I, like many fishermen, thought the worst, that it could never work. But in fact we had been working towards it for many years with ever-improving selectivity. The idea of large amounts of juvenile fish going over the side has always gone against my idea of what fishing should be. However in some fisheries that use smaller mesh there has always been a problem with excessive discards and in this regard I agree totally with the objectives of the landing obligation.

There are two main objectives of the landing obligation. The first one, reduction of catches that are below the minimum conservation reference size, is probably the easiest to deal with. These fish cannot be sold for human consumption, as they are too small. This is the area where fishermen and other stakeholders have been working to reduce the discards of undersized fish and in some fisheries discards of juvenile fish are down to very low levels. This has greatly helped in rebuilding stocks as more fish can now mature and reproduce.

The other objective of the landing obligation, eliminating discards of mature fish, is much more difficult to solve. Demersal fisheries in North West waters are very mixed in nature and this makes the task very difficult. The tools available to us may not solve this problem for all stocks. As I see it the tools currently in place are as follows:

Quota Uplift is from my understanding the increase of TAC to allow for the amount of fish that used to go over the side to be landed, thus reducing discards and increasing landings but not increasing the effort on the stock. However, the quantities of quota uplift have been derisory in most cases around 10%. This I think is because the cart was put before the horse in that the discard problem was not dealt with before the landing obligation was applied to these species, for example Celtic Sea haddock. In my view, the new technical conservation measures should have been introduced at least 2 years before a discard ban was imposed.

The discard problem for some species will be solved by getting a high survival exemption. At the moment work is ongoing on skates and rays, Nephrops and plaice to see if a survivability exemption can allow them to be released back to the sea. This however will not solve the problem for the fisherman who sees good quality fish going over the side.

De minimis is a small quantity of a stock which can be discarded – usually about 5% of the TAC. This is to allow for some discarding of fish below minimum conservation reference size and is to be reduced to zero over time. I

(continued)

believe it will be necessary to have de minimis in place permanently to allow for some small amount of juvenile discards, which inevitably are caught.

Fishermen, scientists and officials have long been working together to solve the problems of discards of fish at sea. While not always agreeing on how to solve a problem, with trials of different selectivity and spatial measures, a compromise can be found. The issue of discards will never be 100% solved but the important thing is to continually strive to improve the situation for the fish and for the fisherman. As I said at the beginning, the idea of a landing obligation is perfectly good but the anomalies of reality have to be considered, and the deadline of first January 2019 for all TAC stocks is a bridge too far."

4.4 Discussion and Conclusions

Here we discuss what the implications of industry perceptions of the Landing Obligation are for management of discards and also broader fisheries governance in the EU. There are some potentially serious consequences arising from the main fishery stakeholders having a negative attitude towards one of the main pillars of EU fisheries policy. We discuss some of the more obvious ones, e.g. compliance, and also some of a more indirect nature but with nevertheless significant implications, such as the quality of scientific data.

The above sections show that a major issue for the fishing industry regarding the LO remains the choke problem. Despite intense efforts to come up with solutions (including a strong industry emphasis on selectivity improvements), Advisory Councils have identified a significant number of fisheries that will have residual choke problems, even after all available mitigation measures are applied. The principle concern that fishers have with the choke problem is the potential for significant negative economic impacts. Some NGOs have proposed specific cases where industry could receive financial compensation but only if they have first implemented effective selectivity measures (SWWAC 2017). A more general application of this approach could incentivize progress while reducing industry fears regarding vessels going out of business due to either choke related fishery closures or the loss of economic efficiency due to the use of more selective gear.

Significant uncertainty still exists among fishers and their representatives regarding how these residual choke problems can be solved without fisheries closures and associated negative economic impacts. Advisory Councils and high-level Member State groups are looking to the Commission for new measures, in addition to those allowed for in Article 15 (the LO) of the 2013 CFP (EU 2013), which could assist in solving this problem. Such measures could include removal of stocks from the TAC process or defining target and bycatch species of which only the target species would be subject to a discard ban (NSAC 2017). Industry proposals have also mentioned the use of overages, or post-landing purchasing of quota as is used in Norway and Iceland to provide some flexibility and reduce choke-type problems. Recent indications from the Commission are that the use of such measures may only be possible when all other available measures under Article 15 have been applied (CEC 2017).

The choke problem has produced some negative outcomes in terms of discard data provision by industry arising from fears that data provided could potentially precipitate choke closures with associated economic impacts. This has resulted in a reported decrease in discard observer coverage in some Member States including Ireland (DAFM 2017) and Sweden (Sverige Radio 2016) and also in regions such as the Eastern Baltic (ICES 2017; Valentinsson et al., this volume). A situation where fishers are not incentivized to provide data represents a backward step in the collaborative process necessary for improved fisheries management. Unless there is a realigning of the incentives for fishers to provide data in support of discard mitigation, this issue is likely to remain a significant barrier to the successful implementation of the LO. The quality of scientific data may also be negatively affected and implementation will be overly reliant on control and enforcement rather than a collaborative approach.

The problem of compromising collaborative research and industry-science relationships is linked also to the more general principle of good governance that involves stakeholder participation in management. These linked issues are all complicated by industry uncertainty regarding the use of data and a poor understanding of the underlying rationale and objectives of the LO. The fact that industry perceives that their views, expressed consistently in the run-up to agreement of the LO, were not really taken on board seem to have created a perception for some of a backward step in terms of partnership and participatory management where the main drivers are bottom-up rather than top-down (see also van Hoof et al., this volume).

This viewpoint was expressed by Poul Degnbol, a former scientific advisor to the Commission, in a 2018 seminar on CFP reform (NWWAC 2018b). He described the LO as an example of top-down fisheries management, which he stated was the wrong way to go about achieving sustainable and effective management and expressed concern that it may have big implications for science. Industry sources have also highlighted that adopting a flexible, adaptive approach to the LO is made more difficult by the co-decision process in Brussels which has proven, in some cases, to be a slow one (Marchal et al. 2016).

The top-down nature of the LO and industry compliance issues are also resulting in more complex management. Each year the Joint Recommendations for regional discard plans and the delegated acts that put these plans on a legal footing become more complex. The Joint Recommendations for 2019, when for the first time all TAC species are subject to the LO, have been drafted and submitted at the time of writing (NSAC 2018b). It can be seen from these that the trend towards complexity has strengthened with a significant increase in both the number of exemptions sought and the number of supporting documents. This increased complexity in a single area of fisheries management is surely contrary to the desire for a less complex and devolved approach that is implicit in the 2013 CFP's move toward regionalization. The complexity is largely due to industry appeals for exemptions but these in turn are driven by the desire to avoid significant choke closures across EU mixed demersal fisheries that a Landing Obligation without exemptions would create. The key to simplifying this policy area most likely lies in finding some improved mechanism for resolving the choke problem.

The additional complexity creates a number of further knock-on effects. It widens the gap between fishers' representatives who are tasked with understanding this complexity and the fishers they represent. Complex rules, combined with uncertainty at the management level, are translated into confusion and inaction at the operational level and create a significant barrier to implementation.

Furthermore the raft of exemptions and selectivity and control measures point to a situation where the implementation timescale will be longer and more complex than envisaged. This is in line with industry statements that the 2019 deadline for full implementation is ambitious. The first amendment of the LO was made in recognition of the fact that it is taking longer to develop multi-year management plans than originally envisaged (EU 2017). Could a similar recognition that LO implementation may likewise take longer than originally hoped take place also? This amendment also shows that difficulties with making changes to Article 15 may be more political than legal in nature.

To conclude, it can be said that there are some, almost unanimously held industry viewpoints on the LO, for example opposition to the requirement to land unwanted catch without any apparent economic value. However, there is also some diversity of views evident among fishers, with some having more proactive views on how to resolve problems posed by the LO while others are more reactive and are simply opposed to it. One side is hoping that the LO will never be implemented while the other is concerned with being as well prepared as possible when it is. The challenge now for fishers is to reconcile their consistently stated goal over the past 20 years, of having fishery-specific discard reduction plans, with a Landing Obligation covering all TAC or MCRS subject species. Proactive industry voices are likely to be much more persuasive than reactive ones in arguing for a discard policy that is both effective in reducing discards and more aligned with industry needs. It remains to be seen how this will play out over the first few years of full LO implementation from 2019 onwards and beyond that in the next reform of the CFP in approximately 2022.

Acknowledgments This work has received funding from the Horizon 2020 Programme under grant agreement DiscardLess number 633680. This support is gratefully acknowledged.

References

Balazuc, A., Goffier, E., Soulet, E., Rochet, M. J., Leleu, K. (2016). EODE – Expérimentation de l'Obligation de DEbarquement à bord de chalutiers de fond artisans de Manche Est et mer du Nord, et essais de valorisation des captures non désirées sous quotas communautaires. http://www.comite-peches.fr/wp-content/uploads/Plaquette_EODE.pdf. Accessed 31 July 2018.

Borges, L. (2015). The evolution of a discard policy in Europe. *Fish and Fisheries*, *16*, 534–540.

Borges, L., & Penas Lado, E. (this volume). Discards in the common fisheries policy: The evolution of the policy. In S.S. Uhlmann, C. Ulrich, S.J. Kennelly (Eds.), *The European Landing Obligation – Reducing discards in complex multi-species and multi-jurisdictional fisheries*. Cham: Springer.

Borges, L., Nielsen, K., Frangoudes, K., Armstrong, C., Borit, M. (2018). Conflicts and trade-offs in implementing the Common Fisheries Policy (CFP) discard policy. DiscardLess Deliverable D7.3. *Zenodo*. http://doi.org/10.5281/zenodo.1238588.

CEC. (1991). Report 1991 from the commission to the council and the European Parliament on the common fisheries policy. SEC(91) 2288 final. Brussels.

CEC. (1992). On the discarding of fish in community fisheries: causes, impact, solutions. Report from the commission to the council. SEC (92) 423 final. Brussels.

CEC. (2001). Green paper on the future of the common fisheries policy. COM(2001) 135 final. Brussels.

CEC. (2007). Communication from the commission to the council and the European Parliament: a policy to reduce unwanted by-catches and eliminate discards in European fisheries. COM(2007) 136 final. Brussels.

CEC. (2009). Commission of the European communities green paper – Reform of the common fisheries policy. COM(2009)163 final. Brussels.

CEC. (2011). Proposal for a regulation of the European Parliament and of the council on the common fisheries policy. COM(2011) 425 final (p. 88). Brussels.

CEC. (2017). EU Commission reply of 16th November 2017 to NSAC Advice Ref. 14–1617. http://nsrac.org/wp-content/uploads/2016/12/Response-to-14-1617-Managing-Fisheries-within-the-Landing-Obligation.pdf. Accessed 31 July 2018.

DAFM. (2017). *Irish report on implementation of the LO in 2016*. Ireland: Department of Agriculture, Food and the Marine. https://www.asktheeu.org/en/request/access_to_member_state_documents_2#incoming-13777. Accessed 31 July 2018.

DFA. (2009). Danish Fishermen's Association response to European Commission Green Paper on CFP reform, 2009. https://ec.europa.eu/fisheries/sites/fisheries/files/docs/body/danish_fishermens_association_en.pdf. Accessed 31 July 2018.

EAPO. (2007). Discard problem in fisheries. http://eapo.com/UserFiles/File/eapo07-17.pdf. Accessed 31 July 2018.

EAPO. (2009). EAPO response to European Commission Green Paper on CFP reform. http://eapo.com/UserFiles/EAPO%20Response%20Green%20Paper%20CFP%20Reform%20-%20Final%20Version.pdf. Accessed 31 July 2018.

EC. (1983). Council regulation (EEC) no 172/83 of 25 January 1983 fixing for certain fish stocks and group of fish stocks occurring in the Community's fishing zone, total allowable catches for 1982, the share of these catches available to the community, the allocation of that share between the member states and the conditions under which the total allowable catches may be fished. *Official Journal of the European Communities*, *L24*, 30–67.

EC. (2002). Council Regulation (EC) No 2371/2002 of 20 December 2002 on the conservation and sustainable exploitation of fisheries resources under the Common Fisheries Policy. *Official Journal of the European Communities*, *L358*, 59–80.

EC. (2017). Fisheries control system inception impact assessment. Ref: Ares(2017)4808152–03/10/2017. http://ec.europa.eu/info/law/betterregulation/initiative/119363/attachment/090166e5b57fb4f6_en. Accessed 31 July 2018.

EC. (2013). Regulation (EU) no 1380/2013 of the European Parliament and of the council of 11 December 2013 on the Common Fisheries Policy. Brussels, Belgium.

EU. (2017). Regulation (EU) no. 2017/2092 of the European Parliament and of the council of 15 November 2017 amending regulation (EU) no 1380/2013 on the Common Fisheries Policy.

Europêche, EAPO, Cogeca. (2012). Press Release European Fishing Industry Organisations launch an effective and workable proposal to address discards. http://eapo.accounts.divinenet.be/UserFiles/20120918%20-%20Press%20Release%20discards%20initiative.pdf. Accessed 31 July 2018.

Federation of Irish Fishermen. (2009). FIF response to the European Commission green paper on CFP reform. https://ec.europa.eu/fisheries/sites/fisheries/files/docs/body/federation_of_irish_fishermen_en.pdf. Accessed 31 July 2018.

ICES. (2017). ICES Advice on fishing opportunities, catch, and effort Baltic Sea Ecoregion cod.27.24–32. Cod (*Gadus morhua*) in subdivisions 24–32, eastern Baltic stock (Eastern Baltic Sea). https://doi.org/10.17895/ices.pub.3096.

Marchal, P., Andersen, J. L., Aranda, M., Fitzpatrick, M., Goti, L., Guyader, O., et al. (2016). A comparative review of fisheries management experiences in the European Union and in other countries worldwide: Iceland, Australia, and New Zealand. *Fish and Fisheries, 17*, 803–824. https://doi.org/10.1111/faf.12147.

NSAC. (2007). North Sea Regional Advisory Council Response to the Commission Communication on reducing unwanted by-catches and eliminating discards in European fisheries. http://www.nsrac.org/wp-content/uploads/2009/09/wd20070702_Position_Paper_on_Discards.pdf. Accessed 31 July 2018.

NSAC. (2016). NSAC Advice Ref. 04–1617 monitoring and control under the landing obligation. www.nsrac.org. Accessed 31 July 2018.

NSAC. (2017). NSAC Advice Ref. 14–1617 managing fisheries within the landing obligation. October 2017. www.nsrac.org. Accessed 31 July 2018.

NSAC. (2018a). North Sea advisory council demersal working group meeting 7th February 2018 report. http://nsrac.org/wp-content/uploads/2017/10/DWG-20180207-London-MReport-3-Approved.docx. Accessed 31 July 2018.

NSAC. (2018b). NSAC Advice Ref. 01–1718 comments on the implementation of the landing obligation in the North Sea demersal fisheries – joint recommendation for a delegated Act for 2019. www.nsrac.org. Accessed 31 July 2018.

NWWAC. (2017). North Western Waters Choke Species Analysis NWW Member States & NWW Advisory Council October 2017. http://www.nwwac.org/publications.26.html. Accessed 31 July 2018.

NWWAC. (2018a). NWWAC response to EC proposals on the EU fisheries control system 29 January 2018. http://www.nwwac.org/publications.26.html. Accessed 31 July 2018.

NWWAC. (2018b). Report on EFARO seminar, Brussels, May 2018. The reformed CFP: an analysis of what went wrong what went well and what should the next CFP look like. http://nwwac.org. Accessed 31 July 2018.

Our Fish. (2017). Thrown away: How illegal discarding in the Baltic Sea is failing EU fisheries and citizens. http://balticsea2020.org/english/press-room/418-new-report-exposes-high-discards-of-cod-in-the-baltic-sea. Accessed 31 July 2018.

Penas Lado, E. (2016). *The common fisheries policy: the quest for sustainability*. West Sussex: Wiley-Blackwell. https://doi.org/10.1002/9781119085676.

Pew. (2018). Pew Charitable Trusts event “countdown to 2020: How far has the EU come in ending overfishing?” 21 February 2018. http://www.pewtrusts.org/en/about/events/2018/countdown-to-2020-how-far-has-the-eu-come-in-ending-overfishing. Accessed 31 July 2018.

Sveriges Radio. (2016). Fishermen do not want to have researchers on the boat. 4th August 2016. https://sverigesradio.se/sida/artikel.aspx?programid=83&artikel=6487184. Accessed 31 July 2018.

Reid, D.G., Calderwood, C., Afonso, P., Fauconnet, L., Pawlowski, L., Plet-Hansen, K.S., et al. (this volume). The best way to reduce discards is by not catching them! In S.S. Uhlmann, C. Ulrich, S.J. Kennelly (Eds.), *The European Landing Obligation – Reducing discards in complex multi-species and multi-jurisdictional fisheries*. Cham: Springer.

Rihan, D., Uhlmann, S.S., Ulrich, C., Breen, M., Catchpole, T. (this volume). Requirements for documentation, data collection and scientific evaluations. In S.S. Uhlmann, C. Ulrich, S.J. Kennelly (Eds.), *The European Landing Obligation – Reducing discards in complex multi-species and multi-jurisdictional fisheries*. Cham: Springer.

Schram, E. & Molenaar, P. (2018). Discards survival probabilities of flatfish and rays in North Sea pulse-trawl fisheries. Wageningen Marine Research (University & Research centre), Wageningen Marine Research report C037/18. 39 pp.

Stockhausen, B. (this volume). How the implementation of the Landing Obligation was weakened. In S.S. Uhlmann, C. Ulrich, S.J. Kennelly (Eds.), *The European Landing Obligation – Reducing discards in complex multi-species and multi-jurisdictional fisheries*. Cham: Springer.

SWWAC. (2017). Opinion 114: Plan for the implementation of the landing obligation. 30 May 2017. http://cc-sud.eu/index.php/en/. Accessed 31 July 2018.

Valentinsson, D., Ringdahl, K., Storr-Paulsen, M., Madsen, N. (this volume). The Baltic cod trawl fishery: The perfect fishery for a successful implementation of the Landing Obligation? In S.S. Uhlmann, C. Ulrich, S.J. Kennelly (Eds.), *The European Landing Obligation – Reducing discards in complex, multi-species and multi-jurisdictional fisheries*. Cham: Springer.

van Hoof, L., Kraan, M., Visser, N.M., Avoyan, E., Batsleer, J., Trapman, B. (this volume). Muddying the waters of the landing obligation: How multi-level governance structures can obscure policy implementation. In S.S. Uhlmann, C. Ulrich, S.J. Kennelly (Eds.), *The European Landing Obligation – Reducing discards in complex, multi-species and multi-jurisdictional fisheries*. Cham: Springer.

Villasante, S., Antelo, M., Christou, M., Fauconnet, L., Frangoudes, K., Maynou, F., et al. (this volume). The implementation of the Landing Obligation in small-scale fisheries of the Southern European Union countries. In S.S. Uhlmann, C. Ulrich, S.J. Kennelly (Eds.), *The European Landing Obligation – Reducing discards in complex multi-species and multi-jurisdictional fisheries*. Cham: Springer.

Zimmermann, C., Kraak, S., Krumme, U., Santos, J., Stotera, S., Nordheim, L. (2015). Research for PECH Committee – Options of handling choke species in the view of the EU landing obligation – the Baltic plaice example. European Parliament (p. 100). https://doi.org/10.2861/808965.

Chapter 5
The Implementation of the Landing Obligation in Small-Scale Fisheries of Southern European Union Countries

Sebastian Villasante, Manel Antelo, Maria Christou, Laurence Fauconnet, Katia Frangoudes, Francesc Maynou, Telmo Morato, Cristina Pita, Pablo Pita, Konstantinos I. Stergiou, Celia Teixeira, George Tserpes, and Vassiliki Vassilopoulou

Abstract In the European Union, discards represent a major source of undocumented mortality, contributing to the overfishing of European fish stocks. However, little attention has been given by the scientific community to discards in the

---

S. Villasante (✉)
Faculty of Political and Social Sciences, University of Santiago de Compostela, Santiago de Compostela, Spain
e-mail: sebastian.villasante@usc.es

M. Antelo · P. Pita
Faculty of Economics and Business Administration, University of Santiago de Compostela, Santiago de Compostela, Spain

M. Christou · K. I. Stergiou
Institute of Marine Biological Resources and Inland Waters, Hellenic Centre for Marine Research, Anavyssos, Greece

Aristotle University of Thessaloniki, Thessaloniki, Greece

L. Fauconnet
Departamento de Oceanografia e Pescas – Okeanos, Instituto do Mar (IMAR), Universidade dos Açores, Horta, Portugal

K. Frangoudes
University of Brest, Ifremer, CNRS, UMR 6308, AMURE, IUEM, Plouzané, France

F. Maynou
Institut de Ciències del Mar, CSIC, Barcelona, Spain

T. Morato
Marine and Environmental Sciences Centre (MARE), Institute of Marine Research (IMAR) and OKEANOS Research Unit, Universidade dos Açores, Horta, Portugal

C. Pita
Department of Environment and Planning & Centre for Environmental and Marine Studies (CESAM), University of Aveiro, Aveiro, Portugal

C. Teixeira
MARE – Marine and Environmental Sciences Centre, Faculdade de Ciências, Universidade de Lisboa, Lisbon, Portugal

S. S. Uhlmann et al. (eds.), *The European Landing Obligation*,
https://doi.org/10.1007/978-3-030-03308-8_5

European Union's small-scale fisheries (SSF). This is mainly due to the fact that discards are mostly generated by industrial fisheries, while SSFs were generally thought to have lower discard rates than industrial fisheries. A Landing Obligation (LO) is being introduced in European waters with the reform of the Common Fisheries Policy (CFP) (Article 15, EU regulation 1380/2013) to limit/reduce discarding. However, management recommendations are required to support its implementation. The reality and challenges to enforce the LO in SSF are analyzed in this chapter, gathering information from different small-scale fisheries and fishers from the Atlantic Ocean and Mediterranean Sea who were asked about their perceptions toward the LO. The objectives of this chapter are to (a) identify the reasons for discarding and (b) investigate the multiple ecological, economic, social, and institutional drivers which act as a barrier toward the implementation of the LO in SSF. Given the high importance of SSF in the southern countries of Europe, different case studies of SSF from France, Greece, Portugal, and Spain coasts are used to illustrate the reasons for discarding, the impacts of the LO on SSF, and the barriers for its implementation.

Keywords Common Fisheries Policy · Discards · Impacts · Landing Obligation · Small-scale fisheries · Southern Europe

5.1 Introduction

In the European Union (EU), discards represent a major source of undocumented (or poorly documented) mortality, contributing to the overfishing of European fish stocks. Discarding levels in EU fisheries vary between locations, gears, species, and fishing grounds (Uhlmann et al. 2013). However, data collection and estimates of discards for all commercial species in EU waters under the CFP are far from being complete and generally have low precision. This reflects the relatively low intensity of discard sampling and the high variability in amounts of fish discarded, even within a single fishery. The omission and/or poor discard data from stock assessments may also result in underestimation of exploitation rates and can lead to biased assessments and policy recommendations, hampering the achievement of resilient and sustainable fishery resources uses (Aarts and Poos 2009).

The implementation of a Landing Obligation (LO) was one of the key elements of the recent reform of the EU Common Fisheries Policy (CFP) (Regulation (EU) No 1380/2013). A phased LO was formally implemented in January 2015, and by 2019

G. Tserpes
Institute of Marine Biological Resources (IMBR), Hellenic Centre of Marine Research, Anavyssos, Greece

V. Vassilopoulou
Institute of Marine Biological Resources and Inland Waters, Hellenic Centre for Marine Research, Anavyssos, Greece

it will be in force in all EU waters, covering all fisheries that capture commercial species covered by the CFP regulation, including SSF. Landings from EU SSF are worth around €2 thousand million euros annually, i.e., 25% of the revenue generated by EU fisheries, and SSF therefore have a high value in the seafood supply chain. Around 80% of EU fishing boats and more than 40% of EU fishers (90,000) are engaged in SSF (Macfadyen et al. 2011), emphasizing that SSF is a sector with great social, economic, and cultural importance for coastal communities, especially in southern Europe.

The small-scale fleet has declined by 20% over the last 10 years, to just over 70,000 vessels. Small-scale vessels are on average between 5 and 7 m in length, weigh 3GT, and have engines with a power of 34 kW (Macfadyen et al. 2011). More than 90% primarily use passive gears (i.e., gears that are not towed or dragged through the water) such as drift and fixed nets, hook and lines, or pots and traps. Despite their importance, for decades, EU fishery policy (e.g., quotas, subsidies, management systems) has focused on large-scale fishing, and there is a lack of knowledge about biological, environmental, socioeconomic, management, and policy aspects of SSF. SSF faces diverse challenges and pressures, not least to establish appropriate governance systems.

However, little research has been done on the impacts of the LO on SSF (Villasante et al. 2015a; Veiga et al. 2016). Therefore, the specific objectives of this chapter are to (i) identify the reasons for discarding among SSF, (ii) determine the factors (ecological, economic, institutional) that act as barriers for the successful implementation of the LO, and (iii) identify the institutional arrangements and/or rules that either inhibit or facilitate an adaptation of the LO.

5.2 The Status of Discards in Small-Scale Fisheries

To examine research gaps regarding discards in SSF, we did a systematic literature search to identify relevant scientific papers published up to August 2018 in Scopus, by searching titles, abstracts, and keywords using the following terms: "fisher*" or "fishing"; "discard*"; and "artisan*" or "small-scale" or "traditional" or "subsistence" or "local" or "industrial" or "commercial" or "large." The results obtained show that the topic of discards in SSFs attracted little attention among the scientific community. A total of 1219 papers have been published on the topic of discards from 1950 to August 2018, of which 952 are related to industrial fisheries (78%) with only 267 papers focused on SSF (21%) (Fig. 5.1). The review also showed that the little attention paid by the scientific community to discards in SSFs is due to the belief that discard problems were mainly concentrated in industrial fisheries, while SSFs generally have lower discard rates (Villasante et al. 2016a).

Discarding occurs not only due to poor gear selectivity and the capture of unwanted "low value" fish but also due to the mismatch between catch composition and regulatory catch or size limits. Undersize fish may be discarded due to the MLS regulations; over-quota fish can be discarded in a multi-species fishery due to quota

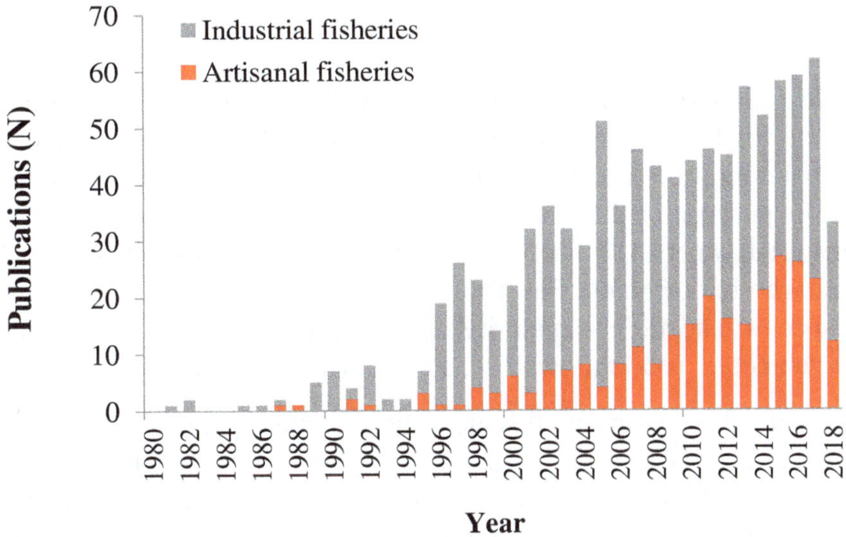

Fig. 5.1 Number of scientific papers published in relation to discards from industrial and artisanal fisheries (1950–2018). (Source: Scopus)

exhaustion of one species, and less valuable size classes of target species may be discarded to make room for more valuable size classes (high grading). Even if high grading has been legally forbidden, it is still known to occur on a regular basis. All these issues are reported to be present in EU SSF (Villasante et al. 2016a, 2016b, 2016c). These different reasons for discarding impact heavily on the willingness to comply with rules and regulations.

5.3 Impacts of the Landing Obligation in Small-Scale Fisheries

The term SSF implies small vessel size and, sometimes, low levels of technology and capital investment per fisher. For the purposes of the European Maritime and Fisheries Fund (Regulation (CE) No 508/2014, "small-scale coastal fishing" was formally defined as fishing done by vessels of an overall length < 12 m and not using towed gear. SSF are thus typically "artisanal" and coastal, using small boats, targeting multiple species using traditional gears.

To investigate the impact of the LO in SSF, we will focus on the impact of this measure on selected SSF in the EU – in France, Greece, Portugal, and Spain. We will describe these fisheries, their discards, the reasons for discarding, impact of the LO, factors that act as barriers for the successful implementation of the LO, and the institutional arrangements and rules that inhibit or facilitate the adaptation to the LO.

5.3.1 France

Small-Scale Fisheries in the British Channel, Celtic Sea, and Bay of Biscay
In France, the small-scale fleet is not legally defined. The number of hours spent at sea is the main criterion to classify the vessels rather than length or use of passive gears. For the purpose of this chapter, only vessels < 12 m of the length operating within territorial waters (12 nm) in the British Channel, Celtic Sea, and Bay of Biscay are taken into account. See Table 5.1 for the characteristics of this fleet and the main species landed. The majority of the target species in the Atlantic Coast/ Ocean are subject to Total Allowable Catch (TAC) regulations.

Interviews were done with small-scale fishers operating in the English Channel, Celtic Sea, and Bay of Biscay as part of the EU DiscardLess project (http://www.discardless.eu). All French fishers interviewed perceived the LO negatively not only because they felt that it will reduce their activity and increase expenditure on their boats but that it also shows that decisions were made at the top level without taking in account the good management practices implemented by the fisheries committees or POs over the last 15 years (van Hoof et al., this volume). That is, French small-scale fishers feel that European decision-makers satisfy claims and interests of lobbies (conservationists, aquaculture) rather those of fishers (De Vos et al. 2016).

According to interviewed fishers, the main reasons for discarding are regulatory, such as quotas, forbidden species, etc. Low market prices and high grading are also given as reasons for discarding (Table 5.2). Damaged fish was also mentioned by netters or long-liners. It is probably that because SSF implemented a quota system later than the larger fleet, and the fact that they have smaller amounts of quota, these fishers discard more. New fishers, who do not own quotas, have to fish under an open national quota system, managed by the national fisheries administration, and because these quotas are rapidly filled, they are obliged to discard all fishes over quota. Young fisherwomen using nets as the main fishing gear say: “As soon as the national quota closes, all fish caught under the quotas system are discarded. We are not members of the local PO, and we cannot access more quotas.” In some regions, POs manage quotas collectively; thus, the quota can be swapped among fishers, with the result that it is easier for them to avoid high discard rates.

Undersize fish, when mentioned, did not really represent a constraint to French small-scale fishers. They all say that the gears they use are more selective than those used by the industrial fleet. But they cannot avoid all undersize fish as “there is not a fishing gear which doesn’t catch undersize fishes.” For those using handline or traps, unwanted fish can easily be returned alive to the sea. The “lack of a good price” for some species, for example, European plaice (*Pleuronectes platessa*) in the Eastern Channel and European hake (*Merluccius merluccius*) or Atlantic horse mackerel (*Trachurus trachurus*) in the Bay of Biscay, is given as another reason for discarding. For fishers, discarding species without commercial value is not perceived as discards. It is the same when it comes to high grading practiced by fishers to

Table 5.1 Case studies characterization

Case study	Country	Fishing fleet	Gear used	Main species landed	Rules and regulations
SSF in the British Channel, Celtic Sea, and Bay of Biscay	France	Vessels less than 12 m operating mainly within territorial waters (12 nm) using mainly passive gear	Gillnets, trammel nets, longlines, handlines, nets, pots and traps, some SSF vessels using dredges or trawls	Common sole, European sea bass, pollack, and monkfish; total landings 80,000 tons (in 2013)	Most target species subject to TAC and MLS
SSF in the Thermaikos Gulf	Greece	Polyvalent passive gears	Nets, pots, longlines, traps	Lands a wide array of species, the most important being hake, common cuttlefish, mullets, annular seabream, saddled seabream, common octopus, common pandora, and scorpionfish; accounted for 18,152 tons (in 2016)	Most target species have minimum landing size (MLS). Plus spatial restrictions and temporal restrictions (e.g., vessels targeting hake and exceeding the limit of >20% of landings are not authorized to fish in February)
SSF in Catalonia	Spain	Polyvalent passive gears operating within 6 nm	Trammel nets, gillnets, boat seines, pots for octopus, and longline	Lands over 200 species, the most important being demersal species (cuttlefish, hake, pandora, sole, golden seabream), sand-eel, octopus, and bonito. Average landings of 3000 ton/year	Some species subject to MLS (e.g., hake, sole, Sparidae, octopus). Technical limitations to the size of fishing gear (e.g., maximum length of nets; maximum number of hooks; maximum number of traps)
SSF in Galicia	Spain	Vessels less than 12 m operating mainly within	Gillnet	Hake, horse mackerel, mackerel, pouting, surmullet.	Most target species subject to TAC, MLS, and fishing effort

(continued)

Table 5.1 (continued)

Case study	Country	Fishing fleet	Gear used	Main species landed	Rules and regulations
		6 nm using passive gear		Average daily catch of 3000 kg (in 2018)	
Deepwater hook-and-line fishery in the Azores	Portugal	Deepwater	Bottom long-lines, handlines	Blackspot seabream, European conger, Forkbeard, silver scabbardfish, bluemouth rockfish, wreckfish. Total landings of 4070 t (in 2014), 15–21 M€ between 2010 and 2017	Target and secondary species subject to TAC (e.g., blackspot seabream, alfonsinos). Deepwater sharks subject to TAC zero. MLS for several species, minimum hook sizes, area and temporal closures, and bans on the use of specific gear
Beach seine	Portugal	Purse seine	Trawling net to the beach; small fishery consisting of solely 143 vessels in the entire country	Small pelagic fish such as mackerel, Atlantic horse mackerel, and sardine	Horse mackerel (*Trachurus* spp.) subject to TAC

Note: SSF, small-scale fisheries; MLS, minimum landing size; nm, nautical miles; TAC, Total Allowable Catch

obtain better prices. Only the biggest individuals are landed; all the others, including those having legal size, are discarded.

The impacts of the LO will be different for fishers using different gears. Netters think that in some seasons they will have high rates of discards (e.g., Atlantic horse mackerel), that they will have to come to the harbor to land before returning back to their fishing areas. Handling and sorting fish will take longer, and they do not know if crew members will do it. For them, the need to employ one more crew member to deal with longer handling times means less income for the crew. All fishers want to know who will pay the different taxes related to auctions, dealing with trash, etc. Netters and long-liners consider that the LO will have a negative economic impact on their activity. But for the more selective handliners, the LO was felt to have little economic impact.

Table 5.2 Reasons for discarding and barriers to implementing the Landings Obligation (LO)

Case study	Country	Main reasons for discarding	Barriers to the implementation of the LO		
			Ecological	Economic	Institutional
SSF in the British Channel, Celtic Sea, and Bay of Biscay	France	Mainly regulations (quotas, MLS, and forbidden species). Also, low market value, lack of commercial value/market, high grading, and damaged catch	Mix fisheries, in some season's abundance of some species are not easy to avoid. Few vessels located in estuary areas deal with undersize fish	The LO will increase operation costs for netters and long-liners (more trips to land all catch, increase in crew to deal with extra work). Worries about who will pay for taxes related to the auction, trash, etc.	Most target species subject to quota and many small-scale fishers operate under the national open quota system, which ends fast
SSF in the Thermaikos Gulf	Greece	Low market value of the landings, damaged catch, mishandling on board, undersize fish, small catch	Many factors, mostly caused by the nature of the Greek SSF (multi-fleet and multi-species). Recent data show that discards have risen and are dominated by alien species	Economic incentives seem to contribute to discarding practices; high local market demand for fish contributes to the regular selling of undersized fish in the black market	Loose enforcement; lack of spatial monitoring system for vessels < 12 m (the majority of SSF fleet). Unknown number of recreational vessels hardens the role of fisheries managers
				Fishers do not perceive an increase in operation costs due to the LO because they have little discards	Fishers oppose the LO because it will decrease their catch. There is the need to decrease the MLS for some species to avoid discards
SSF in Catalonia	Spain	Damaged catch; low market value	Largely mixed fisheries with relatively small quantities of discards of regulated species; very	Increased cost of sorting; inexistence of economic outlet for unwanted catches brought to land	Loose monitoring, control, and enforcement capacity by the fisheries administration; lack of

(continued)

Table 5.2 (continued)

Case study	Country	Main reasons for discarding	Barriers to the implementation of the LO		
			Ecological	Economic	Institutional
			difficult to optimize operations to completely avoid unwanted catches		incentives for compliance
SSF in Galicia	Spain	Lack of quotas for harvested commercial species	Largely mixed fisheries with relatively small quantities of discards of regulated species; very difficult to optimize operations to completely avoid unwanted catch	The hold space on board is currently optimized, and it would not be possible to expand the hold space without affecting the navigability of the fishing vessels. Small-scale fishing vessels hold their catches on board in boxes classified by species and size, and the potential increase on their number would increase insecurity of the vessels	Fishers strongly oppose the LO and the mandatory measure to annotate all catches in the electronic logbook, because it will be very difficult and impractical during the fishing activities
Deepwater hook-and-line fishery in the Azores	Portugal	Undersize fish (< MLS), quota in the case of "alfonsinos," low market value, damaged catch	Difficult to avoid unwanted catch due to mixed resources, especially juveniles of blackspot seabream; fishers perceived high abundance of deepwater sharks	Fishers strongly oppose that unwanted undersize catch cannot be sold for direct human consumption; Representatives of fish auctions concerned about the economic costs of collecting and dealing with the unwanted catch	Fishers strongly oppose that catch will count against quota; limiting quota for "alfonsinos" and TAC zero for deepwater sharks could prematurely choke the fishery

(continued)

Table 5.2 (continued)

Case study	Country	Main reasons for discarding	Barriers to the implementation of the LO		
			Ecological	Economic	Institutional
Beach seine	Portugal	Undersize fish (< MLS), low market value	Difficult to implement the LO due to fishery being carried out on the beach	Fishers strongly oppose the fact that catches of juvenile horse mackerel (*Trachurus* spp.) cannot be sold for human consumption	Fishery carried out in areas of great ecological sensitivity (nursery areas, spawning zones, and/or growing areas) and undersize fish constitute an important part of the catch, but these can survive

Note: SSF, small-scale fisheries; MLS, minimum landing size; LO, Landing Obligation

It was felt that the ecosystem will also likely be negatively impacted by the LO because discards returned to the sea are often eaten by birds, other fish, mammals, or benthic scavengers (Depestele et al., this volume). Small-scale fishers wonder what will happen to the ecosystem if discarding practices are ended. They also prefer to continue discarding as usual rather than supporting the aquaculture sector which they perceive is bound to benefit from the implementation of the LO.

For the moment, SSF avoids unwanted catches, especially undersize fishes, by changing fishing areas. Their main concern is the avoidance of seasonal species like mackerels (*Scomber* spp.) for which they have little or no quota at all. Choke species are the most important constraint because there is always the risk of the fishery to choke, rendering a continuation of operation impossible. Until now, the LO has not been fully implemented, with exemptions having been implemented in all regional seas, but discards are still not landed nor registered officially.

5.3.2 *Greece*

Small-Scale Fisheries in the Thermaikos Gulf

SSF accounts for the majority of SSF vessels operating in Greek waters (94%) with a fleet numbering 12,762 vessels in 2014. They are active along the extensive Greek coastline, using polyvalent passive gears and catching a multitude of species (Stergiou et al. 2002; Gonçalves et al. 2007; Tzanatos et al. 2007; Brodersen et al. 2016), and the SSF métiers exhibit significant spatiotemporal variations in catch composition (Tzanatos et al. 2007; Palialexis and Vassilopoulou 2012a, b;

Table 5.1). Landings are channeled to the market through short supply chains, or directly to restaurants, and sold at an average value of 9 €·kg^{-1}. However, landings per vessel as well as income per fisher are generally very low, and each business has low invested capital.

The Greek SSFs are mostly family-owned vessels with one or two people on board, sometimes the husband and wife together. Based on the Data Collection Framework in 2014, this is the largest fishing fleet in European waters, with a steady decrease since 2008, following the general trend in the overall Greek fleet. This segment had a combined gross tonnage of 24.8 thousand GT and a total power of 238.3 thousand kW (STECF 2016).

SSFs are characterized by their multi-gear nature and the targeting of multiple species, with *Sepia officinalis*, *Mullus surmuletus*, *Diplodus annularis*, *Oblada melanura*, *Octopus vulgaris*, *Pagellus erythrinus*, and *Scorpaena porcus* being landed in high numbers (Stergiou et al. 2002; Gonçalves et al. 2007; Tzanatos et al. 2007; Brodersen et al. 2016) and the SSF metiers exhibit significant spatio-temporal variations in the catch composition (Tzanatos et al. 2007; Palialexis and Vassilopoulou 2012a, b).

In relation to discarding practices, SSF in Greece documents relatively low discarding, with estimates ~10% of the total catch (Tzanatos et al. 2007; Vassilopoulou et al. 2007). More recent data show that discards have risen (17% of the catch in 2014–2016, compared to 7.5% of the catch in 2004–2006) and have been dominated by alien species catches: *Siganus luridus* – which is commercial in some regions – represented 18% of discards in weight, while three more alien species (*Siganus rivulatus*, *Stephanolepsis diaspros*, *Balistes capriscus*) have also been documented in a SSF in the Saronikos Gulf (Brodersen et al. 2016).

SSF discards are a result of (i) low commercial value of the landings (e.g., Atlantic lizardfish (*Synodus saurus*); (ii) fishing practices, i.e., damage to individuals before being brought on board (e.g., European hake); (iii) mishandling on board; (iv) the catch of undersize individuals for species under MLS regimes (e.g., annular seabream (*Diplodus annularis*); and (v) fish having commercial value but not caught in adequate numbers to be sold (Tzanatos et al. 2007; Gonçalves et al. 2007) (Table 5.2). Other factors such as soaking time, depth of the fishing operations, and the mesh used affect considerably the discard numbers in the trammel net fisheries of the Ionian Sea (Vassilopoulou, unpublished data). In Table 5.1 more information is given on the studies dedicated to the investigation of discard practices of SSF in Greece. They all showed that the overall discarded fraction from SSF is considered as far from being negligible.

There are many factors that act as barriers for the successful implementation of the LO, mostly caused by the nature of the Greek SSF (i.e., different gear used with different species being targeted simultaneously) (Table 5.2). Economic incentives to not discard result in undersized fish being sold regularly on the black market (Damalas and Vassilopoulou 2013).

Interviews with small-scale fishers operating in the Thermaikos Gulf were done in 2015 and 2017 as part of two H2020 projects (MINOUW http://minouw-project.eu

and DiscardLess). Small-scale fishers in the Thermaikos Gulf said that they never heard about the LO (Christou et al. 2017; Maynou et al. 2017; Fitzpatrick et al. 2017). But as soon as it was explained to them what the LO means, all of them declared to be against it. The rule is perceived as an additional threat for their activity. Small-scale fishers operating in the area say that they are currently in competition with dolphins which constantly destroy their fishing gear (nets) and damage captured fish. They said that dolphins leave little fish in the nets. "If we want to bring fish home, we have to watch our nets; therefore, we stay on the spot." To avoid nets being destroyed due to the presence of the dolphins, fishers never set their nets for several hours. Sometimes soaking time is less than an hour, and fishers of this region have problems earning a living. Within this short time of operation, discards are very low.

Fishers did not know that the LO is already implemented in Greece and had never heard about the ongoing exemptions already granted to them. For them, the main reason for discarding is regulatory and principally the MLS. The other reasons are damaged fish and lack of market prices for some species. They considered that the discarded quantities are low, and they would not have problems to land them if they had to do so. Nowadays, unwanted catches are often landed for human consumption. For example, undersized fish may be offered as a present to clients, to family members and friends, especially when practicing direct sales. Species used to make fish soups, or undersize fish appreciated by the local market (e.g., surmullet *Mullus surmuletus*) are often given as gifts.

They do not have any problem moving to another fishing area when the quantity of MLS individuals is high because they stay near the nets during the fishing operation "to chase dolphins attacking their gear." If catches contain a lot of small fish, they turn to other fishing grounds. They do not face the same problem when they use pots because unwanted catches remain alive, and as soon as they are put on board, they are released into the sea. They think that live individuals have a high survival rate as soon as they are back into the water. In this way, small-scale fishers think that the LO is not a problem compared to the threat represented by dolphins. For them, the daily struggle against dolphins makes LO a softer constraint. LO doesn't really impact their activity due to their low rates of discards. It is observed that this latter finding contradicts their first negative vision of the LO.

According to fishers, the LO will impact more on trawlers, which generate more discards. For SSF fishers it is a good thing that these boats will have to reduce discards. These two métiers do not operate in the same areas, and little competition for space occurs. But both fleets are targeting the same species, and during the months that trawlers are operating (trawling activity is forbidden in territorial waters between June and the end of September), SSF fishers have problems selling their catch at a good price.

When asked whether they record discards, fishers respond that until the end of 2017 "nobody asked them to record them." And if somebody tells them to do so, they will not comply because they "don't want to complicate their life by adding more administrative tasks." From the interviews, it appears that small-scale fishers of the Thermaikos Gulf are against the LO by principle, but an analysis of their discourse demonstrates the opposite. This is due to the fact that such a rule will have little impact on their activities. In the case of effective implementation, they can easily adapt to the LO. The gears that fishers use are among the most selective, and they do not think that they will need to make more effort under the LO (Table 5.2).

The current conditions of SSF in the Thermaikos Gulf may be different from other areas in Greece, but in terms of discards, it seems to be similar to the results of other studies undertaken in that country. Yet, it is crucial to investigate discard levels specific for each métier and quantify the discards problem among the whole SSF sector, using robust indices (Stergiou et al. 2007). The low discards generated by Mediterranean fisheries (Tsagarakis et al. 2014) and also by other fisheries (including areas under a quota system) should prompt authorities to claim a specific exemption at the EU level, as SSF is an important activity for coastal communities and provides income and employment for local populations in areas with few alternative economic activities (Pita et al. 2010), particularly in small, isolated islands. The ongoing financial recession in Greece has further hardened the socio-economic state of these fisheries. Thus, it is important to safeguard the sector and maintain the social and economic sustainability of the coastal communities.

5.3.3 Portugal

Two examples are provided for the impact of the LO in the Portuguese SSF sector: the beach seine fisheries in mainland Portugal, an ancient activity registered in the National Archive of Intangible Cultural Heritage (e.g., in Costa da Caparica beach seine fishery; Diário da República, 2nd series, N° 34, of 16 February 2017), and the deepwater hook-and-line fisheries in the Azores islands.

5.3.3.1 The Beach Seine Fishery

The beach seine fishery is an ancient commercial fishing activity on the Portuguese coast, with reports dating as far back as the early fifteenth century (Franca and Costa 1979; Martins et al. 2000). Nowadays, the beach seine fleet is composed of 143 vessels, distributed along the Portuguese mainland coast, mainly on the northwest coast (European Commission 2018) (Table 5.1). Each vessel employs ~12 people, 5 working on board the vessel, and 7 working on land. This is a seasonal fishery, typically occurring from March to November. The main target species of the fishery are small pelagic fish such as Atlantic chub mackerel (*Scomber colias*), Atlantic horse mackerel, and European pilchard (*Sardina pilchardus*) (Gaspar and Pereira 2014). In Portugal, official fishing statistics landings are presented by fleet component (divided into trawling, purse seine, and multi-gear); therefore, it is not possible to know the proportion of landings (in volume and value) by the beach seine fishery.

The beach seine fishery operates in a coastal zone of great ecological sensitivity, as this fishing activity occurs in nursery areas, spawning zones, and/or growing areas for many species of high economic interest. As a consequence, juveniles constitute an important fraction of captures, which commercialization is not allowed by law, as individuals are < MLS, and are thus discarded (Jorge et al. 2002). In addition to capture large numbers of juveniles, this fishing technique is not selective and also captures a wide variety of bycatch species (Faltas 1997; Lamberth et al. 1997; Cabral et al. 2003), despite having seasonally target species (Fagundes et al. 2007). The low commercial value of bycatches and legal constraints results in bycatches not being traded and mostly discarded (Cabral et al. 2003). In this fishery, the LO was implemented on January 1, 2015, and applies to catch of horse mackerels (*Trachurus* spp.) and blue whiting (*Micromesistius poutassou*).

5.3.3.2 The Deep-water Hook-and-Line Fishery in Azores

The Azores is a Portuguese oceanic archipelago in the North Atlantic Ocean, with a one million km^2 exclusive economic zone (EEZ), no continental shelf and great depths, with an important demersal fishing around the island slopes and the many seamounts present in the area (Silva and Pinho 2007; Morato et al. 2008). The bottom hook-and-line fishery is the most important fishery in the region, employing about 60% of all professional fishers in the archipelago (Carvalho et al. 2011). This fishery is mostly small-scale, with 92% of the vessels < 12 m (N = 478 in 2016) (Table 5.1). Two main fishing gears are used: (i) bottom longlines targeting mainly deep-sea demersal fishes, such as blackspot seabream (*Pagellus bogaraveo*), alfonsinos (*Beryx* spp.), or blackbelly rosefish (*Helicolenus dactylopterus*), or deeper species such as common mora (*Mora mora*), and (ii) handline targeting mostly blackspot seabream and wreckfish (*Polyprion americanus*). Both gears operate all year round. The bottom longline and handline fishery is by far the most valuable in terms of landed value, with an annual landed value varying from 15 to 21 million € for the period 2010–2017, around 58% of all landed value in the Azores (SREA, http://estatistica.azores.gov.pt).

Unreported catch for this fishery was estimated to amount to 830 t per year on average over the period 2000–2014, i.e., 10.3% of the total catch (Pham et al. 2013). Around half (47%) of this unreported catch is used as bait, kept for crew consumption, or offered, while the remaining was discarded at sea (447 t·year^{-1}) (Fauconnet et al. in press).

Discard practices are believed to be similar between handlines and bottom longlines. Observer data suggest that the catch of individuals smaller than the MLS is the main cause for discarding in this fishery, followed by low market value (Canha 2013). About 90 species are regularly discarded by this fishery, 40%

of which due to low commercial value (Canha 2013). However, 61% of the discards can be attributed to six species of commercially important fish, such as silver scabbardfish (*Lepidopus caudatus*), European conger (*Conger conger*), blackbelly rosefish, splendid alfonsino (*B. splendens*), blackspot seabream, and thornback ray (*Raja clavata*). At least ten species of deepwater sharks are occasionally caught by this fishery. Even if limited, this bycatch is of concern since many deepwater sharks such as *Deania* spp., *Centrophorus* spp., *Etmopterus* spp., *Centroscymnus* spp., and kitefin shark (*Dalatias licha*) are listed in the IUCN red list of endangered species. Due to their vulnerability, the EU has set their TAC to zero in 2010 (EC Reg. N° 1359/2008). Since then, discard of those species has been compulsory. Deepwater sharks accounted for 8% of the discards of the fishery over the period 2010–2014 (Fauconnet et al. 2016).

The implementation of the LO in Azorean demersal fisheries will only take place from January 2019 onward. Several factors were identified, in semi-structured interviews and meeting with stakeholders, as part of the DiscardLess project, to potentially act as barriers for the successful implementation of the LO in the Azores (Table 5.2).

5.3.4 *Spain*

Two cases from Spain illustrate the complexities of the implementation of the LO in different regions of Spain: the case of SSF in Catalonia (NW Mediterranean) and the gillnet fishery in Galicia (NW Atlantic).

5.3.4.1 Small-Scale Fisheries in Catalonia

SSF in Catalonia is carried out by a relatively high number of fishing units (365 out of a fleet of 727 vessels in 2016) operating from 32 fishing harbors. Landings of the SSF fleet oscillate between 1800 and 3800 t·year^{-1} in recent years (average 2960 t·year^{-1} for the period 2000–2016), with a corresponding value of landings around 15 million € (Table 5.1). This fisheries production corresponds to ca. 10% of the production/landings in Catalonia but employs ca. 50% of the fishing fleet and 25% of the labor/fishers. The fleet operates in coastal waters, typically within 6 miles of the coast, uses a multitude of gears, and lands over 200 species, the most important being demersal species and sand eels (*Gymnammodytes* spp.), common octopus (*Octopus vulgaris*), and Atlantic bonito (*Sarda sarda*). Refer to Table 5.1 for detailed information about the fleet, gear and main species landed. In general, the commercial catches of each individual vessel are very low (20–50 kg·day^{-1}) but of high value, with ex-vessel prices of the target species oscillating between 10 and 20 €·kg^{-1}.

Compared to other segments of the fleet, SSF is highly selective, and the amount of discards is relatively low. The fractions of the catch that are discarded are usually noncommercial species, such as epibenthic invertebrates, or damaged fish. Commercial species that could be otherwise sold are discarded when they are damaged due to scavengers preying on the catch. This problem is particularly acute for set net fishing gear (trammel nets and longlines). Undersize fish are usually not discarded but sold on the black market. Field studies carried out in the MINOUW project show that the amount of catches below legal size is generally low, but for certain species and certain gear deployments, the proportion of catches that will fall under the remit of the Landing Obligation can be high. Trammel nets employing inner panels of 40–60 mm mesh can produce a relatively high proportion of undersize European seabass (*Dicentrarchus labrax*), Sand steenbras (*Lithognathus mormyrus*), blackspot seabream (*Pagellus bogaraveo*), or common sole (*Solea solea*). In the case of the blackspot seabream, its large legal size (33 cm TL) results in all catches from all fishing gears studied being undersize. Several barriers for the implementation of the LO have been identified such as low quantities of discards, the lack of capacity to monitor SSF by the regional administration, and increase costs of sorting, among others (Table 5.2).

5.3.4.2 The Gillnet Fishery in Galicia

Fishing is a major contributor to gross domestic product in Galicia (an autonomous community in northwestern Spain), the main fishing region in Spain (Villasante 2012). The artisanal/SSF fleet is comprised mainly of small vessels (on average 6 m long), fishing with a great variety of passive gears, the so-called *artes menores* (traps, hooks and lines, gill and trammel nets, and small seines), and exploiting a diverse range of species, most of which are subjected to TACs.

The fleet using gillnets comprises 1000 fishing vessels, operating in a multispecific SSF, mainly harvesting European hake, pouting (*Trisopterus luscus*), horse mackerels, and surmullet at depths of 30–140 m and up to 8–10 miles from the coast. Based on results from interviews started in 2015 and updated until 2018, the reasons for discarding are the precautionary closure and the full closure of the fishery due to the full harvest of the TAC (Villasante et al. 2016a, b).

Recently, Villasante et al. (2015b) estimated the total removals of fisheries catches (including IUU catches, subsistence catches, and discards for commercial and recreational fisheries) for the 1950–2010 period. The authors demonstrated that the discard rate for SSFs ranges between 5–18% depending on the type of commercial species harvested. However, the authors also found that the discard rate for some sedentary resources (e.g., goose barnacle (*Pollicipes pollicipes*) 74% and razor clams *Ensis* spp. 49%) can be significantly higher than for other SSFs.

However, the species under TAC and quota regulations present high discard rates which ranged between 0 and −50% (European hake, mackerels) and/or 50−200%

(horse mackerels). Catches of horse mackerels and mackerels are highly variable due to migratory movements from Portuguese to Galician waters and can sometimes lead to high discard rates. Harvesting of immature individuals was reported to be very low or nonexistent for all species caught by this fishery (Villasante et al. 2015b, 2016c).

Regarding the compliance to the LO, the expert's opinion and the participatory consultation made with the small-scale fisheries sector show that changing the fisheries management system based on the TAC regulation would be the most important reason to comply with the LO (Villasante et al. 2016a, b) (Table 5.2).

5.4 Conclusion

Despite the increased recognition of SSFs, there is a still need to ensure that policy-makers receive robust scientific data about such fisheries on which to base decisions and thus ensure coherent policy. Our results show that only 21% of 1219 papers that have been published until 2018 focused on the discard problem in SSF.

Key SSFs from around Europe selected to investigate the reasons for discarding, impact, and barriers to implementing the LO illustrate that discard rates vary greatly from fishery to fishery and species to species. However, the main reasons fishers discard are relatively similar from fishery to fishery and are mostly due to regulations (mainly TACs, quotas, and MLS), low market value of some catch components, capture of noncommercial species, high grading, and damaged catch.

Small-scale fishers perceive that it will be difficult to comply with the LO and could identify ecological, economic, and institutional barriers to the implementation of the LO. From an ecological perspective, most fishers are of the opinion that resources are largely mixed, and unwanted catch is very difficult to avoid. For example, the fact that the beach seine fishery in Portugal is carried out in areas of great ecological sensitivity, such as nursery areas, results in the capture of large numbers of juveniles. From an institutional perspective, the lack of monitoring, control, and enforcement capacity by fisheries jurisdictions, combined with lack of incentives for compliance, are critical barriers perceived by fishers for the implementation of the LO in all case studies. Plus, some fishers identified that the implementation of the LO requires the adoption of more selective gear technology (Galicia and Azores). Azores fishers think they are already using one of the most selective gear in European fisheries and as such that the LO should not apply to them.

From an economic perspective, fishers state that the LO will increase the operational costs of fishing activities. They strongly oppose the fact that unwanted undersize catch cannot be sold for human consumption and that this catch will count against their quota. In general, the potential socioeconomic impacts of the LO could be high for SSF. For example, it is estimated that the future yield (catches) under the LO in Galicia (Spain) would be only 50% of catches expected in the absence of the LO, regardless of the total volume of quotas allocated to the fleet.

Acknowledgment The authors acknowledge the financial support from the European COST Action "Ocean Governance for Sustainability – challenges, options and the role of science" and by the ICES Science Fund Project "Social Transformations of Marine Social-Ecological Systems", and MINOUW (Grant Agreement 634495). C. Pita acknowledges FCT/MEC national funds and FEDER co-funding, within the PT2020 partnership Agreement and Compete 2020, for the financial support to CESAM (Grant no UID/AMB/50017/2013). C.M. Teixeira had the support of the Fundação para a Ciência e a Tecnologia (FCT) (Pest-OE/MAR/UI0199/2011); and C.M. Teixeira and C. Pita were supported by the Research Project "LESSisMORE – LESS discards and LESS fishing effort for BETTER efficiency on the small-scale fisheries" (Ref. "LISBOA-01-0145-FEDER-028179"), support by the FEDER Funds through the COMPETE 2020, by the PIDDAC through FCT/MCTES. TM thanks the support from the European Union's Horizon 2020 research and innovation project DiscardLess (Grant Agreement No 633680), the Fundação para a Ciência e Tecnologia (FCT) strategic project UID/MAR/04292/2013 granted to MARE. He is also supported by the Program Investigador FCT (IF/01194/2013/CP1199/CT0002).

References

Aarts, G., & Poos, J. (2009). Comprehensive discard reconstruction and abundance estimation using flexible selectivity functions. *ICES Journal of Marine Science*, *4*, 763–771.

Brodersen, M.M., Haralabous, J., Chalari, Dictyopoulos, C., Dogrammatzi, K., Vassilopoulou, V. (2016). Preliminary comparative study of trammel net fisheries in the Saronikos Gulf a decade apart 16th Panhellenic Conference of Ichthyologists, Kavala, pp. 129–132.

Cabral, H., Duque, J., Costa, M.J. (2003). Discards of the beach seine fishery in the central coast of Portugal. *Fisheries Research*, *63*(1), 63–71. https://doi.org/10.1016/S0165-7836(03)00004-3.

Canha, A. (2013). *Caracterização das rejeições na pescaria de demersais nos Açores*. Master thesis dissertation, University of the Azores, pp 76.

Carvalho, N., Edwards-Jones, G., Isidro E. (2011). Defining scale in fisheries: small versus large-scale fishing operations in the Azores. *Fisheries Research*, *109*, 360–369.

Christou, M., Haralabous, J., Stergiou, K.I., Damalas, D., Maravelias, C.D. (2017). An evaluation of socioeconomic factors that influence fishers' discard behaviour in the Greek bottom trawl fishery. *Fisheries Research*, *195*, 105–115.

Damalas, D., & Vassilopoulou, V. (2013). Slack regulation compliance in the Mediterranean fisheries: a paradigm from the Greek Aegean Sea demersal fishery, modelling discard ogives. *Fisheries Management and Ecology*, *20*, 21–33.

De Vos, B.I., Döring, R., Aranda, M., Buisman, F.C., Frangoudes K., Goti L., et al. (2016). New modes of fisheries governance: Implementation of the landing obligation in four European countries. *Marine Policy*, *64*, 1–8.

Depestele, J., Feekings, J., Reid, D., Cook, R., Gascuel, D., Girardin, R., et al. (this volume). The impact of fisheries discards on scavengers in the sea. In S.S. Uhlmann, C. Ulrich, S.J. Kennelly (Eds.), *The European Landing Obligation – Reducing discards in complex, multi-species and multi-jurisdictional fisheries*. Cham: Springer.

EU [European Commission]. (2018). Fleet register on the NET 2018. Data provided by MS at 01/06/2018. Build of 13/06/2018 http://ec.europa.eu/fisheries/fleet/index.cfm?method=Search.SearchAdvanced&country. Accessed 13 June 2018.

Fagundes, L., Tomás, A., Casarini, L., Bueno, E., Lopes, G., Machado, D., et al. (2007) A pesca de arrasto-de-praianailha de São Vincente, São Paulo, Brasil. *Série Relatórios Técnicos* N° 29.

Faltas, S.N. (1997). Analysis of beach seine catch from Abu Qir Bay (Egypt). *Bulletin of the National Institute of Oceanography and Fisheries, 23*, 69–82.

Fauconnet, L., Pham, C., Canha, A., Afonso, P., Vandeperre, F., Machete, M., et al. (2016). *Estimating total fisheries discards in an oceanic archipelago of the NE Atlantic*. 7th World Fisheries Congress, Busan, South Korea, 23–27th May 2016.

Fauconnet, L., Pham, C., Canha, A., Afonso, P., Diogo, H., Machete, M., Silva, M.A., Vandeperre, F., Morato, T. (in press) *An overview of fisheries discards in the Azores*. Fisheries Research

Fitzpatrick, M., Quetglas, T., Frangoudes, K., Triantaphyllidis, G., Nielsen, K. (2017). *DiscardLess policy brief No 2: year 2 of the landing obligation: key issues in Mediterranean fisheries*. https://doi.org/10.5281/zenodo.573666. Accessed 25 July 2018.

Franca, M., & Costa, F. (1979) Nota sobre as xávegas da Costa da Caparica e Fonte da Telha. *Boletín Instituto Nacional de Investigación das Pescas, 1*, 37–69.

Gaspar, M., & Pereira, F. (2014). Pequena pesca na costa continental portuguesa: caracterização sócio-económica, descrição da actividade e identificação de problemas. *Instituto Português do Mar e da Atmosfera*. pp 268.

Gonçalves, J., Stergiou, K., Hernando, J., Puente, E., Moutopoulos, D., Arregi, L., et al. (2007). Discards from experimental trammel nets in southern European small-scale fisheries. *Fisheries Research, 88*, 5–14.

Jorge, I, Siborro, S., Sobral, M. (2002). Contribuição para o conhecimento da pescaria da xávega da zona centro. *Relatórios Científicos e Técnicos Instituto de Investigação das Pescas e do Mar, 85*, 1–22.

Lamberth, S., Sauer, W., Mann, B., Brouwer, S., Clark, B., Erasmus, C. (1997). The status of the South African beach-seine and gill-net fisheries. *South African Journal of Marine Science*, *18*, 195–202.

Macfadyen, G., Salz, P., Cappell, R. (2011). *Characteristics of small-scale coastal fisheries in Europe*. Policy Department: Structural and Cohesion Policies, European Parliament, Fisheries pp 162 .

Martins, R., Carneiro, M., Rebordão, F., Sobral, M. (2000). A pesca com arte de xávega. *Relatorios Científicos e Técnicos Instituto de Investigacão das Pescas Mar, 48*, pp 32.

Maynou, F., del Gil, M.Mar, Vitale, S., Giusto, G., Foutsi, A., Range, M., et al. (2017). Fishers' perceptions of the European Union discards ban : perspective from south European fisheries. *Marine Policy*, *89*, 147–153.

Morato, T., Machete, M., Kitchingman, A., Tempera, F., Lai, S., Menezes G., et al. (2008). Abundance and distribution of seamounts in the Azores. *Marine Ecology Progress Series, 357*, 17–21.

Palialexis, A., & Vassilopoulou, V. (2012a). *Metier identification in trammel net fisheries in Greece*. Oral paper presented at the 10th Panhellenic Symposium of Oceanography and Fisheries, Athens.

Palialexis, A., & Vassilopoulou V. (2012b). *The local character of trammel net fisheries in Greece and the need of regional spatial approach for management effectiveness*. Oral paper presented at the 6th World Fisheries Congress, Edinburgh 7–11th May 2012.

Pham, C., Canha, A., Diogo, H., Pereira, J., Prieto, R., Morato, T. 2013. Total marine fishery catches for the Azores (1950–2010). *ICES Journal of Marine Sciences, 70*, 564–577.

Pita, C., Dickey, H., Pierce, G., Mente, E., Theodossiou, I. (2010). Willingness for mobility amongst European fishermen. *Journal of Rural Studies, 26*, 308–319.

Silva, H., & Pinho, M. (2007). Exploitation, management and conservation: small-scale fishing on seamounts. In T. J. Pitcher, T. Morato, P. J. B. Hart, M.R. Clark, N. Haggan, R. S. Santos (Eds.), Seamounts: ecology, fisheries & conservation (pp. 333–399). Oxford: Blackwell Publishing.

STECF [Scientific, Technical and Economic Committee for Fisheries]. (2016). The 2016 annual economic report on the EU fishing fleet. (STECF-16-11). 2016. Publications Office of the European Union, Luxembourg, pp 470.

Stergiou, K., Moutopoulos, D., Erzini, K. (2002). Gill net and longlines fisheries in Cyclades waters (Aegean Sea): species composition and gear competition. *Fisheries Research, 57*, 25–37.

Stergiou, K., Moutopoulos, D., Casal, H., Erzini K. (2007). Trophic signatures of small-scale fishing gears: implications for conservation and management. *Marine Ecology Progress Series, 333*, 117–128.

Tsagarakis, K., Palialexis, A., Vassilopoulou, V. (2014). Mediterranean fishery discards: review of the existing knowledge. *ICES Journal of Marine Sciences, 71*, 1219–1234.

Tzanatos, E., Somarakis, S., Tserpes, G., Koutsikopoulos C. (2007). Discarding practices in a Mediterranean small-scale fishing fleet (Patraikos Gulf, Greece). *Fisheries Management Ecology, 14*, 277–285.

Uhlmann, S.S., van Helmond, A.T., Stefánsdóttir, K., Sigurðardóttir, S., Haralabous, J., Bellido, J. M., et al. (2013). Discarded fish in European waters: general patterns and contrasts. *ICES Journal of Marine Sciences, 71*, 1235–1245.

van Hoof, L., Kraan, M., Visser, N.M., Avoyan, E., Batsleer, J., Trapman, B. (this volume). Muddying the waters of the Landing Obligation: How multi-level governance structures can obscure policy implementation. In S.S. Uhlmann, C. Ulrich, S.J. Kennelly (Eds.), *The European Landing Obligation – Reducing discards in complex, multi-species and multi-jurisdictional fisheries*. Cham: Springer.

Vassilopoulou, V., Anastasopoulou, K., Haralambous, C., Christides, G., Glykokokkalos, S., et al. (2007). *Preliminary results of monitoring discards by coastal fishery vessels in Greek waters*. Oral paper presented at the 13th Panhellenic Symposium of Ichthyologists, Mytilene, Greece, Proceedings: 109−116.

Veiga, P., Pita, C., Rangel, M., Gonçalves, J.M., Campos, A., Fernandes, P., et al. (2016). The EU landing obligation and European small-scale fisheries: what are the odds for success? *Marine Policy, 64*, 64–71.

Villasante, S. (2012). The management of the blue whiting fishery as complex social- ecologic system: the Galician case. *Marine Policy, 36*(3), 1301–1308.

Villasante, S., Pazos Guimeráns, C., Garcia Rodrigues, J., Antelo, M., Rivero Rodríguez, S., Da Rocha, J.M., et al. (2015a). *Small-scale fisheries and the zero-discard target*(p. 73). Brussels: European Parliament, Directorate-General for Internal Policies Policy Department B: Structural and Cohesion Policies.

Villasante, S., Macho, G., Isusi De Rivero, J., Divovich, E., Zylich, K., Zeller, D., et al. (2015b). *Estimates of total fisheries removals from the Northwest of Spain (1950–2010)*. Working Paper Series #51, University of British Columbia, Canada, pp 18.

Villasante, S., Pita, C., Pazos Guimeráns, C., Rodrigues, J., Antelo, M., Rivero Rodríguez, et al. (2016a). To land or not to land: How stakeholders perceive the zero-discard policy in European small-scale fisheries? *Marine Policy, 71*, 166–174.

Villasante, S., Pierce, G., Pita, C., Pazos Guimeráns, C., Rodrigues, J., Antelo, M., et al. (2016b). Fishers' perceptions about the EU discards policy and its economic impact on small-scale fishers in Galicia (North West Spain). *Ecological Economics, 130*, 130–138.

Villasante, S., Macho, G., Isusi de Rivero, J., Divovich, E., Zylich, K., Zeller, D., et al. (2016c). Spain (North West). In D. Pauly, & D. Zeller (Eds.), *Global Atlas of marine fisheries: a critical appraisal of catches and ecosystem impacts* (p. 397), Washington, DC: Island Press.

Chapter 6
Potential Economic Consequences of the Landing Obligation

Ayoe Hoff, Hans Frost, Peder Andersen, Raul Prellezo, Lucía Rueda, George Triantaphyllidis, Ioanna Argyrou, Athanassios Tsikliras, Arina Motova, Sigrid Lehuta, Hazel Curtis, Gonzalo Rodríguez-Rodríguez, Hugo M. Ballesteros, Julio Valeiras, and José María Bellido

Abstract To assess the likely economic outcomes to fishing fleets of the Landing Obligation (LO), bioeconomic models covering seven European fisheries, ranging from the North East Atlantic to the Mediterranean, have been applied to estimate the

---

A. Hoff (✉) · H. Frost · P. Andersen
Department of Food and Resource Economics (IFRO), University of Copenhagen, Frederiksberg C, Denmark
e-mail: ah@ifro.ku.dk

R. Prellezo
AZTI, Sukarrieta, BIZKAIA, Spain

L. Rueda
Centro Oceanográfico de Baleares, Instituto Español de Oceanografía, Palma, Spain

G. Triantaphyllidis · I. Argyrou
NAYS Ltd., Kallithea, Athens

A. Tsikliras
Laboratory of Ichthyology, School of Biology, Aristotle University of Thessaloniki University Campus, Thessaloniki, Greece

A. Motova · H. Curtis
Seafish, Edinburgh, UK

S. Lehuta
Ifremer, Fisheries Ecol & Modelling Unit, Nantes, France

G. Rodríguez-Rodríguez · H. M. Ballesteros
Fisheries Economics and Natural Resources Research Unit, Department of Applied Economics, Faculty of Economics and Business Administration, University of Santiago, Santiago de Compostela, A Coruña, Spain

J. Valeiras
Instituto Español de Oceanografía (IEO), Centro Oceanográfico de Vigo, Vigo, Spain

J. M. Bellido
Instituto Español de Oceanografía, Centro Oceanográfico de Murcia (IEO), San Pedro del Pinatar, Murcia, Spain

S. S. Uhlmann et al. (eds.), *The European Landing Obligation*,
https://doi.org/10.1007/978-3-030-03308-8_6

economic performance of fleets before and after implementing the LO. It is shown that for most of the analysed fisheries, their economic outcome will be negatively affected in the long term by the LO, when compared to the expected outcome with no LO. Efficient mitigation strategies (exemptions, quota uplifts, improved selectivity, effort reallocation and others) may, for some of the analysed fisheries, reduce the negative economic effect of the LO. Moreover, the possibility to trade quotas, both nationally and internationally, may also reduce the economic losses caused by the LO. However, even with mitigation strategies and/or quota trade in place, most of the analysed fisheries are worse off under the LO than what could be expected if the LO was not implemented.

Keywords Costs and earnings · Discards · Economic repercussions · Fisheries management · Fleet adjustment

6.1 Introduction

Commercial fisheries in Europe are diverse, with fish being caught for varied purposes ranging from high-value species for human consumption to fish used for fishmeal and fish oil. Technological and biological interactions make it difficult to catch target species completely selectively. For almost a century, landings of immature fish have been prohibited by regulations. Discarding fish below a minimum conservation reference size (MCRS) has been mandatory in European waters since the adoption of the Common Fisheries Policy (CFP) in 1983. The CFP Landing Obligation (LO) of 2013 requires fish under the MCRS to be landed, with implementation being phased in from 2015 to 2019. Similarly, before 2013, it was forbidden to land species for which quota was exhausted, and discarding of catches at or above MCRS was therefore required, a logical practice in mixed species fisheries.

Many businesses expect significant short-term negative economic repercussions of the LO due to increased operating costs, decreased income from landings and underutilisation of quotas (Condie et al. 2014). However, the actual outcomes of the LO will depend on several factors, including (i) the management system in place, (ii) application of exemptions (e.g. *de minimis* allowance of discards up to 5%), (iii) interannual transfers, (iv) catch allowances of stocks without TACs, (v) quota adjustments and quota swaps/movements, (vi) application of selectivity measures, (vii) costs of landing unwanted catch, (viii) prices obtained for unwanted fish and (ix) compliance of the sector. It is hoped that short-term losses could be mitigated by longer-term gains, given the desired reduced pressure on fish stocks and anticipated increases in quota and catch rates.

This chapter considers economic outcomes for fleets by analysing possible economic effects of the LO for seven diverse European case studies comprising (i) UK and Danish North Sea demersal fisheries, (ii) the French demersal trawl fishery in the Eastern English Channel, (iii) the Spanish trawl fishery in the Bay of Biscay, (iv) the Spanish trawl fishery in the Cantabrian-NW region, (v) the Greek trawl and small-scale coastal fishery in the Thermaikos Gulf (Eastern Mediterranean) and (vi) the Spanish demersal trawl fishery in the Western Mediterranean. Common for all these fleets is that they have a history of substantial unwanted catches before the LO; therefore it can be expected that the LO would affect them substantially.

6.2 What Can the Literature of Economics Tell Us?

The literature tells us that fishers tend to use more fishing effort than socially optimal due to market failures such as the tragedy of the commons (Hardin 1968). An unregulated, open-access fishery leads to overexploitation of fish resources; therefore the EU has attempted to prevent this by use of total allowable catches (TACs), limited fishing effort, MCRS and technical specifications for fishing gears, closed areas and seasons, among other measures. However, in mixed fisheries, these restrictions may also, in some cases, encourage a 'race to fish' and may increase incentives to discard because low quotas of some species prevent full exploitation of species with higher quotas. Quotas may also increase incentives to high-grade (discard lower-value fish) to maximise profit.

Although the CFP has a common approach to managing the fishing opportunities of the European Union including rules of compulsory discard, the management and organisation of fleets differ between Member States. Therefore, the economic repercussions of the LO not only depend on the rules of the LO but also on the national management system to which fishing businesses are subjected.

When interest for the discard issue arose in the 1990s, four general types of factors that encouraged discarding were identified (FAO 1996b; Nordic Council of Ministers 2003): (i) *institutional*, e.g. management measures such as quotas, effort restrictions, minimum landing size of fish and mesh size regulations; (ii) *biological*, e.g. species interaction and characteristics of the fish (e.g. gender, and size); (iii) *technological* such as gear selectivity (e.g. prohibited gear, damage to fish); and (iv) *economic*, for example, price and cost relationships determined on the market and high-grading (discarding low-value fish, both regulated and unregulated) to maximise profit by using quota and room on-board for more valuable fish (Batsleer et al. 2015).

Discarding originates primarily from non-selective catch and high-grading practices. These form the basis for the empirical and theoretical economic research that has been done regarding discarding over the last 20 years.

This research began in the 1990s (e.g. Flaaten and Larsen 1991; Frost 1996; Christensen 1996; Pascoe and Revill 2004). Empirical approaches also appeared in conferences and research programmes (FAO 1996a, b; Clucas 1997). In the FAO context, the economics of discarding can be found in Pascoe (1997) with an update in Kelleher (2005). The Nordic Council of Ministers (2003) investigated incentives to discard and options to reduce it. An EU Framework 7 project, NECESSITY, investigated options to reduce discarding by using increased mesh sizes or panels in fishing gear (Frost et al. 2007).

Alongside empirical research, theoretical work based on socio-economic modelling has developed. One approach concerns unwanted catch in open-access and individual transferable quota (ITQ)-managed fisheries (Ward 1994; Ward et al. 2012; Boyce 1996; Turner 1996, 1997). These analyses usually include two species (target and nontarget) and two fleets and deal with the optimal use and allocation of effort subject to a profit- (or resource rent-) maximising objective. In this context, bycatches of nontarget species constitute an endogenous externality, i.e. an outside impact influenced by fishers. In a simple situation where harvest of the target and nontarget species is in fixed proportions, fishing effort used to harvest target species can simply be scaled up and down to reach a first-best optimum. However, harvest of target and nontarget species may take place in variable proportions. Boyce (1996) compares maximisation of welfare in situations governed by open access and ITQs of harvesting two such species by two fleets. He finds that open access leads to excessive bycatches and that an ITQ system can only secure a first-best optimum if imposed on both target species and bycatches. Segerson (2007) extends this analysis to include stochastic bycatches and shows that neither landing fees nor ITQs on both target species and bycatches can secure an expected first-best optimum in which all market failures are corrected in an economically optimal way. A different approach to analyse bycatches is adopted in Abbott and Wilen (2009) where actual regulation, as opposed to estimated economically optimal regulation, is introduced. A given fishery is regulated with quotas for both target and nontarget species combined with limited entry programmes, and this actual regulation generates excessive bycatches and too short harvest seasons.

Another theoretical approach deals with high-grading. High-grading may occur for several reasons, e.g. to extend a quota that is nearly exhausted, to get the best value per tonne of quota or to make room on-board the vessel for more valuable fish. Arnason (1994) and Anderson (1994) show that a traditional ITQ system strengthens the incentive to high-grade. However, Turner (1997) shows that a value-based ITQ system (quotas measured in value instead of volume) secures a welfare optimal level of high-grading in a similar way that open access does. Under open-access or effort management, the distance between fishing grounds and ports of landing affects vessel operators' decisions about catching patterns; limited hold or processing capacity may be increased in the short term for high-priced fish through discarding of low-priced fish, and this discarding can thus pay for one or two more hauls per trip (turnaround cost) (see Vestergaard 1996). In the market policy of the CFP, the suppression of withdrawal prices in 2014 also constituted an incentive to discard, as the removal of a fixed minimum price increases the economic propensity to discard.

Analysing high-grading requires the inclusion of high- and low-priced fish. This can be done by including age-structured fish stocks in the model or simply dividing the stock in two parts: a low-priced and a high-priced part.

Fish sales prices relative to fishing costs also influence the incentives to discard: if the price of fish is lower than the costs of putting the fish on the market, then the fish should be discarded – at least from an economic point of view. However, it may pay to land fish even when handling costs are higher than the total value. That is, if costs of discarding are higher than the loss likely to be incurred by putting the fish on the market, then the fish should be landed.

When it is illegal to discard fish while incentives to discard remain, monitoring and control must be effectively invoked to offset incentives to discard (Sutinen and Andersen 1985; Nuevo et al., this volume). Also, social norms, trust and cooperation play a role (Sutinen and Kuperan 1999; Kraak and Hart, this volume). When it is difficult to monitor vessel operations at sea, vessel operators may decide not to comply with regulation. Jensen and Vestergaard (2002) consider discarding in a moral hazard context, i.e. when fishers hide their actions at sea; when these actions cannot be detected, repercussions are placed on them based on common elements such as estimated changes in target fish stocks.

To discourage non-compliance, measures are required to assist enforcement, including penalties, incentives to adapt to social norms, increased acceptance of management rules (Sutinen and Kuperan 1999; Kraak and Hart, this volume) and a governance structure which addresses diverging perceptions about the legitimacy of discarding in the first place (Fitzpatrick et al., this volume; van Hoof et al., this volume). In theory, a premium can be introduced, e.g. an increase in the price of fish that would otherwise be discarded because of a low price. It could also be invoked as a penalty placed on the (estimated) net benefit from discarding. In such a case, the vessel operator will include the benefit/penalty in their decision function. However, he/she will also consider the probability of being detected and the likelihood and size of any fine. If the risk of being detected and the penalty are low, fish will probably be discarded and vice versa.

6.3 The European Case Study Fisheries

Possible economic implications of the LO are presented through seven diverse European fishery case studies. Characteristics of each case are summarised in Table 6.1. Cases are divided into three groups: (i) demersal fisheries in the North Sea, West of Scotland and English Channel, represented by fleets from Denmark, the UK and France, (ii) Spanish Atlantic fisheries represented by the Basque mixed demersal fishery in the Bay of Biscay and the Galician trawl fleet in the Cantabrian-NW region and (iii) Mediterranean fisheries represented by two mixed demersal trawl fisheries from the Balearic Islands (Spain, Western Mediterranean) and the Greek trawl and small-scale coastal fishery in the Thermaikos Gulf (Eastern Mediterranean).

All cases have different management systems on top of which the LO is imposed. However, all have a certain degree of MCRS regulation, and before the LO, it was compulsory to discard fish below MCRS, with a few derogations in certain pelagic

Table 6.1 Base characteristics of the European case study fisheries with respect to the type of fishery, its target species, a brief description of the management system and key reasons for discarding

	Fishery	Target species	Fleet	Management system	Reasons for discarding
North Sea, West of Scotland, Eastern English channel	Danish North Sea demersal fishery	Cod, plaice, hake, haddock, sole and Norway lobster	Netters and trawlers, with length groups from 12 to 40 metres	TACs allocated in ITQs, MCRS	Quota utilisation optimisation Fish below MCRS High-grading
	UK mixed demersal fisheries in the North Sea, West of Scotland and area 7	73 main UK stocks targeted by different fleets in different areas. Pelagic species and non-quota species representing around 58% of value and 75% of weight landed by UK fleet are excluded	All UK active vessels grouped in 99 producer organisation fleet segments	TACs allocated in fixed quota allocation units that can be pooled within a PO, traded by vessel owners, or can be leased by other vessels in the same or other PO, MCRS	Quota utilisation optimisation Fish below MCRS High-grading
	French demersal fishery in the Eastern English Channel	Sole, scallops, whiting, cephalopods, cod, red mullet, sea bass and plaice	Bottom trawlers, mixed trawlers and trawl-dredgers, with length groups from 12 to 40 metres	TACs, MCRS, seasonal closures for scallops and effort limitation	Quota utilisation optimisation Fish below MCRS High-grading
Mediterranean	Spanish demersal fishery in Western Mediterranean	Four different fishing tactics are used, depending on the main tar get species (Palmer et al. 2009): (1) shallow shelf (striped red mullet), (2) deep shelf (European	Mixed demersal trawl	MCRS and other technical measures	Hake below MCRS High-grading Discard of low-value species

(continued)

Table 6.1 (continued)

	Fishery	Target species	Fleet	Management system	Reasons for discarding
		hake), (3) upper slope (Norway lobster) and (4) middle slope (red shrimp)			
	Greek demersal trawl and small-scale fishery in the Thermaikos Gulf	Mainly hake and red mullet (also surmullet and deep-water rose shrimp)	Bottom trawlers and small-scale coastal vessels using gill nets and trammel nets	Spatial and temporal restrictions, MCRS, other technical measure	Hake and red mullet below MCRS High-grading
Spanish fishery in the Atlantic	Spanish mixed demersal trawl fishery in the Bay of Biscay	Pair trawlers: mainly hake. Otter trawlers: hake, megrims, horse mackerel, blue whiting, mackerel, rays, red mullet, seabass, squids and cuttlefish	Pair and otter trawlers using different métiers	The fleet is managed with fishing rights, TACs and Total allowable Effort, together with mesh and MCRS limitations	Quota utilisation optimisation Fish below MCRS
	Spanish demersal trawl fishery in the Cantabrian-NW region	Hake, megrim, anglerfish, blue whiting, horse mackerel and mackerel	Otter bottom trawlers (average length 28 metres)	The fleet is managed with fishing rights and Total allowable Effort together with mesh and minimum landing size limitations	Quota utilisation optimisation Fish below MCRS

fisheries. Under the LO, it has become obligatory to land these fish, but they cannot be sold for human consumption. On top of this new obligation to land small fish, the North Sea, West of Scotland and English Channel fisheries are regulated by TACs, in some cases combined with effort regulation and technical conservation measures. While TACs are set at the European level, national quotas (i.e. fixed shares of the TACs) are managed differently by the Member States. They are managed as ITQs in Denmark, are distributed between producer organisations (POs) and vessel owners in the UK in a system that is essentially a quasi-ITQ system and are distributed between POs in France. Swaps and quota exchanges are allowed between organisations in the UK and France. The Atlantic Spanish fisheries are regulated with Total Allowable Effort (Prellezo et al. 2016) and TACs. The Mediterranean fisheries are regulated through technical gear specifications and MCRS for the main target species, temporal and spatial closures and other technical measures (Stergiou et al. 2016).

Demersal fishing activities in the North Sea, West of Scotland and in the English Channel have highly mixed catches of species, and therefore it is not possible to fully catch all quotas at the same time in the year, leading to either underutilisation of quota or discarding of fish for which quotas are exhausted first. Under the LO, the risk of a choke situation, i.e. having to stop fishing when the quota of a low-quota stock is exhausted, is a great concern to managers and vessel operators alike (Ulrich et al. 2011). This is especially expected to be a problem for French vessels, operating with fixed quota shares within producer/fishery organisations, while this problem may be less severe for UK and Danish fleets, where quota trade may mitigate the problem to some extent. For Spanish demersal fisheries in the Bay of Biscay, mackerel and horse mackerel are discarded because of low-quota allocations, i.e. to optimise quota utilisation of other species, while hake is primarily discarded because of being below MCRS. Thus, in these fisheries choke situations may also be an issue. In the Mediterranean fisheries, discarding is primarily due to fish below MCRS and to high-grading. As such, all cases face lower revenues under the LO given that previously discarded fish of low value and below MCRS must now be landed, combined with increased handling costs of these unwanted catches.

6.3.1 Mitigation Strategies

Given the different challenges that the selected fishing fleets face under the LO, different scenarios have been analysed, mainly addressing (i) how fleets will respond given the threats faced and (ii) how economic losses can be reduced through mitigation strategies most relevant for that fleet. Table 6.2 gives an outline of the scenarios analysed for each case study.

In all case studies, the economic situation was analysed for the fleet, given the current management system (cf. Table 6.1), i.e. if the LO had not been implemented (named 'business as usual'). This scenario is used as a first benchmark when analysing the effects of the LO. In all case studies the full implementation of the LO with no exemptions was also analysed, i.e. the economic situation for the fleets given their current management system with the LO superimposed. This is a second benchmark against which the effects of introducing mitigation strategies are compared. Application of full implementation in the case study models was based on different assumptions for each case study:

Table 6.2 Scenarios analysed for the European case study fisheries

	North Sea, West of Scotland and English Channel			Spanish Atlantic fisheries		Mediterranean fisheries	
	Denmark	UK	France	Bay of Biscay[1]	Cantabrian-NW	Spain (W. Med)	Greece (E. Med)
Business as usual (no LO)	■	■	■	■	■	■	■
Full LO implementation, no exemptions	■	■	■	■	■	■	■
De minimis	■			■	■		
Year Transfer				■			
Mesh size selectivity				■		■	■
Effort reallocation[2]/Flexibility	■		■		■		
Quota adjustment		■	■				
Decrease minimum landings size	■						
Catch allowance for stocks with zero TACs		■					
Vessel effort movements between métiers		■					
Quota movement (swaps)		■					

Notes: [1]Quota adjustments assumed in all LO scenarios for the Bay of Biscay

[2]Effort reallocation can be seasonal and between fleets (the Danish case) and spatially (the French case) or more efficient effort use (the Cantabrian-NW case)

- In the Danish North Sea demersal case, fish below MCRS must be landed, with gradual implementation from 2016 to 2019 depending on species.
- In the UK mixed demersal fleets, each vessel in a PO has its initial quota available, and by 2019 no demersal species below MCRS can be discarded. The LO is implemented gradually towards 2019 depending on the fish stock.
- In the French mixed demersal case, vessels in métiers are forbidden to continue fishing as soon as the quota of one of their target stocks is reached, and fishing effort is then allocated between the remaining métiers. Fish under MCRS are landed but cannot be sold (price set to zero).
- In the Bay of Biscay Basque mixed demersal trawl case, the fishing activity of a given métier is stopped when the most binding quota share is reached.

- In the Galician mixed trawl case, all catches of species subject to TACs or MCRS must be landed.
- In both Mediterranean cases, a 10% increase in daily variable costs and one more crew member on-board are assumed to reflect the extra effort needed to bring ashore unwanted catches. Three full implementation scenarios were examined for the Greek case (Eastern Mediterranean) based on varying discard rates: (i) 5% increase of daily costs, no extra crew member; (ii) 10% increase of daily costs, 10% extra crew (the original full implementation scenario); and (iii) 20% increase of daily costs, 20% extra crew (based on the discard rates reported in the literature). The reason for the extra full implementation scenarios was that, according to official reports (DCF 2016), the percentage of hake and red mullet discards in Greece had dropped to less than 5% since 2013; thus, this case differs substantially from initial estimates that were based on the literature (e.g. Tsagarakis et al. (2014)).

The analysed mitigation strategies (see Table 6.2) are different for each case, reflecting the specific challenges each fleet faces when the LO is introduced.

In the UK, Danish and French cases, the focus is on maximising quota utilisation. For the Danish demersal fishery, the effect of introducing a 5% *de minimis* exemption is analysed. In addition, economic effects of lowering the MCRS for cod (making it possible to sell some fraction of cod below the previous MCRS) are analysed. For the UK North Sea and West of Scotland mixed demersal fleets, a number of mitigation strategies are analysed: (i) allowance for catching and landing species with zero TAC; (ii) as scenario (i) but with quota adjustment to all TAC species; (iii) as scenario (ii) but with the possibility to reallocate effort to other areas of operation to better utilise producer organisation (PO) quota; (iv) as scenario (iii) but with quota reallocation allowed within the UK to maximise use of quotas; and (v) as scenario (iv) but with international and national swaps at the level of the baseline year incorporated and UK end of year quota reallocated to PO fleets in need of quota. The French mixed demersal fishery in the English Channel case focused on (i) quota adjustments for sole, plaice, cod and whiting and (ii) assuming that fishers can shift to fish in other areas.

The choke situation and having to land fish below MCRS are also issues in the Spanish Atlantic cases. Thus the focus is on quota utilisation optimisation and on fishing gear selectivity. For the Spanish Bay of Biscay mixed demersal fishery, the focus is on investigating the economic effects of implementing (i) 5% *de minimis* exemption, (ii) inter-year quota flexibility, (iii) combining *de minimis* and inter-year flexibility and (iv) selectivity changes for the pair trawlers, given the single-species nature of their catches (90% hake), assuming a change in minimum mesh size from 100 mm to 120 mm. For the Spanish demersal trawl fishery in the Cantabrian-NW region, the focus is on (i) 5% *de minimis* exemption and (ii) effects of improved selectivity, e.g. through effort reallocation or non-compliance, assuming this will reduce unwanted catches by 50%.

The two Mediterranean cases focus predominantly on selectivity issues, given their high catches of unwanted species and fish below MCRS. For the Spanish demersal trawl fishery around the Balearic Islands (Western Mediterranean), several selectivity possibilities for hake are analysed: (i) no fishing mortality for hake at age 0, (ii) no fishing mortality of hake below MCRS (by decreasing the fishing mortality of age 1 individuals by 10%) and (iii) no fishing mortality of immature individuals (through modification of age-selectivity parameters).

For the Greek demersal trawl and small-scale coastal fishery in the Thermaikos Gulf (Eastern Mediterranean), three selectivity scenarios are applied to both hake and red mullet: (i) no fishing mortality at age 0, (ii) no fishing mortality below MCRS (by additionally decreasing the fishing mortality of age 1 individuals by 10%) and (iii) no fishing mortality for hake and red mullet at ages 0 and 1 through modification of age-selectivity parameters.

6.3.2 *The Model Tools*

The analyses were done using different bioeconomic models constructed for the geographical areas of the case study fleets (Table 6.3). Given the level of detail and complexity of each model, model descriptions are not provided in this chapter but can be found in the references listed in Table 6.3. All but one of the models are dynamic, evaluating the development of fleet capacity, economic performance and effort, together with stock dynamics, during the period 2015–2025. The exception is the analysis of the Spanish trawl fishery in the Cantabrian-NW region, which is based on input-output models.

6.4 Results

Analyses of the economic consequences of implementing the LO include two parts, firstly the economic outcome under the LO relative to the outcome if the LO had not been introduced and secondly the LO mitigation scenarios benchmarked against the LO scenario with no exemptions or other mitigation strategies included. These results differ depending on whether they are evaluated in the short or long term. Short term is defined as a period in which only variable inputs can change (e.g. fuel and crew) but not fixed inputs such as vessels, equipment and gear, while in the long term, all inputs can change.

Generally, some short-term negative economic effects of the LO can be expected. The main reasons for this are (i) the choke species issue for fisheries regulated with quotas, whereby catch of some species is constrained once catch of another species has reached its total quota, (ii) that landing of unwanted fish below MCRS and of low market value will replace landings above MCRS and of high value and (iii) the higher costs created by landing instead of discarding. The scale of these short-term losses is case-specific. In the long term, choke situations and displacement of vessels to other areas are expected to reduce fishing pressure, leading to biomass increases and thus improved fishing possibilities. However, ensuing economic improvements

Table 6.3 Model tools applied to evaluate the consequences of the LO for European case fisheries

	Fishery	Model
North Sea, West of Scotland, Eastern English channel: Mixed demersal	Danish North Sea demersal fishery	Fishrent: A bioeconomic profit maximisation model integrating, and allowing feedback between, the economy and the biology of the fishery (Frost et al. 2013)
	UK mixed demersal fisheries in the North Sea, West of Scotland and area 7	SEAFISH: Based on the Fishrent structure, the SEAFISH simulation model is developed to analyse the activity of the total UK fleet (Mardle et al. 2017)
	French demersal fishery in the Eastern English Channel	ISIS-Fish: A spatialised operational simulation model which simulates the dynamics of fish populations and fleets of the mixed fisheries in the Eastern Channel (Pelletier et al. 2009; Lehuta et al. 2015)
Mediterranean	Spanish demersal fishery in the Western Mediterranean	MEFISTO (Mediterranean Fisheries Simulation Tool): A bioeconomic fisheries simulation model with an age-structured biological component (Lleonart et al. 2003, https://mefisto2017.wordpress.com/)
	Greek demersal fishery in the Thermaikos Gulf (Eastern Mediterranean)	
Spanish Atlantic fisheries	Spanish mixed demersal trawl fishery in the Bay of Biscay	FLBEIA: A management strategy evaluation model coupling economic, biological and social dimensions; it shares economic structure with Fishrent but with an age-structured biological component (Garcia et al. 2017)
	Spanish demersal trawl fishery in the Cantabrian-NW region	Input-output analysis: Based on input-output tables for the Galician Fishing and Preserved Fish Sectors 2011 (García-Negro et al. 2016), the function of production of the fleet was recalculated considering the LO and the biological data obtained from IEO (Spanish Institute of Oceanography) campaigns

will differ for individual fleet segments and vessel businesses, depending on catch composition and on whether TACs increase proportionally when biomasses increase. If the latter is not the case, the choke situation may be enhanced.

Here we present a single year view of the economic outcome of the LO for the considered fisheries in 2025 assuming that the LO has been fully implemented (Table 6.4). The exception to this is the Cantabrian-NW case that represents a static view of the impact of the LO in an average year (based on 2014–2016) in the

Table 6.4 This table displays the economic outcomes in 2025 for the LO scenarios relative to the scenario assuming no LO (business as usual)

Mitigation measures[3]	North Sea, West of Scotland and English Channel			Spanish Atlantic fisheries		Mediterranean fisheries	
	Denmark	UK	France	Bay of Biscay	Cantabrian-NW	Spain (W. Med)	Greece (E. Med)
Full LO implementation, no exemptions	P	R	R	P	P	P	P
De minimis	P			P	P		
Year Transfer				P			
Mesh size selectivity				P		P	P
Effort reallocation[2]	P		R		P		
Quotaadjustment		R	R				
Decrease minimum landings size	P						
Catch allowance for stocks with zero TAC		R					
Vessel effort movements between metiers		R					
Quota movement (swaps)		R					

For most scenarios the economic outcome is measured as the total profit in 2025 for the included fleets, while for the UK and French cases, the economic outcome is measured in total revenue for the included fleets. Results at a glance: total economic result (profit='P', revenue='R') in 2025 with LO implemented relative to the business-as-usual case (no LO)

Note: [1]For the Spanish Cantabrian-NW case, the results represent the expected outcome in 2017 given the assumed scenario

[2]Effort reallocation can be seasonal and between fleets (the Danish case) and spatially (the French case) or more efficient effort use (the Cantabrian-NW case)

[3]Yellow indicates less than 5% change, red indicates more than 5% decrease and green indicates more than 5% increase

scenarios considering full implementation and *de minimis*, while the scenario considering flexibility is a longer-term view, assuming a 50% reduction of catches in the long term given improved effort reallocation or other means. Whether 2025 corresponds to a long term will, to some degree, depend on the specific case study, i.e. on whether adjustments are ongoing in the given fleet or whether equilibrium is reached. Theoretically, a better measure of impacts would have been the net present value (NPV) covering the whole period from 2015 to 2025. However, not all models included in the present synthesis are able to provide NPVs over that period, and it has therefore been chosen to present the outcomes for 2025 alone.

In 2025, four of the seven case studies are expected to be negatively affected by the LO, when no exemptions are assumed (see Table 6.4). The exceptions are the Danish North Sea demersal fleet, the Spanish Bay of Biscay fleet and the Eastern Mediterranean fleet. The reasons for the expected economic losses are increased daily and crew costs (Western Mediterranean case), the industry being unable to process the previously discarded fish, and lost landings value due to cessation of fishing after choke situations (the UK and French cases). The assumption of constant TACs in the French case probably exacerbates the problems and results in overly pessimistic scenarios. For the Danish case, where choking on low-quota stocks is the greatest concern, possible negative economic consequences of the LO are reduced through (i) quota trade under the ITQ system in place and (ii) seasonal effort flexibility. In the Spanish mixed demersal fleet in the Bay of Biscay, possible economic losses are reduced by the effects of choke situations reducing mortality and increasing stock size, i.e. under full implementation of the LO, other fleets face choke situations and cease fishing before catching quotas of other stocks, such that the target species stock size increases in the long term, thus increasing catch possibilities (Prellezo et al. 2016). In the Eastern Mediterranean case study, the percentage of discards for hake and red mullet that are officially reported is below 5% for trawlers and even lower for netters (DCF 2016). For that reason, the full LO implementation scenario will result in very low increase (< 5%) in the daily costs and will not necessarily require an extra crew member to handle the extra catch.

Compared with how the case study fisheries would have evolved without the LO, the LO implemented with mitigation measures is, in some cases, expected to make the fisheries equally or better off in 2025. This is so for the Danish North Sea demersal fishery, as the ITQ management system makes it possible for the fleets involved to avoid choke situations through quota trade and seasonal effort flexibility. For the Spanish demersal fishery in the Bay of Biscay, interannual quota flexibility (with a limit of 10% of the initial quota) and increased selectivity (assuming an increase in minimum mesh size from 100 mm to 120 mm) also limit the possible negative economic effects of the LO. However, the application of the *de minimis* exemption has a negative effect in the long term. The application of the *de minimis* exemption increases the fishing mortalities compared to the case with no LO and the harvest control rule will then reduce the advised TAC for the next year (which then happens every year). Thus, the penalty imposed, given increased fishing mortalities, is higher than the flexibility gained by the exemption itself.

Increased selectivity also makes the fishery better off for the Spanish fishery around the Balearic Islands (Western Mediterranean) and for the Greek trawl and small-scale coastal fishery in the Thermaikos Gulf, especially if the catch of immature hake individuals is totally avoided, which raises the profit in 2025 above what could be expected without the LO.

At a glance Table 6.4 shows that for most of the analysed fisheries, their economic outcome will be negatively affected in the long term by the LO. But the possibility to trade quotas, both nationally and internationally, and for some fleets increased selectivity and/or year-transfers may mitigate this effect.

Under the LO, a key question is to what degree the overall negative economic outcome can be avoided through appropriate mitigation measures. Table 6.5 shows the economic outcome in 2025 in the mitigation strategy scenarios for each of the analysed fisheries, relative to the expected situation in 2025, assuming full implementation of the LO with no exemptions.

Half of the mitigation strategies analysed do not significantly improve the economic outcome relative to full implementation of the LO with no exemptions (see Table 6.5). This is the case for the Spanish trawl fishery in the Bay of Biscay when

Table 6.5 Results at a glance: total economic result (profit='P', revenue='R') in 2025 with mitigations relative to full implementation of the LO with no mitigations

Mitigation measures[3]	North Sea, West of Scotland and English Channel			Spanish Atlantic fisheries		Mediterranean fisheries	
	Denmark	UK	France	Bay of Biscay	Cantabrian-NW	Spain, (W. Med)	Greece (E. Med)
De minimis	P			P	P		
Year Transfer				P			
Mesh size selectivity				P		P	P
Flexibility (effort reallocation)	P		R		P		
Quota adjustment		R	R				
Decrease minimum landings size	P						
Catch allowance for stocks with zero TAC		R					
Vessel effort movements between métiers		R					
Quota movement (swaps)		R					

Note: [1]For the Spanish Cantabrian-NW case, the results represent the expected outcome in 2017 given the assumed scenario

[2]Effort reallocation can be seasonal and between fleets (the Danish case) and spatially (the French case) or more efficient effort use (the Cantabrian-NW case)

[3]Yellow indicates less than 5% change, red indicates more than 5% decrease and green indicates more than 5% increase

inter-year quota transfers and increased mesh size selectivity are introduced. Likewise, the *de minimis* exemption in the Spanish fishery in the Cantabrian-NW region does not lead to an increased economic result compared to when no exemptions are applied. In parallel with this, applying the *de minimis* exemption leads to a reduced economic result for the Spanish demersal fishery in the Bay of Biscay, because increased fishing pressure leads to higher mortality and reduced hake and megrim stocks and thus reduced fishing possibilities. Likewise, for the French fishery in the Eastern English Channel, applying increased flexibility through spatial effort reallocation leads to a decreased economic result compared to the LO with no mitigation strategies.

Mitigation strategies that do increase the economic outcome relative to the full implementation of the LO with no exemptions are (i) quota adjustments in the French demersal fishery in the Eastern English Channel, (ii) more efficient effort use leading to reduced unwanted catches in the Spanish fishery in the Cantabrian-NW region, (iii) increased selectivity in the Spanish fishery around the Balearic Islands (Western Mediterranean) and (iv) all mitigation strategies (quota adjustment, catch allowance for zero TAC stocks, vessel movements between metiers and quota swaps) considered for the UK fishing fleets, but the scale of changes depend on the fishing fleets concerned (North Sea and West of Scotland).

Thus, at a glance, Table 6.5 shows that for the analysed fisheries, the most effective mitigation strategies depend on both the fishing fleet and the management system in place. Model structure and the assumptions applied in the models may also influence the results, but all models have been calibrated and tested against the actual situation in each case study. It is thus believed that the relative results provided in each case study are good indicators of the effects of the LO and applied mitigation strategies.

6.5 Summary and Policy Recommendations

To assess the likely fleet economic repercussions of the Landing Obligation, bioeconomic models covering seven European fisheries have been applied to estimate the economic performance of fleets before and after implementing the LO. The selected fisheries cover different species compositions and fishing technologies and different management systems ranging from the North East Atlantic to the Mediterranean.

When the four groups of factors that encourage discarding, i.e. institutional, biological, technological and economical, are combined, the main issues to address are (i) that certain stocks cause a choke species situation for some fleets; (ii) landings of small or damaged fish, which have low market values; and (iii) illegal high-grading as a consequence of the two former issues when vessel operators seek to maximise their profits. Consequently, it is important to improve catch selectivity through gear changes, changes to fishing patterns and effort reallocation and to apply management measures that decrease effects of choke situations, such as enabling quotas to be traded or reallocated. Finally, the use of price measures (deemed value) that consider the differences between market prices and the social value of the fish should be considered to reduce the relative benefits of high-grading (Pascoe 1997).

In the short term, when fleet structure has not adapted to the new situation, the introduction of the LO will generally result in decreasing profits for all selected fleets mainly because of choke situations constraining the catch of other species in TAC-regulated fisheries and because of lower catches of higher-value larger fish given the requirement to land undersize fish in MCRS-regulated fisheries. Obviously, this is of concern for vessel operators, who see the risk that their economic performance will deteriorate.

In the long term, economic repercussions will differ as the four factors mentioned above interact in different ways for each fishery and the type of management affects the options for businesses to adjust. In the Mediterranean, which is managed with MCRS and has a wide variety of species, the main anticipated issue is the cost of dealing with undersized fish which cannot be sold for human consumption and for which there is a lack of processing facilities to make it into fishmeal or other non-food products. This issue also applies to the northern fisheries, but here the choke issue also plays a role. Countries that have tradable quota systems, such as Denmark and the UK, can, to some degree, avoid or delay choke situations through quota trade. While trading is possible however, there are no mechanisms to ensure or require trading of quota units to mitigate choke situations. The choke issue could be more severe for stocks managed by non-transferable quota shares such as in France and Spain. Although long-term profits are expected to increase, some vessel businesses may not have the financial resources to overcome the severe economic losses predicted during the first years of implementation. Some governments might find it appropriate to implement measures to ensure that businesses do not fail as a result of short-term impacts of a fully implemented Landing Obligation.

What defines the short and long term depends on the individual fisheries and how fast these are able to adjust to the new situation. It must be expected that the fleet structure will have adapted in ~10 years for most of the analysed fishing fleets, which is why it has been chosen to monitor the results in 2025 in the present context.

Mitigation strategies such as selectivity changes, *de minimis* exemptions and quota adjustments equal to previous discarded quantities could enable fishing businesses to increase profits with the implementation of the LO. But for the fisheries analysed in this chapter, the profits are generally lower than or equal to profit with no LO. For the North East Atlantic fisheries, regulated with TACs and quotas, a useful policy could be to further develop a system to mitigate the problem of choke stocks. Such a policy is already in place in the EU (cf. Frost 2010) through the annual setting of TACs, when single-species assessments show recommended total removals that are adjusted to take account of multispecies interactions and fleets' technological characteristics. Reducing differences between stock TACs and fleets' catch compositions could mitigate the choke problem and allow individuals, producer organisations or fleet segments to land and sell fish and decrease the inherent incentive to discard. However, this approach would also to some extent negate the purpose of the Landing Obligation, which is to encourage more selective fishing by creating incentives to avoid catching species with lower quotas. To create incentives to avoid catching fish below MCRS, price measures could be used to correct the difference between the sale price and the estimated social value of the fish. The difference must be sufficient to cover handling costs of landing the fish, and thus create an incentive to do so, but not high enough to incentivise targeting the fish beyond the quota, and vice versa to reduce prices for fish species with unused quotas.

Generally, the modelled economic outcomes for the seven selected fisheries under the LO suggest that fishing businesses may have incentives not to comply with the LO. Monitoring and enforcement are generally considered to be currently insufficient to motivate compliance. Therefore, high-grading and continuing to discard will still be an issue that must be addressed. The success of the LO is likely to require either larger investment in monitoring and enforcement or implementation of policies that create incentives for compliance or at least weaken incentives for non-compliance.

Acknowledgements This work received funding from the Horizon 2020 Programme under grant agreement DiscardLess number 633680 and from the LIFE+Environmental Program of the European Union under grant agreement iSEAS project, Ref. LIFE13 ENV/ES/000131. This support is gratefully acknowledged.

References

Abbott, J.K., & Wilen, J.E. (2009). Regulation of fisheries bycatch with common-pool output quotas. *Journal of Environmental Economics and Management 57*, 195–204.

Anderson, L.G. (1994). An economic analysis of high grading in ITQ fisheries regulation programs. *Marine Resource Economics 9*(3), 209–226.

Arnason, R. (1994). On catch discarding in fisheries. *Marine Resource Economics*, *9(3)*, 189–207.

Batsleer, J., Hamon, K.G., van Overzee, H.M.J, Rijnsdorp, A.D., Poos J.J. (2015). High-grading and over-quota discarding in mixed fisheries. *Reviews in Fish Biology and Fisheries*, *25*, 715–736.

Boyce, J.R. (1996). An economic analysis of the fisheries bycatch problem. *Journal of Environmental Economics and Management*, *31*, 314–336.

Christensen, S. (1996). Potential Bio-economic impact of reduced mortality of cod end escapees in the shrimp fishery in the Davis Strait. In A.V. Soldal (Ed.), *Bidødelighed i nordiske trawlfiskerier. Volum 2: Konsekvensudredninger*. Nord 1996:17. Nordic Council of Ministers, Copenhagen.

Condie, H.M., Catchpole, T.L., Grant, A. (2014) The short-term impacts of implementing catch quotas and a discard ban on English North Sea otter trawlers. *ICES Journal of Marine Science 71*, 1266–1276. doi: https://doi.org/10.1093/icesjms/fst187

Clucas, I. (1997). A study of the options for utilization of bycatch and discards from Marine capture fisheries. FAO Fisheries Circular No. 928 FIIU/C928. FAO, Rome.

DCF. (2016). On the Greek National Fisheries Data Collection Programme for 2015. Annual report, Directorate General for Fisheries, Ministry of Reconstruction of Production, Environment and Energy, Athens, Greece (p. 456).

FAO. (1996a). A global assessment of fisheries bycatch and discards and the technical consultation on reduction of wastage in fisheries (Tokyo, Japan, 28 October – 1 November 1996). Fisheries technical paper 339, Rome.

FAO. (1996b). Report of the technical consultation on reduction of wastage in fisheries. Tokyo, Japan, 28 October – 1 November 1996. FAO Fisheries report no. 547, Rome.

Fitzpatrick M., Frangoudes K., Fauconnet L., Quetglas A. (this volume). Fishing industry's perspectives on the EU Landing Obligation. In S.S. Uhlmann, C. Ulrich, S.J. Kennelly (Eds.), *The European Landing Obligation – Reducing discards in complex, multi-species and multi-jurisdictional fisheries*. Cham: Springer.

Flaaten, O., Larsen, N.-J. (1991). Sorting panels or normal trawl in cod fishery. An economic analysis. Working paper. University of Tromsø. (In Norwegian).

Frost, H. (1996). Economic impact of changes in by-mortality. In A.V. Soldal (Ed), *Bidødelighed i nordiske trawlfiskerier. Volum 2: Konsekvensudredninger*. Nord 1996:17. Nordic Council of Ministers, Copenhagen.

Frost, H., Boom, J.T., Buisman, E., Innes, J., Metz, S., Rodgers, P., Taal, K. (2007). Economic impact assessment of changes in fishing gear. NECESSITY. FOI report no. 194. Institute of Food and Resource Economics, Copenhagen.

Frost, H. (2010). European Union Fisheries Management. In R. Quentin Grafton, R. Hilborn, D. Squires, M. Tait, M.J. Williams (Eds.), *Handbook of marine fisheries conservation and management* (Ch. 35). Oxford University Press.

Frost, H., Andersen, P., Hoff, A. (2013). Management of complex fisheries: Lessons learned from a simulation model. *Canadian Journal of Agricultural Economics*, *6*, 283–307.

Garcia, D., Sánchez, S., Prellezo, R., Urtizberea, A., Andrés, M. (2017). FLBEIA: A simulation model to conduct Bio-Economic evaluation of fisheries management strategies. *SoftwareX*, *6*, 141–147.

García-Negro, M.doC., Rodríguez-Rodríguez, G., Ballesteros, H.M., Sálamo Otero, P. (2016). Táboas Input-Output da Pesca-Conservas Galega 2011. Consellería do Medio Rural e do Mar. Xunta de Galicia, Santiago de Compostela.

Hardin, G. (1968). The tragedy of the commons. *Science*, *162*, 1243–1248.

Jensen, F., & Vestergaard, N. (2002). Moral hazard problems in fisheries regulation: The case of illegal landings and discard. *Resource and Energy Economics, 24(4)*, 281–299.

Kelleher, K. (2005). Discards in the world's marine fisheries. An update. FAO fisheries technical paper no. 470. Rome, FAO (p. 131).

Kraak, S.B.M, & Hart, P.J.B. (this volume). Creating a breeding ground for compliance and honest reporting under the Landing Obligation: Insights from behavioural science. In S.S. Uhlmann, C. Ulrich, S.J. Kennelly (Eds.), *The European Landing Obligation – Reducing discards in complex, multi-species and multi-jurisdictional fisheries*. Cham: Springer.

Lehuta, S., Youen, V., Marchal, P. (2015). A spatial model of the mixed demersal fisheries in the Eastern Channel. In *Marine productivity: Perturbations and resilience of socio-ecosystems*. Proceedings of 15'th French-Japanese Oceanographic symposium, pp. 187–195.

Lleonart, J., Maynou, F., Recasens, L., Franquesa, R. (2003). A bioeconomic model for Mediterranean fisheries, the hake off Catalonia (western Mediterranean) as a case study. *Scientia Marina*, *67*, 337–351.

Mardle, S., Russel, J., Motova, A.. (2017). Seafish bioeconomic modelling – methodology report. Seafish Report No. SR702.

Nordic Council of Ministers. (2003). Report from a Workshop on discarding in Nordic fisheries. Editor: John Willy Valdemarsen, Fangstseksjonen, Havforskningsinstituttet, Bergen. Sophienberg Slot, København, 18–20. november 2002. TemaNord 2003:537. Nordic Council of Ministers, Copenhagen.

Nuevo, M., Morgado, C., Sala, A, (this volume). Monitoring the implementation of the Landing Obligation: Last Haul programme. In S.S. Uhlmann, C. Ulrich, S.J. Kennelly (Eds.), *The European Landing Obligation – Reducing discards in complex, multi-species and multi-jurisdictional fisheries*. Cham: Springer.

Palmer, M., Quetglas, A., Guijarro, B., Moranta, J., Ordines, F. Massutí, E. (2009). Performance of artificial neural networks and discriminant analysis in predicting fishing tactics from multispecific fisheries. *Canadian Journal of Fisheries and Aquatic Sciences*, *66*(2), 224–237.

Pascoe, S. (1997). Bycatch management and the economics of discarding. FAO Fisheries technical paper 370. FAO, Rome.

Pascoe, S., & Revill, A. (2004) Costs and benefits of bycatch reduction in European Brown Shrimp fisheries. *Environmental and Resource Economics 27*, 43–64.

Pelletier, D., Mahevas, S., Drouineau, H., Vermard, Y., Thebaud, O., Guyader, O., Poussind, B. (2009). Evaluation of the Bioeconomic sustainability of multi-species multi-fleet fisheries under a wide range of policy options using ISIS-Fish. *Ecological Modelling, 220*, 1013–1033.

Prellezo, R., Carmona, I., Garcia, D. (2016). The bad, the good and the very good of the landing obligation implementation in the Bay of Biscay: A case study of Basque trawlers. *Fisheries Research, 181*, 172–185.

Segerson, K. (2007). Reducing stochastic sea turtle bycatch: An efficiency analysis of alternative policies. Working paper, Department of Economics, University of Connecticut, Storrs.

Stergiou, K.I., Somarakis, S., Triantafyllou, G., Tsiaras, K.P., Giannoulaki, M., Petihakis, G., et al. (2016). Trends in productivity and biomass yields in the Mediterranean Sea large marine ecosystem during climate change. *Environmental Development, 17*(1), 57–74

Sutinen, J.G., & Andersen, P. (1985). The economics of fisheries law and enforcement. *Land Economics, 61*(4), 387−397.

Sutinen, J.G., & Kuperan, K. (1999). A socio-economic theory of regulatory compliance. *International Journal of Social Economics, 26*, 174−193.

Tsagarakis, K., Palialexis, A., Vassilopoulou, V. (2014). *ICES Journal of Marine Science 71*, 1219–1234.

Turner, M.A. (1996). Value-based ITQs. *Marine Resource Economics* 11, 59–69.

Turner, M.A. (1997). Quota-induced discarding in heterogeneous fisheries. *Journal of Environmental Economics and management, 33*, 186−195.

Ulrich, C., Reeves, S.A., Vermard, Y., Holmes, S.J., Vanhee, W. (2011). Reconciling single-species TACs in the North Sea demersal fisheries using the Fcube mixed-fisheries advice framework. *ICES Journal of Marine Science, 68*(7), 1535–1547.

van Hoof L., Kraan M., Visser N.M., et al. (this volume). Muddying the waters of the Landing Obligation: How multi-level governance structures can obscure policy implementation. In S.S. Uhlmann, C. Ulrich, S.J. Kennelly (Eds.), *The European Landing Obligation – Reducing discards in complex, multi-species and multi-jurisdictional fisheries*. Cham: Springer.

Vestergaard, N. (1996). Discard behavior, highgrading and regulation: The case of the Greenland shrimp fishery. *Marine Resource Economics, 11*(4), 247−266.

Ward, J.M. (1994). The bioeconomic implications of bycatch reduction devise as a stock conservation management measure. *Marine Resource Economics, 9*(3), 227–240.

Ward, J.M., Benaka L.R., Moore C.M., Meyers S. (2012). Bycatch in Marine fisheries, *Marine Fisheries Review* 74, No. 2. United States Department of Commerce.

Chapter 7
The Impact of Fisheries Discards on Scavengers in the Sea

Jochen Depestele, Jordan Feekings, David G. Reid, Robin Cook, Didier Gascuel, Raphael Girardin, Michael Heath, Pierre-Yves Hernvann, Telmo Morato, Ambre Soszynski, and Marie Savina-Rolland

Abstract A scavenger is an animal that feeds on dead animals (carrion) that it has not killed itself. Fisheries discards are often seen as an important food source for marine scavengers so the reduction of discards due to the Landing Obligation may affect their populations. The literature on scavenging in marine ecosystems is considerable, due to its importance in the trophic ecology of many species. Although discards undoubtedly contribute to these species' food sources, few can be seen to be solely dependent on carrion (including discards). Ecosystem models predicted that discards contributed very little to the diet of scavengers at a regional scale. A reduction in discards through the Landing Obligation may therefore affect

---

Electronic supplementary material The online version of this chapter (https://doi.org/10.1007/978-3-030-03308-8_7) contains supplementary material, which is available to authorized users.

J. Depestele (✉)
Flanders Research Institute for Agriculture, Fisheries and Food (ILVO), Oostende, Belgium
e-mail: jochen.depestele@ilvo.vlaanderen.be

J. Feekings
National Institute of Aquatic Resources, DTU Aqua, Technical University of Denmark, Hirtshals, Denmark

D. G. Reid
Marine Institute, Oranmore, County Galway, Ireland

R. Cook · M. Heath
Department of Mathematics and Statistics, University of Strathclyde, Glasgow, UK

D. Gascuel · P.-Y. Hernvann
Université Bretagne Loire, Agrocampus Ouest, UMR 985 Ecology and Ecosystem Health, Rennes, France

R. Girardin
Ifremer, Channel and North Sea Fisheries Research Unit, Boulogne sur Mer, France

T. Morato · A. Soszynski
Marine and Environmental Sciences Centre (MARE), Institute of Marine Research (IMAR) and OKEANOS Research Unit, Universidade dos Açores, Horta, Portugal

M. Savina-Rolland
Ifremer, Fishery Technology and Biology Laboratory, Lorient, France

S. S. Uhlmann et al. (eds.), *The European Landing Obligation*,
https://doi.org/10.1007/978-3-030-03308-8_7

populations for a few species in some areas, but generally this is unlikely to be the case. But it is challenging to identify how important discards might be to scavengers, as they are taxonomically diverse and vary in the role they play in scavenging interactions.

Keywords Carrion · Discard consumption · Food subsidies · Food web models · Scavengers

7.1 Introduction

Trophic interactions are increasingly recognized as an important driver of ecosystem change (Pikitch et al. 2004; Möllman et al. 2015). Foraging relationships primarily focus on predator-prey interactions while the consumption of carrion, as a high-quality form of dead animal matter, has received far less attention in the ecosystem context. There is considerable literature at the experimental level, but very few studies have brought this information to the higher level of assessing the actual role of discards/carrion in the marine food web. Carrion consumption should however have different consequences for the structure and functioning of food webs than predation, because it does not cause direct mortality or demographic changes (Wilson and Wolkovich 2011).

A scavenger is an animal that feeds on dead animals (carrion) that it has not killed itself (Dictionary 2018). Obligate scavengers are those that rely on carrion for survival and reproduction. Facultative scavengers are those species that will scavenge, but do not depend solely on carrion for their survival or reproduction (Beasley et al. 2015). On land, vultures are believed to be the only obligate vertebrate scavengers (DeVault et al. 2016). In the sea some benthic scavengers (e.g., hagfish, Lysianassidae amphipods) may also be obligate scavengers (Kaiser and Moore 1999; Smith and Baco 2003; Beasley et al. 2012). Facultative scavengers, in contrast, are widely present in the marine environment and range from nematodes to crustaceans, echinoderms, molluscs, fish and marine mammals (Jensen 1987; Luque et al. 2006).

Apart from whale deaths, recordings of naturally-occurring marine carrion in the sea are limited (Britton and Morton 1994; Smith and Baco 2003). The lack of naturally-occurring carrion may be due to few animals dying from natural senescence (Britton and Morton 1994) or to their rapid consumption by scavengers (Kaiser and Moore 1999). Fisheries produce non-discarded carrion due to mortality in the tow path or organisms escaping the capture process (Broadhurst et al. 2006; Collie et al. 2017; Hiddink et al. 2017). Marine carrion from fisheries discards was globally estimated to have been less than 5 million tonnes in 1950, rising to a peak of 18.8 million in 1989 and now less than 10 million tons per year (Kelleher 2005; Zeller et al. 2018).

Fisheries discards are often perceived as an important source of food for marine scavengers that may be lost if discarding stops. A decline in European fisheries discards is currently envisaged with the gradual phasing in of the Landing Obligation (EU 2013). Such a lack of discards may have knock-on effects on seabird populations and communities (Votier et al. 2004; Bicknell et al. 2013) and this may also be true for non-avian marine scavengers, depending on their position along the obligate-to-facultative scavenging continuum.

This chapter focuses on scavengers in the sea rather than scavenging by seabirds. Two approaches were used to assess the impact of discards on scavengers in the sea: (1) a review of knowledge from field observations and (2) modelling studies. Noticeably, the review of field observations (Sect. 7.2) is less directly dealing with the EU Landing Obligation itself compared to other chapters of this book. But this study represents the first published synthesis on the fundamental knowledge on the biological and ecological processes involved in scavenging, and thus provides a very comprehensive and novel overview on the resilience of scavengers in the sea and on their potential ability to switch to other food sources if fisheries discards are reduced. This improved understanding of the processes involved will then be useful for a more accurate parameterization of the ecosystem models presented in the second part of this chapter. The final section of this chapter 'Synthesis and outlook' summarizes information about discard-scavenger interactions and how to progress knowledge on this topic.

7.2 Field Observations of Discard-Scavenger Interactions

Fisheries discards are considered to be an important food source for marine scavengers in the sea (e.g. Link and Almeida 2002; Fondo et al. 2015). Several field studies have been conducted using various observational techniques in various locations and seasons (Yamamura 1997; Groenewold 2000). Here we review these studies and empirical information with the overall objective of (1) identifying scavenger taxa and (2) scaling scavenger taxa along the continuum of obligate to facultative scavenging. In Sect. 7.2.4, we discuss the relevance of these observational studies to the assessment of discard-consumer candidates and highlight current knowledge gaps.

7.2.1 Methodological Approach

7.2.1.1 List of Observational Studies

We listed observational studies that identified organisms that are attracted to discards or to marine carrion (presented as a proxy for discards). Most of the reviewed studies used baited traps or lines ($N = 16$) or baited video frames ($N = 16$; Fig. 7.1)

Fig. 7.1 Images taken from a baited camera trial in the Kattegat. (Photo courtesy Feekings, unpublished data; Feekings and Krag 2015)

to observe interactions between scavengers and marine carrion. Alternative observational techniques made use of laboratory observations ($N = 8$), stomach analyses ($N = 6$) or divers ($N = 3$). We also included seven studies that investigated aggregations, increased abundances or stomach contents of scavengers after the passage of a trawl, including the reaction of scavengers to carrion produced in the tow path and/or carrion from discards (Kaiser and Hiddink 2007). Three field studies on scavenging at ghost fishing nets have also been listed (Gilman 2015). These observations were complemented with studies of interactions between surface scavengers and discards from spatial interaction analysis in combination with direct observations by eye or using remote sensing systems with GPS-referenced data (Bicknell et al. 2016). The extensive series of observational studies are documented in Table 7.A of the Electronic supplementary material[1] to this chapter.

The majority of the studies were done in Europe and thus have a direct relevance to the EU discard policy (30 out of 43 studies). Most studies were conducted at or near the seabed ($N = 39$, with one in the intertidal zone), with three studies in the mesopelagic zone and only four studies at the water surface; 20 studies were done on

[1]Table 7.A in the Online Supplement refers to an extensive overview of studies investigating interactions between discards and scavengers.

the continental shelf and 6 in deep sea areas (> 500 m). The number of studies in the various depth zones of the sea (surface, mesopelagic and bentho-demersal zone) suggest how much aquatic scavenging activity is occurring in each zone. The limited number of studies in the mesopelagic zone is partly due to high sinking rates of the carrion, although it may also reflect the difficulty in studying carrion-scavenger interactions in that zone.

7.2.1.2 Review of Empirical Information and Observational Studies

Empirical information and observational studies were reviewed in two steps to address the objectives of: (1) scavenger identification and (2) scavenger scaling along the obligate to facultative scavenging continuum.

Scavenger Identification

Primary scavenger taxa from observational studies were identified (Table 7.A in the Online Supplement). This list of scavenger taxa complements the species enumeration from Britton and Morton (1994) with recent studies (1990–2017) but requires caution in determining the relative importance of different scavenger groups as consumers of marine carrion, because the different techniques used have technical constraints in determining species composition. The retention efficiency and mesh sizes of the different types of traps, for instance, may cause bias in the observed abundances.

The list of scavenger species from observational studies were complemented using data from commercial baited fisheries. Species were listed as scavenger species when annual landings between 2003 and 2016 exceeded 1 ton in ICES Subdivisions IV and VII a, d, f and VIIg. This list of species excluded discarded taxa. Discard data were available from French and UK longline fisheries (Da Silva 2009; Cornou et al. 2015). A list of species which are highly discarded (more than half of the catch) was provided for the French and UK commercial baited fisheries. These species generally comprise < 10% of the total catch of all species combined (in weight).

Assessing Scavenger Abilities

The variety of species covered in the overview of field studies (Table 7.A in the Online Supplement) and the species from commercial baited fisheries highlighted recurrent characteristics across all scavengers in the sea: from the surface into the seabed and from marine mammals to infaunal invertebrates. The overview showed that a vast range of marine aquatic taxa utilize a scavenging lifestyle to a lesser or greater extent along the facultative-obligate scavenging continuum.

Species that rely on scavenging to sustain substantial portions of their diets must encounter a sufficient amount of carrion and must be able to out-compete potential competitors and efficiently assimilate carrion to meet their energetic requirements (Ruxton et al. 2014; Kane et al. 2017). We summarized the suite of biological and functional traits that increase carrion discovery and monopoly (Table 7.A in the Online Supplement) because these traits can be used to assess any organism's ability to scavenge, i.e. to scale an organism along the obligate to facultative scavenging continuum.

A scale of scavenging was developed to predict the qualitative importance of carrion in a scavenger's diet (DeVault et al. 2003; Kane et al. 2017) and to partition the encountered carrion between scavenging taxa. This approach is analogous to the partitioning of fisheries discards between aerial and aquatic scavengers, which is presented in Depestele et al. (2016).

Two principal parameters of optimal foraging strategy were used to this end: (1) encounter probability and (2) handling tactics.

1. Encounter probability is the likelihood to come across carrion or fisheries discards. Carrion availability is, in general, relatively unpredictable, ephemeral and of short duration (Britton and Morton 1994; Kaiser and Moore 1999; DeVault et al. 2003; Yang et al. 2010). Scavengers can either scan vast areas at low metabolic cost to detect carrion or have limited home ranges where carrion is regularly produced and rapidly detected and encountered (Schlacher et al. 2013; Moleón et al. 2014). Four traits are presented to assess whether an organism is likely to find carrion quickly: home range, detection ability, locomotion and metabolism, the latter is the rate at which animals expend energy in relation to its acquisition (via feeding) (Brown et al. 2004). Scavengers are expected to reduce energetic maintenance costs to allow for longer inter-feeding periods (Kane et al. 2017).
2. Upon encountering carrion, scavengers require handling tactics to overcome competitors and maximize nutrient gains during feeding so that they can replenish their reserves until the next, unpredictable discovery of carrion (Ruxton and Houston 2004). Two traits, i.e. competitive abilities and capacity to facilitate the consumption of the encountered carrion, are presented to evaluate whether an organism is well fitted to scavenge.

Both parameters must occur to promote an organism's ability to find and consume carrion like fisheries discards (Depestele et al. 2016). The traits within each parameter do not necessarily work multiplicatively or additively, but taken together, they make up a qualitative scale of scavenging abilities that can be applied to any species to assess the likely relative encounter and consumption of carrion in relation to other scavenging taxa (Greene, 1986; Kane et al. 2017).

7.2.2 *Identification of Scavenging Taxa*

7.2.2.1 Observational Studies

A few studies highlight the importance of marine mammals as surface scavengers (killer whales, dolphins and seals), but most studies focused on the identification of taxa in association with the seabed. Demersal fish scavengers that were identified in more than ten field studies in Table 7.A in the Online Supplement included the orders Gadiformes and Perciformes. Interestingly, both orders also occurred frequently in longline landings (Sect. 7.2.2.2). Gadiformes covered several families: Gadidae (e.g. *Gadus morhua*, *Merlangius merlangus*), Merluccidae (e.g. *Merluccius merluccius*), Lotidae (e.g. *Molva molva*), Moridae (e.g. *Antimora rostrata*) and Macrouridae (e.g. *Macrourus holotrachys*). Perciformes also occurred frequently and were represented by several families: Sparidae, Labridae, Callionymidae, Zoarcidae, Trachinidae, Scombridae, Gobiidae. Fish taxa that were observed in fewer studies ($N = 4–10$) or in longline discards (see below) were sharks belonging to the Carcharhiniformes (e.g. *Scyliorhinus canicula*), or taxa from the orders Myxiniformes (hagfishes), Anguilliformes (e.g. *Conger conger*), Pleuronectiformes (e.g. *Limanda limanda*) and Scorpaeniformes (e.g. *Triglidae*). Rajiformes and Squaliformes occurred in less than four studies.

Invertebrate taxa were dominated by Decapoda, in particular Brachyura (>10 studies) like *Cancer pagurus*, *Hyas araneus*, *Maja* spp. and portunid crabs (e.g. *Carcinus maenas*, *Liocarcinus* spp. and *Necora puber*). Amphipoda (*Orchomene* spp., *Scopelocheirus hopei*), Isopoda (e.g. *Natatolana borealis*), Asteroidea (e.g. *Asterias rubens*), Ophiuroidea and Neogastropoda (in particular Buccinidae and Nassariidae) were also attracted to bait in at least ten studies. Taxa encountered in fewer studies ($N = 3–10$) were Cephalopoda, as well as hermit crabs, lobsters and shrimps, i.e. Decapoda belonging to the Anomura (Paguroidea, e.g. *Pagurus bernhardus*), Nephropoidea (e.g. *Homarus gammarus* and *Nephrops norvegicus*) or the Caridea (e.g. *Crangon crangon*). Polychaeta and Nemertea were identified in less than three studies, which is likely due to the mesh sizes used in the sampling methods. Landings from pot fisheries highlighted the importance of brachyuran Decapoda, whelks (*Buccinum undatum*) and Cephalopods.

7.2.2.2 Commercial Baited Fisheries

Scavenger species from landings data of commercial baited fisheries are listed in Table 7.1. Species which were discarded in French and UK commercial baited fisheries were: *Conger conger*, sharks and ray species (*Galeus melastomus*, *Mustelus* spp., *Raja undulata*, *Scyliorhinus* spp.) and quota-limited species like *Brosme brosme*, *Helicolenus dactylopterus*, *Phycis blennoides* or *Micromesistius poutassou*.

7.2.3 Assessing Scavenger Abilities

7.2.3.1 Encounter Probability: Home Range

Most marine organisms exhibit site fidelity, i.e. their movements are directed and confined to a smaller area rather than random. Routine activities like resting, spawning and feeding are done in established areas, defined as their home range (Pittman and McAlpine 2003). Home ranges are related to foraging strategies which

Table 7.1 Landings of commercially valuable scavenger species from pot and longline fisheries between 2003–2016, extracted from fisheries data collection website (JRC 2018)

	Scientific name	Common name	Mean annual landings (tons)
Pot fisheries	*Cancer pagurus*	Edible crab	233
	Buccinum undatum	Common whelk	232
	Homarus gammarus	European lobster	27
	Maja squinado	Spinous spider crab	12
	Necora puber	velvet swimming crab	9
	Sepiidae, Sepiolidae	Cuttlefishes,	6
	Sepia officinalis	Common cuttlefish	2
	Carcinus maenas	Shore crab	2
Longline fisheries	*Scomber scombrus*	Mackerel	22
	Merluccius merluccius	Hake	12
	Gadus morhua	Atlantic cod	4
	Dicentrarchus labrax	Seabass	3
	Molva molva	Ling	2
	Pollachius pollachius	Pollack	2
	Squalus acanthias	Spiny dogfish	2
	Conger conger	Conger eel	2

are broadly categorized as (1) scavengers employing a sit-and-wait strategy and (2) those actively foraging in search of carrion, the free-ranging strategy (Greene 1986; Higginson and Ruxton 2015).

The nature of the sit-and-wait strategy confines the home range of the scavenger to a smaller radius of activity. Carrion detection of 'sit-and-wait' scavengers typically requires well-developed sensory capabilities (Sect. 7.2.3.2) but offers in return the advantage of refuge from predation and low metabolic costs (Løkkeborg et al. 2000; Bailey and Priede 2002). Hagfishes are a primary example of fish taxa with a sedentary life style and low metabolic requirements (Lesser et al. 1997). The 'sit-and-wait' strategy of scavengers has been observed in field trials as they rapidly arrive at carrion e.g. snake eels (*Ophichthus rufus*) and amphipods (Sect. 7.2.3.3; Table 7.A in the Online Supplement). These rapid responses, reflected in short arrival times in baited experiments, have been used to model fish scavenger abundances (Bailey et al. 2007). Several amphipod and isopod species with small home ranges (Sainte-Marie 1986; Sainte-Marie and Hargrave 1987; Groenewold and Fonds 2000; Johansen 2000) are candidates that quickly respond to odour plumes (Tamburri and Barry 1999).

Free-ranging scavengers typically occupy larger areas to search for food and may apply a range of search techniques to encounter carrion. Bailey and Priede (2002) described two techniques in the deep sea: (1) drifting on ambient currents and detecting carrion by the sound produced by other animals feeding and (2) cross-current swimming which increases the probability of detecting carrion, albeit at a high-energy cost. Free-ranging facultative scavengers in continental shelf areas actively forage with little, if any, knowledge of resource availability (Sims et al. 2008). When resources are sparse and patchily distributed, several species, including sharks and teleost fish, exhibit Lévy flights (where area-restricted search (ARS) behaviour is alternated with movements across longer distances to detect resources – Fauchald and Tveraa 2003), while Brownian movements occur when resources are abundant (Humphries et al. 2010; Sims et al. 2012). ARS-behaviour has been observed in facultative free-ranging scavengers like plaice *Pleuronectes platessa* (Hill et al. 2000), cod *Gadus morhua* (Løkkeborg and Fernö 1999) and ling *Molva molva* (Løkkeborg et al. 2000) and over larger areas by Atlantic cod than by ling, which increases the probability of carrion encounters. Highly mobile generalist species like cod and sharks generally occupy larger home ranges than less mobile specialist such as gobies, *Conger conger* and epibenthic invertebrates like *Cancer pagurus* and *Maja squinado* (Pittman and McAlpine 2003; Pita and Freire 2011; Abecasis et al. 2014; Carlson et al. 2014). The combination of ARS-behaviour and larger home ranges increases the probability of encountering carrion that does not normally constitute part of their diet. When accounting for the spatial extent of home ranges, one should acknowledge that free-ranging scavengers may establish different areas for spawning and feeding. And feeding areas may vary further diurnally, monthly, seasonally and by animal personality, which can affect their interest in scavenging (Yamamura 1997; Hunter et al. 2005; Humphries et al. 2017; Villegas-Ríos et al. 2018).

7.2.3.2 Encounter Probability: Detection Ability

Foraging by fish and marine invertebrates typically consists of various phases with distinctly different characteristics (Hara 2011; Kamio and Derby 2017). The first phase, finding food, involves food detection whereby scavengers are alerted by a stimulus. Initial arousal is followed by a search, involving orienting and tracking the food source. The second phase, food selection and consumption, typically involves other mechanisms, which are discussed in Sect. 7.2.3.5.

The sensory mechanisms for distant food detection in aquatic environments are primarily olfactory (smell) and gustatory (taste), rather than vision, electro- or mechanoreception (Løkkeborg et al. 2010). Whereas terrestrial scavengers may discriminate themselves from non-scavengers by well-developed olfactory senses (Kane et al. 2017; Verheggen et al. 2017), the distinction is much less pronounced in benthic and demersal aquatic environments where visibility is much lower and olfactory search behaviour is more predominant across a variety of taxa and foraging types (Seibel and Drazen 2007; Paul et al. 2011; Puglisi et al. 2014). Also, the chemical composition of the cues does not distinguish between marine scavenging and non-scavenging taxa, as non-nutritious metabolic waste from urine or other tissues of living organisms might also signal the presence of prey. The presence of odour in water is long-lasting (hours) over long distances (hundreds of meters) and has promoted evolutionary chemoreception beyond the exclusive domain of scavengers. Nearly all demersal fish and benthic invertebrates use olfaction for distant food detection and respond to similar threshold levels (Kohn 1961; Croll 1983; Zimmer-Faust 1987; Hara 1994; Derby and Sorensen 2008; Hay 2011; Løkkeborg et al. 2014; Kamio and Derby 2017).

While the detection thresholds of amino acids in live prey may not be species-specific (Hara 1994), various species show different behaviours when presented with choices between living, damaged or dead prey (Jenkins et al. 2004; Brewer and Konar 2005). Two deep-sea scavengers for instance, the hagfish *Eptatretus stouti* and the amphipod *Orchomenella obtusa*, began searching for dead and decaying food within seconds while they did not for odours that reflected live prey (Tamburri and Barry 1999). The ratio of amino acids to ammonia decreases with increasing carrion age, which may reflect relatively decreasing nutritional quality and elicit different responses across scavenging taxa (Zimmer-Faust 1987). Fisheries discards constitute a variety of living, damaged and dead organisms (Depestele et al. 2014) and may, as such, attract scavengers along a continuum of those preferring freshly killed carrion to those depredating on merely damaged and/or live prey (Table 7.A in the Online Supplement). Dead discards could be colonized by bacteria in a few days, which will then deter certain higher-order scavengers (Burkepile et al. 2006; Hussain et al. 2013). These processes may lead to a clear succession of scavengers in deep-sea environments and for larger carcasses (Smith and Baco 2003; Quaggiotto et al. 2016). In continental shelf areas, it is more likely that dead discards will not reside in the benthic environment for longer than a few days (Groenewold and Fonds 2000), and therefore primarily attract higher-order scavengers which will consume the freshly killed carrion before it can fragment by physical forces and deteriorate. Even the scraps created by wasteful feeders that macerate carrion may be fed upon by indirect feeders before becoming available to detritivores (Davenport et al. 2016).

It is therefore not so much the detection threshold of carrion which will determine whether an organism will encounter carrion or not, but rather the combination of response time following detection (see arrival times in Table 7.A in the Online Supplement) and the area of attraction. Groenewold and Fonds (2000) estimated, for instance, that the attraction areas of gadoids reached up to 1200 m^2, of amphipods, hermit crabs and swimming crabs to 100 m^2, of starfish, whelks, dab, shrimp and brittle stars between 10 and 100 m^2 and attraction areas were less than 10 m^2 for gobies, solenette, sea urchins and sandstars. Both response time and attraction area are thought to be a function of swimming abilities (Sect. 7.2.3.3), intrinsic motivation (e.g. hunger status, Moore and Howarth 1996; Laidre and Elwood 2008), local enhancement (e.g. through mucus secretion by conspecifics Lee et al. 2004) or social learning behaviour (Ryer and Olla 1992; Brown and Laland 2003) as well as a function of environmental variability (e.g. currents, substratum – Bailey and Priede 2002; Stoner 2004), diel rhythm (Bozzano and Sardá 2002) and fishery-related stimuli (e.g. acoustic cues from fishery-generated noise – Thode et al. 2007).

7.2.3.3 Encounter Probability: Locomotion

Locomotion performance is another important driver in finding carrion, and can be considered as a trade-off between extensive searching to find carrion fast versus limited searching with extended resting times between feeding bouts (Ruxton and Bailey 2005). Sit-and-wait scavengers will typically reduce searching and invest in rapid reaction upon carrion detection. Hagfish, for instance, respond quickly to carrion (Tamburri and Barry 1999) and move fast towards it despite their limited swimming capabilities (Collins et al. 1999). In contrast to other burrowing scavengers like *Nephrops norvegicus* which only forage upon carrion in the vicinity of their safe burrows (Nickell and Atkinson 1995), it is likely that hagfishes aggregate close to a previous feeding area to increase the probability of carrion encounters (Martinez et al. 2011). Free-ranging scavengers, in contrast, invest to a greater extent in scanning large areas to increase probabilities of encountering carrion. This foraging strategy is efficient when they can cover vast areas at a low energetic cost (Ruxton and Bailey 2005). The high energetic cost of swimming for marine mammals for instance (Williams 1999) means that they need to trade-off metabolic costs and carrion encounter between natural foraging behaviours and scavenging upon discards following fishing vessels. Whether marine mammals adapt to a scavenging lifestyle may also vary between individuals (see examples in Table 7.A in the Online Supplement). Other large marine organisms, like sharks, have been proposed as good scavenging candidates, as their large pectoral fins are adapted to cruising swimming (Carrier et al. 2004), much like typically obligate aerial scavengers, such as vultures (Kane et al. 2017). Indeed, cruising specialists like sharks and scombrids (Videler and He 2010) appeared regularly in field trials (Sect. 7.2.2). Similarly, taxa with increased swimming abilities like Gadiformes and Perciformes occurred in a larger number of studies than orders with lower swimming speed such as Pleuronectiformes, Rajiformes and Anguilliformes (Videler and He 2010; van Weerden et al. 2014). Several field studies also showed that fish arrived first, followed by fast-moving invertebrates like decapod crustaceans and lastly by slow-moving foragers such as starfish and whelks (Table 7.A in the Online Supplement). Swimming abilities are therefore an important driver to determine an organism's ability to move towards carrion fast.

7.2.3.4 Encounter Probability: Metabolism

Animal biology and ecology depends on metabolism to fuel vital activities, such as foraging (Glazier 2014; Harrison 2017). Metabolic rate (MR) provides an objective measure to attribute cost to their activities like locomotion, predator-prey interactions and to assess what animals do compared to some optimal behaviour, i.e. a behaviour that maximizes one or more biological characteristics such as growth or reproductive success (Metcalfe et al. 2016a, b) or a scavenger versus a predator strategy. Body size and temperature are primary determinants of respiration (Brown et al. 2004) but once accounted for, metabolic rate also reveals much about activities like foraging and the risk of being eaten (Hirst and Forster 2013; Glazier 2014). In an interplay with locomotive performance, carrion detection and foraging time, scavengers tend to reduce metabolic requirements in contrast to their predatory counterparts. A striking example is the higher MR of shallow-living pelagic predators like gadoids and tunas using visual stimuli to locate their food and spending energy to pursuit them (Farrell 1991; Seibel and Drazen 2007). The limited likelihood of a scavenging foraging strategy in pelagic organisms is further reinforced by the division of discards into floating versus rapidly sinking carrion, reducing their incidental encounter in midwater (Harris and Poiner 1990; Croxall and Prince 1994; Bergmann et al. 2002).

Scavengers exhibit low metabolic requirements so that they can survive long periods without food and benefit from efficient assimilation of carrion when it becomes available. Sharks, for instance, have the ability to fast for weeks and focus on energy-rich carrion to replenish their expenditures (Fallows et al. 2013). Other large organisms like marine mammals have higher energy expenditures to maintain their body temperature (Williams 1999; Hirt et al. 2017) and need to trade-off gains and costs of scavenging. Ansmann et al. (2012) showed, for instance, that dolphins preferred associations with trawlers as a reliable, easily located and large food source when fishing effort was high, but they returned to group living to find food when fishing effort reduced and costs of depredation became too high. In contrast, scavenging hagfish, amphipods, starfish and gastropods may survive long periods (4 weeks to 13 months) without food (Vahl 1984; Tamburri and Barry 1999). Lysianassoid amphipods, in particular, exhibit a sit-and-wait strategy, withstanding long periods of starvation, followed by a rapid response and localisation followed by high rates of consumption and efficient carrion utilisation (Smith and Baldwin 1982; Sainte-Marie 1992). Shallow-living demersal scavenging fish (Gadiformes, Perciformes and Pleuronectiformes) have standard metabolic rates within the same order of magnitude, while energy consumption of Anguilliformes and Myxiniformes is generally much lower (Lesser et al. 1997; Clarke and Johnston 1999; Drazen et al. 2011). Lower mass-specific standard respiration rates of slowly moving (crawling) invertebrate scavengers (including crustaceans and echinoderms) were also suggested in Brey (2010).

Besides taxonomic differences in metabolic rates, a significant decline of metabolic rate with depth was demonstrated for organisms whose activities depend on light and vision (e.g. benthopelagic fish like cod), even after adjustments for temperature and body size (Seibel and Drazen 2007; Drazen and Seibel 2007). Comparisons of the deep-sea grenadier *Coryphaenoides armatus* with the facultative scavenger *Gadus morhua*, for instance, showed the increased scavenging abilities of the grenadier moving at slow swimming speeds and with a low metabolic rate, i.e. 15–30% lower, than similar-sized cod at similar temperatures (Ruxton and Bailey 2005). Such depth-related differences in metabolic rates were not found between shallow and deep-living echinoderms, benthic fish, crustaceans and cephalopods – organisms which depend less on visual interactions as do their benthopelagic counterparts.

7.2.3.5 Handling Tactics: Competitive Abilities

Scavenging typically involves a large number of individuals at the carrion and leads to a complex network of direct competition (this Section), facilitation (Sect. 7.2.3.6) and indirect processes e.g. increased or decreased predation (Oro et al. 2013; Moleón et al. 2014). Assessing an organism's competitive abilities is particularly predictive for marine scavengers, because of the short generation times of carrion pulses (Nowlin et al. 2008; Beasley et al. 2012). Indeed, baited experiments show how numerous interactions between scavengers dictate the opportunities to consume the encountered carrion (see Table 7.A in the Online Supplement and references therein). Observations from baited field trials focus primarily on competition related to small amounts of carrion as used in field studies, rather than the potentially greater amounts of carrion which become available from commercial discarding practices.

The competitive abilities of marine taxa are positively correlated to their willingness to expose themselves to risk (boldness), which greatly influences their opportunities for food consumption and, ultimately, survival (Ward et al. 2006; Hamilton 2018). Several baited experiments highlighted the trade-off between feeding preferences (e.g. by damage level of discards), nutritional status (starvation level), the risk of death by predation (e.g. risk aversion of dead conspecifics) and carrion density and distribution (McKillup and McKillup 1994; Davenport and Moore 2002; Collins and Gerald 2009; Tanner et al. 2011; Yeh and Drazen 2011). Overall, organisms with larger body sizes have a higher efficacy as scavengers than smaller individuals, which is not only because they arrive first at the carrion (Sect. 7.2.3.3), but also because they ingest carrion faster, have access to larger bodied carrion, are less prone to hyper predation and are more powerful in outcompeting with smaller organisms (Juanes 1992; Collins et al. 2005; Ward et al. 2006; Nordström et al. 2015). In response, smaller organisms have developed alternative strategies to compensate for their small body size. Small scavengers can aggregate at the carrion and outnumber competitors, e.g. abundance increases of hagfish and amphipods (Table 7.A in the Online Supplement). They may also discourage competitors, e.g. by slime excretion in hagfish (Zintzen et al. 2011) or by threat displays in crabs and fish (Davenport et al. 2016). Competition may also lead to temporal niche partitioning and successional stages of carrion feeding, which is not only driven by the differential reaction speed to carrion but also by an organism's ability to ingest it (Sect. 7.2.3.6).

7.2.3.6 Handling Tactics: Facilitation

Once the food has been encountered and competitors excluded, carrion can be selected (or rejected) for consumption. The selection and consumption of carrion as a food source is facilitated when overlapping with dietary preferences, and when food processing abilities of the organism are adapted to the pulsed availability of the resource. These two characteristics, (1) dietary preferences and (2) food processing, jointly determine which taxa are best able to select and consume carrion and fisheries discards.

Scavengers in baited experiments were not indiscriminately attracted to any bait type but showed differential attraction according to dietary preferences (e.g. Groenewold and Fonds 2000; Jenkins et al. 2004 in Table 7.A in the Online Supplement). Starfish, for instance, showed higher abundances and attraction to mollusc bait types, with a preference for damaged carrion (Jenkins et al. 2004; Brewer and Konar 2005). The amphipod *Tryphosa nana* preferred dead crustaceans (Kaiser and Moore 1999). Dietary preferences are not dictated by olfactory sensory systems only, but are influenced by gustatory, visual, electro- and/or mechanosensory stimuli in species-specific ways (Carrier et al. 2004; Derby and Sorensen 2008; Løkkeborg et al. 2014; Kamio and Derby 2017). Information on sensory systems are scarce in scavenging observations, but lessons can be drawn from observations of predators. Plaice, dab and flounder, for instance, use visual stimuli more than Dover sole to target prey (de Groot 1969). Another predator, ling, targets mobile prey more often than cod, potentially showing increased dependency on visual stimuli (Løkkeborg 1998). Elasmobranch species feeding on (live) benthic prey use electrosensory means to locate food prior to its consumption (Desender et al. 2017). The dependency on other sensory systems like electrosensory location or the dependency on the mobility of food items may reduce the relevance of marine carrion (and 'dead' discards) as a targetable food source for organisms using these feeding mechanisms.

Whether scavengers will feed upon marine carrion will in part depend on these dietary portfolios and the sensory systems to locate their food, but they will also be dictated by a species' dietary plasticity. Several examples have illustrated the dietary flexibility of birds, dolphins, fish and amphipods in response to carrion availability (Whitehead and Reeves 2005; Ansmann et al. 2012; Sinopoli et al. 2012; Oro et al. 2013; Johnson et al. 2015; Seefeldt et al. 2017).

Carrion size is another potential driver of a species' ability to consume the encountered carrion. Food size is positively related with mouth gape size in fish (Scharf et al. 2000; Johnson et al. 2012), while brachyurans prefer medium-sized food items as a trade-off between energetic gains from larger food amounts with the mechanical costs of predation on larger species (Juanes 1992; Kaiser et al. 1993).

While these studies suggest that larger scavengers are more likely to consume larger carrion, a series of studies demonstrated various strategies and techniques that scavengers deploy to profit from carrion overcoming the gape limitation. Small scavengers like amphipods, gobies and hagfish may immerse themselves into cavities of decaying animals and forage from inside out ensuring consumption of rich nutrient sources first (Kaiser and Moore 1999; Bucking et al. 2011; Polačik et al. 2015). Scavengers may also jerk or shake carrion to tear it to pieces or may, in case of larger carrion, spin around their longitudinal axis, as observed in eels (Helfman and Clark 1986) and Atlantic cod (Svendsen 2018). Sharks and brachyurans scavengers are particularly able to exhibit high biting or crushing forces to cut larger food items into digestible pieces (Preston et al. 1996; Huber and Motta 2004; Lucifora et al. 2009). Macerating practices are generally wasteful practices, generating easy access to food for indirect feeders (Sainte-Marie 1992), which is another frequently applied scavenging tactic (Sainte-Marie 1992; Britton and Morton 1994).

Small-sized carrion used in baited experiments is expected to be consumed within days (Table 7.A in the Online Supplement). These experiments presented small amounts of small-sized carrion in comparison to fisheries discards, where larger amounts are released at once. It is likely that not all carrion is consumed within days, leading to decay and subsequent changes in palatability of the discarded carrion. While some species like gastropods do not clearly discriminate between fresh and old carrion (Morton and Jones 2003), others (*Eptatretus stouti, Pycnopodia helianthoides* and *Orchomene* spp.) show clear preferences for dead or damaged organisms (Moore and Wong 1995; Tamburri and Barry 1999; Brewer and Konar 2005). Successional stages have been documented for large-sized carrion, like whale falls in the deep sea (Smith and Baco 2003) and in shallow waters (Glover et al. 2010; Quaggiotto et al. 2016).

The limited time window for carrion consumption (due to, for example, competition and decay) has also stimulated morphological and metabolic adaptations to maximise energy gain from carrion while minimizing handling time. Deep-sea lysianassoid amphipods, for instance, can be divided into two groups: highly specialised necrophagivores (e.g. *Eurythenes gryllus*) that have mandibles and guts which process food in batches, while *Orchomene* spp. process food in a more continuous way, and have small guts and mandibles which are not well suited for rapid food ingestion (Sainte-Marie 1992; Jones et al. 1998). Another typical scavenger, Pacific hagfish (*Eptatretus stoutii*), can ingest large amounts of protein which would result in high rates of post-feeding ammonia and urea excretion. Indeed, Pacific hagfish has a wide scope for ammonia and urea excretion, which allows it not only to process decomposing tissue but also to process large amounts of protein which it acquires during short-term feeding bouts (Wilkie et al. 2017).

7.2.4 Towards Identification of the Most Likely Discard-Consumer Candidates

Field observation studies have elucidated how various traits, and trade-offs between these traits, enable organisms to scavenge upon marine carrion. A trait-based assessment enables scaling of marine scavengers along the obligate-facultative scavenging continuum is given in Fig. 7.2. This scaling can pinpoint the scavenging taxa that will primarily make use of marine carrion (including discards) and therefore have the potential to experience a population level effect when discard levels change. As discussed in more details in Sect. 7.4, the next step of this work will thus be to perform the actual ranking of the taxa identified in Sect. 7.2.2 into the continuum, on the basis of their life history.

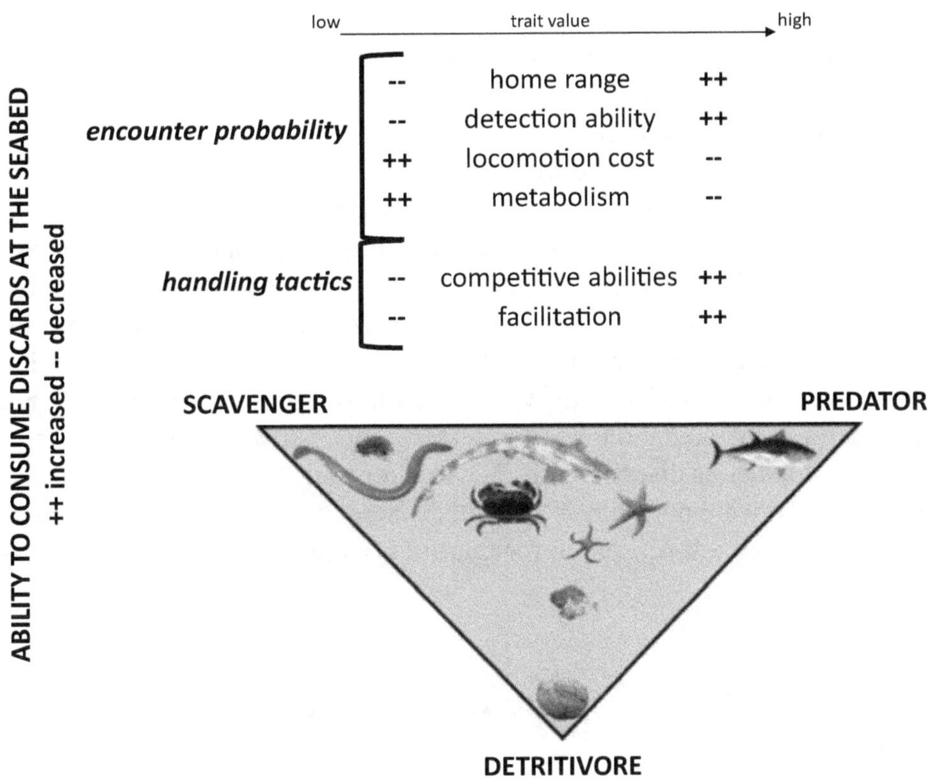

Fig. 7.2 Encounter probability and handling tactics determine an organism's scavenging ability. The traits of these two parameters make up a qualitative scale of scavenging that can be applied to any given species to assess the relative importance of carrion, including dead discards, in its diet. The ranking of species in the triangle is for illustrative purposes but does not represent a formal analysis as such. (Scavengers from left to right: hagfish and amphipod spp., dogfish and edible crab. In the centre: starfish, ophiurids and hermit crabs. Predator: tuna. Detritivore: Echinocardium spp). NB: The probability that facultative scavengers also consume fisheries discards (with more or less focus upon predation or detritivorous strategies – e.g. injured or stressed discards) was not estimated in field observation trials and requires further research. Inspired by Kane et al. (2017)

Most observational studies used marine carrion as a proxy for discards in their field tests. In general, these trials did not reflect commercial discarding practices. We acknowledge, for instance, that not all discarded organisms are dead when they are returned to the sea after commercial fishing. Foraging upon a mixture of dead, stressed, injured and potentially undamaged discarded organisms implies that discards are not only suitable as a food source to obligate scavengers, but also to predators. The consumption of stressed and injured organisms following discarding practices were less frequently examined in the observational studies but may affect foraging by facultative scavengers in commercial discarding practices.

Observational studies have also not focused on mimicking the total amount of carrion that is discarded in commercial fishing practices, nor the way it is spread or lumped on the seabed or the frequency of discard availability. Although observational studies indicated that discards are consumed in a short time window (within days), we are not aware of any observational studies underpinning these assumptions in relation to the recurrent high amounts of discards that become available following commercial discarding. When the assumption of rapid consumption of all discarded material does not hold in certain circumstances, we may expect that not all fisheries-induced carrion will be consumed by scavengers. A fraction of the discards may be fragmented and become available to detritivores later.

In conclusion, field observation studies improve the qualitative assessment of the scavenging abilities of taxa on marine carrion (Fig. 7.2, upper panel). To evaluate how organisms with scavenging abilities profit from fisheries discards requires additional information, notably on the total amount of dead versus living discards and how discards are consumed by organisms with various foraging strategies. The impact of discards on marine taxa with a facultative scavenging strategy requires information on how predation and/or detritivorous strategies may influence the consumption of fisheries discards (Fig. 7.2, lower panel). Predation and detritivorous foraging strategies were not addressed in this chapter but cannot be ignored when the impact of discards as a food source for marine organisms is evaluated.

7.3 Modelling Approaches to Discard-Scavenger Interactions

Discarding is considered in existing ecosystem models evaluating the effects of various fisheries management measures (Fulton et al. 2014; Kaplan et al. 2014; Libralato et al. 2015; Mackinson et al. 2018), but very few have focused on exploring the effects of varying discard flows on marine food webs (Lauria 2012; Heath et al. 2014; Fondo et al. 2015). Here, we used five ecosystem models implemented across Europe to illustrate the lessons learned and challenges still ahead in modelling discard flows in marine food webs.

7.3.1 Materials and Methods

Three ecosystem modelling frameworks have been used: Ecopath with Ecosim (EwE), StrathE2E and Atlantis (Table 7.2). They differ in their representation of space, time and processes but all of them aim at modelling the whole food web by using functional groups (representing species sharing similar life history traits).

Five existing models were considered in this study. They are generally implemented at the scale of fisheries management regions (Fig. 7.3). The food web description in each region is specific to its general ecology, fisheries and conservation stakes (see Table 7.3 for details). However, in the outputs presented here, biomasses of functional groups were pooled in aggregated categories for the purpose of comparison. Discard flows are specific to each model and have been parameterised using data from fisheries monitoring programs.

Table 7.2 Main features of the three modelling frameworks used

Model type	Principle	Functional groups	Space	More details
EwE	Mass balanced	Biomass pool (with life stanza possible) Functional groups and singled-out species	No	Heymans et al. (2016)
StrathE2E	Tracks the flows of nitrogen in groups of functionally similar taxa and materials, spanning the entire ecosystem from biogeochemistry to seabirds and marine mammals	Biomass pools (with life stanza possible)	Vertical layers	Heath (2012)
Atlantis		Age structured population (abundance) and growth (average individual weight) dynamic for vertebrates and biomass pols for invertebrates Functional groups and singled-out species	Irregular horizontal polygons and regular vertical layers	Fulton (2010)

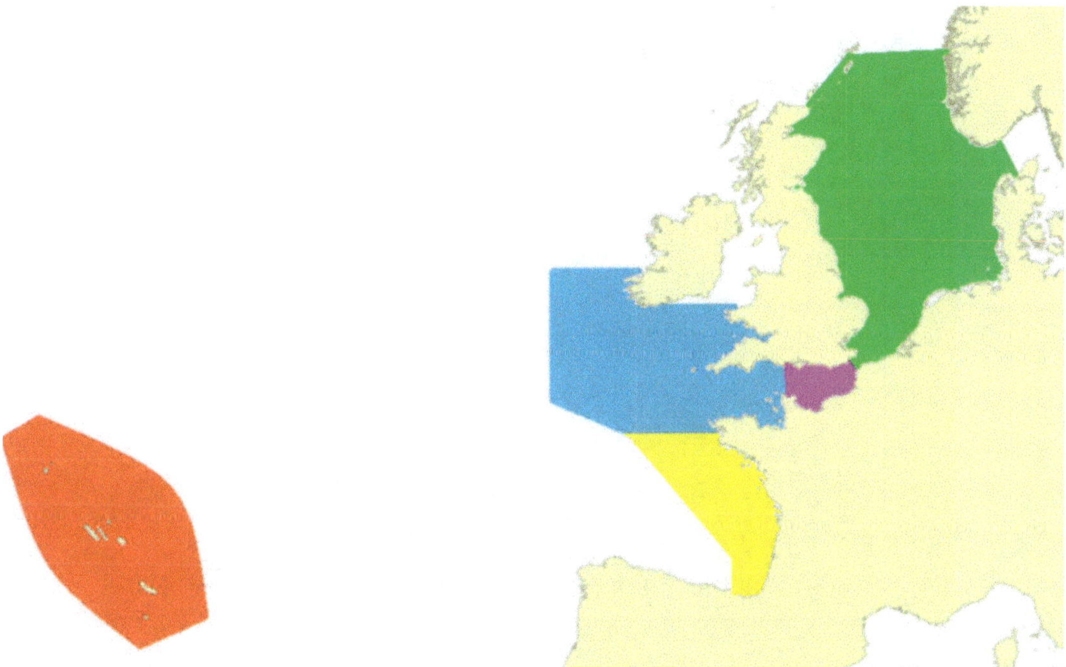

Fig. 7.3 Geographic implementation of the 5 models used: the Azores EEZ (red), the Bay of Biscay (ICES regions VIIIa and b, yellow), the Celtic Sea (VIIe, h, f, g and j, blue), the Eastern Channel (VIId), and the North Sea (IV c, b and most of a)

We evaluated two scenarios using the design in Table 7.3:

1. "Discard as Usual": fishing mortalities and discard rates are fixed per group and constant throughout the simulation, the discard survival rates are all assumed to be nil;

Table 7.3 Main features of the ecosystem models used in five areas

Area	Models	Groups	Historical run	Projections	Outputs	References
Azores (open ocean and deep-sea environments)	EwE	46 groups 2 primary producers, 2 zooplankton, 5 benthic and 1 pelagic invertebrates, 18 fish, 1 seabird, 1 turtle, 3 mammals functional groups 11 individual fish species 1 discards and 1 detritus groups	Ecosim fitted on 1997–2013	Run for 17 years (from 2013 to 2030) using an average F over the last 3 years of the historical run	Average of the last 3 years	Morato et al. (2016)
Bay of Biscay	EwE	43 groups 3 primary producers, 1 bacteria, 3 zooplankton, 7 benthic and 1 pelagic invertebrates, 7 fish, 2 mammals and 2 birds functional groups 13 individual fish species (1 with 2 life stanza) and 1 benthic invertebrate species 1 discards and 1detritus groups	Ecosim fitted on 1980–2013	Run for 17 years (from 2013 to 2030) using an average F over the last 3 years of the historical run	Average of the last 3 years	Moullec et al. (2017)
Celtic Sea	EwE	48 groups 3 primary producers, 1 bacteria, 3 zooplankton, 7 benthic and 1 pelagic invertebrates, 7 fish, 2 mammals and 2 birds functional groups 17 individual fish species (2 with 2 life stanza) and 1 benthic invertebrate species 1 discards and 1detritus groups	Ecosim fitted on 1980–2013	Run for 17 years (from 2013 to 2030) using an average F over the last 3 years of the historical run	Average of the last 3 years	Moullec et al. (2017)

(continued)

Table 7.3 (continued)

Area	Models	Groups	Historical run	Projections	Outputs	References
North Sea	StrathE2E	23 groups 2 primary producers, 2 zooplankton, 2 benthic invertebrates, 2 fish (with each 2 life stanza) and 1 birds and mammals functional groups 3 detritus, and 2 discards groups 6 nutrient groups	Model calibrated on averaged biomass/ catch/effort over 1970–1999	Historical run = baseline "No discard" scenario based on the historical run (no projection)	Annual average once stationary annual cycle reached	Heath et al. (2014)
Eastern English Channel (spatialised into 32 polygons)	Atlantis	43 groups 1 primary producers, 2 bacteria, 3 zooplankton, 8 benthic and 1 pelagic invertebrates, 11 fish, 2 mammals and 1 seabird functional groups 7 individual fish species and 1 benthic invertebrate species 2 detritus and 1 discards groups 3 nutrient groups	Model calibrated on averaged biomass/ catch/effort over 2002–2011	"No discard" scenario based on the historical run (no projection)	Annual average once stationary annual cycle reached	Girardin et al. (2018)

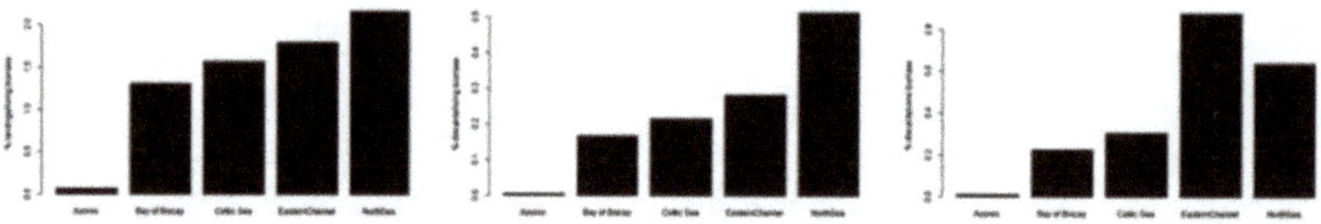

Fig. 7.4 Living biomass extracted from the ecosystem (landings) over the total biomass (*left*), biomass returned to the ecosystem as discards over the total living biomass (*middle*) and discards over total potential consumer biomass (*right*). The total living biomass includes vertebrates, benthic and pelagic invertebrates, benthic primary producers, as well as phytoplankton and zooplankton. The total consumer biomass includes vertebrates, benthic and pelagic invertebrates (excluding zooplankton and phytoplankton and benthic primary producers)

2. "No Discards": total fishing mortalities per group remain the same, but the discard rates are all set to 0, leading to no discards being released in the system[2], which allows one to test the 'pure' effect of the discard flow into the marine food web.

The scenarios represent extreme cases to bracket the range of possible realities and allow us to isolate the ecosystem response to the flow of discards. In the baseline model, zero survival of discarded fish is extreme, as is complete cessation of discarding in "No Discards" (especially without changes in the fishing pressure).

7.3.2 *Results*

7.3.2.1 The Flow of Discards Into the Environment

The fishing pressure varies considerably across the different modelled ecosystems, but it is apparent from Fig. 7.4 that the flow of discards into the marine ecosystem is extremely low (< 1%), when compared to the biomass of potential consumers of discards (all living groups except plankton and primary producers).

In the five models used, the actual proportion of discards in the diet of a scavenger depends on its (parameterized) preference for this type of food, and on the abundance of discards available to feed on (together with other factors depending on the model, such as spatial overlap or consumer clearance rate). Fig. 7.5 shows the average calculated diet of three groups of scavengers in the Eastern English Channel Atlantis model, among those with the highest parametrized preference for discards. The discards never accounted for more than 0.4% of the total diet, which is due to the very low abundance of discards available to consume.

[2]In the case of the North Sea StrathE2E model however, a residual discard flow remains for the benthos, corresponding to 4% of the baseline discard flow (Heath et al. 2014).

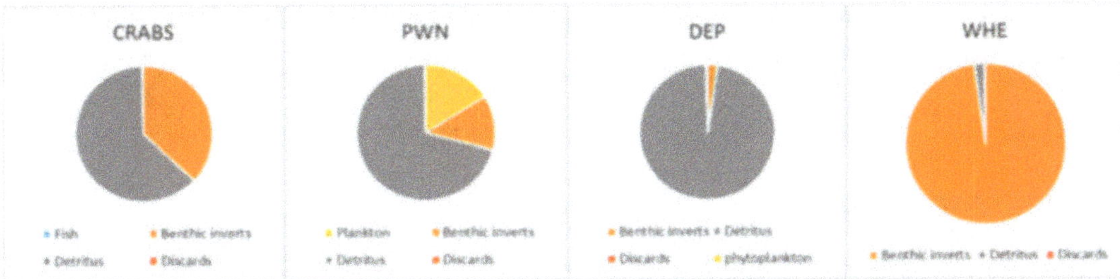

Fig. 7.5 Average diet composition of four groups of scavengers in the Atlantis Eastern English Channel model: crabs, prawns (PWN), deposit-feeders (DEP) (mainly amphipods and some polychaetes) and whelks (WHE), as calculated by the model. All the consumed food items appear in the legend, but the ones representing less than 1% of the diet do not appear on the pie charts. Benthic inverts: benthic invertebrates

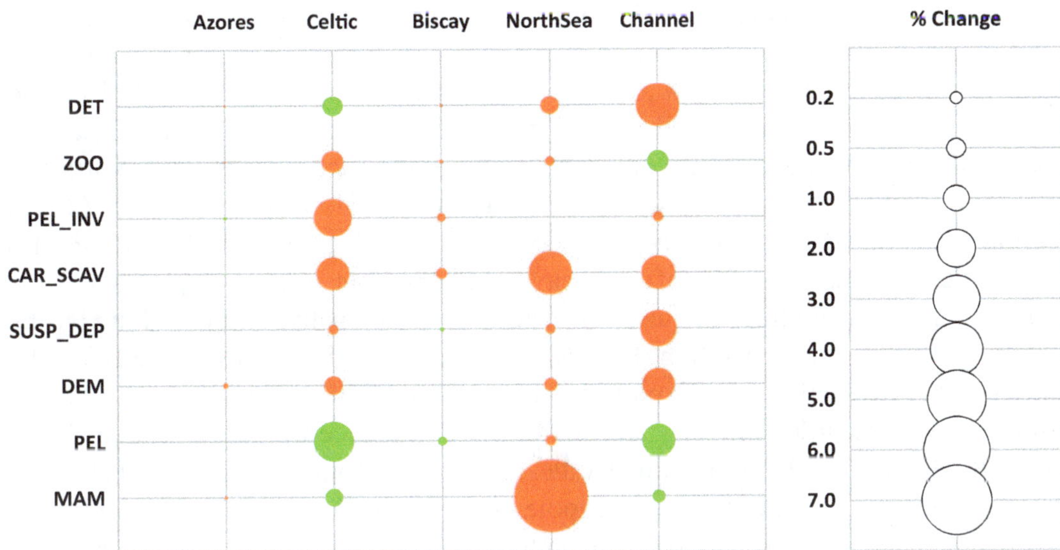

Fig. 7.6 Percentage changes in the biomass of aggregated categories between the "No Discards" scenario and the "Discard as usual" baseline. Data is shown as absolute values colored in red for negative change and green for positive change. The size of the circle is proportional to the absolute % changes in the biomass. Aggregated categories: DET: detritus, ZOO: zooplankton, PEL_INV: pelagic invertebrates, CAR_SCAV: carnivorous and scavenging benthos, SUSP_DEP: suspension and deposit feeding benthos, DEM: demersal and bentho-pelagic fish, PEL: pelagic fish, MAM: mammals

7.3.2.2 Effect of Removing Discards on the Ecosystem

Removing discards from the ecosystem had an extremely low effect on the evaluated ecosystems (Fig. 7.6), with virtually no effect (< 0.2%) on the Azores ecosystem, where the discard flow was the smallest. In the other case studies, the effect of removing discards was small but noticeable on the benthic carnivores and scavengers (−1.3% in the Celtic Sea and −2.4% in the North Sea). The largest negative effect of removing discards was observed in the North Sea on the mammals/seabirds group. Not surprisingly, the effect of removing discards is dependent on the importance of discards in the diets of the different groups.

7.3.3 Discussion

Discard flows may seem large when considered as absolute values, and with regards to the waste of natural resources it represents. When considered as a potential food source, however, discard flows appear to be low relative to the living biomass in the ecosystem, and when compared with other sources of food (natural carrion).

The models found little effect of a full discard ban (with no changes in the fishing pressure) on the ecosystem. These results are in agreement with other modelling studies. Sánchez and Olaso (2004) confirmed the low importance of discards as a food source in the Cantabrian Sea in comparison to detritus, primary producers or other low trophic levels (0.07% of the total food intake, EwE model). Kaiser and Hiddink (2007) and Collie et al. (2017) estimated that fisheries-generated carrion could only sustain benthic carnivores for 3 days per year at the scale of the North Sea and scavenging fish for approximately 6 days per year. Depestele et al. (2016) similarly estimated that discards in the Grande Vasière (northern Bay of Biscay) contributed to less than 2% of the scavenging benthic community's total food requirements. Our findings are also consistent with the spatial analysis done by the Working Group on Ecosystem Effects of Fishing Activities (WGECO) of the International Council for the Exploration of the Sea (ICES) who found no relationship between discarded biomass and scavenger abundance in the North Sea (ICES 2016; ICES 2017). Other modelling studies, in contrast, found an influential effect of discards on benthic scavengers (Catchpole et al. 2006; Fondo et al. 2015) but at a smaller spatial scale (respectively Moreton Bay, and the North Sea *Nephrops* English fishing grounds). Ecosystem modelling can help extrapolate available data and empirical knowledge (to a certain extent) on the fate of discards in the environment at the regional scale, and in testing a full Landing Obligation scenario. However, some discrepancies between modelling and empirical knowledge should be noted. First, the discards are all pooled into one group in the models, while observational studies showed that the consumption of discards differs between the provided types of discards and dietary preferences of the scavengers (Table 7.A in the Online Supplement). Second, benthic and demersal scavengers tend to be aggregated into coarse functional groups in ecosystem models. This aggregation does not allow one to fully account for varying degrees of importance of discards to specific taxa (e.g. Sect. 7.2 of this chapter; Kaiser and Hiddink 2007; Mackinson and Daskalov 2017). Further splitting the scavenger groups would help, but only if sufficient data are available for parametrization, which is unlikely at the scale of the presented models.

A closer interaction between empirical and modelling approaches could increase our understanding of the importance of discards in the marine food web. For example, observational studies may inform modelling exercises regarding the differences between discard consumption ratios of taxa within one functional group. Observational studies may also set the boundaries to conduct sensitivity analyses on the existing groups, and help explore whether a different partitioning of discards as food sources across taxa results in a significant food subsidy for some species and not for others.

A major source of uncertainty in the modelling work was due to the quality of the discards data. The limited fleet coverage of discard monitoring programs hampers the possibility of discard estimations at a small spatial scale. Furthermore, current discard monitoring programs focus primarily on commercial species and have been designed to estimate discard rates per fleet and stock rather than to estimate discard flow into the environment. Reliable quantitative information on the ratio of commercial versus non-commercial discards is relatively sparse (Uhlmann et al. 2013; Depestele 2015), but existing data could be used to speculate on the orders of magnitude changes that non-commercial discards may cause in the energy flow in the ecosystem. Here again, sensitivity analyses will be necessary to complete the present work.

7.4 Synthesis and Outlook

There is a considerable literature about the ecology of scavenging in marine ecosystems, showing that scavenging is an important aspect of the trophic ecology of many marine species. Discards undoubtedly contribute to this with a global maximum of close to 20 million tonnes per year at its peak. Despite this, it is challenging to identify how important discards might be to the scavenging community, as they are very diverse taxonomically and in the role they play in scavenging interactions.

When marine scavengers are explicitly included as groups in ecosystem models, the impact of stopping discards is shown as minimal (Sect. 7.3). These findings corroborate earlier empirical and analytical studies where it was suggested that discards only deliver a small-scale food subsidy in large areas. They contrast, however, with a few studies (Catchpole et al. 2006) where fisheries discards may have more significant effects in certain locations with high discards

One problem with our conclusion regarding the low impact of discards is that we treated the scavenger community as a few functional groups over large spatial scales. It is more than likely that there will be particular species that are more dependent on a discard subsidy than others (Table 7.A in the Online Supplement). This was the case with small-spotted catsharks (*Scyliorhinus canicula*) and smooth skate (*Dipturus inominatus*) which made significant use of fisheries discards but not with starry rays (*Raja asterias*) (Forman and Dunn 2012; Navarro et al. 2016). This highlights one way in which modelling studies could be taken forward; to differentiate within the models between scavenger species that might be more, or less, dependent on discards. This would integrate improved knowledge regarding these species' relative dependencies on carrion (Sect. 7.2). It is easy to speculate, for instance, that hagfish, which consume 82% of discards would be more affected than more opportunistic species by banning discards. That is, it may be possible to include more categories with low to high discard dependencies within the functional groups. The trait-based approach discussed in this chapter is intended to stimulate such an assessment of scavenging abilities.

Scavenging is not only an activity of epibenthic invertebrate scavenger communities as we know that not all discards reach the seabed. Significant amounts will be consumed by marine mammals and seabirds, depending on the type of discard (Depestele et al. 2016). An unknown but likely smaller quantity will also be taken by mid water scavengers. Many demersal fish species will also likely scavenge to some extent (Table 7.A in the Online Supplement). Finally, discards may often be in large quantities locally, and may simply exceed the capacity of scavengers to utilise them all, and will then likely become available to a detritivore community. Empirical field evidence should be the main source of information to parameterise discard partitioning and usage, and therefore suggest scenarios for the models.

It is, however, important to also remain aware of the limitations of this evidence. Empirical parameterisation should be based on more than one sampling technique given the selectivity issues with observation within particular benthic and demersal environments. While numerous studies in continental European seas have highlighted what this means for benthic invertebrate scavengers, few experiments have focused on implications to fish. Also, while baited traps or cameras are excellent tools for studying which species might utilise discards and what discard types they prefer, they may not be representative of the real situation, for instance in the amount, condition or species mix of actual discards. Most importantly, these *in situ* methods are expensive and difficult and only provide information that is specific to a place, time and particular conditions. Extrapolating those data to population levels can be risky (Levin 1992; Dickey-Collas et al. 2014).

This brings us back to the value of the models. The models could be used in a number of scenarios to test their sensitivity to assumptions about the scavenger group. These could include giving the scavenger group a much greater dependency on discards than might otherwise be expected, and possibly finding at what level of dependency a discard ban would likely start to have effects on populations. Alternatively, we could assume a much higher level of discarding, prior to the Landing Obligation, than is suggested from observer data. Then the models could be run to show the sensitivity to the assumptions of such a high, pre-Landing Obligation discard volume. The sensitivity testing should give us some idea of what we still need to know to understand the importance of discards in the marine food web, and so the likely effects of eliminating them. This information could then be used to design subsequent field studies and analytical assessments. To date most field work has had a broad natural history approach – what do we see coming to some discarded fish? The model sensitivities could be used to define more hypothesis-based field work, focusing on particular species, discard types, places and seasons.

In conclusion, thus far, modelling and empirical studies suggest that carrion is used by many species, but few species are solely dependent on carrion (including discards). For a few species and areas, a reduction in discards due to the Landing Obligation may have a population level effect, but generally this is unlikely to be the case. The way forward is to determine the sensitivities in the models to certain assumptions, particularly in the importance of discards to some species in a broader scavenging community, and the actual volume of discards. That information can then be used to focus the field work on testing hypotheses which are tailored to particular species and/or discard volumes in particular locations, periods and fisheries where modelling predicts higher impacts of discards on scavengers.

Acknowledgments Part of this work has received funding from the Horizon 2020 Programme under grant agreement DiscardLess number 633680. This support is gratefully acknowledged. TM and AS acknowledge support by the Fundação para a Ciência e Tecnologia (FCT) strategic project UID/MAR/04292/2013 granted to MARE. TM is supported by the Program Investigador FCT (IF/01194/2013/CP1199/CT0002).

References

Abecasis, D., Afonso, P., Erzini, K. (2014). Combining multispecies home range and distribution models aids assessment of MPA effectiveness. *Marine Ecology Progress Series, 513,* 155–169.

Ansmann, I.C., Parra, G.J., Chilvers, B.L., et al. (2012). Dolphins restructure social system after reduction of commercial fisheries. *Animal Behaviour, 84,* 575–581.

Bailey, D.M., & Priede, I.G. (2002). Predicting fish behaviour in response to abyssal food falls. Marine Biology, 141: 831–840.

Bailey, D.M., King, N.J., Priede, I.G. (2007). Cameras and carcasses historical and current methods for using artificial food falls to study deep-water animals. *Marine Ecology Progress Series, 350,* 179–192.

Beasley, J.C., Olson, Z.H., Devault, T.L. (2012). Carrion cycling in food webs: Comparisons among terrestrial and marine ecosystems. *Oikos*, *121*, 1021–1026.

Beasley, J.C., Olson, Z.H., Devault, T.L. (2015). Ecological role of vertebrate scavengers. In M.E. Benbow, J.K. Tomerlin, A.M. Tarone (Eds.), Carrion ecology, evolution and their applications, (pp. 107–127). Boca Raton/London/New York: CRC Press/Taylor & Francis Group.

Bergmann, M., Wieczorek, S.K., Moore, P.G., et al. (2002). Utilisation of invertebrates discarded from the Nephrops fishery by variously selective benthic scavengers in the west of Scotland. *Marine Ecology Progress Series, 233,* 185–198.

Bicknell, A.W.J., Oro, D., Camphuysen, K., et al. (2013). Potential consequences of discard reform for seabird communities. *Journal of Applied Ecology, 50,* 649–658.

Bicknell, A.W.J., Godley, B.J, Sheehan, E.V. et al. (2016) Camera technology for monitoring marine biodiversity and human impact. *Frontiers in Ecology and the Environment, 14,* 424–432.

Bozzano, A., Sardá, F. (2002). Fishery discard consumption rate and scavenging activity in the northwestern Mediterranean Sea.*ICES Journal of Marine Science, 59,* 15–28.

Brewer, R., & Konar, B. (2005). Chemosensory responses and foraging behavior of the seastar *Pycnopodia helianthoides*. *Marine Biology*, *147*, 789–795.

Brey, T. (2010). An empirical model for estimating aquatic invertebrate respiration. *Methods in Ecology and Evolution, 1*, 92–101.

Britton, B.C., & Morton, B. (1994). Marine carrion and scavengers. *Oceanography and Marine Biology: An Annual Review*. *32*, 369–434.

Broadhurst, M.K., Suuronen, P., Hulme, A. (2006). Estimating collateral mortality from towed fishing gear. *Fish and Fisheries*, *7*, 180–218.

Brown, C., & Laland, K.N. (2003). Social learning in fishes: A review. *Fish and Fisheries, 4*, 280–288.

Brown, J.H., Gillooly, J.F., Allen, A.P., et al. (2004). Toward a metabolic theory of ecology. *Ecology, 85,* 1771–1789.

Bucking, C., Glover, C.N., Wood, C.M. (2011). Digestion under duress: Nutrient acquisition and metabolism during hypoxia in the Pacific hagfish. *Physiological and Biochemical Zoology: Ecological and Evolutionary Approaches, 84,* 607–617.

Burkepile, D.E, Parker, J.D., Woodson, C.B. et al. (2006). Chemically mediated competition between microbes and animals: Microbes as consumers in food webs. *Ecology, 87*, 2821–2831.

Carlson, A.E., Hoffmayer, E.R., Tribuzio, C.A., et al. (2014). The use of satellite tags to redefine movement patterns of spiny dogfish (*Squalus acanthias*) along the u.S. East coast: Implications for fisheries management. *PLoS ONE, 9,* e103384.

Carrier, J.C., Musick, J.A., Heithaus, M.R. (2004). *Biology of sharks and their relatives.* Boca Raton: CRC Press.

Catchpole, T.L., Frid, C.L.J., Gray, T.S. (2006). Importance of discards from the English *Nephrops norvegicus* fishery in the North Sea to marine scavengers. *Marine Ecology Progress Series, 313*, 215–226.

Clarke, A., & Johnston, N.M. (1999). Scaling of metabolic rate with body mass and temperature in teleost fish. *Journal of Animal Ecology, 68*, 893–905.

Collie, J., Hiddink, J.G., van Kooten, T., Rijnsdorp, A.D., Kaiser, M.J., Jennings, S., Hilborn, R. (2017). Indirect effects of bottom fishing on the productivity of marine fish. *Fish and Fisheries, 18,* 619–637.

Collins, A.M., & Gerald, G.W. (2009). Attraction of flatworms at various hunger levels toward cues from an odonate predator. *Ethology, 115*, 449–456.

Collins, M.A., Yau, C., Nolan, C.P., et al. (1999). Behavioural observations on the scavenging fauna of the Patagonian slope. *Journal of the Marine Biological Association of the United Kingdom, 79*, 963–970.

Collins, M.A., Bailey, D.M., Ruxton, G.D. et al. (2005). Trends in body size across an environmental gradient: A differential response in scavenging and non-scavenging demersal deep-sea fish. *Proceedings of the Royal Society B: Biological Sciences, 272,* 2051–2057.

Cornou, A.S., Diméet, J., Tétard, A. et al. (2015). Observation à bord des navires de pêche professionnelle. Bilan de l'échantillonnage 2013, 1–381

Croll, R.P. (1983). Gastropod chemoreception. *Biological Reviews, 58*, 293–319.

Croxall, J.P., & Prince, P.A. (1994). Dead or alive, night or day: How do albatrosses catch squid? *Antarctic Science, 6*, 155–162.

Da Silva, J.F. (2009). Elasmobranchs & Commercial Fisheries around the British Isles: Spatial and Temporal Dynamics. Universidade do Porto.

Davenport, J., & Moore, P.G. (2002). Behavioural responses of the netted dogwhelk *Nassarius reticulatus* to olfactory signals derived from conspecific and nonconspecific carrion. *Journal of the Marine Biological Association of the United Kingdom, 82,* 967–969.

Davenport, J., McCullough, S., Thomas, R.W. et al. (2016). Behavioural responses of shallow-water benthic marine scavengers to fish carrion: In *A preliminary study. Marine and freshwater behaviour and physiology*, (pp. 1–15). Taylor & Francis.

de Groot, S.J. (1969). Digestive system and sensorial factors in relation to the feeding behaviour of flatfish (Pleuronectiformes). *ICES Journal of Marine Science, 32,* 385–394.

Depestele, J. (2015). *The fate of discards from marine fisheries*. PhD thesis Ghent University. Gent, Belgium. 286pp.

Depestele, J., Desender, M., Benoît, H.P., et al. (2014). Short-term survival of discarded target fish and non-target invertebrate species in the "eurocutter" beam trawl fishery of the southern North Sea. *Fisheries Research, 154*, 82–92.

Depestele, J., Rochet, M.J., Dorémus, G., et al. (2016). Favorites and leftovers on the menu of scavenging seabirds: modelling spatiotemporal variation in discard consumption. *Canadian Journal of Fisheries and Aquatic Sciences, 73,* 1–14.

Derby, C.D., & Sorensen, P.W. (2008). Neural processing, perception, and behavioral responses to natural chemical stimuli by fish and crustaceans. *Journal of Chemical Ecology, 34,* 898–914.

Desender, M., Kajiura, S., Ampe, B., et al. (2017). Pulse trawling: Evaluating its impact on prey detection by small-spotted catshark (*Scyliorhinus canicula*). *Journal of Experimental Marine Biology and Ecology, 486,* 336–343.

DeVault, T.L., Rhodes, O.E., Shivik, J.A. (2003). Scavenging by vertebrates: Behavioral, ecological, and evolutionary perspectives on an important energy transfer pathway in terrestrial ecosystems. *Oikos, 102,* 225–234.

DeVault, T.L., Beasley, J.C., Olson, Z.H., et al. (2016). Ecosystem services provided by avian scavengers. In D.G.W.C.H. Şekercioğlu, C.J. Whelan (Eds.), *Why birds matter: Avian ecological function and ecosystem services* (pp. 235–270). Chicago: University of Chicago Press.

Dickey-Collas, M., Payne, M.R., Trenkel, V.M., et al. (2014). Hazard warning: Model misuse ahead. *ICES Journal of Marine Science, 71,* 2300–2306.

Dictionary, o. (2018). Scavenger Meaning in the Cambridge English Dictionary. [online] Dictionary.cambridge.org. Available at: https://dictionary.cambridge.org/dictionary/english/scavenger. Accessed 20 June 2018.

Drazen, J.C., & Seibel, B.A. (2007). Depth-related trends in metabolism of benthic and benthopelagic deep-sea fishes. *Limnology and Oceanography, 52*, 2306–2316.

Drazen, J.C., Yeh, J., Friedman, J., et al. (2011). Metabolism and enzyme activities of hagfish from shallow and deep water of the Pacific Ocean. *Comparative Biochemistry and Physiology Part A: Molecular & Integrative Physiology, 159*, 182–187.

EU. (2013). European Regulation No. 1380/2013 of the European parliament and of the Council of 11 December 2013 on the Common Fisheries Policy, amending Council Regulations (EC) No 1954/2003 and (EC) No 1224/2009 and repealing Council Regulations (EC) No 2371/2002 and (EC) No 639/2004 and Council Decision 2004/585/EC. Official Journal of the European Union, L354: 22.

Fallows, C., Gallagher, A.J., Hammerschlag, N. (2013). White Sharks (*Carcharodon carcharias*) scavenging on whales and its potential role in further shaping the ecology of an apex predator. *PLoS ONE, 8*, e60797.

Farrell, A.P. (1991). From hagfish to tuna: A perspective on cardiac function in fish. *Physiological Zoology, 64*, 1137–1164.

Fauchald, P., & Tveraa, T. (2003). Using first-passage time in the analysis of area-restricted search and habitat selection. In *Ecology* (pp. 282–288). Ecological Society of America.

Feekings, J., & Krag, L.A. (2015). Fate of discard trial [online]. Available at: http://www.discardless.eu/video/entry/fate-of-discard-trial. Accessed 20 June 2018.

Fondo, E.N., Chaloupka, M., Heymans, J.J, et al. (2015). Banning fisheries discards abruptly has a negative impact on the population dynamics of charismatic marine megafauna. *PLoS ONE, 10*, e0144543.

Forman, J.S, & Dunn, M.R. (2012) Diet and scavenging habits of the smooth skate *Dipturus innominatus*. *Journal of Fish Biology, 80,* 1546–1562.

Fulton, E.A. (2010). Approaches to end-to-end ecosystem models. *Journal of Marine Systems, 81*, 171–183.

Fulton, E.A., Smith, A.D.M., Smith, D.C., et al. (2014). An integrated approach is needed for ecosystem based fisheries management: Insights from ecosystem-level management strategy evaluation. *PLoS ONE, 9,* e84242.

Gilman, E. (2015). Status of international monitoring and management of abandoned, lost and discarded fishing gear and ghost fishing. *Marine Policy, 60*, 225–239.

Girardin, R., Fulton, E.A., Lehuta, S., et al. (2018). Identification of the main processes underlying ecosystem functioning in the Eastern English Channel, with a focus on flatfish species, as revealed through the application of the Atlantis end-to-end model. *Estuarine, Coastal and Shelf Science, 201*: 208–222.

Glazier, D. (2014). Metabolic scaling in complex living systems. *Systems, 2,* 451.

Glover, A.G., Higgs, N.D., Bagley, P.M., et al. (2010) A live video observatory reveals temporal processes at a shelf-depth whale-fall. *Cahiers de Biologie Marine, 51*, 375–381.

Greene, C.H. (1986). Patterns of prey selection: Implications of predator foraging tactics. *The American Naturalist, 128*, 824–839.

Groenewold, S. (2000). *The effects of beam trawl fishery on the food consumption of scavenging epibenthic invertebrates and demersal fish in the southern North Sea*. PhD thesis Hamburg University. Hamburg, Germany. 158pp.

Groenewold, S., & Fonds, M. (2000). Effects on benthic scavengers of discards and damaged benthos produced by the beam-trawl fishery in the southern North Sea. *ICES Journal of Marine Science, 57,* 1395–1406.

Hamilton, S.L. (2018). From a sea of phenotypic traits, fast reaction and boldness emerge as the most influential to survival in marine fish. *Functional Ecology, 32,* 856–857.

Hara, T.J. (1994). The diversity of chemical stimulation in fish olfaction and gustation. *Reviews in Fish Biology and Fisheries, 4*, 1–35.

Hara, T.J. (2011). Smell, taste, and chemical sensing | Chemoreception (Smell and Taste): An Introduction A2 – Farrell, Anthony P. In *Encyclopedia of Fish Physiology* (pp. 183–186). San Diego: Academic Press.

Harris, A.N., & Poiner, I.R. (1990). By-catch of the Prawn Fishery of Torres Strait; Composition and Partitioning of the discards into components that float of sink. *Marine and Freshwater Research, 41*, 37–52.

Harrison, J.F. (2017). Do performance-safety tradeoffs cause hypometric metabolic scaling in animals? *Trends in Ecology & Evolution, 32*, 653–664.

Hay, M.E. (2011). Crustaceans as powerful models in aquatic chemical ecology. In T. Breithaupt, and M. Thiel (Eds.), *Chemical Communication in Crustaceans* (pp. 41–62). New York: Springer.

Heath, M.R. (2012). Ecosystem limits to food web fluxes and fisheries yields in the North Sea simulated with an end-to-end food web model. *Progress in Oceanography, 102*, 42–66.

Heath, M.R., Cook, R.M., Cameron, A.I., et al. (2014). Cascading ecological effects of eliminating fishery discards. *Nature Communications, 5,* 1–8.

Helfman, G.S., & Clark, J.B. (1986). Rotational feeding: Overcoming gape-limited foraging in anguillid eels. *Copeia, 1986,* 679–685.

Heymans, J.J., Coll, M., Link, J.S., et al. (2016). Best practice in Ecopath with Ecosim food-web models for ecosystem-based management. *Ecological Modelling, 331,* 173–184.

Hiddink, J., Jennings, S., Sciberras, M., et al.. (2017). Global analysis of depletion and recovery of seabed biota following bottom trawling disturbance. *Proceedings of the National Academy of Sciences, 114,* 8301–8306.

Higginson, A.D., & Ruxton, G.D. (2015) Foraging mode switching: the importance of prey distribution and foraging currency. *Animal Behaviour, 105,* 121–137.

Hill, S., Burrows, M.T., Hughes, R.N. (2000). Increased turning per unit distance as an area-restricted search mechanism in a pause-travel predator, juvenile plaice, foraging for buried bivalves. *Journal of Fish Biology, 56*, 1497–1508.

Hirst, A.G., & Forster, J. (2013) When growth models are not universal: Evidence from marine invertebrates. *Proceedings of the Royal Society B: Biological Sciences, 280,* 20131546.

Hirt, M.R., Jetz, W., Rall, B.C. et al. (2017). A general scaling law reveals why the largest animals are not the fastest. *Nature Ecology & Evolution, 1*(8), 1116–1122.

Huber, D.R., & Motta, P.J. (2004). Comparative analysis of methods for determining bite force in the spiny dogfish *Squalus acanthias*. *Journal of Experimental Zoology Part A: Comparative Experimental Biology, 301A*, 26–37.

Humphries, N.E., Queiroz, N., Dyer, J.R.M., et al. (2010). Environmental context explains Lévy and Brownian movement patterns of marine predators. *Nature, 465*, 1066–1069.

Humphries, N. E, Simpson, S.J., Sims, D.W. (2017). Diel vertical migration and central place foraging in benthic predators. *Marine Ecology Progress Series, 582*, 163–180.

Hunter, E., Buckley, A.A., Stewart, C., et al. (2005). Migratory behaviour of the thornback ray, *Raja clavata*, in the southern North Sea. *Journal of the Marine Biological Association of the United Kingdom, 85*, 1095–1105.

Hussain, A., Saraiva, L.R., Ferrero, D.M., et al. (2013). High-affinity olfactory receptor for the death-associated odor cadaverine. *Proceedings of the National Academy of Sciences, 110,* 19579–19584.

ICES. (2016). Report of the Working Group on the Ecosystem Effects of Fishing Activities (WGECO), 6–13 april 2016. 110 pp.

ICES. (2017). Report of the Working Group on the Ecosystem Effects of Fishing Activities (WGECO), 5–12 april 2017. 122 pp.

Jenkins, S.R., Mullen, C., Brand, A.R. (2004). Predator and scavenger aggregation to discarded by-catch from dredge fisheries: Importance of damage level. *Journal of Sea Research, 51,* 69–76.

Jensen, P. (1987). Feeding ecology of free-living aquatic nematodes. *Marine Ecology Progress Series,* 35, 187–196.

Johansen, P-O. (2000). Bait attraction studies on the scavenging deepwater isopod *Natatolana borealis* (Crustacea, Isopoda). *Ophelia, 53,* 27–35.

Johnson, A.F., Valls, M., Moranta, J., et al. (2012). Effect of prey abundance and size on the distribution of demersal fishes. *Canadian Journal of Fisheries and Aquatic Sciences, 69,* 191–200.

Johnson, A.F., Gorelli, G., Jenkins, S.R. et al.. (2015) Effects of bottom trawling on fish foraging and feeding. *Proceedings of the Royal Society of London B: Biological Sciences, 282*(1799), 20142336.

Jones, E.G., Collins, M.A., Bagley, P.M., et al. (1998). The fate of cetacean carcasses in the deep sea: observations on consumption rates and succession of scavenging species in the abyssal north-east Atlantic Ocean. *Proceedings of the Royal Society of London. Series B: Biological Sciences, 265,* 1119–1127.

JRC. (2018). Joint Research Centre (JRC) fisheries data collection web site [online]. Available at: datacollection.jrc.ec.europa.eu. Accessed 14 Feb 2018.

Juanes, F. (1992). Why do decapod crustaceans prefer small-sized molluscan prey? *Marine Ecology Progress Series, 87,* 239–249.

Kaiser, M.J., & Hiddink, J.G. (2007). Food subsidies from fisheries to continental shelf benthic scavengers. *Marine Ecology Progress Series, 350,* 267–276.

Kaiser, M.J., & Moore, P.G. (1999) Obligate marine scavengers: Do they exist? *Journal of Natural History, 33*, 475–481.

Kaiser, M.J., Hughes, R.N., Gibson, R.N. (1993). Factors affecting diet selection in the shore crab, *Carcinus maenus* (L.). *Animal Behaviour, 45*, 83–92.

Kamio, M., & Derby, C.D. (2017). Finding food: how marine invertebrates use chemical cues to track and select food. *Natural Product Reports, 34*, 514–528.

Kane, A., Healy, K., Guillerme, T., et al. (2017). A recipe for scavenging in vertebrates – the natural history of a behaviour. *Ecography, 40,* 324–334.

Kaplan, I.C., Holland, D.S., Fulton, E.A. (2014). Finding the accelerator and brake in an individual quota fishery: Linking ecology, economics, and fleet dynamics of US West Coast trawl fisheries. *ICES Journal of Marine Science, 71*, 308–319.

Kelleher, K.(2005). Discards in the worlds marine fisheries. An update. FAO Fisheries Technical Paper 470. 131 pp.

Kohn, A.J. (1961). Chemoreception in Gastropod Molluscs. *American Zoologist, 1*, 291–308.

Laidre, M.E., & Elwood, R.W. (2008). Motivation matters: Cheliped extension displays in the hermit crab, Pagurus bernhardus, are honest signals of hunger. *Animal Behaviour, 75*, 2041–2047.

Lauria, V. (2012). Impacts of climate change and fisheries on the Celtic Sea ecosystem. In *Faculty of marine science and engineering* (p. 249). Plymouth: University of Plymouth.

Lee, C.G., Huettel, M., Hong, J.S., et al. (2004). Carrion-feeding on the sediment surface at nocturnal low tides by the polychaete *Phyllodoce mucosa*. *Marine Biology, 145,* 575–583.

Lesser, M. P, Martini, F.H., Heiser, J.B. (1997). Ecology of the hagfish, *Myxine glutinosa* L. in the Gulf of Maine I. Metabolic rates and energetics. *Journal of Experimental Marine Biology and Ecology, 208,* 215–225.

Levin, S.A. (1992). The problem of pattern and scale in ecology: The Robert H. MacArthur Award Lecture. *Ecology, 73,* 1943–1967.

Libralato, S., Caccin, A., Pranovi, F. (2015). Modeling species invasions using thermal and trophic niche dynamics under climate change. *Frontiers in Marine Science, 2,* 29.

Link, J.S., & Almeida, F.P. (2002). Opportunistic feeding of longhorn sculpin (*Myoxocephalus octodecemspinosus*): Are scallop fishery discards an important food subsidy for scavengers on Georges Bank? *Fishery Bulletin, 100,* 381–385.

Løkkeborg, S. (1998). Feeding behaviour of cod, *Gadus morhua*: Activity rhythm and chemically mediated food search. *Animal Behaviour, 56,* 371–378.

Løkkeborg, S., & Fernö, A. (1999). Diel activity pattern and food search behaviour in cod, *Gadus morhua*. *Environmental Biology of Fishes, 54,* 345–353.

Løkkeborg, S., Skajaa, K., Fernö, A. (2000). Food-search strategy in ling (*Molva molva* L.): Crepuscular activity and use of space. *Journal of Experimental Marine Biology and Ecology, 247,* 195–208.

Løkkeborg, S., Fernö, A, Humborstad, O.B. (2010). Fish Behavior in Relation to Longlines. In P. He (Ed.), *Behavior of Marine Fishes: Capture processes and conservation challenges* (pp. 105–141). Oxford: Wiley-Blackwell.

Løkkeborg, S., Siikavuopio, S. I, Humborstad, O.B., et al. (2014). Towards more efficient longline fisheries: Fish feeding behaviour, bait characteristics and development of alternative baits. *Reviews in Fish Biology and Fisheries, 24,* 985–1003.

Lucifora, L.O., García, V.B., Menni, R.C. et al (2009). Effects of body size, age and maturity stage on diet in a large shark: Ecological and applied implications. *Ecological Research, 24*: 109–118.

Luque, P.L., Davis, C.G., Reid, D.G., et al.. (2006). Opportunistic sightings of killer whales from Scottish pelagic trawlers fishing for mackerel and herring off North Scotland (UK) between 2000 and 2006. *Aquatic Living Resources, 19,* 403–410.

Mackinson, S., & Daskalov, G. (2007). An ecosystem model of the North Sea to support an ecosystem approach to fisheries management: description and parameterisation. ICES Document Sci. Ser. Tech Rep., Cefas Lowestoft, 142. 196 pp.

Mackinson, S., Platts, M., Garcia, C., et al. (2018). Evaluating the fishery and ecological consequences of the proposed North Sea multi-annual plan. *PLoS ONE, 13,* e0190015.

Martinez, I., Jones, E.G., Davie, S.L., et al. (2011). Variability in behaviour of four fish species attracted to baited underwater cameras in the North Sea. *Hydrobiologia, 670,* 23.

McKillup, S.C., & McKillup, R.V. (1994). The decision to feed by a scavenger in relation to the risks of predation and starvation. *Oecologia, 97,* 41–48.

Metcalfe, N.B., Van Leeuwen, T.E., Killen, S.S. (2016a). Does individual variation in metabolic phenotype predict fish behaviour and performance? *Journal of Fish Biology, 88,* 298–321.

Metcalfe, J.D., Wright, S., Tudorache, C., et al. (2016b). Recent advances in telemetry for estimating the energy metabolism of wild fishes. *Journal of Fish Biology, 88,* 284–297.

Moleón, M., Sánchez-Zapata, J.A., Selva, N., et al. (2014). Inter-specific interactions linking predation and scavenging in terrestrial vertebrate assemblages. *Biological Reviews, 89,* 1042–1054.

Möllmann, C., Folke, C., Edwards, M., et al. (2015). Marine regime shifts around the globe: Theory, drivers and impacts. *Philosophical Transactions of the Royal Society of London. Series B, Biological Sciences, 370,* 20130260.

Moore, P.G., & Howarth, J. (1996). Foraging by marine scavengers: Effects of relatedness, bait damage and hunger *Journal of Sea Research, 36,* 267–273.

Moore, P.G., & Wong, Y.M. (1995). *Orchomene nanus* (Krøyer) (Amphipoda: Lysianassoidea), a selective scavenger of dead crabs: Feeding preferences in the field. *Journal of Experimental Marine Biology and Ecology, 192,* 35–45.

Morato, T., Lemey, E., Menezes, G., et al. (2016). Food-web and ecosystem structure of the open-ocean and deep-sea environments of the Azores, NE Atlantic. *Frontiers in Marine Science, 3*, 245.

Morton, B., & Jones, D.S. (2003). The dietary preferences of a suite of carrion-scavenging gastropods (Nassariidae, Buccinidae) in Princess royal harbour, Albany, Western Australia. *Journal of Molluscan Studies, 69,* 151–156.

Moullec, F., Gascuel, D., Bentorcha, K., et al. (2017) Trophic models: What do we learn about Celtic Sea and Bay of Biscay ecosystems? *Journal of Marine Systems, 172*: 104–117.

Navarro, J., Cardador, L., Fernández, Á.M., et al. (2016) Differences in the relative roles of environment, prey availability and human activity in the spatial distribution of two marine mesopredators living in highly exploited ecosystems. *Journal of Biogeography, 43*, 440–450.

Nickell, L. A, & Atkinson, R.J.A. (1995). Functional morphology of burrows and trophic modes of three thalassinidean shrimp species, and a new approach to the classification of thalassinidean burrow morphology. *Marine Ecology Progress Series, 128*, 181–197.

Nordström, M.C., Aarnio, K., & Törnroos, A. et al. (2015). Nestedness of trophic links and biological traits in a marine food web. *Ecosphere*, 6, art161.

Nowlin, W.H., Vanni, M.J., Yang, L.H. (2008). Comparing resource pulses in aquatic and terrestrial ecosystems. *Ecology, 89*, 647–659.

Oro, D., Genovart, M., Tavecchia, G., et al. (2013). Ecological and evolutionary implications of food subsidies from humans. *Ecology Letters, 16*, 1501–1514.

Paul, V.J., Ritson-Williams, R., Sharp, K. (2011). Marine chemical ecology in benthic environments. *Natural Product Reports, 28*, 345–387.

Pikitch, E.K., Santora, C., Babcock, E.A., et al. (2004) Ecosystem-based fishery management. *Science, 305*, 346–347.

Pita, P., & Freire, J. (2011). Movements of three large coastal predatory fishes in the northeast Atlantic: A preliminary telemetry study. *Scientia Marina, 2011*(75), 12.

Pittman, S.J., & McAlpine, C.A. (2003). Movements of marine fish and decapod crustaceans: Process, theory and application. In A.J. Southward, P.A. Tyler, C.M. Young, L.A. Fuiman (Eds.), *Advances in Marine Biology* (Vol 44, pp. 205–294).

Polačik, M., Jurajda, P., Blažek, R., et al. (2015). Carcass feeding as a cryptic foraging mode in round goby *Neogobius melanostomus*. *Journal of Fish Biology, 87*, 194–199.

Preston, S.J, Revie, I.C, Orr, J.F, et al. (1996). A comparison of the strengths of gastropod shells with forces generated by potential crab predators. *Journal of Zoology, 238,* 181–193.

Puglisi, M.P., Sneed, J.M., Sharp, K.H., et al. (2014). Marine chemical ecology in benthic environments. *Natural Product Reports, 31*, 1510–1553.

Quaggiotto, M.M., Burke, L.R., McCafferty, D.J., et al. (2016). First investigations of the consumption of seal carcasses by terrestrial and marine scavengers. *Glasgow Naturalist, 26*, 33–52.

Ruxton, G.D., & Bailey, D.M. (2005). Searching speeds and the energetic feasibility of an obligate whale-scavenging fish. *Deep Sea Research Part I: Oceanographic Research Papers, 52*, 1536–1541.

Ruxton, G.D., & Houston, D.C. (2004). Energetic feasibility of an obligate marine scavenger. *Marine Ecology Progress Series, 266,* 59–63.

Ruxton, G.D., Wilkinson, D.M., Schaefer, H.M., et al.. (2014). Why fruit rots: Theoretical support for Janzen's theory of microbe-macrobe competition. *Proceedings of the Royal Society B 281*, 20133320.

Ryer, C.H., & Olla, B.L. (1992). Social mechanisms facilitating exploitation of spatially variable ephemeral food patches in a pelagic marine fish. *Animal Behaviour 44,* 69–74.

Sainte-Marie, B. (1986). Feeding and swimming of lysianassid amphipods in a shallow cold-water bay. *Marine Biology, 91*, 219–229.

Sainte-Marie, B. (1992). Foraging of scavenging deep-sea lysianassoid amphipods. In G.T.P.V. Rowe (Ed.) *Deep-sea food chains and the global carbon cycle.* Berlin/Heidelberg/New York: Springer.

Sainte-Marie, B., & Hargrave, B.T. (1987). Estimation of scavenger abundance and distance of attraction to bait. *Marine Biology, 94*, 431–443.

Sánchez, F., & Olaso, I. (2004). Effects of fisheries on the Cantabrian Sea shelf ecosystem. In *Ecological modelling placing fisheries in their ecosystem context* (pp. 151–174).

Scharf, F.S., Juanes, F., Rountree, R.A. (2000). Predator size – prey size relationships of marine fish predators: Interspecific variation and effects of ontogeny and body size on trophic-niche breadth. *Marine Ecology Progress Series, 208*, 229–248.

Schlacher, T.A, Strydom, S., Connolly, R.M. (2013). Multiple scavengers respond rapidly to pulsed carrion resources at the land–ocean interface. *Acta Oecologica, 48*, 7–12.

Seefeldt, M.A., Campana, G.L., Deregibus, D., et al. (2017). Different feeding strategies in Antarctic scavenging amphipods and their implications for colonisation success in times of retreating glaciers. *Frontiers in Zoology, 14*, 59.

Seibel, B.A., & Drazen, J.C. (2007). The rate of metabolism in marine animals: Eanvironmental constraints, ecological demands and energetic opportunities. *Philosophical Transactions of the Royal Society B: Biological Sciences, 362*, 2061–2078.

Sims, D.W., Southall, E.J., Humphries, N.E., et al. (2008). Scaling laws of marine predator search behaviour. *Nature, 451,* 1098.

Sims, D.W., Humphries, N.E., Bradford, R.W. et al. (2012). Lévy flight and Brownian search patterns of a free-ranging predator reflect different prey field characteristics. *Journal of Animal Ecology, 81*, 432–442.

Sinopoli, M., Fanelli, E., D'Anna, G., et al. (2012). Assessing the effects of a trawling ban on diet and trophic level of hake, *Merluccius merluccius*, in the southern Tyrrhenian Sea. *Scientia Marina, 76,* 677–690.

Smith, C.R., & Baco, A.R. (2003). Ecology of whale falls at the deep-sea floor. In R.N. Gibson, R.J.A. Atkinson (Eds.), *Oceanography and marine biology*, (Vol. 41, pp. 311–354).

Smith, K.L, & Baldwin, R.J. (1982). Scavenging deep-sea amphipods: Effects of food odor on oxygen consumption and a proposed metabolic strategy. *Marine Biology, 68*, 287–298.

Stoner, A.W. (2004). Effects of environmental variables on fish feeding ecology: Implications for the performance of baited fishing gear and stock assessment. *Journal of Fish Biology, 65*, 1445–1471.

Svendsen, J.C. (2018). *Cod death roll.* https://www.youtube.com/watch?v=CE_lmtIX3Bg. Accessed 20 June 2018.

Tamburri, M.N., & Barry, J.P. (1999). Adaptations for scavenging by three diverse bathyla species, *Eptatretus stouti*, *Neptunea amianta* and *Orchomene obtusus*. *Deep Sea Research Part I: Oceanographic Research Papers, 46*, 2079–2093.

Tanner, C.J., Salah, G.L., Jackson, A.L. (2011). Feeding and non-feeding aggression can be induced in invasive shore crabs by altering food distribution. *Behavioral Ecology and Sociobiology, 65*, 249–256.

Thode, A., Straley, J., Tiemann, C.O., et al. (2007). Observations of potential acoustic cues that attract sperm whales to longline fishing in the Gulf of Alaska. *The Journal of the Acoustical Society of America 122*, 1265–1277.

Uhlmann, S.S., Coers, A., van Helmond, A.T.M., et al. (2013). Discard sampling of Dutch bottom-trawl and seine fisheries in 2012. CVO Report 13.015. 76pp.

Vahl, O. (1984). The relationship between specific dynamic action (SDA) and growth in the common starfish, *Asterias rubens L. Oecologia, 61*, 122–125.

van Weerden, J.F., Reid, D.A.P., Hemelrijk, C.K. (2014). A meta-analysis of steady undulatory swimming. *Fish and Fisheries 15*(3), 397–409.

Verheggen, F., Perrault, K.A., Megido, R.C., et al. (2017). The odor of death: An overview of current knowledge on characterization and applications. *Bioscience, 67*, 600–613.

Videler, J.J, & He, P. (2010). Swimming in marine fish. In P. He (Ed.), *Behavior of marine fishes: Capture processes and Conservation challenge* (pp. 3–24). New York: Wiley-Blackwell.

Villegas-Ríos, D., Réale, D., Freitas, C., et al. (2018). Personalities influence spatial responses to environmental fluctuations in wild fish. *The Journal of Animal Ecology*. https://doi.org/10.1111/1365-2656.12872

Votier, S.C., Furness, R.W., Bearhop, S., et al. (2004). Changes in fisheries discard rates and seabird communities. *Nature*, *427*, 727–730.

Ward, A.J.W., Webster, M.M, Hart, P.J.B. (2006). Intraspecific food competition in fishes. *Fish and Fisheries, 7,* 231–261.

Whitehead, H., & Reeves, R. (2005). Killer whales and whaling: The scavenging hypothesis. *Biology Letters, 1*, 415–418.

Wilkie, M.P, Clifford, A.M., Edwards, S.L, et al. (2017). Wide scope for ammonia and urea excretion in foraging Pacific hagfish. *Marine Biology, 164*, 126.

Williams, T.M. (1999). The evolution of cost efficient swimming in marine mammals: limits to energetic optimization. *Philosophical Transactions of the Royal Society of London. Series B: Biological Sciences, 354*, 193–201.

Wilson, E.E., & Wolkovich, E.M. (2011). Scavenging: How carnivores and carrion structure communities. *Trends in Ecology & Evolution, 26*, 129–135.

Yamamura, O. (1997). Scavenging on discarded saury by demersal fishes off Sendai Bay, northern Japan. *Journal of Fish Biology, 50,* 919–925.

Yang, L.H., Edwards, K.F., Byrnes, J.E., et al. (2010). A meta-analysis of resource pulse-consumer interactions. *Ecological Monographs, 80*, 125–151.

Yeh, J., & Drazen, J.C. (2011). Baited-camera observations of deep-sea megafaunal scavenger ecology on the California slope. *Marine Ecology Progress Series, 424*, 145–156.

Zeller, D., Cashion, T., Palomares, M., et al.. (2018). Global marine fisheries discards: A synthesis of reconstructed data. *Fish and Fisheries, 19*: 30–39.

Zimmer-Faust, R.K. (1987). Crustacean chemical perception: Towards a theory on optimal chemoreception. *The Biological Bulletin, 172,* 10–29.

Zintzen, V., Roberts, C.D., Anderson, M.J., et al. (2011). Hagfish predatory behaviour and slime defence mechanism. *Scientific Reports, 1*, 131.

Part III
Cultural, Institutional and Multi-Jurisdictional Challenges

Chapter 8
How the Implementation of the Landing Obligation Was Weakened

Björn Stockhausen

Abstract This chapter covers the development of provisions and exemptions to the Landing Obligation in the years following the adoption of the Common Fisheries Policy in December 2013. It focuses on the processes leading to certain changes in Article 15, the development of discard plans, and describes reasons for the slow implementation of the Landing Obligation. It provides further insight into why the intention of the objective of the discard ban, the reduction of unwanted catches, has not yet been achieved to its maximum possible extent.

Keywords Data collection · Discard plans · High survival · Implementation · Omnibus Regulation

8.1 Introduction

Prior to the reform of the Common Fisheries Policy (CFP), European policy makers were aware of the issue of discarding fish and considered it to wasting a natural resource. The European Commission described ways forward in a green paper (European Commission 2009) and outlined provisions to improve that situation in its final legislative proposal in 2011 (European Commission 2011).

This initial proposal, while it contained already a gradual phasing-in of the new rules per fish stock, was substantially changed by both co-legislators, the European Parliament and the European Council of Ministers (hereafter referred to as "Council"). Amongst those changes were a delayed timeline for the introduction of the new Landing Obligation, a range of exemptions based on scientific evaluations, and the possibility for Member States to jointly propose, on a regional basis, so-called discard plans, comprising details of the implementation of the Landing Obligation for respective sea basins.

A political agreement between the European Parliament, the Council and the Commission was found in May 2013 at the end of the trilogue where all three entities

B. Stockhausen (✉)
Seas At Risk, Brussels, Belgium

S. S. Uhlmann et al. (eds.), *The European Landing Obligation*,
https://doi.org/10.1007/978-3-030-03308-8_8

negotiated a compromise between their respective positions they had adopted before. After the publication of the basic regulation of the Common Fisheries Policy, the new legal text and its provisions came into force on January 1st, 2014.

However, the political developments and discussions in the following years after the legal text was adopted changed the impact on fisheries management again, and in many cases weakened the positive impact on selectivity and fisheries management a full implementation might have had.

This chapter covers the development of respective provisions in the Landing Obligation (CFP Article 15) in the years following its legal adoption. It focuses on the processes leading to certain changes in Article 15, the development of discard plans and describes reasons for the slow implementation of the Landing Obligation. It provides further insight into why the intention of the objective of the discard ban, the reduction of unwanted catches, has not yet been achieved to the maximum possible extent.

8.2 The Omnibus Regulation 2013–2015

One week before Christmas' Eve 2013, and two weeks before the adopted basic regulation of the reformed CFP came into force, the Commission released a proposal for the so-called 'Omnibus regulation' (2013/0436 [COD]).

The reason for this proposal was that the new reformed CFP contained several key objectives and provisions that meant that existing legislation was no longer in line with the new policy framework. In particular, the phasing in of the Landing Obligation from January 1st, 2015 onwards meant that there was not enough time to completely revise other framework legislation containing detailed rules that made the application of the new paradigm regarding discards difficult. This is exemplified by the Commission stating that "*legislation is required to remove any legal and practical impediments to implementation on a transitional basis while this new framework is being developed*" (European Commission 2013).

In its proposal for the Omnibus regulation, the Commission therefore proposed amendments to Council regulations (EC) No 850/98 ('Technical Measures Regulation), (EC) No 1224/2009 (the 'Control Regulation'), and other associated regulations. Provisions within the Technical Measures Regulation such as the Minimum Landing Size (MLS) and catch composition rules meant challenges for fishers, as they would stand in contradiction to the Landing Obligation that foresaw that over time unwanted catches should be avoided and reduced as far as possible. From a control perspective the implementation of the Landing Obligation would mean a shift from monitoring and controlling of landings to catches, this meant that, amongst others, the respective rules on data recording in the Control Regulation would have to be adapted.

Apart from the content of the proposal, the timing of the adoption of the reformed CFP played a key role. The political process was initially hampered by the fact that the Commission only published its proposal in December 2013, while a political agreement between the co-legislators on the reformed CFP had been found in May 2013 and the final agreed text existed since October 2013. Further, despite the approaching entry into force of the first phase of the Landing Obligation by January 1st, 2015, the co-legislators were not able to agree on their position on the Omnibus regulation proposal and subsequently on a joint compromise that would have allowed amending the existing regulations in time, before January 2015.

While the Council agreed on its position regarding the Commission proposal on the Omnibus Regulation in June 2014, the Fisheries Committee of the European Parliament was facing upcoming European Elections in May 2014. The Fisheries Committee would only resume its meetings in July – after the election. This implied that the composition of members of parliament could change and potentially further delay discussions. In January 2014, the report on the legislative file was attributed (European Parliament 2014a) to the Fisheries Committee president who would table a working document. In February (European Parliament 2014b), the Commission was invited to present the proposal, while a foreseen exchange of views on the dossier in March was cancelled (European Parliament 2014c). For the April committee meeting, a hearing (European Parliament 2014d) had been organised as agreed in February to provide insights to members on the Landing Obligation on both technical measures and control aspects.

It was therefore unfortunately not possible for the Fisheries Committee to discuss the proposal further prior to the end of its legislative term and to facilitate and accelerate the legislative procedure towards an earlier adoption in 2014 than ultimately achieved in 2015.

After the European Elections in May 2014, the new members of the Fisheries Committee reconvened in July. A new Chair of the Fisheries Committee was appointed, who continued discussions based on previous agreements. For the July 2014 meeting, a working document had been prepared summarising the presentations and discussion of the April hearing including additional studies by the European Parliament Policy Department B (European Parliament 2014e). As the Fisheries Committee had been partially reshuffled and new members had to become familiar with the file, it took still until December for the Fisheries Committee to adopt its position. It was only in January 2015 that a political agreement with the Council was reached. Following approval by Council and Parliament plenary, the final legislative act came into force in June 2015 – 6 months after the Landing Obligation had been introduced for the first fisheries (European Union 2015).

At this point, the content of the Omnibus proposal had changed significantly. What was originally intended as a regulation to align existing legislation with the new CFP and in particular to facilitate the implementation of the Landing Obligation, had in some respects weakened the recently agreed CFP.

Amongst the changes was, for example, an additional exemption to the Landing Obligation. The originally adopted Article 15 of the CFP foresaw exemptions from the Landing Obligation such as for high survivability or for *de minimis* in case that increases in selectivity are difficult to achieve or to avoid disproportionate costs for handling the previously discarded fish (Rihan et al., this volume). This means that as a last resort, when despite other efforts to reduce the amount of unwanted catches such as swaps between Member States and temporal and spatial measures, there still remains a certain quantity of unwanted fish, a *de minimis* exemption can be granted based on scientific evaluations, to provide for an outlet and allow for the discarding of those fish. Similar for the exemption for high survivability: In case that scientific studies clearly show that certain species survive to a certain extent, the decision maker can grant the exemption for this species from the obligation to land it.

The Omnibus regulation amended the respective paragraph by adding an exemption for "*fish which shows damage caused by predators*", without specifying details of what damage means or how it would be controlled. This amendment was only inserted after the vote in the European Parliament Fisheries Committee during the trilogue negotiations, which meant there were fewer opportunities to discuss its implications or to decide on required provisions to guarantee the control of the new exemption.

On control and enforcement, the Commission's initial proposal for the Omnibus regulation foresaw that violating the Landing Obligation would constitute a serious infringement under the current Control Regulation (EC) 1224/2009. The proposal was amended by the co-legislators to include a delay of 2 years, meaning that one could violate the rules of the Landing Obligation for 2 additional years during which this would not constitute a serious infringement from the Landing Obligation.

Further, the Commission had proposed to delete the threshold of 50kg of species caught and retained on board for compulsory recording in the logbook. This became necessary for the reason that Article 15.1 of the CFP required that all catches of species that are subject to catch limits (or that are subject to minimum sizes in the Mediterranean) "*shall be... recorded*" and was therefore an essential part of the legal alignment of new and old legislation. Especially, as with the CFP reform, the management shifted from landings to catches, the monitoring and control of all quantities of the latter was fundamental. Nevertheless, Parliament and Council agreed to reinstate the threshold in their political agreement leading to the final adopted text, maintaining a 50 kg threshold, potentially leaving, if added up across the EU, large quantities of removals unaccounted for.

Lastly, amendments were introduced by the Parliament which required the Commission to annually report on the implementation of the Landing Obligation, based on information provided from Member States and Advisory Councils from 2016 onwards. These reporting requirements were subsequently specified by the Scientific, Technical and Economic Committee for Fisheries, STECF (STECF 2016b; see Rihan et al., this volume). These reports have provided useful insight into the process and progress of the implementation of the Landing Obligation.

In summary, despite the pressing timeframe, the co-legislators did not manage to adopt the required legal changes in the Omnibus Regulation before the first phase of the Landing Obligation came into force in January 2015. Further, the Omnibus Regulation introduced amendments to the recently reformed CFP and to some key aspects of the Landing Obligation which weakened the overarching policy that was agreed only 18 months earlier.

In summary, one of the elements that weakened the Landing Obligation policy was the Omnibus regulation (EU 2015/812) which was supposed to legally align existing legislation such as the 'Fisheries Control Regulation' (EC 1224/2009) or the 'Technical Measures Regulation' (EC 850/98). It weakened the recently reformed basic regulation of the CFP by adding additional exemptions to the Landing Obligation in Article 15.

8.3 High Survival – A Concept Undefined by Decision Makers

Amongst the exemptions from the Landing Obligation was Article 15.4(b), stating that the Landing Obligation shall not apply to "*species for which scientific evidence demonstrates high survival rates, taking into account the characteristics of the gear, of the fishing practices and of the ecosystem*".

Unfortunately, the term 'high survival' was not further described or clarified by the co-legislators in the whole legislative document. The exemption was introduced by both the Parliament and the Council in their report (European Parliament 2013) and general approach (Council of the European Union 2013), respectively. Following the political agreement on the reformed CFP, the STECF was asked in September 2013 for scientific support, and the request by the European Commission already mentioned that "*There is currently no objective means to define 'high survival rates'*". The STECF was asked to (i) develop guidelines or identification of best practice for undertaking discard survival studies, (ii) develop an objective framework to define high survival which will provide managers with a range of the likely impacts of different options depending on the definition used; (iii) assess the impacts if a proportion of the landed catch that would have been discarded might otherwise have survived and how this may affect estimates of fishing mortality, spawning stock biomass (SSB) and associated reference points; and (iv) if possible define a predefined list of species and fisheries that could be considered for exemption on the basis of high survival (STECF 2013).

Following this initial request, a series of STECF meetings provided recommendations on the different requests. Already the first report of the September 2013 meeting, the STECF Expert Working Group EWG 13-16, stated "*that the term 'high survival' is somewhat subjective*", and that "*defining a single value* [of survival] *cannot be scientifically rationalised and therefore EWG 13-16 advises that assessing proposed exemptions on the basis of 'high survival' need to be considered on a case-by-case basis taking* [into] *account the specificities of the species and fisheries involved*". This important remark was re-iterated in the following years by stating that "*without clear definitions of the terms, [...] 'high survival', there are no objective scientific criteria to judge whether any proposed exemptions from the Landing Obligation are merited.*" The STECF therefore assessed the scientific basis for the different exemptions requested, but referred the judgement "*whether such proposals are merited using relevant subjective criteria*" back to the decision-makers, in this case the European Commission – as being responsible for adopting specific discard plans through delegated acts (Rihan et al., this volume).

Over subsequent years the STECF provided much needed clarity and scientific support to facilitate the implementation of the Landing Obligation. For example, apart from developing guidelines for survivability studies, the STECF EWG 13–17 recommended guidelines for the regional discard management plans "*where an exemption under high survival is being proposed*". In relation to high survival exemptions, the EWG recommended specific information that should be included, e.g. species name, respected stock(s) and management unit, descriptions of gear types used, catch composition and other operational characteristics, the discard profile, selectivity measures developed and effects of the Landing Obligation on the stocks (STECF 2014a, b).

Nevertheless, when the STECF was requested to evaluate Joint Recommendations submitted to the Commission by respective regional Member States groups in the following years, these recommendations seem not to have been taken on board by Member States.

The STECF in 2015 reports of "*inconsistencies in the definition of the fleets*", "*it is unclear … to which fleets such de minimis volumes will be accessible*" or that exemptions "*appear to be intended to cover residual discards and as such essentially equate to 'business as usual*'" (STECF 2015b). In many cases, additional supporting information had to be requested from Member States after the submission of the Joint Recommendations to improve the justification provided. Still, the STECF noted in several cases that the additional information provided referred only to single Member States rather than corresponding to the fishery for which the exemption was requested.

This situation continued in 2016 when the STECF identified "*a number of general issues and limitations*" in the Joint Recommendations (STECF 2016a). Even after additional responses were received from Member States and regional groups, information was often insufficient or lacking. "*None of the JR [Joint Recommendations] received contain any concrete measures for the documentation of catches*", which was a violation of the requirements of Article 15.1 CFP (see previous part of this chapter).

Since 2015, discard plans have been adopted for all major sea basins of the European Union, i.e., the Baltic Sea, the North Sea, North- and South-Western Waters, the Mediterranean and the Black Sea. By 2017, Joint Recommendations had generally improved, but still contained "*limitations in the information provided*" and "*often information is provided for one fleet but not for other fleets using similar gears and which would be also affected*", as noted by the STECF (STECF 2017b). The number of species and fisheries included in the discard plans and therefore subject to the Landing Obligation, has increased only slowly. Until June 2017, from a total number 174 stocks subject to the Landing Obligation, there were still 77 stocks not covered at all while 97 were either fully or partially included in discard plans.

The legal wording adopted by policy makers on the Landing Obligation put a lot of burden on the STECF that was requested to evaluate the Joint Recommendations submitted by regional Member State groups (Rihan et al., this volume). The STECF had, from the very beginning, provided guidelines on the information necessary to conduct a satisfactory evaluation, which provided Member States with the tools and information for a gradual implementation. Despite the ambitious timeframe, Member States could have coordinated data collection in a better way than eventually happened, especially as every year the STECF provided information on where information was missing and how the justifications should be improved which could have been used for the JR of the next year. Regarding the justification provided, the Commission had to send throughout the years additional requests to Members States after receiving the respective Joint Recommendations, and even in these cases, Member States did not always provide sufficient information. Finally, the number of stocks introduced to the discards plans was increased only at a slow pace, which will mean that in the final year of the implementation of the Landing Obligation, in 2019, a large number of remaining stocks still have to be implemented into the Landing Obligation through the respective discard plans.

In summary, the adopted wording in Article 15 of the Common Fisheries Policy regarding 'high survival' had not been defined in the legal text. As it is an important threshold for whether or not a species can be exempted from the Landing Obligation, its application had to be discussed. This section reviewed how this was clarified by respective scientific bodies, highlighting how these became even more entangled in the political process.

8.4 The Drafting of Discard Plans

Article 15.6 of the CFP foresees that so-called discard plans containing the specifications for the Landing Obligation in a specific sea basin shall be adopted through delegated acts by the European Commission, following the legislative procedure outlined in Article 18 of the CFP. Article 18 foresees that in line with the regionalisation concept, Member States with a direct management interest can submit Joint Recommendations to achieve the respective objective of the conservation measure, in this case the Landing Obligation.

In this context, groups of Member States have submitted since 2014 so-called Joint Recommendations for the respective sea basins, outlining amongst others the fisheries to include in the discard plans, closures, and their requests for exemptions from the Landing Obligation and including evidence to justify these exemptions.

As the quality of the justifications provided should be used by the European Commission to determine whether exemptions are granted subsequently, it should be obvious that Member States apply their greatest efforts to provide complete information based on the best available scientific advice, and that subsequently the

European Commission executes the greatest scrutiny when deciding on whether to grant exemptions (or not) to ensure that the corresponding objective of the CFP, the avoidance and reduction of unwanted catches, is not unnecessarily and unjustifiably weakened. Member States also should have a strong interest in providing well-founded Joint Recommendations because the requested exemptions ultimately lead to facilitating the implementation of the Landing Obligation for their fleets and fishers.

As outlined above, these joint recommendations are evaluated by the STECF by scrutinising the supporting information and providing corresponding recommendations to the European Commission (Rihan et al., this volume). And as documented in several STECF reports, the underlying data and studies provided were often insufficient for various reasons (e.g., limited research funding for a well-replicated study). The question to ask is whether the European Commission has followed the scientific recommendations. For example, in cases where the STECF indicated that the justification provided for certain exemptions had been insufficient, why did the European Commission adopt the discard plans including the respective exemption?

As outlined in the previous section, in several cases the initial information provided by Member States was insufficient and the Commission had to request additional information after the meeting of the respective Expert Working Group, putting additional burden on the subsequent STECF plenary meeting. In addition to that, less time could be spent analysing the new information, compared to a dedicated stand-alone EWG.

It was also clear that the timing of the subsequent scientific evaluation and legislative process required an early submission of the Joint Recommendations to allow for the discard plans to come into force on January 1st of the following year. The submission date had been established by the European Commission to be the end of May every year for the Joint Recommendations of the following year to allow for sufficient time for both processes. This date was also often not respected and additional information was sent to the EWG only days before, or in some cases even during the meeting itself, which further impeded the necessary analysis. Therefore, it would have been completely justified to not give Member States the additional possibility to provide information after the deadline or after the EWG meeting. Even in 2017, three years after the process of providing supporting information to justify exemptions had initiated and despite guidelines and templates provided by the STECF, the European Commission still found it necessary to request additional information from Member States.

Additionally, the final adopted legislative act for a discard plan often granted Member States the possibility to provide information until May of the following year to be analysed by the STECF at the following July plenary meeting. See for example the 2016 discard plans with respect to certain fisheries of hake in south-western waters (European Commission 2015a) and whiting in north-western waters (European Commission 2015b); the 2017 discard plans with respect to certain fisheries of hake in south-western waters (European Commission 2016a) and sole in north-western waters (European Commission 2016b), and the 2018 discard plans with respect to certain fisheries of sole in north-western waters (European Commission 2018).

Looking into one of these examples in detail, it is important to note that for hake in south-western waters, additional information was requested twice in subsequent years and that both requests were for the same fisheries. The report of the STECF expert group in 2015 stated that that no relevant information as well as limited and non-quantitative information were presented (STECF 2015b). The following STECF plenary (STECF 2015a) meeting in July 2015 analysed additional information provided by the respective Member States upon request of the European Commission following the process described above, and the report still stated that the additional supporting information is still "*rather generic*" and that "*further selectivity studies should be carried out*". Nevertheless, the adopted discard plan for 2016 granted the exemption including the obligation to provide additional information. In 2016, the next STECF expert group (STECF 2016a) was requested to evaluate whether the required additional information was provided. The STECF concluded that the information provided related to only one Member State involved in the fishery and that no specific information for other Member State fleets was provided that were (in 2016) and would be (in 2017) subject to an exemption. The report continued stating that unclear catch information was supplied and that the Joint Recommendation was missing the level of *de minimis*. The report concludes that "*it is still not currently possible to evaluate whether the arguments of disproportionate costs are well founded*". Again, the subsequent STECF summer plenary had to evaluate the additional information provided by Member States upon request by the European Commission, and the report again stated that little additional information was provided and that additional selectivity information provided does not contain any additional evidence. In the same procedure as the previous year, the exemption was granted in the discard plan for 2017 as outlined above, with the requirement to provide additional supporting information.

This means, that by the end of 2017, for 2 years, a fishery had been granted an exemption, despite repeated scientific analysis stated each time that inconsistencies and insufficient information prevailed, and despite that respective Member States had multiple chances to provide the missing information.

In hindsight, this example shows that the European Commission could have applied stricter scrutiny regarding the requirements for Member States to provide information that supported exemptions. Exemptions should have been adopted only when the required information was complete. The multiple STECF evaluations resulting from subsequent weakly supported exemptions not only placed an extra burden on the STECF, but might have given the impression to Member States that no serious efforts are necessary to obtain exemptions from the Landing Obligation.

In summary, the Commission seems not to have followed the evaluation by the STECF when transposing the JRs into discard plans, because justifications were often incomplete or information was not provided – despite additional requests or the possibility to provide information until the following year.

8.5 Data Collection and Reporting Requirements Under the Landing Obligation

The CFP states in Recital 46 that "*Fisheries management based on the best available scientific advice requires harmonised, reliable and accurate data sets*" and that Member States should collect data on discards under exemptions. With regard to the Landing Obligation, this is further specified in Article 15.1 that requires that all catches of species which are subject to catch limits (or that are subject to minimum sizes in the Mediterranean) "*shall be … recorded*". With regard to *de minimis* exemptions, it is further specified in Article 15.5(c) (ii) that even though these catches "*shall not be counted against the relevant quotas; however, all such catches shall be fully recorded*".

This shows that the co-legislators were fully aware of the importance of a comprehensive data collection system to provide the respective scientific entities such as the STECF and ICES with the best available data for scientific evaluation. The corresponding provisions to outline the data collection provisions, also referred in Article 15.5.d ("*provisions on documentation of catches*") were supposed to be included in multiannual plans or, where these were not adopted, in discard plans, as specified in Article 15.6.

As outlined previously, quite a number of discard plans have been adopted and amended since the phasing in of the Landing Obligation in 2015 (19 as of May 2018) – with no provisions for collecting and recording data. The Joint Recommendations submitted by regional groups of Member States to the European Commission have been thoroughly analysed by respective STECF expert groups and plenary meetings. Before the phasing in of the Landing Obligation, the STECF reported "*From a control perspective, the fact that catches discarded under the de minimis provisions do not count against quota, creates a significant risk of non-compliance around de minimis*" (STECF 2014b). Therefore, stronger legal requirements to fully record these catches would have been important, especially as the previously discarded amounts of fish are added as additional quota (so-called "top-ups") while deducting the *de minimis* amounts.

Looking at the analyses done by the STECF, this seems not to have been the case. The STECF recognised early in the process, in 2014, in one of the first Mediterranean Joint Recommendations, that while a rationale to collect data was mentioned, the document lacked indication how the data will be collected. In 2015, the assigned STECF EWG reported a general lack of relevant data associated for fleets and vessels falling under the Landing Obligation, "*which is necessary to estimate of their relative contribution to the overall catches of the stocks concerned and the potential volumes of de minimis catches that may be attributed/allocated to them*". The STECF added that "*that no specific provisions have been included in the JR's* [Joint Recommendations]" for the recording of *de minimis* catches and discards. In 2016, the STECF EWG reiterated the missing respective provisions in Joint Recommendations regarding the requirement to fully record *de minimis* catches following Article 15.5 (c) (ii) of the CFP (STECF 2016a), as did likewise the respective STECF EWG in 2017 (STECF 2017b). In addition, the 2017 STECF spring plenary notes that "*Member States also indicate a lack of reporting by vessel operators of fish discarded under exemptions (i.e., de minimis and high survivability), discards of fish currently not subject to the Landing Obligation and catches of fish below MCRS*" (STECF 2017a), which is iterated in 2018, adding the remark that "*Based on the Member States reports, the quantities of discards and unwanted catches being recorded in logbooks are extremely low and do not match information from observer trips or from last observed haul analyses carried out by inspectors. Inaccurate or incomplete catch data will compromise the provision of scientific advice*" (STECF 2018).

The consistency outlined in these reports regarding the monitoring requirement agreed to in the 2013 CFP reform shows that the implementation of the Landing Obligation has been suffering from a fundamental disrespect and mis-understanding for the importance of this key aspect. Despite the continuous and annual remarks by the respective scientific bodies, Member States seem to not have taken the requirement seriously, potentially because it might have shown that actual improvements in selectivity have not been implemented or achieved to the point necessary to facilitate the full implementation of the Landing Obligation. Better data collection is of utmost importance during this time of changing policies to accompany the political process. Also, the scientific assessments during this period have potentially suffered. The new CFP with its paradigm shift from requiring to land previously discarded fish, requires adjusting data collection to have the best available scientific knowledge on how many fish area actually returned to sea.

In summary, the change in the policy objective from the requirement to discard unwanted catches to the obligation to avoid and reduce as far as possible unwanted catches also came with the requirement to monitor and record the respective amounts of catches, both for legal and for stock assessment reasons.

Acknowledgments Disclaimer: The views and opinions expressed in this chapter are those of the author and do not reflect the official position of any employer.

References

Council of the European Union. (2013). Revised general approach on the proposal for a regulation of the European Parliament and of the Council on the Common Fisheries Policy 6108/1/13 REV1 http://register.consilium.europa.eu/doc/srv?l=EN&f=ST%206108%202013%20REV%201. Accessed 27 August 2018.

European Commission. (2009). Green paper on the reform of the common fisheries policy. COM (2009)163 final. https://eur-lex.europa.eu/LexUriServ/LexUriServ.do?uri=COM:2009:0163:FIN:EN:PDF. Accessed 27 August 2018.

European Commission. (2011). Proposal for a regulation of the European Parliament and of the Council on the common fisheries policy. COM(2011) 425 final. https://eur-lex.europa.eu/legal-content/EN/TXT/PDF/?uri=CELEX:52011PC0425&from=EN. Accessed 27 August 2018.

European Commission. (2013). Proposal for a REGULATION OF THE EUROPEAN PARLIAMENT AND OF THE COUNCIL amending Council Regulations (EC) No 850/98, (EC) No 2187/2005, (EC) No 1967/2006, (EC) No 1098/2007, No 254/2002, (EC) No 2347/2002 and (EC) No 1224/2009 and repealing (EC) No 1434/98 as regards the landing obligation. https://eur-lex.europa.eu/resource.html?uri=cellar:017cc877-67c5-11e3-a7e4-01aa75ed71a1.0004.01/DOC_1&format=PDF. Accessed 27 August 2018.

European Commission. (2015a). Commission Delegated Regulation (EU) 2015/2439 of 12 October 2015 establishing a discard plan for certain demersal fisheries in south-western waters. https://eur-lex.europa.eu/legal-content/EN/TXT/PDF/?uri=CELEX:32015R2439&from=EN. Accessed 27 August 2018.

European Commission. (2015b). Commission Delegated Regulation (EU) 2015/2438 of 12 October 2015 establishing a discard plan for certain demersal fisheries in north-western waters. https://eur-lex.europa.eu/legal-content/EN/TXT/PDF/?uri=CELEX:32015R2438&from=EN. Accessed 27 August 2018.

European Commission. (2016a). Commission Delegated Regulation (EU) 2016/2374 of 12 October 2016 establishing a discard plan for certain demersal fisheries in South-Western waters. https://eur-lex.europa.eu/legal-content/EN/TXT/PDF/?uri=CELEX:32016R2374&from=EN. Accessed 27 August 2018.

European Commission. (2016b). Commission Delegated Regulation (EU) 2016/2375 of 12 October 2016 establishing a discard plan for certain demersal fisheries in North-Western waters. https://eur-lex.europa.eu/legal-content/EN/TXT/PDF/?uri=CELEX:32016R2375&from=EN. Accessed 27 August 2018.

European Commission. (2018). Commission Delegated Regulation (EU) 2018/46 of 20 October 2017 establishing a discard plan for certain demersal and deep sea fisheries in North-Western waters for the year 2018. https://eur-lex.europa.eu/legal-content/EN/TXT/PDF/?uri=CELEX:32018R0046&from=EN. Accessed 27 August 2018.

European Parliament. (2013) European Parliament legislative resolution of 6 February 2013 on the proposal for a regulation of the European Parliament and of the Council on the Common Fisheries Policy (COM(2011)0425 – C7-0198/2011 – 2011/0195(COD)). P7_TA(2013)0040. http://www.europarl.europa.eu/sides/getDoc.do?pubRef=-//EP//NONSGML+TA+P7-TA-2013-0040+0+DOC+PDF+V0//EN. Accessed 27 August 2018.

European Parliament. (2014a). European Parliament Fisheries Committee January 2014, minutes. PECH_PV(2014)0122_1. http://www.europarl.europa.eu/meetdocs/2009_2014/documents/pech/pv/1015/1015544/1015544en.pdf. Accessed 27 August 2018.

European Parliament. (2014b). European Parliament Fisheries Committee February 2014, draft agenda. PECH(2014)0211_2. http://www.europarl.europa.eu/meetdocs/2009_2014/documents/pech/oj/1017/1017048/1017048en.pdf. Accessed 27 August 2018.

European Parliament. (2014c). European Parliament Fisheries Committee March 2014, minutes. PECH_PV(2014)0318_1 http://www.europarl.europa.eu/meetdocs/2009_2014/documents/pech/pv/1023/1023382/1023382en.pdf. Accessed 27 August 2018.

European Parliament. (2014d). European Parliament Fisheries Committee. Hearing on 7 April 2014 on "Implementing the discard ban". http://www.europarl.europa.eu/committees/en/pech/events-hearings.html?id=20140331CHE82053. Accessed 27 August 2018.

European Parliament. (2014e). European Parliament Fisheries Committee working document on the Committee hearing "Implementing the discard ban". PE536.149v01-00. http://www.europarl.europa.eu/sides/getDoc.do?pubRef=-%2f%2fEP%2f%2fNONSGML%2bCOMPARL%2bPE-536.149%2b01%2bDOC%2bPDF%2bV0%2f%2fEN. Accessed 27 August 2018.

European Union. (2015). Regulation (EU) 2015/812 of the European Parliament and of the Council of 20 May 2015 amending Council Regulations (EC) No 850/98, (EC) No 2187/2005, (EC) No 1967/2006, (EC) No 1098/2007, (EC) No 254/2002, (EC) No 2347/2002 and (EC) No 1224/2009, and Regulations (EU) No 1379/2013 and (EU) No 1380/2013 of the European Parliament and of the Council, as regards the landing obligation, and repealing Council Regulation (EC) No

1434/98. OJ L133, 29.5.2015. https://eur-lex.europa.eu/legal-content/EN/TXT/PDF/?uri=CELEX:32015R0812&from=EN. Accessed 27 August 2018.

Rihan, D., Uhlmann, S.S., Ulrich, C., Breen, M., Catchpole, T. (this volume). Requirements for documentation, data collection and scientific evaluations. In S.S. Uhlmann, C. Ulrich, S.J. Kennelly (Eds.), *The European Landing Obligation – Reducing discards in complex, multi-species and multi-jurisdictional fisheries*. Cham: Springer.

STECF. (2013). STECF EXPERT WORKING GROUP EWG 13-16 on a Landing Obligation in EU Fisheries, Terms of Reference. https://stecf.jrc.ec.europa.eu/c/document_library/get_file?uuid=f222e1e5-2f0c-420b-9adb-2fa6c25814df&groupId=43805. Accessed 27 August 2018.

STECF. (2014a). Scientific, Technical and Economic Committee for Fisheries (STECF) – Landing Obligation in EU Fisheries – part II (STECF-14-01). 2014. Publications Office of the European Union, Luxembourg, EUR 26551 EN, JRC 88869, 67 pp. https://stecf.jrc.ec.europa.eu/documents/43805/633247/STECF+14-01++Landing+obligations+in+EU+fisheries+-+p2.pdf. Accessed 27 August 2018.

STECF. (2014b). Scientific, Technical and Economic Committee for Fisheries (STECF) – 46th Plenary Meeting Report (PLEN-14-02). 2014. Publications Office of the European Union, Luxembourg, EUR 26810 EN, JRC 91540, 117 pp. https://stecf.jrc.ec.europa.eu/documents/43805/812327/STECF+PLEN+14-02.pdf. Accessed 27 August 2018.

STECF. (2015a). Scientific, Technical and Economic Committee for Fisheries (STECF) – 49th Plenary meeting Report (PLEN-15-02).2015. Publications Office of the European Union, Luxembourg, EUR 27404EN, JRC 97003, 127 pp. https://stecf.jrc.ec.europa.eu/c/document_library/get_file?uuid=78be1880-f78e-446b-895b-41d4bca7dd8c&groupId=43805. Accessed 27 August 2018.

STECF. (2015b). Scientific, Technical and Economic Committee for Fisheries (STECF) – Landing Obligation – Part 5 (demersal species for NWW, SWW and North Sea) (STECF-15-10)2015. Publications Office of the European Union, Luxembourg, EUR 27407 EN, JRC 96949, 62 pp. https://stecf.jrc.ec.europa.eu/documents/43805/999871/STECF+15-10+-+Landing+obligations+-+p5.pdf. Accessed 27 August 2018.

STECF. (2016a). Scientific, Technical and Economic Committee for Fisheries (STECF) – Evaluation of the landing obligation joint recommendations (STECF-16-10); Publications Office of the European Union, Luxembourg; EUR 27758 EN. https://doi.org/10.2788/59074. https://stecf.jrc.ec.europa.eu/documents/43805/1471816/STECF+16-10+-+Evaluation+of+LO+joint+recommendations.pdf. Accessed 27 August 2018.

STECF. (2016b). Scientific, Technical and Economic Committee for Fisheries (STECF) – Methodology and data requirements for reporting on the Landing Obligation (STECF-16-13). 2016. Publications Office of the European Union, Luxembourg, EUR 27758 EN. https://doi.org/10.2788/984496. https://stecf.jrc.ec.europa.eu/documents/43805/1399955/STECF+16-13+-+Methods+and+data+requirements+for+LO.pdf. Accessed 27 August 2018.

STECF. (2017a). Scientific, Technical and Economic Committee for Fisheries (STECF) – 54th Plenary Meeting Report (PLEN-17-01); Publications Office of the European Union, Luxembourg; EUR 28569 EN. 10.2760/33472. https://stecf.jrc.ec.europa.eu/documents/43805/1672821/STECF+PLEN+17-01.pdf. Accessed 27 August 2018.

STECF. (2017b). Scientific, Technical and Economic Committee for Fisheries (STECF) – Evaluation of the landing obligation joint recommendations (STECF-17-08). Publications Office of the European Union, Luxembourg, 2017. ISBN: 978-92-79-67480-8. doi:10.2760/149272, JRC107574. https://stecf.jrc.ec.europa.eu/documents/43805/1710831/STECF+17-08+-+Evaluation+of+LO+joint+recommendations.pdf. Accessed 27 August 2018.

STECF. (2018). Scientific, Technical and Economic Committee for Fisheries (STECF) – 57th Plenary Meeting Report (PLEN-18-01), Publications Office of the European Union, Luxembourg, 2018. ISBN: 978-92-79-85804-8. doi:10.2760/088784, JRC111800. https://stecf.jrc.ec.europa.eu/c/document_library/get_file?uuid=91be62f0-3aa7-4151-8a0c-b595444a8458&groupId=43805. Accessed 27 August 2018.

Chapter 9
Muddying the Waters of the Landing Obligation: How Multi-level Governance Structures Can Obscure Policy Implementation

Luc van Hoof, Marloes Kraan, Noor M. Visser, Emma Avoyan, Jurgen Batsleer, and Brita Trapman

Abstract The 2013 reform of the European Common Fisheries Policy (CFP) included an increased drive for regionalisation of the policy implementation and the introduction of the Landing Obligation (LO). The process of implementing the LO takes place at multiple levels of governance within the EU. We use the case of the implementation of the LO in the Netherlands, where policymakers and the fishing industry cooperate towards a workable policy implementation. In this paper, we argue that the EU's complex and unconsolidated implementation structure hampers a fair and clear implementation process. Three main causes can be distinguished: first, a lack of a shared understanding of the goal of the Landing Obligation within and between the different governance levels that are involved in the implementation process. Second, no meaningful discussions are taking place between concurrent resource users, resource managers and supporters of the LO regarding the need and usefulness of the measure, as there is no arena in the governance system for them to meet. With the introduction of the Regional Advisory Councils in the 2002 CFP reform, a platform for discussion between fishers and NGOs was created, but this platform has only an advisory role and does not include the Member States. Third, the relationship between different decision-making bodies is unclear, as is the manner in which stakeholder input will be included in decision-making about

---

L. van Hoof · J. Batsleer · B. Trapman
Wageningen Marine Research, IJmuiden, The Netherlands

M. Kraan (✉)
Wageningen Marine Research, IJmuiden, The Netherlands

Environmental Policy Group, Wageningen University, Wageningen, The Netherlands
e-mail: marloes.kraan@wur.nl

N. M. Visser
University of Amsterdam, Amsterdam, The Netherlands

E. Avoyan
Institute for Management Research, Radboud University Nijmegen, Nijmegen, The Netherlands

S. S. Uhlmann et al. (eds.), *The European Landing Obligation*,
https://doi.org/10.1007/978-3-030-03308-8_9

implementing the LO. The result of this implementation process has been a diluted policy where the goal, its execution and its effectiveness remain unclear.

Keywords Common Fisheries Policy · Landing obligation · Multi-level governance · Regionalisation · Subsidiarity

9.1 Introduction

Since its inception in the 1980s, the EU Common Fisheries Policy (CFP) has been reformed more or less every 10 years (1992, 2002, 2013). The latest reform of the CFP came into force in 2014 (European Commission 2013). A Green Paper presented the analysis on which the reform was based (European Commission 2009). That paper described five main structural failures of the CFP: a deep-rooted problem of fleet overcapacity, imprecise policy objectives resulting in insufficient guidance for decisions and implementation, a decision-making system that encourages a short-term focus, a framework that does not give sufficient responsibility to the industry and a lack of political will to ensure compliance and thus poor compliance by the industry. Among the proposals to reform the CFP mentioned at the time (European Commission 2011a) were the introduction of a discard ban; the application of maximum sustainable yield as a leading principle for the management of fish stocks; regionalisation, addressing social issues such as labour conditions, employment and wages; and the introduction of transferable fishing concessions (European Commission 2011b).

Starting from these justifications for reform, the last reform of the CFP introduced two major changes: the introduction of the Landing Obligation and the regionalisation of CFP implementation. Article 15 includes the obligation for European fisheries to land all discards of commercial species for which catch restrictions apply. The term 'discards' refers to catches that are discarded either because they consist of unwanted catch, or catch that cannot be landed due to rules of catch composition and minimum landing size, lack of quota or catches that have (relatively) low or no economic value (Mytilineou et al. 2018; Zeller et al. 2018). Article 18 of the CFP describes the regional cooperation on conservation measures (European Commission 2013). Throughout the Regulation, the process of regionalisation is mentioned as an important instrument in CFP implementation.

In this chapter we will examine how a new policy instrument (i.e. the Landing Obligation, LO) and a new governance setting (regionalisation) are working in practice, using the case of Dutch fisheries. The LO is introduced in a complex process that takes place at multiple levels in the EU governance system: (1) the general policy has been described in the reformed CFP in Article 15; (2) the details have to be filled in at the regional level in multi-annual plans or discard plans by groups of Member States (such as the 'Scheveningen group' for the North Sea); (3) for each Member State, the implementation process is discussed with stakeholders (De Vos et al. 2016), and research projects are performed by national research institutes to provide the necessary information (van Hoof and Kraus 2017). These national processes then feed into the regional process, and (4) the regional discard plans consequently have to be presented to the European Commission and the Council of Fisheries Ministers for approval (Borges and Penas-Lado, this volume; Rihan et al., this volume). The regional groups are advised by the Advisory Councils, representing stakeholders, and attended by members of the fishing industry and NGO representatives.

To analyse this process, we use the case of the introduction of the LO in Dutch fisheries. We use the perspective of multi-level governance to analyse institutional changes and the outcomes of the process. The inputs for this study were obtained from literature analyses, interviews with relevant policymakers, industry representatives, fishers and NGO representatives, by observing and analysing meetings between the fishing sector and the government and by attending meetings of the Dutch Landing Obligation steering group.

In Sect. 9.2, within the context of European fisheries management and its governance setting, the principle of subsidiarity and multi-level governance will be introduced. In Sect. 9.3, a short history of the evolution of the Landing Obligation is presented, in relation to both society's call to reduce fish discards and the policy process leading to the inclusion of the LO into the CFP. In Sect. 9.4, we investigate how the LO is being operationalised in Dutch fisheries within the context of regional North Sea cooperation. In Sect. 9.5, we reflect on the implementation of the LO and aspects of multi-level governance, and Sect. 9.6 presents the conclusions.

9.2 Fisheries Management, Subsidiarity and Multi-level Governance

Fisheries management in the EU is one of only five areas in which the Commission has sole competence (Hawkins 2005; van Hoof 2010b). This implies that policy proposals are developed by the European Commission, and the sole responsibility of the Council of Fisheries Ministers (hereby referred to as 'Fisheries Council') is to approve or reject the proposal. Since the Lisbon Treaty in 2007, this responsibility has been shared between the Fisheries Council and the European Parliament (Symes 2012).

This setting stems from the principle of subsidiarity: according to Spicker (1991), 'subsidiarity' is based on a view of society in which responsibilities are conditioned by the closeness of people's relationships. Intervention at higher levels of society shall be seen as subsidiary to the obligations of smaller social units. Applied more narrowly in the context of the EU, subsidiarity refers to a functional division of administrative responsibilities, although sometimes the principle is referred back to its wider usage where it implies an emphasis on decentralisation and diversity

(Spicker 1991). Kickert (2003) defines subsidiarity as the principle under which the state shall interfere only if the autonomous lower parts seriously fail to fulfil their tasks. In EU fisheries management, subsidiarity is defined in the European Treaty as fisheries management being the sole competence of the EU (CFP) (Raakjear et al. 2010; van Hoof 2010a). Hence, for the CFP, the principle of subsidiarity implies that the EU formulates the general principles of the CFP and is responsible for its general outline, while the Member States are responsible for the implementation of the general policies (Frost and Andersen 2006). Consequently under the CFP the Commission can only command the management level but does not control implementation (Sissenwine and Symes 2007).

With the 2013 reform of the CFP (EU 2013) and the introduction of Article 18, the institutional setting has changed slightly by introducing the regional level into the CFP implementation. Article 18 includes the provision that for some of the conservation objectives (such as the LO), Member States affected by these measures may present recommendations to the European Commission on how to achieve the objectives. In the development of their proposal, the Member States are required to consult the relevant Advisory Council. However, little authority is formally transferred to the regional level and the stakeholders, as the Commission is not accountable for how it handles the inputs from the regional Member States, which themselves are not accountable for how they handle the inputs from the Advisory Councils (Eliasen et al. 2015:227).

The rather complex institutional setting of European Fisheries management consists of the Council of Ministers, the European Commission (EC) and the European Parliament (EP) at the supranational EU level and the Member States at the national level. Therefore fisheries management is shaped to a large extent during a continuous process of draft legislation being prepared by the European Commission, legislative acts being jointly decided by the Council of the European Union and the European Parliament acting together through the co-decision procedure and adopted legislative acts being implemented by the EU Member States (Hegland et al. 2012). One problem with the regionalisation of the CFP, however, is that the European Treaties (which govern the policy-making conduct of EU) do not recognise regions as part of the executive process (Symes 2012).

European fisheries policies are prepared in a way that resembles neither traditional international politics nor policy-making by nation-states. The process takes place in a complex nested setting in which Member States, the European Council of Ministers, the European Commission and the European Parliament play a distinct role. With the concept of multi-level governance, it is possible to capture the shifting locus of governance from the traditional state level to subnational and supranational levels. More specifically, multi-level governance points to sharing policy-making competencies in a complex system of negotiation. The many actors involved in the discussion are located at several levels of nested governmental institutions (supranational, regional, national and local) as well as private actors (i.e. NGOs, producers, consumers, citizens, scientists) (van Tatenhove 2003). Using the perspective of multi-level governance for the analysis of a complex system shows how the EU is clearly still a treaty-based polity and its Member States keep the ultimate authority to approve all decisions (Smith 2004). Nevertheless, responsibility for implementing the CFP still belongs to the Member States.

Within the realm of the LO, we can distinguish three layers of policy-making, policy operationalisation and implementation, i.e. the European level, the Member State level and, in between, the newly created regional level of Member States who discuss regional implementation of the LO. Using the case of Dutch fisheries, we will look at how the LO came about and how it is being implemented across this institutional setting, both from the perspective of the societal call to address discarding of fish and the policy process of incorporating the LO into the CFP.

9.3 A Short History of the EU Landing Obligation

In 2007, the EC published a document about the policy to reduce unwanted bycatches and eliminate discards in European fisheries (European Commission 2007). It suggested a gradual elimination of discards by changing fishers' behaviour and technologies, followed by a discard ban, thus placing the discard problem on the European fisheries managers' agenda (Borges 2015).

In 2008, a British vessel was captured on camera discarding tonnes of fish. This sparked widespread societal awareness of the discard problem. Discarding was perceived by people as morally reprehensible, leading to a waste of marine resources and contributing to fish mortality (Self 2015). This societal call to action, coupled with the slow rate of implementation of discard reduction, substantially increased political pressure on the EU to reconsider the discard policy and to promptly address the discard problem (Borges 2015). Following this development, the EC started reviewing the CFP by publishing the Green Paper proposal on the CFP reform (Commission of the Europe an Communities 2009) and launching a public consultation to discuss the proposed reforms (European Commission 2016). The proposal stated that 'the future CFP should ensure that discarding no longer takes place' (Johnsen and Eliasen 2011).

The debate on banning discards took great momentum when British chef Hugh Fearnley-Whittingstall launched his 'Hugh's Fish Fight' public campaign against fisheries discards (Fish Fight 2014; Borges 2015). The campaign gained a large support across Europe from celebrities, major retailers and the public in general; it also received substantial funding and support of a number of environmental NGOs (Fish Fight 2014). Just before the start of the 'Hugh's Fish Fight' campaign, the new EC had been installed, and the new Commissioner for Maritime Affairs and Fisheries, in the process of reforming the CFP, adopted the idea of an LO (Schwägerl 2013). Involved in the fisheries decision-making process for the first time (European Commission 2016), the European Parliament responded to the expectations of the European public by 'holding the line and fighting hard in the negotiations... for a radical overhaul of EU fisheries policies' (Schwägerl 2013:1–2).

As a result, the previously dismissed 'discard ban' was included in the 2013 reform of the CFP as the 'Landing Obligation'. The introduction to the CFP 2013 explains some of the aims of this LO. Consideration 26 states: '[...] Unwanted catches and discards constitute a substantial waste and negatively affect the sustainable exploitation of marine biological resources and marine ecosystems and the financial viability of fisheries. [...].' (European Commission 2013:24–25). This we perceive as a 'best use perspective', as the simplest explanation of the obligation to land in this perspective: one must land unwanted bycatches, therefore making 'best use' of these catches. In addition to the best use perspective, consideration 29 states: '[...] In the management of the landing obligation, it is necessary that Member States do their utmost to reduce unwanted catches. To this end, improvements of selective fishing techniques to avoid and reduce, as far as possible, unwanted catches must have high priority. [...].' (European Commission 2013). This we perceive as a 'selectivity perspective'. In this perspective, landing discards is a tool to create an incentive for fishers to improve their species and size selectivity, as landing unwanted bycatch would be a burden for them.

It must be noted that the intended effect of the Landing Obligation differs between the two perspectives. In the 'best use' perspective, catches are perceived to be landed and to be used; in the latter, the 'selectivity perspective', the burden of landing fish is used to stimulate selectivity, through which a decrease in the amount of discarded fish is established. The relation between these two intended effects and/or the hierarchy of these intents is not clear, as consideration 29 gives the selectivity perspective 'high priority': improvements of selective fishing techniques to avoid and reduce, as far as possible, unwanted catches must have high priority (European Commission 2013). As for the 'best use' consideration, the CFP states that the discards landed are not allowed to be used for direct human consumption. Article 15 Section 11 states that for the species subject to the Landing Obligation as specified in paragraph 1, the use of catches of species below the minimum conservation reference size (MCRS) shall be restricted to purposes other than direct human consumption, including fish meal, fish oil, pet food, food additives, pharmaceuticals and cosmetics. Although the intention of Article 15 Section 11 is clear, as the regulation limits the use of discards to prevent the development of a market for undersized or over-quota fish because this could lead to an incentive to catch undersized or over-quota fish, it does prevent a large proportion of discards from being used for human consumption.

Besides the fundamental difference in perception of the core objective of the LO (best use versus selectivity), some of the discards are a direct result of previous CFP legislation. Measures such as total allowable catch (TAC) or minimum landing sizes forbid landing and hence selling over-quota catches (Copes 1986; Poos et al. 2010) and small fish (Harley et al. 2000; Rochet and Trenkel 2005). The natural result is discarding of these catches (Mytilineou et al. 2018; Zeller et al. 2018). Hence, the LO conflicts with certain provisions within current legislation, which lead fishers to discard fish that they are not allowed to keep on board (Sardà et al. 2015). To remove the legal and practical impediments to the LO and ensure a legally sound and consistent basis for the introduction of the LO, the EC proposed in 2014 to adjust eight EU fisheries laws/regulations that were not yet adapted to the Landing Obligation (Popescu 2015; Borges and Penas-Lado, this volume).

Exemplary of this is the debate on the so-called 'Omnibus regulation' proposal, initiated at the beginning of 2014. This debate resulted in Regulation (EU) 2015/812 of the European Parliament and of the Council of May 20, 2015 that amended several Council and Parliament regulations[1] which became in fact a technical road map for the implementation of the LO. That document reflected a period of additional discussions, meetings and hearings at the EU level on the core objectives and paths of implementation, separating the involved institutions into different 'camps'. Some members of the European Parliament, together with the European Commission, signalled a desire to put this legal adjustment in place as soon as possible without undermining the main provisions of the initial CFP. Some other members of the European Parliament, together with the Council, put forward a number of changes in the proposal (Naver 2014).

The 'Omnibus regulation' is one area where the discourse of the LO reflects differences in opinions on the 'how' and 'when' of the Landing Obligation. Article 15 of the CFP does include provisions allowing some flexibility under the LO. Paragraph 4 of the reformed CFP (Art. 15 §4; (European Commission 2013) specifies exemptions for prohibited species, species having high survival rates after being discarded, and the so-called de minimis exemption (further elaborated in §5 (c)). The latter implies that 5% of the catch can be discarded, considering that it is very difficult to improve the selectivity of a gear or when using selective fishing gear, handling of unwanted catches on board leads to 'disproportionate costs'. Interspecies flexibility is allowed to a maximum of 9% (§8) and year-to-year flexibility is allowed up to 10% (§9). However, flexibilities provided in Article 15 are associated with great uncertainty, particularly about how the provisions will be implemented in practice: the wording leaves room for subjective interpretation, e.g. 'scientific evidence demonstrating high survival rates' (Batsleer 2016).

Paragraph 5 (Art. 15 §5) notes that details of the implementation of the Landing Obligation must be specified in multi-annual plans (§8–9), in the context of the Landing Obligation (also called 'discard plans'). These details include the high survival[2] exemptions and the de minimis exemptions. Between October 2014 and beginning of 2018, the Commission adopted 15 of these documents in preparation of the Landing Obligation. Exemptions must be specified according to the type of

---

[1]Council Regulations (EC) No 850/98, (EC) No 2187/2005, (EC) No 1967/2006, (EC) No 1098/2007, (EC) No 254/2002, (EC) No 2347/2002 and (EC) No 1224/2009, and Regulations (EU) No 1379/2013 and (EU) No 1380/2013 of the European Parliament and of the Council, as regards the landing obligation, and repealing Council Regulation (EC) No 1434/98 (European Commission 2015).

[2]Fishers should be allowed to continue discarding species which, according to the best available scientific advice, have a high survival rate when released into the sea (European Commission, considerations (27):25; Chaps. 2, 3 and 4).

fishing (small pelagic, large pelagic or demersal), the species of fish, the specific type of fishing gear and the fishing ground and may apply only for a limited amount of time. This results in highly detailed plans.

9.4 The Implementation of the Landing Obligation in Dutch Fisheries

Based on the Landing Obligation proposed by the European Council and agreed in the Council of June 2012, the Dutch ministry started preparing for the Landing Obligation in autumn 2012. A large part of the Dutch fleet fishes in a mixed demersal fishery that involves substantial amounts of discards (Catchpole et al. 2008). Reduction of discards was therefore seen as a major challenge. The topics that the government wished to address were thus the economic impact of the Landing Obligation for the fishing sector and its implications for control and surveillance, and for the scientific monitoring of catches.

The Dutch government launched a LO working group (*werkgroep Aanlandplicht*). Scientists, NGOs and industry representatives were invited to attend monthly meetings, together with civil servants from the Ministry and the control agency. The Ministry deemed it necessary to cooperate with the industry and aimed to garner industry support for necessary measures. In reality the opposite occurred; the LO triggered an immense negative response in the Dutch fishing fleet and resulted in the industry representatives refusing the offer to participate in the working group (Pastoors et al. 2014).

Internationally, other fishing sectors (EAPO, Europêche and Cogeca) opposed the plans and developed their 'alternative proposal' with a focus on stock sustainability.[3] Over the course of 2013, resistance grew in the Netherlands with one of the two fisheries organisations in the Netherlands, the *Vissersbond*, launching a campaign entitled 'Landing Obligation NO' with a manifesto.[4] The actual content of the Landing Obligation document was as yet unknown because the European Parliament was still preparing its opinion regarding the Commission's CFP proposal. This hindered the Dutch government's communication regarding the expected impact. The fisheries representatives and the government did meet to discuss the draft policy, however. In the summer of 2013, it was decided that brainstorming sessions would be organised with the fishers, scientists and policy officers to discuss possible ideas for research towards improved selectivity. However, during the first brainstorming session (in Urk on August 31, 2013), the fishers expressed marked pessimism (Kraan and Verweij in press) because they saw no

[3]http://nsrac.org/wp-content/uploads/2013/01/Paper-5.1-EAPO-Alternative-proposal-discards-approach-by-EU-Industry.pdf

[4]https://www.goeree-overflakkee.nl/document.php?m=46&fileid=14793&f=84374857b90e60c39d03f7a7aef57228&attachment=0&c=22055

ecological and economic benefits resulting from the LO and they feared economic loss. This precluded a constructive or productive discussion about the LO.

From an ecological perspective, the fishers argued that a large part of the discarded catch could survive the discarding process. As such, they expected that by having to land the entire catch under the LO, mortality of the stocks would only increase, reducing potential future yield. In addition, they feared cascading effects in the ecosystem to arise as many other (scavenging) species, i.e. birds and benthic fauna, feed off discards.

From an economic point of view, undersized fish are not allowed to be sold for human consumption. In this context, fishers will obtain a low price for such fish, lower than the cost of the additional workload of having to handle, sort and store the unwanted catches on board. In addition, capacity problems in the fish hold as well as choke species[5] may cause an early cessation of fishing activities resulting in a loss of potential catches and yield. Both the ecological and economic impacts were still poorly understood when the Landing Obligation came into force. This lack of understanding was met with suspicion and fear for the future by the industry. In turn, this uncertainty led to an attitude towards the LO that was likely to undermine compliance (Batsleer 2016).

Besides the Dutch fishers' fear of how the LO would affect the ecosystem, they also feared the market-driven consequences of landing undersized fish (Trapman and Kraan 2015:28). In the 1980s and 1990s, there had been a major black market for undersized sole, and the fishing sector had fought hard to ban this practice. Many fishers feared that the LO would bring this practice back. To prevent a black market from developing, the government had summoned to destroy undersized fish immediately at the auction by pouring red dye over the fish.

As a result of the fishers' resistance to brainstorming about how to cope with a potential LO, future sessions were cancelled and replaced with sessions where the Minister explained to the fishers the LO and listened to their concerns. These meetings, one held in Scheveningen and one in Urk, have been characterised by Pastoors et al. (2014) as 'parallel monologues' with three 'talking lines' (Table 9.1).

Table 9.1 Parallel lines of reasoning between fishers and the Dutch government

	Fishers	Minister
1	The Landing Obligation is impossible to implement	The Landing Obligation is a fact
2	We want a principled discussion about the goals and measures of the LO	We will not have a principled discussion about the goals and measures of the LO
3	Emphasising impossibilities, concerns and dilemmas	Seeking room to manoeuvre in the implementation of the Landing Obligation

Source: Pastoors et al. (2014)

[5]Choke species are species for which quota are so restrictive that quota will be fully exploited at an early stage in the year, resulting, in a mixed fisheries, that the fishery has to be completely stopped, although for other species there may well still be sufficient quota.

The two perspectives inherent in the Landing Obligation (best use and selectivity) have caused confusion during implementation of LO in the Netherlands. From the discussions it became clear that fishers and policy officers had fundamentally opposing perceptions of what the ecological consequences of the Landing Obligation would be (see Kraan and Verweij in press). This disconnect did not become obvious as the Minister refused to have a principled discussion. The 'best use' perspective conflicts strongly with the perception of the fishers; they think that the Landing Obligation will conversely result in massive 'waste', as the discards (hardly reduced) cannot be used and will have to be destroyed on land (Kraan and Verweij in press). This undermines their willingness to cooperate and think constructively about how to implement the measure.

The third talking line, however, provided a bridge between fishers and government. The government created a research agenda based on all of the (perceived) impossibilities, concerns and dilemmas the fishers mentioned and provided funding in 2013 to implement research based on proposals from the fishing sector. The Minister emphasised from the start that she would seek 'room to manoeuvre' for the fleet. During 2014 and 2015, meetings were organised between the Ministry, fishers and researchers to discuss research directions and findings. Research was directed at developing what was labelled as 'best practices' by gaining a better understanding of amounts of discards of different fisheries, possible gear innovations, the survival of discarded fish (and ways to improve survival), camera monitoring on board and processing and valorisation of discards.

It became clear from interviews with fishers that they thought it was impossible to prepare for the LO because of the long preparation trajectory, while the final details about the measure itself still remained uncertain. When asked in 2014/2015 about possibilities to fish more selectively by the choices they make on where and when to fish, fishers expressed uncertainty and indicated that they could not envision how the (draft) rules would influence their behaviour (Trapman and Kraan 2015). Again, their fundamental resistance to the LO was linked to their (lack of) willingness and ability to think constructively about how to live with it.

In light of deeply entrenched resistance of most Dutch fishers to the Landing Obligation, the Dutch therefore prepared for it by cooperating between fisheries representatives, policy officers and scientists to search for the best way to implement the new rules. The participants involved in the process built up interactional expertise (Stange 2017) as they were frequently and directly in contact with each other. Fisheries representatives combined their knowledge of the fleet and fishing practices with the knowledge of the political arena and science and thereby thought constructively about what was technically feasible, scientifically researchable and politically achievable. A small group of fishers did participate in the cooperative research projects[6] and thus also got some of these insights as well; however, most of the individual fishers were left out of these developments and their resistance to the LO remained high.

---

[6]See http://cvo-visserij.nl/projecten-2/best-practices/ for some of these projects.

This development illustrated the need to improve communication with and within the fishing community, since all discussions and negotiations at the national level had to be explained and brought to acceptance in the fleet by the fisheries representatives. The attempts to improve communication were not fully successful, resulting in the 2016 establishment of a fisher protest movement called 'EMK' (*Eendracht Maakt Kracht*; Unity Creates Strength). EMK opposed the LO as well as other measures impacting the fishers, such as the development of windmill parks in the North Sea and the establishment of marine protected areas. EMK has organised annual demonstrations since 2016: in Rotterdam (August 2016; an armada of 50 vessels sailed on the Maas to the city centre, in conjunction with 'World Harbour Day'), in Brussels (May 2017; EMK delivered a petition to the European Parliament) and in Amsterdam (June 2018; an armada of 15 vessels and other fishers who came by bus to deliver a petition to a member of the European Parliament).[7] The EMK fishers aim for repeal of the LO. The fishing industry as a whole (consisting of different organisations and groups using different avenues) has also tried to bring its opinion on the LO to different arenas at different levels. The demonstrations of the EMK group were aimed at the Dutch general public and Dutch EP members; but the Dutch parliament was also mobilised from time to time, resulting in a motion to the Minister in December 2015 to go to Brussels to ask for a delay in the implementation of the Landing Obligation.

Parallel to this development, the Ministry had to (re-)negotiate at the regional level with the other North Sea Member States in the 'Scheveningen' group. In that process, Member States present their ideas for policy measures in the discard plans at the Scheveningen group meeting, the Scheveningen group drafts a joint recommendation for the annual discard plans. These are then sent to the STECF[7] for evaluation, who then advises the EC (see Rihan et al., this volume). The EC then sends it as a Delegated Act to the European Parliament, which then also has to agree before the plans can be finally accepted. This is a lengthy process, in which (highly) technical matters and politics get mixed (see also Stockhausen, this volume). In this process, the Ministry balances the expectations regarding the other Member States and 'room to manoeuvre' of the fishing fleet in consultation with the fisheries representatives (fisheries representative June 6, 2018). This results in some policy proposals from the Dutch fleet representatives being filtered out at national level and no longer discussed at the regional level.

9.5 The Muddy Waters of Multi-level Governance

Starting from the very beginning of the CFP reform process, considerable disparity in the views articulated by different European institutions and EU Member States was observed regarding the LO and its implications (Self 2015). The position taken

[7]The Scientific, Technical and Economic Committee on Fisheries of the European Commission, the highest Scientific Advisory body for fisheries of the EU

by the European Parliament, representing the interests of European citizens, was very much influenced by public discontent with the fishing practices of discarding with discards seen as unacceptable behaviour (Frangoudes et al. 2015). The 'Hugh's Fish Fight' campaign played an important role in portraying the image of waste, and the image of mature cod being discarded in British small-scale fisheries was instrumental. The European Parliament adopted the LO with the aim of changing current fishing practices. However, they had failed to consider all of the factors driving discarding (Borges 2015). The driving factors behind the discarding of North Sea cod (i.e. lack of quota) are completely different than those behind the discarding of undersized plaice (i.e. CFP regulation; minimum landing size) in a fishery such as the Dutch mixed fisheries.

In contrast, the Council of Ministers has been much more reluctant to embrace changes in the CFP (Schwägerl 2013). Among the national governments, there are differing interpretations of the LO (Viðarsson et al. 2015; Borges et al. 2016) which results in a situation in which '[a]t the level of the Member States, the rules of the game seemed not yet to be fixed, and open for change' (de Vos et al. 2016).

The general public, fuelled by 'Hugh's Fish Fight' and its commercial and NGO supporters, regards discarding as a waste of living resources (Heath et al. 2014; Borges 2015). During the public consultation process preceding the introduction of the LO, environmental NGOs strongly supported the principles of a discard ban. To them, discarding is a wasteful practice that damages marine life and is consequently an unsustainable practice (Villasante et al. 2016a, b).

The LO has been met with strong resistance by the European fishing industry. The fishers' response to the new CFP reform and the LO was initially influenced by their exclusion from the decision-making process that preceded introduction of the Regulation (De Vos et al. 2016). The fishers therefore perceive the introduction of the LO as a top-down process (SOCIOEC 2013; Frangoudes et al. 2015; De Vos et al. 2016; Fitzpatrick et al., this volume). Moreover, they do not generally comprehend the LO's relevance, its biological meaningfulness nor its overall objective (Hedley et al. 2015). They also emphasise that there are different reasons for discarding, including some that result from the very rules of the same CFP (Kraan et al. 2015; Borges and Penas-Lado, this volume). This has resulted in a lack of legitimacy and acceptance not only among Dutch fishers but among most European fishers, weakening their willingness to comply with the LO requirements.

The reform of the CFP met with three major shifts in the governance set-up: co-decision, regionalisation and the LO. The co-decision procedure for the first time provided a strong voice to the European Parliament in policy development.

Parallel to the implementation of the Marine Strategy Framework Directive, using a regional seas approach in which in drawing up marine strategies for the waters within each marine region, Member States are required to cooperate closely (van Hoof and van Tatenhove 2009; van Hoof et al. 2012; Van Leeuwen et al. 2012). And, due to the 2013 reform, the CFP now has a regional component. This regional approach creates a new layer in the institutional setting of fisheries policy between the European institutions and Member States.

Finally, the LO is a Policy, yet it carries many characteristics of a Directive. In the CFP, the wider goal of the LO is formulated, but the operationalisation of the policy into concrete measures is developed over time. This will allow for adaptation of the LO to local and regional circumstances and practices.

In principle these three developments added to a multi-level governance structure, contributing to improved subsidiarity; in theory it emphasises decentralisation and diversity resulting in policy implementation that is as close to fishing practices as possible (Spicker 1991). However, in practice, bottom-up proposals from stakeholders to improve policy do not make it up the chain of multi-level governance due to political considerations (i.e. assessing that the proposals will probably not get support from the other Member States). Also, as a consequence of this new institutional setting, it becomes clear that there is no one single overarching discourse steering the operationalisation of the LO. At different levels and across institutions, the main driver and hence the main objective of the LO ('best use' perspective or 'selectivity' perspective) are interpreted differently.

From the 'best use' perspective, the perspective adhered to by the general public and the main driver of the anti-discards campaign, human consumption of bycatch fish is stimulated. From this perspective, it is hard to understand why the unwanted bycatches, and hence former discards, cannot be made available for consumption. Parallel to this perspective is the fishers' opinion that when unwanted bycatches are discarded, they still fulfil a role in the marine ecosystem, yet once landed, they are subtracted from the ecosystem. If, after landing, the catches are then destroyed, this seems to contrast with a 'best use' perspective.

From the 'selectivity' perspective, fishers find it hard to see ways of increasing selectivity. As a result, we see today that across Europe exemptions are increasingly sought under the de minimis regulation, 'high survival' exemption (Rihan et al., this volume) and the Omnibus regulation, to find ways to exclude certain fish species from the LO. A new discourse seems to present itself in which, at national levels, efforts are being made to prove that species have a high survival rate after being caught and discarded.

Hence, instead of developing a clear understanding of the aim of the LO, its operationalisation and implementation at the most relevant level of the multi-governance structure of fisheries policy in the EU, it appears that the current governance setting of the CFP is obscuring the discussion, with different interpretations of the aims at different levels of governance and different perceptions of the principles behind the LO and of the consequences when implemented.

9.6 Conclusions

The introduction of the LO and regionalisation of the CFP have reshaped the multi-level setting of fisheries policy-making in Europe. However, when looking at implementation and operationalisation of the LO, one can wonder whether these changes actually have resulted in a more open, transparent and participatory manner of policy-making. Has it shifted from the subsidiary level to the lowest possible institutional level?

The debate on the objective of the LO between the 'best use' and 'selectivity' perspectives, resulting in different interpretations at different levels, obscures the debate. It highlights the lack of a shared goal of the LO within and between the different governance levels that are involved in the implementation process.

As we have seen in the case of the Netherlands, this partly stems from the lack of discussion between opposing resource users and supporters of the Landing Obligation about the need and usefulness of the measure because there is no arena in the governance system where they meet. In addition, in the new institutional setting, two things are unclear: the relation between different decision-making bodies and the way in which stakeholders' input will be incorporated in decision-making regarding the implementation of the Landing Obligation. The multi-level setting of fisheries policy in Europe has made the debate more opaque instead of more transparent and open. The result of this implementation process has been a diluted policy whose goal, execution and effectiveness are all equally muddy.

The multi-level governance setting could have been instrumental using the subsidiarity principle of having discussions and decisions made at the most relevant policy and institutional level. The regional level is added to the Member States and the European institutions of the Council, Parliament and Commission. The effectiveness of the setting would have benefitted greatly from clearly defined roles for industry and other stakeholders at each level of governance. Participation requires a clear platform; and clearly defining and communicating policy goals are essential to the process.

Besides this lack of a principled discussion about the motivation and objective of the LO, implementation also appears to be muddled. Testing for high survival rates of discarded fish is in fact not so much implementing the LO but trying to escape from the LO. This can be traced to the inconsistency in interpretation of the objectives of the LO, a lack of thorough debate linking objectives with fishing practice (i.e. what is more wasteful – discarding fish back into the ecosystem with an unclear fate or landing them and thus clearly choosing for fish mortality?) and from the lack of reflection in an earlier stage on the driving forces behind the discards.

Surely, the fishing industry does see benefits of reducing unwanted catches, but does not see any justification for having to land all catches of quota-managed species. Indeed the 'force' of the LO has resulted in improved selectivity in sub-sections of the fishing industry, which otherwise might not had been accomplished. However, the final goal of a total ban on discards is far from being attained. Perhaps the major change that is needed to reduce discards requires a transition to a whole new system of fisheries, including different fishing practices, different fish consumption patterns and different fisheries management. Implementing rigorous mitigation measures, such as the LO, should be supported by a common understanding of the context in which the problem occurs and how best to address it. If not, it may lead to unexpected and unwanted consequences and undermine acceptance and compliance. At times, it is clear that a complex multi-level governance structure may only muddy the waters instead of resolving the debate.

References

Batsleer, J. (2016). Fleet dynamics in a changing policy environment. PhD thesis. Wageningen, Wageningen University.

Borges, L. (2015). The evolution of a discard policy in Europe. *Fish and Fisheries, 16*, 534–540.

Borges, L., Cocas, L., Nielsen, K.N. (2016). Discard ban and balanced harvest: A contradiction? *ICES Journal of Marine Science*, *73*, 1632–1639.

Borges, L., & Penas Lado, E. (this volume). Discards in the common fisheries policy: The evolution of the policy. In S.S. Uhlmann, C. Ulrich, S.J. Kennelly (Eds.), *The European Landing Obligation – Reducing discards in complex, multi-species and multi-jurisdictional fisheries*. Cham: Springer.

Catchpole, T., Keeken, O.v., Gray, T., Piet, G. (2008). The discard problem – A comparative analysis of two fisheries: The English *Nephrops* fishery and the Dutch beam trawl fishery. *Ocean & Coastal Management*, *51*, 772–778.

Copes, P. (1986). A critical review of the individual quota as a device in *Fisheries Management. Land Economics*, *62*, 278–291.

De Vos, B., Döring, R., Aranda, M., Buisman, F., Frangoudes, K., Goti, L., Macher, C., et al. (2016). New modes of fisheries governance: implementation of the landing obligation in four European countries. *Marine Policy*, *64*, 1–8.

Eliasen, S.Q., Hegland, T.J., Raakjær, J. (2015). Decentralising: the implementation of regionalisation and co-management under the post-2013 Common Fisheries Policy. *Marine Policy*, *62*, 224–232.

European Commission. (2007). Communication From the Commission to the Council and the European Parliament on a policy to reduce unwanted by-catches and eliminate discards in European Fisheries. Commission of the European Communities, COM(2007) 136 final, 8 pp.

European Commission. (2009). Green Paper reform of the common fisheries policy, COM163 final.

European Commission. (2011a). Communication from the Commission to the European Parliament, the Council, the European Economic and Social Committee and the Committee of the Regions; reform of the common fisheries policy COM 417 final. Brussels.

European Commission. (2011b). Reform package.

European Commission. (2013). Regulation (EU) No 1380/2013 of the European Parliament and of the Council of 11 December 2013 on the common fisheries policy, amending Council Regulations (EC) No 1954/2003 and (EC) No 1224/2009 and repealing Council Regulations (EC) No 2371/2002 and (EC) No 639/2004 and Council Decision 2004/585/EC.

European Commission. (2015). Regulation (EU) 2015/812 of the European Parliament and of the Council of 20 May 2015 amending Council Regulations (EC) No 850/98, (EC) No 2187/2005, (EC) No 1967/2006, (EC) No 1098/2007, (EC) No 254/2002, (EC) No 2347/2002 and (EC) No 1224/2009, and Regulations (EU) No 1379/2013 and (EU) No 1380/2013 of the European Parliament and of the Council, as regards the landing obligation, and repealing Council Regulation (EC) No 1434/98.

European Commission. (2016). Fact sheets on the European Union: the common fisheries policy, origins and development. Available at: http://ec.europa.eu/fisheries/cfp/index_en.htm. Last accessed: August 30, 2018.

Fish Fight. (2014). The fish fight story. Available at: http://www.fishfight.net/story.html. Last accessed: August 22, 2018.

Fitzpatrick, M., Frangoudes, K., Fauconnet, L., Quetglas, A. (this volume). Fishing industry perspectives on the EU Landing Obligation. In S.S. Uhlmann, C. Ulrich, S.J. Kennelly (Eds.), *The European Landing Obligation – Reducing discards in complex multi-species and multi-jurisdictional fisheries*. Cham: Springer.

Frangoudes, K., Leleu, K., Rochet M-J., Trekel, V. (2015). Vision of French fishers about the European Union regulation on Landing Obligation (LO): which ecological and economical impacts and which strategies to cope with it? Extended abstract ICES ASC. Available at: http://www.ices.dk/sites/pub/ASCExtendedAbstracts/Shared%20Documents/L%20-%20Science-industry%20partnership.%20The%20value%20of%20cooperative%20research%20in%20fisheries%20and%20marine%20management/L2115.pdf. Last accessed: August 30, 2018.

Frost, H., & Andersen, P. (2006). The Common Fisheries Policy of the European Union and fisheries economics. *Marine Policy*, *30*, 737–746.

Harley, S.J., Millar, R.B., McArdle, B.H. (2000). Examining the effects of changes in the minimum legal sizes used in the Hauraki Gulf snapper (*Pagrus auratus*) fishery in New Zealand. *Fisheries Research*, *45*, 179–187.

Hawkins, T. (2005). The role of partnerships in the governance of fisheries within the European Union. In T.S. Gray (Ed)., *Participation in fisheries governance* (pp. 27–44). Dordrecht: Springer.

Heath, M.R., Cook, R.M., Cameron, A.I., Morris, D.J., Speirs, D.C. (2014). Cascading ecological effects of eliminating fishery discards. *Nature Communications*, *5*, 3893.

Hedley, C., Catchpole, T., Santos, A. (2015). The landing obligation and its implications on the control of fisheries. The European Parliament: Brussels, Belgium.

Hegland, T.J., Ounanian, K., Raakjær, J. (2012). Why and how to regionalise the Common Fisheries Policy. *Maritime Studies*, *11*, 7.

Johnsen, J.P., & Eliasen, S. (2011). Solving complex fisheries management problems: what the EU can learn from the Nordic experiences of reduction of discards. *Marine Policy*, *35*, 130–139.

Kickert, W. (2003). Beneath consensual corporatism: traditions of governance in the Netherlands. *Public Administration*, *81*, 119–140.

Kraan, M., Verkempynck, R., Steins, N. (2015). Technical Measures in the Atlantic and the North Sea – working with Stakeholders towards meaningful revision. Workshop – in depth analysis. In: *A new technical measures framework for the new Common Fisheries Policy.* Marcus Breuer and Carmen-Paz Marti. Available at: Brussels EU report. Available at: http://www.europarl.europa.eu/studies. Last accessed: August 30, 2018.

Kraan, M.L., & Verweij, M. (in press). Implementing the landing obligation in the Netherlands; an analysis of the gap between fishery and the ministry. In P. Holm, M. Hadjimichael, S. Mackinson (Eds.), *Bridging the gap: collaborative research practices in the fisheries.* MARE Publication Series. Springer.

Mytilineou, C., Herrmann, B., Mantopoulou-Palouka, D., Sala, A., Megalofonou, P., Handling editor: Finbarr, O.N. (2018). Modelling gear and fishers size selection for escapees, discards, and landings: a case study in Mediterranean trawl fisheries. *ICES Journal of Marine Science*, fsy047-fsy047.

Naver, A. (2014). Dispute in Parliament creates confusion over discard ban. CFP Reform Watch. Available at: http://cfp-reformwatch.eu/2014/12/dispute-in-parliament-creates-confusion-over-discard-ban/.

Pastoors, M.A., Buisman, E., van Oostenbrugge, H., Kraan, M.L., van Beek, F.A., Uhlmann, S.S., van Helmond, A.T.M. (2014). Fasering discard ban. IMARES report C070/14.

Poos, J.J., Bogaards, J.A., Quirijns, F.J., Gillis, D.M., Rijnsdorp, A.D. (2010). Individual quotas, fishing effort allocation, and over-quota discarding in mixed fisheries. *ICES Journal of Marine Science, 67*, 323–333.

Popescu, I. (2015). Adapting EU fisheries legislation to the landing obligation. European Parliament Research Service, EU Legislation in Progress, Briefing.

Raakjear, J., Abreu, H., Armstrong, C., Hegland, T., van Hoof, L., Ounanian, K., Ramirez, P., et al. (2010). Exploring the option of regionalising the Common Fisheries Policy. Making the European Fisheries Ecosystem Plan Operational (MEFEPO): Work package 4 technical document.

Rihan, D., Uhlmann, S.S., Ulrich, C., Breen, M., Catchpole, T. (this volume). Requirements for documentation, data collection and scientific evaluations. In S.S. Uhlmann, C. Ulrich, S.J. Kennelly (Eds.), *The European Landing Obligation – Reducing discards in complex multi-species and multi-jurisdictional fisheries*. Cham: Springer.

Rochet, M.-J., & Trenkel, V.M. (2005). Factors for the variability of discards: assumptions and field evidence. *Canadian Journal of Fisheries and Aquatic Sciences* 62, 224–235.

Sardà, F., Coll, M., Heymans, J.J., & Stergiou, K.I. (2015). Overlooked impacts and challenges of the new European discard ban. *Fish and Fisheries*, *16*, 175–180.

Schwägerl, C. (2013). 'Will reform finally end the plunder of Europe's fisheries?. *Yale Environment*, 360.

Self, E. (2015). Who speaks for the fish: the tragedy of Europe's Common Fisheries Policy. *Vand. J. Transnat'l L.*, *48*, 577.

Sissenwine, M., & Symes, D. (2007). Reflections on the common fisheries policy. In report to the general directorate for fisheries and maritime affairs of the European commission. http://www.greenpeace.org/raw/content/denmark/press/rapporter-og-dokumenter/reflections-on-the-common-fish.pdf. Brussel.

Smith, M. (2004). Toward a theory of EU foreign policy-making: Multi-level governance, domestic politics, and national adaptation to Europe's common foreign and security policy. *Journal of European Public Policy*, *11*, 740–758.

SOCIOEC. (2013). Report on governance and stakeholder involvement in fisheries and analysis of EU policy framework. Socio-economic effects of management measures of the future CFP (Socioec) project Deliverable. Marine Law and Ocean Policy Research Services Limited.

Spicker, P. (1991). The principle of subsidiarity and the social policy of the European Community. *Journal of European Social Policy*, *1*, 3–14.

Stange, K. (2017). Knowledge production at boundaries: An inquiry into collaboration to make management plans for European fisheries. Wageningen University and Research. Environmental Policy Group. PhD thesis.

Stockhausen, B. (this volume). How the implementation of the Landing Obligation was weakened. In S.S. Uhlmann, C. Ulrich, S.J. Kennelly (Eds.), *The European Landing Obligation – Reducing discards in complex, multi-species and multi-jurisdictional fisheries*. Cham: Springer.

Symes, D. (2012). Regionalising the Common Fisheries Policy: Context, content and controversy. *Maritime Studies*, 11: 6.

Trapman, B., & Kraan, M.L. (2015). Aanpassingen visserijgedrag en -techniek in de tongvisserij in verband met de aanlandplicht. IMARES report C142/15.

van Hoof, L. (2010a). Co-management: an alternative to enforcement? *ICES Journal of Marine Science*, 67.

van Hoof, L. (2010b). Who rules the waves? Governance and new institutional arrangements in Dutch Fisheries Management in the context of the European Common Fisheries Policy. PhD Thesis. Environmental Policy Group. Wageningen University, Wageningen.

van Hoof, L., & Kraus, G. (2017). Is there a need for a new governance model for regionalised Fisheries Management? Implications for science and advice. *Marine Policy*, *84*, 152–155.

van Hoof, L., van Leeuwen, J., van Tatenhove, J. (2012). All at Sea; regionalisation and integration of Marine Policy in Europe. *Maritime Studies, 11, 9.*

van Hoof, L., & van Tatenhove, J. (2009). EU marine policy on the move: the tension between fisheries and maritime policy. *Marine Policy*, *33*, 726–732.

Van Leeuwen, J., van Hoof, L., van Tatenhove, J. (2012). Institutional ambiguity in implementing the European Union Marine Strategy Framework Directive. *Marine Policy*, *36*, 636–643.

van Tatenhove, J. (2003). Multi-Level Governance and the 'institutional void': The interplay between front stage and backstage politics. In Workshop 21 'Assessing Emergent Forms of Governance: European Public Policies Beyond the 'Institutional Void'', ECPR Joint Sessions of Workshops. University of Edinburgh.

Viðarsson, J., Guðjónsson, Þ., Sigurðardóttir, S. (2015). Report on current practices in the handling of unavoidable, unwanted catches. Project Deliverable, Strategies for the gradual elimination of discards in European fisheries, Discardless project.

Villasante, S., Pierce, G.J., Pita, C., Guimeráns, C.P., Rodrigues, J.G., Antelo, M., et al. (2016a). Fishers' perceptions about the EU discards policy and its economic impact on small-scale fisheries in Galicia (North West Spain). *Ecological Economics*, *130*, 130–138.

Villasante, S., Pita, C., Pierce, G.J., Guimeráns, C.P., Rodrigues, J.G., Antelo, M., et al. (2016b). To land or not to land: How do stakeholders perceive the zero discard policy in European small-scale fisheries? *Marine Policy*, *71*, 166–174.

Zeller, D., Cashion, T., Palomares, M., Pauly, D. (2018). Global marine fisheries discards: a synthesis of reconstructed data. *Fish and Fisheries*, *19*, 30–39.

Chapter 10
The Baltic Cod Trawl Fishery: The Perfect Fishery for a Successful Implementation of the Landing Obligation?

Daniel Valentinsson, Katja Ringdahl, Marie Storr-Paulsen, and Niels Madsen

Abstract The cod fisheries in the Baltic Sea were among the first EU fisheries with a full implementation of the EU Landing Obligation (LO) or so-called 'discard ban', phased in from 2015 onwards. This chapter describes key aspects for the successful management of Baltic cod such as the long history of scientific data collection for stock assessment and cod management as well as a well-documented history of work aimed at increased selectivity in cod trawls. We then analyse how the scientific data used for stock assessment has been affected by the LO and how the knowledge of Baltic cod selectivity has been used and developed since its introduction. We conclude that in spite of many good prerequisites, the introduction of the LO in Baltic cod fisheries has been unsuccessful and has failed to deliver any of the expected benefits. Data quality for stock assessments has deteriorated, discarding of cod has not decreased despite a reduced minimum size and there are no indications of increased gear selectivity in the fishery. Finally, we propose potential explanations for this failure and recommend actions that may be needed to make the Landing Obligation more successful.

Keywords Baltic Sea · Common fisheries policy · Discard ban · Discards · European Union · Regionalisation · Selectivity · Trawl

---

D. Valentinsson (✉) · K. Ringdahl
Department of Aquatic Resources, Institute of Marine Research, Swedish University of Agricultural Sciences, Lysekil, Sweden
e-mail: daniel.valentinsson@slu.se

M. Storr-Paulsen
National Institute of Aquatic Resources, DTU-Technical University of Denmark, Lyngby, Denmark

N. Madsen
Department of Chemistry and Bioscience – Section of Biology and Environmental Science, Aalborg University, Aalborg, Denmark

S. S. Uhlmann et al. (eds.), *The European Landing Obligation*,
https://doi.org/10.1007/978-3-030-03308-8_10

10.1 Introduction

Baltic Sea cod was the first demersal species with a full implementation of the EU Landing Obligation (LO) in 2015. The LO meant that the dominant fisheries that use trawls and nets were phased-in at once, with only a survival exemption for the negligible cod catches from pots and traps valid between 2015 and 2017 (Commission Delegated Regulation (EU) No 1396/2014). Catches from cod trawls in the Baltic, in contrast to other European demersal trawl fisheries, are relatively clean with only flounder (non-quota species) and some plaice as notable bycatches. In 2017, plaice was also included in the Baltic Sea LO. This could potentially lead to challenges for some Baltic countries as the relative stability (i.e. the fixed allocation key of TAC between countries) means that few countries in the Baltic hold a relatively large share of the total plaice quota. This could potentially be problematic for the countries with no quota or with a very low quota share if they cannot avoid plaice in their cod catches.

Because bottom trawls are by far the main gear used to catch Baltic cod and because both gear selectivity research and discard observers programmes have a long history in the Baltic, the setting and background of this fishery seem perfect for a successful LO implementation. This chapter focuses on the prerequisites and outcomes of the LO in the Baltic trawl fishery targeting cod. Our aim is to describe the backdrop to the LO implementation and subsequent indications of the outcomes. We first describe the discard sampling using scientific observers and how these data are used in stock assessments. We also depict the long history of trawl selectivity research in Baltic cod trawls and then discuss the outcomes of these aspects in light of the articulated aims of the Landing Obligation (Anon 2018a), i.e., reduced unwanted catches, improved quality of scientific data and industry adaption through the adoption of more selective fishing practices.

10.1.1 The Baltic Cod Stocks: Stock Development and Current Status

The Baltic Sea is inhabited by two genetically distinct cod populations, i.e. eastern and western Baltic cod. There is evidence supporting the difference between the two populations, based on differences in spawning time, otolith shape (Hüssy et al. 2016) and genetics (Nielsen et al. 2003; Nielsen et al. 2005). The cod in the Baltic Sea live in brackish water characterized by low salinity and an oxygenated surface layer with a hypoxic deeper saline layer that gives a permanent halocline. This influences egg buoyancy and thereby egg survival, especially in the more eastern part of the Baltic Sea. The water volume suitable for egg fertilisation and development of eastern Baltic cod (i.e., the reproductive volume) is defined as the volume of water with a PSU >12 and oxygen content >2 ml/l (Köster et al. 2003).

Cod fisheries in the Baltic Sea have a long history (MacKenzie et al. 2002). More intensive exploitation only started in the 1950s (Bagge and Thurow 1994; Eero et al. 2007). The cod fishery on the eastern stock increased further in the early 1980s as a result of increased stock biomass due to favourable reproductive conditions and strong year classes (Eero et al. 2011). As the fishery intensified during the 1980s, a period of low productivity due to deteriorated environmental conditions began which, together with intensive exploitation, rapidly reduced the stock. In a 10 year period, the stock size was reduced by more than 85% (ICES 2013), and in 2005 the spawning stock biomass was estimated to be at a historic low, close to 10% of the high levels observed in the 1980s. However, after 2005 the stock started to rebuild again and the positive developments were partly assigned to effective management measures and increased reproductive success. In 2010, the stock was thought to have recovered (Eero et al. 2012a). Shortly after, it was evident that although recruitment was estimated to be good, large cod were missing in scientific surveys and commercial catches. The nutritional condition of adult cod has been continuously declining since the 1990s. However, since the mid-2000s, when cod abundance started to increase again, the proportion of eastern cod with a very low condition index increased rapidly (ICES 2017a). The reasons for deteriorating nutritional condition and possibly reduced growth are not fully understood. Different hypotheses are suggested including: (i) low prey availability in the area where cod are mainly found (Eero et al. 2012b); (ii) increased size selectivity in commercial fisheries which may have led to high mortality of large cod, (iii) a greater proportion of small-sized fish in the stock and their contribution to density-dependent effects and (iv) increased extent of low oxygen areas that could affect cod growth either directly via physiological processes or indirectly via affecting benthic prey availability (Chabot and Dutil 1999; Svedäng and Hornborg 2014; Eero et al. 2015; Casini et al. 2016). In 2017 the eastern Baltic cod stock was estimated to be at a very low level and cod larger than 45 cm were very scarce (ICES 2018).

The western Baltic cod spawns in specific areas in saline waters deeper than 20–40 m, with the main spawning areas being the Little and the Great Belt, the Sound and the Kiel and Mecklenburg Bays. The western Baltic cod mixes with both the eastern Baltic cod stock in the Arkona Basin (Subdivision 24) and with the Kattegat cod stock in the southern part of Kattegat (ICES 2018). The western Baltic cod stock has also experienced large fluctuations in stock development over time. In the mid-1980s, landings were close to 50,000 t in the western Baltic management area to below 6000 t in 2017. Unlike the eastern Baltic cod, there is no documentation of decreased condition or impairment by reduced growth of western Baltic cod. The western Baltic cod has experienced high fishing pressure and shown poor recruitment for several years and was assessed to be well below reference points at the onset of the LO in 2015. However, although the spawning stock was at a historically low level in 2016, a new large year class was observed, which is likely to influence the development of the stock in the following years (ICES 2018).

10.2 Data Collection and Assessment of Baltic Cod

Internationally coordinated management of the two Baltic cod stocks stems back to the early 1970s. Before many of the countries around the Baltic Sea joined the EU, the management authority was the International Baltic Sea Fishery Commission (IBSFC), which ceased to exist in 2007. The two cod stocks were regulated together by a total Baltic TAC (total allowable catch) until 2003, but have since then been regulated separately. The introduction of a TAC in the late 1980s was prompted by the drastic decline in stock size and catches in the mid 1980s, just after the peak in stock size when catches exceeded 400,000 t (Bagge and Thurow 1994; Suuronen et al. 2007). The first EU management plan for Baltic cod was introduced in 2007 (Council Reg (EC) No 1098/2007). The cod plan was repealed in the summer of 2016 when a new multi-annual plan for all Baltic stocks of cod, herring and sprat was introduced (Reg (EU) No 2016/1139).

The assessment of the eastern Baltic cod stock was conducted for many years as a full age-based analytical assessment. In 2014, however, ICES could no longer accept the assessment. There were several factors that had changed the prerequisite for the assessment: Ageing discrepancies between countries, reduced growth and the unexplained disappearance of larger cod despite the appearances of good recruitment. Therefore, a reliable analytical assessment could not be produced. The advice since 2015 has been based on the so-called ICES "category three" assessments. These assessments mainly use trends in an index of spawning stock or total biomass from research surveys. The current assessment approach for the eastern stock uses the numbers of cod above 30 cm (as a proxy for Spawning Stock Biomass, SSB) from the survey where the average abundance for the two most recent years is divided by the average for the three previous years ("2 over 3 rule"). Thus, the provision of high quality management advice is challenged by a number of changes in cod biology and the ecological conditions affecting cod (ICES 2017a). The change of assessment methodology drastically changed the perception of the eastern stock and resulted in a large reduction in the advised quota in contrast to quota increases in previous years. Since 2015 TACs have been set well above advised quotas but have not been fully utilised (ICES 2018), meaning that quotas have been non-restrictive in some Member States. Before 2015, individual cod were assigned to their stock of origin (western or eastern cod) only according to the area where they were caught. This was changed in 2015 when a split on the cod caught in the transition area (SD 24) was applied, assigning the individual cod in the transition area to either stock based on otolith shape analyses and genetics. A consequence of the split was that it was not possible to use the full time series back to 1970 as before, but the time series had to be truncated to 1994. In 2013, substantial recreational catches were also included in the western Baltic cod stock assessment (ICES 2013). This was one of the first stocks in Europe to take recreational catches into account in fisheries management advice. The western stock assessment is still conducted as a full age-based analytical assessment, because it is not facing the severe uncertainties, such as the absence of large cod and impaired growth, observed among eastern cod.

The quantities of cod discarded by Baltic cod trawlers have been estimated based on data collected by on-board observers on commercial fishing vessels since 1996 (early from an EU perspective). Since 2002 the monitoring has been done under the EC data collection regulations (EU No 199/2008, Commission Decision 2010/93/EU, EU No 2017/1004, Commission Decision 2016/1251/EU). The objective of the monitoring programme is exclusively scientific and the observers have no role in control and enforcement. The observers measure the volume of discarded catches and species and size composition of all the catch (discards and retained). All relevant biological information concerning discards is also recorded. Discard estimates for the eastern Baltic cod stock were included in the assessment of the stock between 2001–2014, until the assessment method changed. Since 2015 discards are included in the advice as part of the harvest rate and ICES advice is based on total catches (landings and discards). For the western Baltic cod discard data have been included in the assessment since 2002.

For a long time the discards of cod in the Baltic were considered relatively low compared with other areas. The average discard rate in the western Baltic cod stock was 8% (for the time period 1994–2017), and for the last three years (2015–2017) has been estimated at lower than 5% (ICES 2018). For the eastern stock, the discard rate was low at the beginning of the time-series but has increased considerably over time. In 2017 the discard rate was estimated at 11% (ICES 2018). Moreover, after a peak in 2013–2014, just before the introduction of the LO, the estimated discard rate fell but is still above long-term average for the stock (Fig. 10.1). Important to note is that MCRS (minimum conservation reference size) was reduced from 38 to 35 cm when the LO was introduced, which is a probable reason for the drop in discard rate

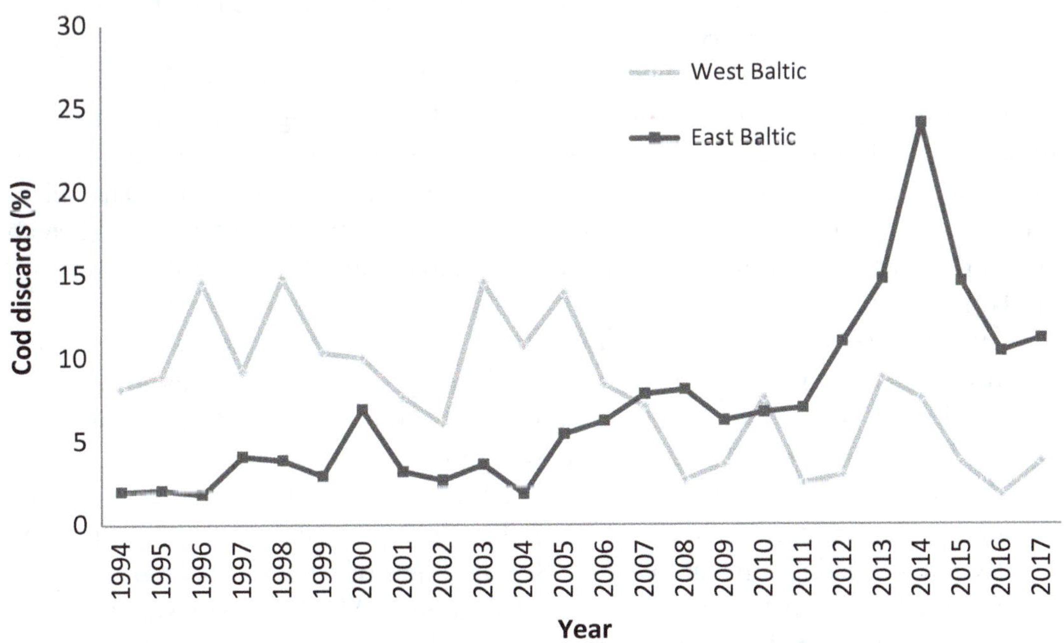

Fig. 10.1 Weight percentage of cod discarded between 1994 and 2017 for the Eastern and Western Baltic cod stocks. Catches combined from all gears (passive and active). (Data from ICES 2018)

in 2015. Furthermore, problems in gaining observer access in some countries led ICES to point out that the discard estimate is highly uncertain and is also considered to be an underestimate (ICES 2017c).

The observer coverage of at-sea sampling programmes is usually low (Uhlmann et al. 2014). In the Swedish and Danish programmes, typically only 0.5–1.5% of the fishing trips are covered. Discard data are also known to be highly variable in terms of quantities and proportions (Kelleher 2005), which makes estimates of overall volumes of discarded fish uncertain. In Sweden, for example, the proportion of cod below MCRS ranged between 2% and 43% in observed trips during 2012–2017.

10.3 Research to Improve Selectivity in Baltic Cod Trawls

Technical conservation measures have a long history in the Baltic Sea. Documented scientific trials with escape windows to reduce catches of young fish in trawls started over 100 years ago (Ridderstad 1915), and the Baltic Sea is an area where many documented selectivity experiments have been carried out (see Madsen 2007). Improving the yield of Baltic cod by increasing size selectivity (length of first capture) and reducing discards have been a management strategy for decades. A series of scientific experiments have followed with the aim of developing and identifying new concepts to improve selectivity in the trawl fishery. As such, the L50 (50% retention length) in a traditional diamond mesh codend can be increased by increasing the mesh size. However, a major focus in the Baltic Sea has been to make the selection curve steeper, which can be assessed by a lowered selection range (SR: L75-L25). By lowering SR, the escape of fish below a defined length (minimum landing or minimum conservation reference size; MLS or MCRS, respectively) is increased and the loss of marketable cod is reduced. This is important for the fishers because they do not want to lose any marketable fish. Fig. 10.2 illustrates the influence of SR for the retention of small and legal sized cod. Furthermore, to guarantee optimal selection, it is important that the selectivity is stable. The selectivity in traditional diamond mesh codends is normally reduced with increased catch weight due to closure of the mesh opening (Robertson and Stewart 1988; Tschernij and Holst 1999).

A common solution to improve trawl selectivity is to fit the net with an escape window ("window"), i.e., a panel with meshes (type and size) suitable to allow unwanted catch to escape. Typically, square mesh (netting turned 45°) panels are inserted into the net (mainly in the codend region), which ensures that meshes stay open in this region. The remaining part of the trawl is most often made of traditional diamond mesh netting. An advantage of the window is that it can be mounted directly in the existing trawl at low cost. Several types of windows have been tested in the Baltic Sea cod fishery over the years (Madsen 2007). The first tested selective codends were equipped with two side windows (known as the "Swedish window" codend), positioned close to the end of the codend (Tschernij et al. 1996). This codend had a higher selection factor (SF: 50% retention length L50/mesh size) and

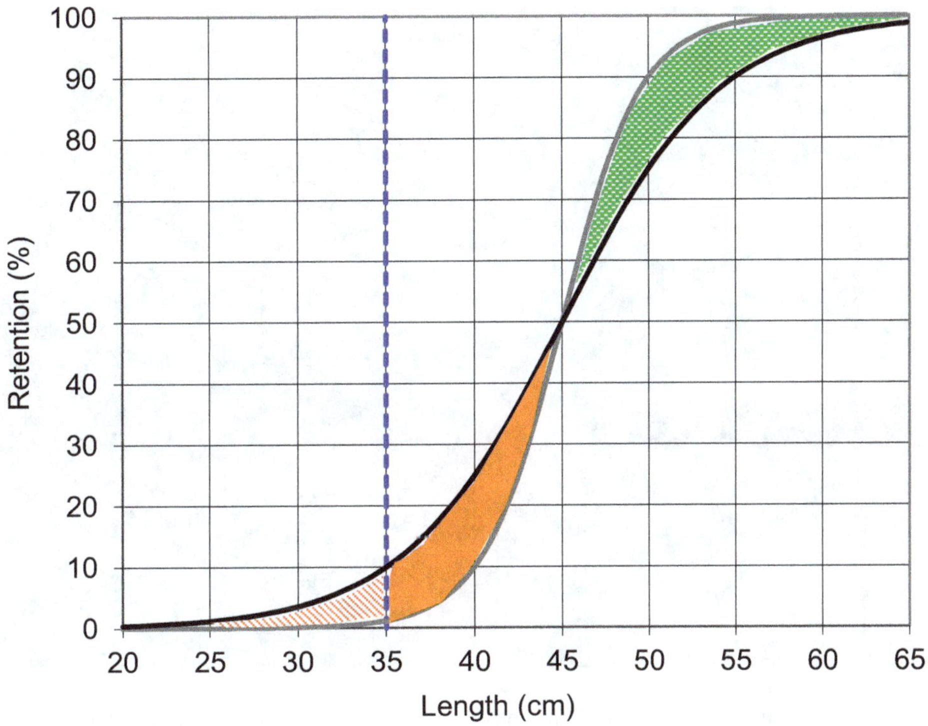

Fig. 10.2 Two hypothetical selection curves with identical L50 (45 cm) but with different selection ranges (SR). The grey curve has a lower SR (5 cm) and the black one, a higher SR (10 cm). Minimum size (35 cm) is indicated by the vertical dashed line. The red shaded area indicates differences in retention of undersized cod (higher retention with higher SR), the orange area indicates differences in retention of cod just above minimum size (higher retention with higher SR) and the green area indicates the difference in retention for larger sizes (higher retention with lower SR)

lower relative selection range (SRA: selection range SR/L50) than a standard diamond mesh codend (Madsen 2007). This was also true when compared to a window codend having side windows ending in front of the lifting strop (known as the "Danish window" codend; Lowry et al. 1995; Madsen et al. 1998). Later experiments demonstrated that the selectivity is improved when the window is moved all the way back to the end of the codend (Madsen 2000) and even further using a codend with a single top window (Madsen 2007). These findings formed the basis for the "Bacoma window" (Fig. 10.3) used today. The idea behind the Bacoma window was to change the selectivity by changing the mesh size in the escape window only (Madsen et al. 2002). This was expected to make stepwise changes easier and to reduce costs compared to extensive changes to the entire gear. Experiments indicated that the SF was relatively constant when mesh size was changed in the window (Madsen et al. 2002). An additional advantage of the Bacoma codend was that it is more flexible and easier to handle than a full square mesh codend. However, the knotless netting used in the window is relatively expensive and difficult to mend (Tschernij and Suuronen 2002).

Fig. 10.3 The standard codend used before introducing selective codends and the T90 and Bacoma codend used today with a simulated 500 kg catch, tested inside the flume tank in Hirtshals, Denmark. In the left column the full codend, in the middle column the last part of the codend where the catch is stored and in the right column the meshes in the area around the lifting strops. (Details about the codends are found in Madsen et al. 2015)

During the last decade scientists have focused on the selective properties of T90 meshes in the Baltic Sea cod fishery (ICES 2011). The T90 codend is a very simple and cheap way to improve the size selectivity of a trawl, because standard conventional netting material can be used. In T90 codends, the standard (T0) diamond mesh netting is turned 90°, which provides a more open mesh under pull (Fig. 10.3). The netting knot size, netting material and codend circumference will influence selectivity (Herrmann et al. 2007; Wienbeck et al. 2011).

In recent years there has been a change in some EU countries where the typical top-down focus (from managers or scientists) of gear research initiatives has become less pronounced, and focus has shifted to bottom-up initiatives where the fishing industry identifies gear issues and develop solutions to better suit their fisheries (see O'Neill et al., this volume). Some of these initiatives are described in ICES (2017b). There are several explanations for this trend. Apart from the obvious ones such as increased transparency, and industry expert knowledge regarding gears and practical know-how regarding their fisheries and catches, another important factor is changes in EU policy. Examples are increased regionalisation and the introduction of the Landing Obligation with its associated choke issues (Borges and Penas Lado, this volume) and a movement to more results-based management with less prescriptive legislation in the reformed CFP. Bottom-up initiatives can also increase industry buy-in and thus help to alleviate some of the mistrust of top-down science and legislation and circumvention of rules and recommendations (Suuronen et al. 2007; Catchpole and Gray 2010; Krag et al. 2016).

Examples of recent or ongoing bottom-up gear related initiatives in the Baltic region are projects like MINIDISC and FastTrack in Denmark and the Selective Fisheries Secretariat in Sweden. The MINIDISC project evaluated the effectiveness of reducing unwanted catches through free gear choice and full documentation (Mortensen et al. 2017). Interestingly, the study reported that although the chosen gears indicated a lowered selectivity, the discard ratio decreased due to either higher landings or lower discards. However, as the technical descriptions of gears and fishing behaviour was limited, it is possible that changes other than gear selectivity might have influenced the results of the study. The FastTrack project and the Selective Fisheries Secretariat are very similar in terms of objectives, set-up and work procedures (Valentinsson 2016, www.fast-track.dk). Both involve an initial selection process of industry ideas where managers, industry and scientific representatives are involved. Accepted projects start with an iterative test and development phase of the gear modification in commercial practice led by the industry participants where they collect some data. A second phase is a scientific trial and evaluation. The large majority of projects within these two initiatives is focused on an improvement and documentation of the available toolbox of gears available for use by fishers to adapt to the LO (Valentinsson 2016).

Three of the projects under the Selective Fisheries Secretariat have focused on reducing unwanted catches in Baltic cod trawl fishery (Nilsson et al. 2018). One project was unsuccessful in reducing the catch of small cod; another showed increased selectivity but included a technically complex gear that cannot be used due to several conflicts with current regulations. The third project studied a modified T90 codend (slightly reduced mesh size, increased codend length and increased circumference). Those results showed that the modified codend reduced the catches of undersized cod (Nilsson 2018). Based on these results, a joint recommendation from BALTFISH in May 2017 to COM suggested allowing this T90 codend as an alternative. After evaluation by STECF, COM approved the recommendation and the new T90 codend was introduced February 1st, 2018 (Commission Delegated Regulation (EU) 2018/47).

10.3.1 Technical Conservation Measures – the Baltic History

The intensive research on improving the selectivity in Baltic cod trawls has led to several legislative changes since 1995. Before that, regulations for cod trawls stipulated a minimum mesh size of 105 mm and a minimum landing size (MLS) of 33 cm. In 1994 the IBSFC changed the technical measures, which was followed by a series of changes over the next decade. From 1995, the baseline mesh size was increased to 120 mm and MLS to 35 cm in order to protect young cod. By derogation a 105 mm codend remained allowed if one of two designs of exit windows was used. This was one of the first EU regulations where selective devices were adopted into legislation (Council EC regulation No 2250/95; Madsen 2007; Feekings et al. 2013). The technical descriptions of the two windows (Danish and Swedish) were very detailed; this was the start of a practice that continued in the following Baltic trawl specifications. In 1998 these gear rules were transposed into EU legislation (Council Reg (EC) No. 88/98). Based on advice from a large international scientific project (the Bacoma project, Suuronen et al. 2000), starting in 2002, IBSFC changed both exit windows with two alternative codends. This meant that either a 130 mm codend or a new codend with 105 mm mesh size and a knotless square mesh window of 120 mm (i.e., the Bacoma codend) had to be used. Use of the Bacoma codend was widespread in early 2002 but due to the increased L50 (~10 cm), initial catch losses for trawlers that used the Bacoma codend were substantial (Tschernij et al. 2004). Therefore, most trawlers rapidly switched to the alternative 130 mm diamond mesh codend or manipulated their Bacoma codends to decrease selectivity (Suuronen et al. 2007), with increased discarding as a result. This led to an EU emergency closure of Baltic cod fisheries in April 2003. When the fishery reopened in August 2003, the Bacoma panel mesh size was reduced to 110 mm, conventional diamond mesh codends were prohibited and cod MLS was increased to 38 cm. The changed mesh size was supposed to better match the new 38 cm MLS (Valentinsson and Tschernij 2003). In 2005 the current technical regulation (Council Reg (EC) No 2187/2005) was introduced and the 110 mm Bacoma codend was allowed. For a few years, the 110 mm Bacoma codend was the only legally approved gear. In 2006, a T90 codend was introduced as an alternative after an evaluation of existing data by ICES, which did not find any difference in selectivity between the 110 mm T90 codend and the 110 mm Bacoma (ICES 2007; Suuronen et al. 2007). The next major change occurred in 2010 when the mesh size of the T90 codend and the Bacoma window was increased from 110 mm to 120 mm to further decrease catches of juvenile cod. The length of the Bacoma window was also extended. The latter measure was to prevent selectivity from decreasing at high catch rates (Madsen et al. 2010). A follow-up analysis of Danish discard data demonstrated that these improvements in selectivity contributed to a reduction of cod discards (Feekings et al. 2013).

In 2014 the regionalisation within the reformed CFP gave Member States more power over technical regulations as part of temporary discard plans in accordance with articles 15(6) and 18(3) of the basic Regulation (Regulation (EU) No 1380/2013). The Member States around the Baltic Sea, organised in the regional group called Baltfish, suggested in their first joint recommendation to the Commission that the minimum conservation reference size (MCRS) should be lowered from 38 to 35 cm as an "efficient and speedy way to minimise cod discards" when the Landing Obligation was introduced 2015 (Anon 2014a). The commission adopted this proposal in their discard plan for 2015–2017 and prolonged it via a delegated act from 2018 (delegated regulation No. 2018/306). The new alternative 115 mm T90 codend was introduced on first February 2018 after a joint recommendation from Baltfish (Commission delegated regulation (EU) 2018/47). This was the first change to Baltic cod trawl regulations since 2010.

Expected changes in size selectivity of Baltic cod trawls over the time period 1994–2018 is shown in Fig. 10.4, where the estimated L50 values and minimum

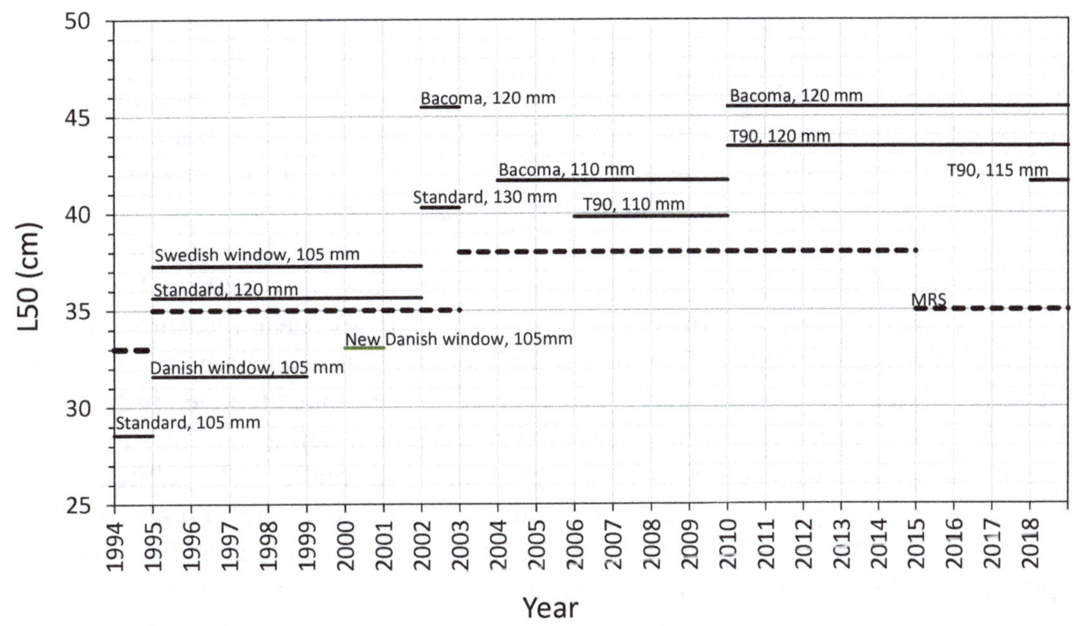

Fig. 10.4 Changes in estimated codend selectivity (L50) in Baltic cod trawls during the past 25 years (some information from Feekings et al. 2013). The minimum mesh opening was set to 120 mm (Bacoma) from January 1st, 2010 in subdivisions 22–24 (western Baltic) and from March 1st, in sub-Divisions 25–32 (Eastern Baltic). The L50 for the T90 120 mm codend is taken from Wienbeck et al. (2011). All other L50 values are taken from Madsen (2007). The same selectivity estimates are assumed for the Bacoma 120 mm and the new Bacoma 120 mm introduced in 2010

sizes over time are indicated. The estimated effect from the changes in selectivity is a continuous increase in L50, giving a total increase of about 15 cm since the early 1990s. This seems to be in conflict with the high discard rates reported in recent years. One reason could be that the estimated selectivity obtained from scientific sea trials does not necessarily reflect the realised selectivity in commercial practice because vessel type, engine power, gear choice, −design and -configuration may all influence selectivity. More information about commercial fishing practice and compliance is necessary to clarify this. In fact, the manipulation of the selectivity in trawls by fishers has been documented (ICES 2017c). Another potential reason is the considerable change in size structure of (mainly Eastern) Baltic cod; small individuals now dominate the stock and very few individuals are larger than 45 cm (ICES 2018). Thus, although the trawls used may be size selective, the catches still have a large proportion of small cod due to the truncated size structure of the fished population.

10.3.2 Technical Conservation Measures – Since the Introduction of the Landing Obligation

The dynamic period of many changes regarding technical measures for Baltic cod trawls in the 2000s was followed by a calmer period after 2010 (Fig. 10.4). However, renewed calls for changes to the technical measures emerged with the LO introduction for Baltic cod in 2015. Industry representatives in the Baltic Sea advisory council have repeatedly declared that the current detailed gear measures have hampered the implementation of the LO, and that a new technical framework with more flexibility is urgently needed (BSAC 2017). In 2015 industry representatives of the Baltic Sea Advisory Council (BSAC) had already recommended amending Reg. 2187/2005 by deleting the reference to the specifications of Bacoma and T90 to allow the use of 105 mm codends as under pre-1995 law (BSAC 2015). Despite all of the existing science and historical regulatory changes, gear measures are still today hotly debated in regional forums and Member States, especially in light of the Landing Obligation and unabated discarding. As mentioned above, information from the industry and observers suggests that trawls are being modified to reduce selectivity, leading to catching a higher proportion of smaller fish (ICES 2017c). This is not a new phenomenon in the Baltic (Valentinsson and Tschernij 2003; Suuronen et al. 2007). ICES (2007) also mentioned that considerable differences in opinion prevailed among Member States and that scientific arguments have gotten lost in a largely emotional debate. This may be one explanation for the lack of cooperation on this issue. Strikingly, in spite of >3 years with the LO and its increased mandate for regional proposals via discard plans, only one joint recommendation for an alternative trawl has been proposed. For control and documentation measures, joint recommendations are equally scarce from the Baltfish regional group of Member States.

A new technical framework regulation is now being negotiated in the EU. Available draft texts seem to consolidate the current two codends (120 mm Bacoma or 120 mm T90) as minimum requirements, with options to adopt joint recommendations from regional groups for alternative gear measures, provided they are at least equivalent in terms of limiting unwanted catches as compared to the baseline gears. In essence, the revision of the technical measures framework does not represent a great change, as regional gear proposals (and control measures) have already been possible via discard plans since the introduction of the LO (art. 15(6) of Regulation (EU) No. 1380/2013).

10.4 Effects of the Landing Obligation on Scientific Data for Stock Assessments

When the Landing Obligation went into effect, it was expected to generate more accurate data because all catches of species that came under this obligation should have been documented and landed. During the years preceding the LO, there were even discussions about whether scientific sea-sampling programmes would still be needed (Anon 2014b; STECF 2014), as it appeared to be more cost-efficient to sample landed catches than board vessels at sea. Proper documentation and landing unwanted catches was also the underlying assumption for increased quotas, as landing quotas would be turned into catch quotas, based on historical discard estimates.

In reality, however, 3 years of the LO in the Baltic Sea trawl fisheries for cod indicate that a majority of the unwanted catches of cod are neither brought ashore nor documented properly. In Sweden, for example, 2–4% of the total cod landings in the fisheries statistics were reported as cod below MCRS (recorded as below minimum size or 'BMS' in log books) between 2015 and 2017 (Fig. 10.5a). The estimated

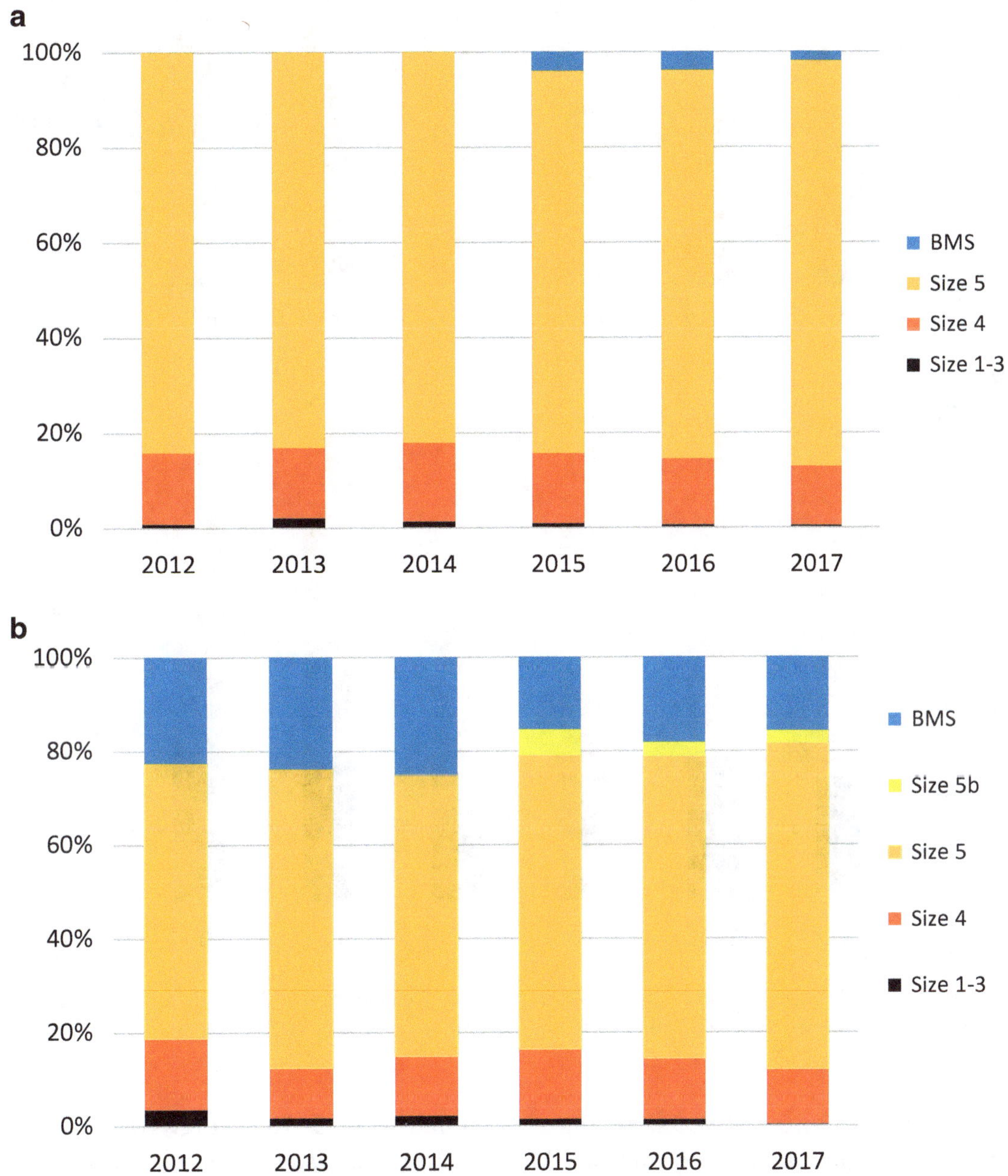

Fig. 10.5 Relative size composition of cod catches from Swedish Baltic trawl fisheries between 2012 and 2017 from (**a**) reported catches (logbooks and sales slips) and (**b**) data from the scientific observer programme. Size 1–3 are cod with an individual weight > 2 kg, size 4 are cod in the range of 1–2 kg and size 5 are cod with an individual weight between 0.3 and 1 kg. 5b are cod with a length 35–37 cm. BMS are cod smaller than the minimum conservation reference size (MCRS)/ minimum landing size (MLS). MLS was 38 cm until 2015 and MCRS 35 cm from 2015 onwards

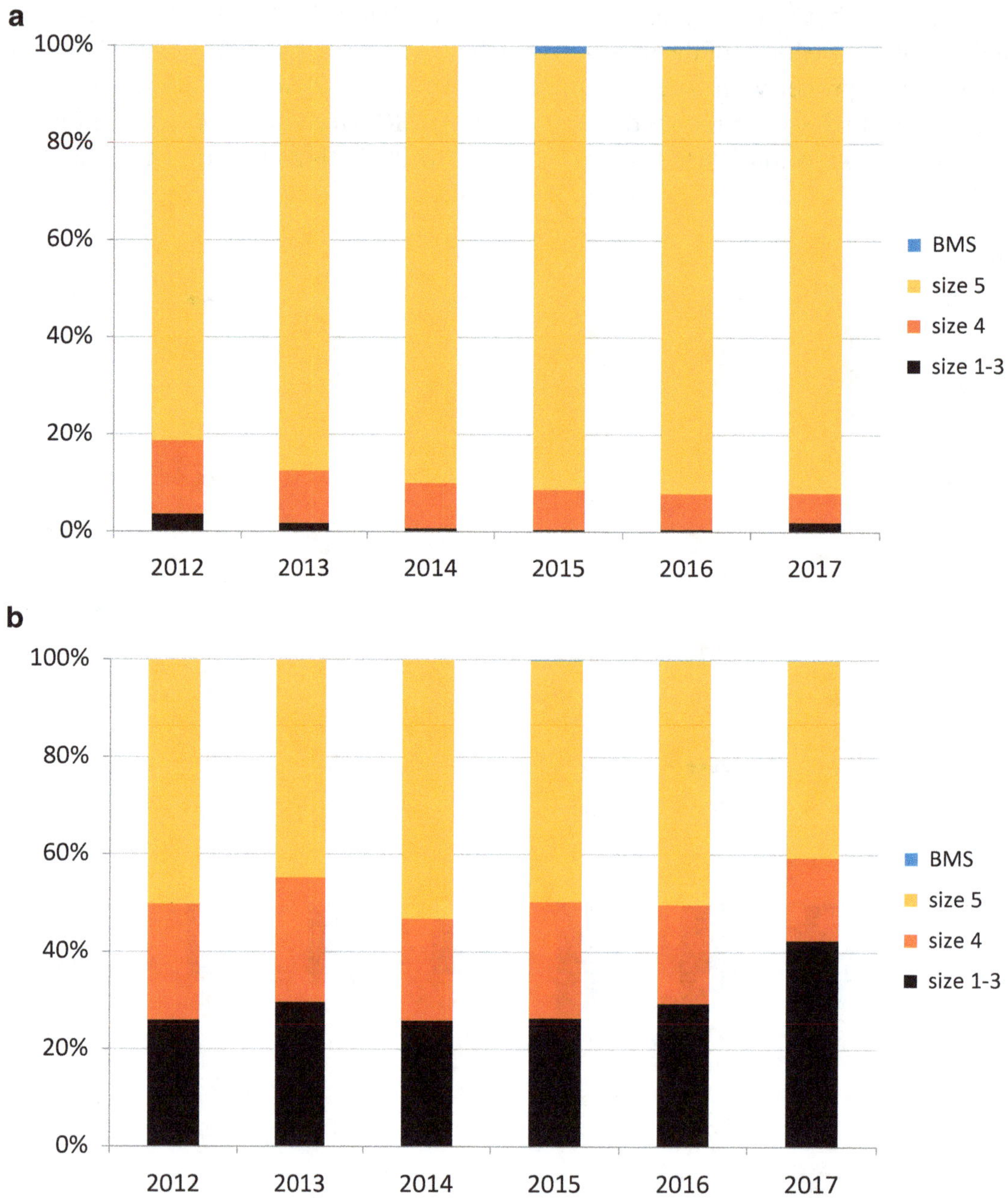

Fig. 10.6 Relative size composition of cod catches from Danish trawl fisheries based on logbooks and sale slips between 2012 and 2017 in the (**a**) Eastern and (**b**) Western Baltic Sea

discard rate from the at-sea sampling programme was 15–18% (Fig. 10.5b), indicating substantial underreporting of BMS cod. In reality, seagoing observers noted that most of the BMS cod were discarded. Similarly, 0.5–1.5% of the total landings in Denmark are reported as BMS for the eastern Baltic cod (Fig. 10.6a) and less than 1% for the western Baltic cod (Fig. 10.6b). Average discard rates from observer trips for the same years vary between 23 and 9%, respectively, indicating that most of the BMS cod is discarded and is also unreported by Danish vessels. Several reasons may explain why the BMS fraction of the catch may not be landed: resistance to changing well-established commercial fishing practices (i.e. from legally discarding to being forced to land small fish); and primarily economic reasons, because it is illegal to sell the BMS fish for human consumption purposes and valorisation options may be underexplored (see also Iñarra et al., this volume). Other reasons may include a lack of buyers and inconvenient handling processes of small cod. In some fisheries, high-grading is still considerable. In these fisheries it seems rather unlikely that the BMS fraction will be landed as long as the enforcement is weak. In the Baltic, however, high-grading has not been considered to be a major problem because there is a market for smaller cod.

The amount of unwanted catches of cod decreased somewhat after the LO was introduced in 2015 (Fig. 10.5b). This decrease is partly a consequence of the reduction in MCRS from 38 cm to 35 cm that occurred at the same time (the 35–38 cm cod is labeled as 'size 5b' in Fig. 10.5b). Landings of the commercial sizes of cod from the Swedish observer trips are also shown. Failure to land unwanted cod catches and report them is widespread and observed far beyond Sweden and Denmark. The proportion of BMS cod landings as reported to ICES by all countries was less than 1% (0.7% for eastern Baltic and 0.5% for western Baltic cod). The assessment working group could thereby not rely on these reported figures as total catches and have therefore used data on unwanted cod from observer programmes instead (ICES 2018).

The European Fisheries Control Agency (EFCA) coordinates the Member States in a control monitoring programme (i.e. last haul; see Nuevo et al., this volume). The objectives are to evaluate compliance with the LO, to compile catch composition data for use in a risk management strategy and to provide information about where and when discards are expected in a particular fishery. Last haul data are not easily available for scientific use. Data acquired from the last haul inspections do not provide all needed biological information and vessels are not selected at random. However, last haul data are an important source of catch information that can be used to validate data from the scientific observer programmes (e.g. by ICES WGBFAS). Better availability of detailed last haul data is therefore needed for scientific purposes. In both Sweden and Denmark, the observed catches below MCRS from the last haul controls have been similar to those estimated from the observer programmes. For example, Swedish last haul data for 2016/17 indicated 12.5% unwanted catches of cod, which is similar to the estimate from the observer data (Fig. 10.5b). Note that both these estimates are considerably larger than self-reported BMS catches by fishers (Fig. 10.5a). Data obtained from the last haul inspections are not easily available nor are they made public due to the confidentiality and potential implications of compliance breaches. Knowledge of unwanted cod catches and exploitation pattern on the stock thereby still heavily rely on data from the scientific sea-sampling programmes that are publically available, primarily through different scientific reports (STECF 2017; ICES 2018), but which may suffer from a significant observer bias (see below).

The introduction of the Landing Obligation has impacted the quality of scientific observer data as well. EU vessels are legally required to allow observers on board. In practice, however, scientific observers only board vessels after permission by the skipper when minimum safety standards are met. In Sweden, the initial response from the fishing industry was strongly opposed to the policy in 2015 when Baltic cod was phased into the LO. This opposition led to denial of observer access to many of the vessels. Sweden was only able to conduct 5 out of 24 planned observer trips on cod trawlers in the Baltic (Anon 2015), leading to the first time that discard estimates could not be provided to ICES. In response, in 2016 the Swedish Agency for Marine and Water Management (the responsible agency) changed the national legislation to fine vessels selected for sampling since 2016 if they refuse to take observers without a valid reason. Denmark did not have the same problem as Sweden initially but the sensitivity of the skippers' willingness to take observers became apparent after a national media debate during early 2018. A news story reported on the discrepancy between quantities of landed and self-reported BMS cod by fishers in contrast to those from scientific observer estimates from Danish trips (Anon 2018b). The examples from both countries highlight the risks for scientific data collection. STECF has expressed concern that increased refusal rates are causing a deterioration in data quality for scientists, and has requested more information on observer refusal rates from Member States (STECF 2016).

Introduction of the LO might introduce other types of biases than restricted access to all active vessels. Before 2015, discarding was not only legal but also even mandatory for undersized catches. When this act became illegal in 2015, the role of observers thus changed from documenting a legal act to documenting an illegal act. This could lead to an increased observer effect, i.e. that fishers change their behaviour when observers are aboard (Kelleher 2005; Anon 2014b; STECF 2016). Fishers can change their behavior in the presence of an observer in several ways: by changing fishing grounds for places were small fish occur less frequently; by changing gear; and by not discarding fish that they normally would discard. Introduction of the LO thereby requires more validation of observer data to control for these potential biases.

10.5 Conclusions – Lessons from the Landing Obligation in the Baltic Cod Trawl Fishery and Future Prospects

We have described that the intentions and expected benefits of the EU Landing Obligation have not been fulfilled after more than 3 years of the LO in the Baltic cod trawl fishery:

(i) Discarding of unwanted catches still occurs at rates roughly comparable to the years before the Landing Obligation, in spite of a lowered MCRS at the onset of the LO in 2015. In fact, there are even indications that discarding may have increased.

(ii) Coupled to the continued discarding of cod, there is also an important element of bad timing and bad luck. The timing of the introduction of the Landing

Obligation for Baltic cod was fixed in the basic Regulation (Reg (EU) No 1380/2013) but coincided with a period of negative developments for the larger eastern Baltic cod stock. The reduced growth and condition and a truncated size distribution without larger cod forced the fishery to target cod around the minimum size. This inevitably resulted in unwanted catches due to the inherent selection range in trawls. The western cod stock was also in a depleted state with continuous TAC reductions since before the LO introduction in 2015. These factors have caused severe economic implications for the fisheries, affected the industry perspective of the future, and thus most likely complicated the LO implementation for the Baltic cod stocks.

(iii) The aim of improved data quality for stock assessment has not been met. The required documentation and landing of unwanted catches is at least an order of magnitude lower than estimated volumes from independent estimates by scientific observer programmes and last haul inspections. Fishermen consistently underreport catches, thus the stock assessments done by ICES cannot rely on official catch data only, but needs to include observer data. At the same time, the role of observers has most likely changed with the LO as they are now supposed to observe and quantify an illegal act. This change has resulted in data shortages due to increased refusal rates of access to vessels and may also have changed the representativeness of the estimates due to a changed observer effect. These trends therefore indicate that scientific data quality has not improved but rather worsened. As unwanted catches are underreported, continued catch sampling from scientific observers is a prerequisite for reliable data. Given the uncertainty of observer data in the LO context, the EFCA last-haul inspections can potentially also provide independent information of unwanted catches although care must be taken due to certain limitations of the sampling methods. The data from these inspections are not easily available however, even for scientific purposes. This needs to change: to make evaluations of observer data possible, to (potentially) increase the quality of stock assessments, and to make the best use of limited public funding. One issue related to stock assessment quality effects of the LO is also the untimely discovery of major biological uncertainties for the eastern stock that negatively affected preparedness for the LO. The uncertainties led ICES to downgrade the assessment from a full analytical assessment to an index based assessment, which also changed the perception of stock status (and advised TAC) negatively just when the LO was introduced in 2015. The discontinuation of the analytical assessment negatively affected the understanding of stock status and development (including year class variability). Furthermore, the index based assessment currently applied does not use discard data, although these are still important for the catch advice from ICES. Large research initiatives are currently occurring in the Baltic countries in order to understand the biological uncertainties and to enable a rollback to full analytical assessment for eastern Baltic cod (ICES 2017a). The western stock assessment is still conducted as a full age-based analytical assessment. Still, implementation of the LO was most likely aggravated by reduced TACs, due to several years of reduced stock size and poor recruitment, before and after the onset of the LO in 2015.

(iv) There are no indications of increased gear selectivity since the introduction of the LO. In fact, there are even anecdotal reports of gear manipulation by fishers to reduce selectivity (ICES 2017c). The increased regional mandate to propose modifications to the current detailed and prescriptive gear regulations via joint recommendations has only been used once so far. This is surprising given the long and voluminous history of scientific selectivity trials, the more recent bottom-up gear research initiatives on Baltic cod trawls, the heated debate around continued discarding and the persistent criticism of current trawl specifications. More constructiveness and responsibility from Member States and regional groups is needed to create positive change and to facilitate the industry to adapt their trawls to the LO requirements. As mentioned above (Conclusion, ii), the lack of larger cod has resulted in increased unwanted catches due to the selection range in trawls. If selectivity (as reported) is further negated by gear manipulation to reduce losses of legal sized cod in this fishery, that struggles with economic viability, this of course further increases the unwanted catches.

(v) So what is needed to make the Landing Obligation work? The expectation that the LO would encourage fishers to avoid unwanted fish is based on an elegant idea but will most likely remain a pious hope until enforcement and incentives are aligned with that goal. The theoretical mechanism behind the expectation is that the LO will result in less unwanted catch via increased gear selectivity (or avoidance) based on the idea that unwanted catches costs quota and are less lucrative than wanted catch. The cost of unwanted fish is thereby supposed to be internalised for the fisher (Catchpole et al. 2017). This internalisation of costs will, however, not be realised if the risk of being caught discarding remains as insignificant as it is today. One of the key challenges is thus to shift the control and monitoring focus to what actually happens at sea, including the use of technologies like remote electronic monitoring (EM; see James et al., this volume). However, given that available control resources are not infinite, it is also essential to develop strong incentives to encourage best practice mitigation methods or behaviours (see Kraak and Hart, this volume). Based on the Baltic experiences so far, the calls for increased flexibility to allow fishers to choose gear solutions probably need to be coupled to increased documentation responsibility for fishers that opts in on freer gear choice (cf. Mortensen et al. 2017). An example of a sound supportive incentive could be to create a twin-tier structure and only allocate the estimated discard share of the quota to vessels that opt to use gears with a proven higher selectivity or that have full documentation via EM, and at the same time subtract the estimated discards from the quota for vessels that opt out of these measures. Such a structure would also be in accordance with Member States' responsibilities for the allocation of quotas in the basic regulation (art. 17 of Reg (EU) No 1380/2013). Other forms of incentives worth exploring may be to stimulate uptake of trawls with desirable selectivity by granting exclusive access to fishing locations (e.g. Real Time Closures RTCs and/or permanent areas) or time periods for fishers who opt to increase selectivity (Madsen and Valentinsson 2010; Condie et al. 2014). Experiences from other areas with discard bans indicate that such additional management measures are required to incentivise a move towards more selective fishing under a discard ban (MRAG 2007; Condie et al. 2014).

References

Anon. (2014a). Baltfish Joint Recommendation No 1, 27 May, 2014. 28 pp. http://www.bsac.dk/

Anon. (2014b). Report of the regional co-ordination meeting for the Baltic (RCM Baltic). 151 pp. https://datacollection.jrc.ec.europa.eu/docs/rcm/2014

Anon. (2015). Annual report for the Swedish national programme for collection of fisheries data 2015. 60 pp. https://datacollection.jrc.ec.europa.eu/ars/2015

Anon. (2018a). Discarding and the landing obligation. https://ec.europa.eu/fisheries/cfp/fishing_rules/discards_en

Anon. (2018b). http://www.fiskerforum.dk/erhvervsnyt/a/dtu-smidt-i-land-af-vrede-fiskere-20022018

Bagge, O., & Thurow, F. (1994). The Baltic cod stock: Fluctuations and possible causes. *ICES Marine Science Symposia 198*, 254–268.

Borges, L., & Penas Lado, E. (this volume). Discards in the common fisheries policy: The evolution of the policy. In S.S. Uhlmann, C. Ulrich, S.J. Kennelly (Eds.), *The European Landing Obligation – Reducing discards in complex multi-species and multi-jurisdictional fisheries*. Cham: Springer.

BSAC. (2015). The Baltic Sea Advisory Council's recommendations on technical measures. 29th September 2015. Ref: BSAC/2015/6. 4 pp. http://www.bsac.dk/BSAC-Resources/BSAC-Statements-and-recommendations

BSAC. (2017). BALTFISH / BSAC / EFCA joint workshop on monitoring, control and enforcement of the Landing Obligation. 9th March 2017. Final Report. 6 pp. http://www.bsac.dk/meetings

Casini, M., Käll, F., Hansson, M., Plikshs, M., Baranova, T., Karlsson, O., et al. (2016). Hypoxic areas, density dependence and food limitation drive the body condition of a heavily exploited marine fish predator. *Royal Society Open Science*, *3*, 160416.

Catchpole, T.C., & Gray, T.S. (2010). Reducing discards of fish at sea: A review of European pilot projects. Journal of Environmental Management, *91*(3), 717–723.

Catchpole, T.L., Riberio-Santos, A., Mangi, S.C., Hedley, C., Gray, T.S. (2017). The challenges of the landing obligation in EU fisheries. *Marine Policy, 82*, 76–86.

Chabot, D., & Dutil, J.D. (1999). Reduced growth of Atlantic cod in non-lethal hypoxic conditions. *Journal of Fish Biology, 55*, 472–491.

Condie, H.M., Grant, A., Catchpole T.L. (2014). Incentivising selective fishing under a policy to ban discards; lessons from European and global fisheries. *Marine Policy, 45*, 287–292.

Eero M., Köster F.W., Plikshs M., Thurow F. (2007). Eastern Baltic cod (Gadus morhua callarias) stock dynamics: Extending the analytical assessment back to the mid-1940s. *ICES Journal of Marine Science*, *64*, 1257–1271.

Eero, M., Mackenzie, B.R., Köster F.W., Gislason, H. (2011). Multi-decadal responses of a cod (*Gadus morhua*) population to human-induced trophic changes, fishing, and climate. *Ecological Applications*, *21*(1): 214–226.

Eero M., Köster F.W., Vinther, M. (2012a). Why is the Eastern Baltic cod recovering? Marine Policy, 36 (1): 235–240.

Eero, M., Vinther, M., Haslob, H., Huwer, B., Casini, M., Storr-Paulsen, M., et al. (2012b). Spatial management of marine resources can enhance the recovery of predators and avoid local depletion of forage fish. *Conservation Letters*, *5*, 486–492.

Eero, M., Hjelm, J., Behrens, J., Buchmann, K., Cardinale, M., Casini, M., et al. (2015). Eastern Baltic cod in distress: Biological changes and challenges for stock assessment. *ICES Journal of Marine Science*, *72* (8), 2180–2186.

Feekings, J. P., Lewy, P., Madsen, N. (2013). The effect of regulation changes and influential factors on Atlantic cod discards in the Baltic Sea demersal trawl fishery. *Canadian Journal of Fisheries and Aquatic Sciences*, *70*, 534–542.

Herrmann, B., Priour, D., Krag, L. (2007). Simulation-based study of the combined effect on cod-end size selection of turning meshes by 90° and reducing the number of meshes in the circumference for round fish. *Fisheries Research*, *84*, 222–232.

Hüssy, K., Mosegaard, H., Albertsen, C.M., Nielsen, E.E., Hemmer-Hansen, J., Eero, M. (2016). Evaluation of otolith shape as a tool for stock discrimination in marine fishes using Baltic Sea cod as a case study. *Fisheries Research*, *174*, 210–218.

ICES. (2007). Report of the ICES advisory committee on fishery management, advisory committee on the marine environment and advisory committee on ecosystems. ICES advice. Book 8, 147 pp. Section 8.3.3.3: ICES response to EU on selectivity of active gears targeting cod in the Baltic Sea.

ICES. (2011). Report of the study group on turned 90° codend selectivity, focusing on Baltic cod selectivity (SGTCOD). 4-6 May 2011, ICES Headquarters, Copenhagen. ICES CM 2011/SSGESST:08. 44 pp.

ICES. (2013). Report of the Baltic Fisheries Assessment Working Group (WGBFAS), 10–17 April 2013, ICES Headquarters, Copenhagen. ICES CM 2013/ACOM:10. 747 pp.

ICES (2017a). Report of the Workshop on Biological Input to Eastern Baltic Cod Assessment (WKBEBCA), 1–2 March 2017. ICES CM 2017/SSGEPD: 19. 40 pp.

ICES (2017b). Interim report of ICES-FAO Working Group on Fishing Technology and Fish Behaviour (WGFTFB), 4-7 April 2017, Nelson, New Zealand. ICES CM 2017/SSGIEOM: 13.

ICES. (2017c). Cod (*Gadus morhua)* in subdivisions 24-32, eastern Baltic stock (eastern Baltic Sea). ICES advice on fishing opportunities, catch and effort. Version 4: 8 March 2018. 10 pp. https://doi.org/10.17895/ices.pub.3096.

ICES. (2018). Report of the Baltic fisheries assessment working group (WGBFAS), 6–13 April 2018, ICES Headquarters, Copenhagen, Denmark. ICES CM 2018/ACOM:11.

Iñarra, B., Bald, C., Cebrián, M., Antelo, L.T., Franco-Uría, A., Vázquez, J.A., Pérez-Martín, R., Zufía, J. (this volume). What to do with unwanted catches: Valorisation options and selection strategies. In S.S. Uhlmann, C. Ulrich, S.J. Kennelly (Eds.), *The European Land-ing Obligation – Reducing discards in complex, multi-species multi-juridictional fisheries*. Cham: Springer.

James, K.M. Campbell, N., Viðarsson, J.R., et al. (this volume). Tools and technologies for the monitoring, control and surveillance of unwanted catches. In S.S. Uhlmann, C. Ulrich, S.J. Kennelly (Eds.), *The European Landing Obligation – Reducing discards in complex multi-species and multi-jurisdictional fisheries*. Cham: Springer.

Kelleher, K. (2005). Discards in the world's marine fisheries. An update. FAO Fisheries Technical Paper. 470. 131 pp.

Köster, F.W., Schnack, D., Möllmann, C. (2003). Scientific knowledge of biological processes that are potentially useful in fish stock predictions. *Scientia Marina*, *67* (1), 101–127.

Kraak, S.B.M, & Hart, P.J.B. (this volume). Creating a breeding ground for compliance and honest reporting under the Landing Obligation: Insights from behavioural science. In S.S. Uhlmann, C. Ulrich, S.J. Kennelly (Eds.), *The European Landing Obligation – Reducing discards in complex multi-species and multi-jurisdictional fisheries*. Cham: Springer.

Krag, L.A., Herrmann, B., Feekings, J., Karlsen, J.D. (2016). Escape panels in trawls – A consistent management tool? *Aquatic Living Resources*, *29*, 306.

Lowry, N., Knudsen, L.H., Wileman, D.. (1995). Selectivity in Baltic cod trawls with square mesh codend windows. ICES CM, 1995/B:5.

MacKenzie, B.R., Alheit, J., Conley, D.J., Holm, P., Kinze, C.C. (2002). Ecological hypotheses for a historical reconstruction of upper trophic level biomass in the Baltic Sea and Skagerrak. *Canadian Journal of Fisheries and Aquatic Sciences*, *59*, 173–190.

Madsen, N. (2000). Experimental adjustments of the escape window position in trawl codends – implications for Baltic Sea cod fishery. *Meddelanden från Havsfiskelaboratoriet i Lysekil*, *329*, (ISSN: 1103-4777).

Madsen, N. (2007). Selectivity of fishing gears used in the Baltic Sea cod fishery. *Reviews in Fish Biology and Fisheries*, *17*, 517–544.

Madsen, N., & Valentinsson, D. (2010). Use of selective devices in trawls to support recovery of the Kattegat cod: A review of experiments and experience. *ICES Journal of Marine Science 67*(9), 2042–2050.

Madsen, N., Moth-Poulsen, T., Lowry, N. (1998). Selectivity experiments with window codends fished in the Baltic Sea cod (*Gadus morhua*) fishery. *Fisheries Research*, *36*, 1–14.

Madsen, N., Holst, R., Foldager, L. (2002). Escape windows to improve the size selectivity in the Baltic cod trawl fishery. *Fisheries Research 57*, 223–235.

Madsen, N., Tschernij, V., Holst, R. (2010). Improving selectivity of the Baltic cod trawl fishery: Experiments to assess the next step. *Fisheries Research 103*, 40–47.

Madsen, N., Hansen, K., Madsen, N.A.H. (2015). Behavior of different trawl codends concepts. *Ocean Engineering 108*, 571–577.

Mortensen, L.O., Ulrich, C., Eliasen, S., Olesen, H.J. (2017). Reducing discards without reducing profit: free gear choice in a Danish result-based management trial. *ICES Journal of Marine Science*, *74* (5), 1469–1479.

MRAG. (2007). Impact assessment of discard policy for specific fisheries. European commission studies and pilot projects for carrying out the common fisheries policy no. FISH/2006/17 –Lot 1, Brussels. 289 pp.

Nielsen, E.E., Hansen, M.M., Ruzzante, D.E., Meldrup, D., Grønkjær, P. (2003). Evidence of a hybrid-zone in Atlantic cod (*Gadus morhua*) in the Baltic and the Danish Belt Sea revealed by individual admixture analysis. *Molecular Ecology*, *12*, 1497–1508.

Nielsen, E.E., Grønkjær, P., Meldrup, D., Paulsen, H. (2005). Retention of juveniles within a hybrid zone between North Sea and Baltic Sea Atlantic cod (*Gadus morhua*). *Canadian Journal of Fisheries and Aquatic Sciences, 62*, 2219–2225.

Nilsson, H.C. (Ed). (2018). Sekretariatet för selektivt fiske-Rapportering av 2016 och 2017 års verksamhet. Aqua reports 2018:4. Swedish University of Agricultural Sciences, Department of Aquatic Resources, Lysekil, 211 pp. ISBN: 978-91-576-9557-4 (electronic version).

Nilsson, H.C., Andersson, E., Hedgärde, M., Königson, S., Ljungberg, P., Lunneryd, S-G., et al. (2018). Projects accomplished by the selective fisheries secretariat 2014–2017: A synthesis report, Aqua reports 2018:13, Swedish University of Agricultural Sciences, Department of Aquatic Resources, Lysekil, 26 pp, ISBN: 978-91-576-9576-5 (electronic version).

Nuevo, M., Morgado, C., Sala, A. (this volume). Monitoring the implementation of the Landing Obligation: The last Haul programme. In S.S. Uhlmann, C. Ulrich, S.J. Kennelly (Eds.), *The European Landing Obligation – Reducing discards in complex, multi-species and multi-jurisdictional fisheries*. Cham: Springer.

O'Neill, F.G., Feekings, J., Fryer, R.G., Fauconnet, L., Afonso, P. (this volume). Discard avoidance by improving fishing gear selectivity: Helping the fishing industry help itself. In S.S. Uhlmann, C. Ulrich, S.J. Kennelly (Eds.), *The European Landing Obligation – Reducing discards in complex, multi-species and multi-jurisdictional fisheries*. Cham: Springer.

Ridderstad, G. (1915). A new construction of trawl-net intended to spare under-sized fish. *Svenska Hydrografisk-Biologiska Kommisionens skrifter*, *6*, 1–21.

Robertson, J.H.B, & Stewart, P.A.M. (1988). A comparison of size selection of haddock and whiting by square and diamond mesh codends. *ICES Journal of Marine Science 44*:148–161.

Scientific, Technical and Economic Committee for Fisheries (STECF). (2014). 45th Plenary meeting report for Scientific, Technical and Economic Committee for Fisheries (PLEN-14-01). 86 pp. https://stecf.jrc.ec.europa.eu/reports/plenary

Scientific, Technical and Economic Committee for Fisheries (STECF). (2016). Methods and data requirements for LO. (STECF-16-13). 95 pp. https://stecf.jrc.ec.europa.eu/reports/discards

Scientific, Technical and Economic Committee for Fisheries (STECF). (2017). Fisheries Dependent Information – Classic (STECF-17-09). 846 pp. https://stecf.jrc.ec.europa.eu/reports/effort

Suuronen, P., Kuikka, S., Lehtonen, E., Tschernij, V., Madsen, N., Holst, R. (2000.) Improving technical management in Baltic cod fishery (BACOMA). Final report (EC 4th framework programme, contract FAIR CT 96–1994). 106 pp.

Suuronen, P., Tschernij, V., Jounela, P., Valentinsson, D., Larsson, P.O. (2007). Factors affecting rule compliance with mesh size regulations in the Baltic cod trawl fishery. *ICES Journal of Marine Science, 64*(8), 1603–1606.

Svedäng, H., & Hornborg, S. (2014). Fishing induces density-dependent growth. *Nature Communications*, *5*, 4152.

Tschernij, V., & Holst, R. (1999). Evidence of factors at vessel-level affecting codend selectivity in Baltic cod demersal fishery. ICES CM 1999/R: 02.

Tschernij, V, Larsson, P.O., Suuronen, P., Holst, R. (1996). Swedish trials in the Baltic Sea to improve selectivity in demersal trawls. ICES CM 1996/B:2.

Tschernij, V., & Suuronen, P. (2002). Improving trawl selectivity in the Baltic. *TemaNord* 2002: 512. 56 pp.

Tschernij, V, Suuronen, P, Jounela, P. (2004). A modelling approach for assessing short-term catch losses as a consequence of a mesh size increase. *Fisheries Research*, *69*, 399–406.

Uhlmann, S.S., van Helmond, A.T.M., Stefánsdóttir, E.K., Siguroardóttir, S., Haralabous, J., Bellido, J.M., et al. (2014). Discarded fish in European waters: General patterns and contrasts. *ICES Journal of Marine Science*, *71*, 1235–1245.

Valentinsson, D. (Ed). (2016). Sekretariatet för selektivt fiske- Rapportering av 2015-års verksamhet. Aqua Reports 2016:8. Swedish University of Agricultural Sciences, Department of Aquatic Resources, Lysekil, 126 pp. ISBN: 978-91-576-9403-4.

Valentinsson, D., & Tschernij T. (2003). An assessment of a mesh size for the "Bacoma design" and traditional diamond mesh cod-end to harmonize trawl selectivity and minimum mesh size. ICES CM 2005/B:05.

Wienbeck, H., Herrmann, B., Moderhak, W., Stepputtis, D. (2011). Effect of netting direction and number of meshes around on size selection in the codend for Baltic cod (*Gadus morhua*). *Fisheries Research*, *109*, 80–88.

Chapter 11
Creating a Breeding Ground for Compliance and Honest Reporting Under the Landing Obligation: Insights from Behavioural Science

Sarah B. M. Kraak and Paul J. B. Hart

Abstract Fisheries regulations aim to maintain fishing mortality and fishing impacts within sustainable limits. Although sustainability is in the long-term interest of fishers, the regulations themselves are usually not in the short-term interest of the individual fisher because they restrict the fisher's economic activity. Therefore, as is the case with all regulations, the temptation exists for non-compliance and dishonest reporting. In the EU and elsewhere, top-down, complex regulations, often leading to unintended consequences, with complex and non-transparent governance-science interactions, may decrease the credibility and legitimacy of fisheries management among fishers. This, in turn, may decrease the motivation to comply and report honestly. The Landing Obligation may make things worse because following the regulation to the letter would often strongly and negatively impact the individual fishers' economic situation. Behavioural science suggests factors that may influence compliance and honesty. Compliance is not necessarily a function of the economic benefits and costs of rule violation: compliance may be more or less, depending on intrinsic motivations. An increased level of self-decision may lead to greater buy-in to sustainable fishing practices and voluntary compliance to catch limits and the Landing Obligation. All else being equal, people in small and self selected groups are inherently more likely to behave "prosocially". In this chapter, some key recommendations based on behavioural science are given for changes in institutional settings that may increase voluntary compliance and sustainable fishing practices. However, transition to a system allowing for more freedom from top-down regulation, with more self-governance, may be difficult due to institutional and cultural barriers and therefore may take many years.

S. B. M. Kraak (✉)
Thünen Institute of Baltic Sea Fisheries, Rostock, Germany
e-mail: sarah.kraak@thuenen.de

P. J. B. Hart
Department of Neuroscience, Psychology and Behaviour, University of Leicester, Leicester, UK

S. S. Uhlmann et al. (eds.), *The European Landing Obligation*,
https://doi.org/10.1007/978-3-030-03308-8_11

Keywords Behavioural economics · Being watched · Carrot or stick · Hidden cost of control · Honest reporting · Lack of trust · Loss aversion · Moral priming · Nonmonetary incentives · Self-decision · Self-selected groups · Voluntary compliance

11.1 Introduction

Prior to the Landing Obligation, the main instrument to control fishing pressure in the European Union (EU) has been the setting of total allowable landings quotas (which have been called Total Allowable Catches, TACs). This instrument has allowed for implementation error because landings quotas do not limit catches. Under landings quotas, unlimited over-quota catches are allowed as long as they are not landed; in other words, they must be discarded at sea. The implementation error occurs because, when over-quota fish are caught and discarded dead, the intended level of exploitation set by the quotas is overshot, sometimes by large and often unknown amounts (Kelleher 2005). Reasons for discarding may be:

1. Fish smaller than the minimum landing size (MLS) have been caught and are not allowed to be landed.
2. Fishers may discard lower-quality fish and utilise their landings quota to land better-quality and higher-priced fish (high-grading) – this practice is forbidden in the EU since 2002, but no offenders have been caught and sanctioned (Schou 2015).
3. In mixed fisheries, fishers may catch over-quota fish when they continue fishing for other species whose quota is not yet exhausted – these fish, which may be (unavoidable, incidental) bycatch species or part of the targeted assemblage, must be discarded at sea (see BBC 2007; Borges and Penas-Lado, this volume).

The Landing Obligation, with total allowable *catch* quotas (limiting actual catches rather than only landings), attempts to make an end to the implementation error caused by the landings quota system. However, it is expected that the EU will experience problems in fully implementing the Landing Obligation if the incentives for discarding continue to exist. For example, although the MLS is abolished under the Landing Obligation, fish smaller than a minimum conservation reference size (MCRS) are not permitted to be sold for human consumption and thus have a lower value. Therefore, incentives for illegally discarding fish below MCRS and high-grading may continue to exist. In addition, in several fishing areas, the problem of "choke species" may arise in mixed fisheries or fisheries with unavoidable bycatch (Prellezo et al. 2016; Fitzpatrick et al., this volume). In such cases the fishery for important commercial species is "choked", i.e. must be closed, before its quota is fished up, because the quota for another species caught in the same fishery is already exhausted. This situation creates incentives to illegally discard over-quota catches of the choke species and continue fishing.

Thus, the EU's Common Fisheries Policy (CFP) can only be fully implemented if the Landing Obligation is fully complied with and catch limits are not exceeded. However, it is not yet clear how to achieve this: it is not specified how the catch limits and Landing Obligation will be enforced and how catches will be verified. The regulation leaves the documentation and compliance monitoring to the Member States (EU 2013, Article 15.13):

> For the purpose of monitoring compliance with the Landing Obligation, Member States shall ensure detailed and accurate documentation of all fishing trips and adequate capacity and means, such as observers, closed-circuit television (CCTV) and others. In doing so, Member States shall respect the principle of efficiency and proportionality.

Assuming that catches cannot be completely observed, counted, documented and verified, the regulation leaves room for implementation error, in two ways:

1. Directly: Fishers may still catch (and discard) in excess of the quotas.
2. Indirectly: Removals will be known to stock assessment scientists with a certain degree of uncertainty. This will lead to imprecise estimates, advice and management measures, for example, too low or too high a TAC. Prior to the Landing Obligation, this has been commonly the case. Scientists have attempted to estimate the removals (including discards) through sampling trips with observers on board. Problems of reliability and representativeness of the samples may, however, increase under the Landing Obligation, because fishers may fear that when scientists are aware of law violations this information will be passed on to the enforcement authorities. To mitigate this problem, data collection for science (i.e. stock assessment) should be kept strictly separated from data collection for enforcement purposes (Mangi et al. 2013). For example, it has been found that Norwegian fishers report bycatch of rare species in scientific programmes, but not in their logbooks, although such reporting is obligatory (J. Vølstad, personal communication).

The implementation error can only be avoided if full trustworthiness or full proof of the catches can be ensured. Full documentation (i.e. proof) of the catch can be achieved by Remote Electronic Monitoring (REM or EM) systems such as camera and sensor systems (for more details, see James et al., this volume), leading to a fully documented fishery (FDF), but these systems are costly, and fishers are often averse to being watched. Some EU Member States also expressed a dismissive attitude towards the use of cameras for monitoring purposes, presumably fearing to turn an anti-FDF-minded sector against their ministry.

Persuading fishers to follow regulations has been one of the toughest problems to be solved in fisheries. In an effort to deal with the difficulty, co-management has been introduced and become more common than it once was (Jentoft 1998; Wilson et al. 2003; Jacobsen et al. 2012). Co-management allows fishers to contribute to the formulation of regulations and gives them a say in how they are applied. Studies in behavioural economics have shown that a feeling of ownership encourages participants to value more highly something that they have put effort into achieving (Norton et al. 2012). Unfortunately, at the level of the EU, the co-management approach is not yet widespread, and the implementation of the Landing Obligation is a top-down regulation that requires fishers to follow a rule that can result in economic outcomes for the individual that could lead to resistance.

11.2 The Commercial Pressures Influencing Fisher Decisions

If management agencies want their rules to be followed, then it is essential that those rules are devised with regard to the commercial pressures on fishers. Rules are more likely to fail if they make it difficult for a fisher to make ends meet financially. This is why the TAC approach, when applied to a mixed fishery, often failed. If her/his trawl emerges from the depths with a catch of a species for which she/he has no quota, then a fisher can be faced with a dilemma. Under the old TAC system, if the fish have market value as is shown in Fig. 11.1, this situation creates the temptation to keep the fish and sell them illegally, and this temptation would be particularly acute if the fisher's economic situation is poor (Booker 2007; BBC 2007). Recognising that under the old TAC system fishers could be regularly faced with this dilemma does not prove that many fishers submitted to the temptation, but gathering evidence for illegal sales of non-quota fish is difficult by its nature although the problem of illegal, unreported and unregulated (IUU) fishing has been extensively examined (Le Gallic and Cox 2006; Sumaila et al. 2006). In certain parts of the world, there is no doubt that illegal fish are landed and sold, and there is also evidence that this happens within EU waters (Couper et al. 2015). Under the new Landing Obligation, the fisher has to stop fishing, even if she/he has quota left for other species, and this will also have economic consequences. For this reason, both management regimes are vulnerable to abuse if the fisher's economic circumstances are poor. It may be necessary in this situation to find alternatives to top-down regulations, which would focus on giving fishers greater power to devise regulations.

11.3 Behavioural Economics: A Discipline Providing Guidance for Addressing the Problem

In the rest of this chapter, we aim to give some guidance on how fishers can be encouraged to stay within the catch limits and to report catches accurately by using insights from the behavioural sciences, specifically the discipline of behavioural economics. Behavioural economics studies the effects and consequences of psychological, social, cognitive and emotional factors on the economic decisions of individuals and institutions, challenging the assumption of human rationality that prevails in classical economic theory (Dhami 2016). Behavioural economics is an empirical science based on experimental work: the behaviour and choices of human

MAGP IV proposals called for Holland (12% tonnage and ters which 'weighted' capacity

Saithe dumping scandal

THIS 180 box haul of big saithe was taken by the Shetland trawler *Guardian Angell* on 21 April this year some 20 miles north of Muckle Flugga. Because of a lack of quota only 10 boxes were gutted and the remaining 170 boxes of big, grade 1 saithe were dumped.

Grade 1 saithe currently makes around £20 a box, meaning fish worth more than £3000 were dumped – and this was just one haul. Trawlers working in these waters regularly and unavoidably catch varying amounts of saithe while fishing for cod, haddock and other roundfish.

The vast majority of it has to be dumped because the UK has a very small quota, leading to a huge waste of a valuable resource for nothing.

■ Open letter to ministers – page 4.

Fig. 11.1 An example of how, under the old TAC system, fishers could be tempted to land illegally dead fish which could not legally be landed. (From Fishing News, UK)

subjects are observed under controlled experimental conditions. The value of behavioural economics in the creation of policy has been given its highest profile by Thaler and Sunstein (2009). They used the terms "nudge" and "nudging" to characterise measures they proposed to encourage people to make decisions that comply with policy or achieve long-term goals. They made suggestions, for example, for how to design procedures to encourage people to submit their tax returns on time or to save more for retirement. The British Government in 2010 was so impressed by the approach that it set up the Behavioural Insights Team which is tasked to "...use insights from behavioural science to encourage people to make better choices for themselves and society" (Behavioural Insights Team 2018). In 2015, Barack Obama, then US president, established the Social and Behavioral Sciences Team charged with a similar task (Congdon and Shankar 2018). Much of what we offer in this chapter is in the same vein.

To find ways of improving the chances of fishers following the rules, we discuss how the following selection of ideas from behavioural economics could contribute to better compliance to the Landing Obligation:

- Determinants of honesty and respect for the law
- Crowding out of voluntary compliance
- The effects of being watched
- Loss aversion

11.3.1 Determinants of Honesty and Respect for the Law

In most countries, the standard approach to obtaining fisher compliance is to deter rule violations through investments in enforcement activities, including at-sea patrols, dockside monitoring and observer programmes (see also James et al., this volume). This approach is built on the assumption that the occurrence of fishery offences is solely a function of the perceived benefits and costs of an offence, such as the gains derived from rule violation, the likelihood of detection and the severity of the penalties (Becker 1974; Hart 1997). However, modern criminology (e.g. Tyler 2006) and behavioural economics (Mazar et al. 2008) recognise that many people comply with rules because they believe it is the right thing to do. In this context, tax compliance is much higher than deterrence models would predict (Frey and Torgler 2007). Individuals are also much influenced by the majority view of the group they are part of (e.g. Aronson and Aronson 2012). When we witness unethical behaviour, our own morality erodes (Ariely 2012). Cheating can be socially contagious (Gino et al. 2009; Mann et al. 2014): as long as we see members of our own social groups behaving in ways that are dishonest, it is likely that we too will recalibrate our internal moral compass and adopt their behaviour as a model for our own. Tax compliance, for example, varies widely across European countries, and a high correlation has been found between perceived tax evasion and tax morale (Frey and Torgler 2007). Similarly, experimentally measured, individual intrinsic honesty is stronger in the subject pools of countries with lower levels of corruption, tax evasion and fraudulent politics than in those of countries where the latter are higher (Hermann et al. 2008; Gächter and Schulz 2016). Individuals may even feel pride about breaking the rules. A study among Danish fishers (Nielsen and Mathiesen 2003) reported that they "feel they are taken hostage by an illegitimate management system, and thus feel it is morally correct not to comply". Within a fishery where crews know each other through the use of a common port, it may be possible to enhance compliance by fostering pride (see Panagopoulos 2010; Harth et al. 2013) about sustainable fishing practices by publishing stories in the fishing press about fishers that have complied.

In laboratory experiments, Mazar et al. (2008) found that (1) the amount of dishonesty is largely insensitive to either the expected external benefits or the costs associated with the deceptive acts; (2) causing people to become more aware of their internal standards for honesty by moral priming decreases their tendency for deception; and (3) increasing the "degrees of freedom" that people have to interpret their actions increases their tendency for deception. For instance, Mazar et al. (2008) found that nonmonetary crime targets (i.e. property rather than money) can increase economically incentivised dishonesty in a laboratory setting. Similar laboratory studies by Mead et al. (2009) found that mental tiredness also increases cheating. These two studies suggest that violation of fishing regulations could at least in part be exacerbated by a lack of moral reminders, the opportunity to "steal" a nonmonetary asset (i.e. fish) and the mental tiredness of fishers. A further complication in a fishery is that the fisher and his crew are often on their own at sea, so that social pressures regarding compliance are distanced from the act of disobedience. Mazar et al. (2008) suggest that understanding dishonesty has important implications for designing effective methods to curb it. The costs of obtaining a particular level of fisheries compliance through enforcement could potentially be reduced through increased investments in activities that improve voluntary compliance.

Studies (e.g. Mazar et al. 2008) have indicated that honesty can be enhanced by moral reminders, such as simply asking people to sign a statement in which they declare their commitment to honesty before taking part in a task rather than after (e.g. signing the honesty statement on the income tax declaration form at the top rather than at the bottom). In fisheries this finding could be applied by making fishers sign the logbook just before logging the information (Kraak et al. 2015). The e-log system could have a confirmation screen which requires the operator to acknowledge that they are filling the form out accurately before the electronic system can receive data input.

A recent experiment showed how priming can affect honest reporting by fishers (Drupp et al. 2016). Using a coin-tossing task, the authors tested whether truth-telling of German fishers, who are known to dislike their EU regulator, is affected by various treatments. Fishers misreported coin tosses to their economic advantage more strongly in a treatment where they were faced with the EU logo. Fishers were more honest in an additional treatment where the source of research funding, namely the EU, was revealed. These apparently contradictory findings suggest that lying is increased towards a disliked regulator, but perhaps decreased when it is made clear from whom the money is "stolen". The implication is that regulators can affect truth-telling behaviour by the nature and communication of their policies.

11.3.2 Crowding Out of Voluntary Compliance

Compliance with the rules is more likely when fishers, whose behaviour is to be regulated, buy-in to those rules. Fisheries management is in many cases a top-down bureaucratic process with centralised control (Daw and Gray 2005). The regulations are viewed by the fishers as opposing rather than supporting their interests, and this manifests itself as a reduced compliance to "the letter" as well as "the spirit" of the regulations (Kuperan and Sutinen 1998; Hatcher et al. 2000; Nielsen 2003; Nielsen and Mathiesen 2003; Kraak 2011). Evidence suggests that the willingness to obey regulations voluntarily depends on whether one is controlled or not (reviewed in Bowles 2008; Richter and van Soest 2012). Counterintuitively, the control imposed by an outside institution undermines – "crowds out" – any intrinsic motivations an individual may have to comply voluntarily. As a result, there is a hidden cost of control, as pointed out by Falk and Kosfeld (2006). The implication is that control can "crowd out" intrinsic motivations, calling for even stronger control, leading to a vicious cycle of mistrust and strong controls. Behavioural economics has established that regulations that are chosen by the individuals (e.g. via voting) are obeyed more, as they are perceived to be more legitimate (Vyrastekova and van Soest 2003). It is apparent from the work of Ostrom that self-imposed rules, self-imposed monitoring and self-imposed sanctions work better (Ostrom 2009). One of the reasons is that control by an outside institution signals mistrust, which directly affects motivational factors, such as cooperation, reciprocity or being a good citizen.

Indeed, fisheries systems can be characterised by mutual mistrust, between regulators and fishers, between scientists and fishers and among fishers themselves. Fishers have lost respect for the rules and regulations because many of them do not seem to make sense (including the Landing Obligation), seem too complex, seem contradictory or seem to provide perverse incentives (Jentoft 1998; Jacobsen et al. 2012). At the same time, fishers are usually not expected to voluntarily take action to fish more sustainably as this would be perceived to reduce rather than maximise catches. Often the institutional set-up is such that fishers are perceived as the adversaries of the management establishment. The key challenge for European fisheries is how one can "crowd in" desirable behaviour by establishing trusting relationships. The problem seems to be how to make such a transition from the current situation – rebuilding of mutual trust cannot be done simply on a short time scale.

Large group size and anonymity may be among the causes of the apparent lack of trust. Social capital, trust and intrinsic motivation to cooperate tend to be higher in small groups of people who regularly interact with each other in non-anonymous ways (Henrich 2004). For example, in mixed fisheries, where vulnerable bycatch species effectively become the choke species, it is advantageous to join in groups and share the individual small bycatch quota (Holland and Jannot 2012; O'Keefe and DeCelles 2013). In the case of the US West Coast (Holland and Jannot 2012), groups of fishers pooling their quotas can set their own rules, not necessarily encoded in law. These people were not necessarily connected in communities before; they came together because they have a common problem that can best be solved by collective action. Also in Europe there are examples of fishers voluntarily pooling their quotas: in a Danish village, boat owners and fishers have established a cooperative company where they have bought quotas jointly, with the aim of securing the community of its present and future catch rights (Schou 2011). In that way, the cooperative company replaces the Danish state as provider of fishing rights. In Toyama Bay, Japan, some of the inshore fishers pool their catches and their costs, others do not. Experiments demonstrated that the fishers who pooled costs and catches were more likely to show cooperative behaviours in laboratory experiments than were the fishers who did not share fishery costs and catches (Carpenter and Seki 2011). Several economic experiments have established that group choice is a key point to facilitate cooperative behaviour. If individuals can self-select into groups, there is a larger tendency to act in the group's interest and also to coordinate on a common cooperative strategy (Brekke et al. 2011; Gürerk et al. 2006).

Operating in groups/cooperatives may be more or less attractive to fishers: a perceived advantage may be the sharing of risk but a perceived disadvantage may be that the individual surrenders her/his own decision-making for the sake of democratic group decision-making. Kraak et al. (2015) proposed that fisheries management could set up a structure in which several levels of organisation are offered to which individual fishers can opt in (e.g. voluntary pooling of quotas); each level has its benefits and costs, but because the individuals can choose themselves, there would be greater acceptance of the disadvantages of the chosen setting. There are significant costs of managing a group/cooperative that need to be covered. To the extent that social behaviour of fishers in small groups decreases the negative externalities to society, e.g. those caused by non-compliance, policies can be designed that effectively subsidise those groups/cooperatives. This can be done in the form of setting aside a portion of the Member State's quota for such social initiatives or else by financial instruments.

Fishers often distrust managers and scientists and vice versa. The co-management movement which has been taking hold in various locations, particularly for small-scale fisheries, is designed to address this lack of trust between fishers, scientists and policy-makers (Wilson et al. 2003; and see papers in Chuenpagdee 2011). To varying degrees, co-management arrangements involve fishers in gathering data, translating their local ecological knowledge to a form which can be incorporated into stock assessments, determining policy and setting regulations. These systems give the fisher a sense of belonging and break down the barriers between the fisher and the regulator. In order to build mutual trust between fishers and scientists, industry-science collaborative projects could be established (e.g. Mangi et al. 2018), for example, in which fishers could try new practices and scientists explore the consequences. In the USA as well as in Europe, various scientist-facilitated initiatives are arising where scientists process and display information that fishers provide to share among themselves, for example, on CPUE hotspots or bycatch rates of species that need to be avoided so that fishers can catch their quotas at a lower impact to the ecosystem (e.g. O'Keefe and DeCelles 2013; Hetherington 2014; Eliasen and Bichel 2016). In collaboration with scientists, the fishing industry can create fishery management plans which comply with management policies. In the Netherlands and the UK, fishing organisations have started to hire scientists that were previously employed by the government to help them check assessments and advice and develop plans.

There are also initiatives to improve the knowledge of fish biology, fishery science and policy-making in fishers, which may further contribute to a more trusting relationship between fishers on the one hand side and regulators and scientists on the other. For example, in the UK, Fishing into the Future (2018) is run by fishers and, among other things, is running courses to broaden the education of fishers. A similar project has run in the Netherlands (Wageningen University and Research 2018), where fishers set up "knowledge groups" around themes and then interacted with fisheries scientists to access scientific knowledge. In the USA, the Marine Resource Education Programme, run by the Gulf of Maine Research Institute in Boston, similarly runs courses in fishery science for fishers.

The "crowding out" hypothesis does not only state that control may undermine intrinsic motivations to comply, but also that monetary incentives may undermine such motivations. In experiments and in the field it has been found that sometimes financial incentives induced more self-interested behaviour, even after they were withdrawn (Bowles 2008). For example, in a study by Cardenas et al. (2000), experiments were run with people in rural Colombia who are confronted with a common pool problem in their daily life. In the experiment subjects were asked to decide how much timber to extract from a forest. The scenario presented was that harvesting had an adverse effect on water quality (as is actually the case in the study region), posing a cost to everyone in the group. The game was played first without any regulations in place, while at a later stage, an extraction norm was introduced that was enforced by a mild probabilistic fine. Cardenas et al. (2000) found that subjects reduced their extraction level immediately after the regulation was introduced but started extracting more aggressively after realising that consequences were rather mild. Strikingly, in the last rounds, extraction levels were higher with the regulation than without. As a result, payoffs were significantly lower when individuals were confronted with a formal rule than in its absence; the weak official rule interacted with the internal norms of the subjects and destroyed their intrinsic motivation to cooperate (Cardenas et al. 2000). Richter and van Soest (2012) reviewed similar experiments, such as the one where imposing a fine on parents arriving late to collect their children at day care increased the number of late-coming parents or the one where small honoraria for seminar speakers may increase the probability of declining the invitation. More generally, in experiments investigating the psychological consequences of money, subjects exposed to the concept of money subsequently showed a more self-reliant but also more self-centred approach to problem-solving than subjects exposed to neutral concepts (Vohs et al. 2006). These results suggest that the application of nonmonetary incentives in fisheries

management should be explored, along with other factors enhancing intrinsic compliance motivation such as moral reminders, non-anonymity, small group size and face-to-face communication. Nevertheless, Bowles (2008) and Richter and van Soest (2012) warn that the loss of social capital may, to a large extent, be irreversible and that from the reviewed experiments, it cannot simply be concluded that regulations or sanctions should be abolished.

11.3.3 The Effects of Being Watched

Technologies have recently emerged for monitoring fishers, such as AIS (automatic identification system), VMS (vessel monitoring system) and drones (Toonen and Bush 2018), but also Electronic Monitoring with cameras on board as a tool for a fully documented fishery (FDF) (James et al., this volume). As mentioned above, some EU Member States expressed a dismissive attitude towards the use of cameras for the purpose of monitoring. The aversion to being watched is in agreement with the notion that too much monitoring may have the result that individuals feel they are not trusted and as a consequence become less trustworthy (Ostrom 1998). In contrast, it has been well documented that people will be more likely to behave "prosocially" (e.g. cooperate, comply, be honest) when being watched in non-anonymous situations (reviewed in Kraak 2011). This occurs because it opens possibilities of direct and indirect reciprocity as well as reputation building – the psychological mechanism may be that good behaviour instils pride or, conversely, it can be driven by the fear of social exclusion when damaging one's reputation (Ouwerkerk et al. 2005). Recent investigations have shown that subtle cues of being watched, such as two stylised eye-like shapes on a computer screen suffice to change human behaviour and reduce selfishness; these eye-shaped cues seem to elicit unconscious, biologically hardwired reactions (Milinski and Rockenbach 2007). Perhaps a way to exploit this human propensity, without the disadvantage of eroding trust due to too much monitoring, is to display a picture of "watching eyes" on the e-logbook screen (Kraak et al. 2015) or anywhere on the vessel.

To persuade fishers to deploy a fully documented fishery, its advantages could be emphasised more clearly to the fishers. Full documentation of the fish supply chain (from net to plate) could bring strong market incentives through information on sustainability of the species, traceability and documentation on how the fish has been caught and treated on board (Mangi et al. 2013). The concepts of traceability and transparency could also be used in more innovative ways. As mentioned above, humans are not only subject to an aversion of being watched, but people may also like being watched when they are proud of what they are doing within the context of a peer group. In the UK, the Moshi Moshi sushi restaurant chain labels fish dishes with Quick Response (QR) codes printed with squid ink on rice paper so that customers can see where the MSC-certified fish comes from (SeafoodSource 2018). In Canada an organisation called This Fish (This Fish 2018) is setting up a system whereby consumers can use QR codes to identify the fisher who caught the

product that is just about to be purchased or eaten. In other markets (meat products), this form of information promotes trust from the consumer, but in the fisheries case, it may also promote compliance from the fishers by instilling in them a greater sense of ownership of the final product (Kraak et al. 2015).

Further research on the "being-watched" effect should be done with experiments that are relevant to the specific settings encountered in fisheries management.

11.3.4 Loss Aversion and the Use of "Carrot" or "Stick" Approaches

In some Member States, pilot projects offered extra quota to participating fishers in return for providing full documentation of their catches for monitoring purposes (fully documented fishery) (van Helmond et al. 2015; Needle et al. 2015; Mortensen et al. 2017). This could be extended in, for example, a tiered approach stipulating that the fishers opting for a fully documented fishery would be subjected to less prescriptive rules and hence have more perceived freedom and flexibility in running their business, while the fishers opting for less stringent monitoring would have to bear the burden of more uncertain catch documentation and be subjected to more restrictive rules and/or a larger reduction of their quotas (Prellezo et al. 2016). Such approaches are framed as a "carrot", where a reward is given for the desired behaviour. The approach, however, can also be framed as a "stick", where a penalty is given when the desired behaviour is not chosen (e.g. where quota would be deducted or restricting rules would be imposed if fishers do not take up the fully documented fishery option). The response to "carrots" versus "sticks" should be carefully considered (Kubanek et al. 2015). Human beings are known to be subject to loss aversion (Tversky and Kahneman 1974, 1991): it is thought that the pain of losing is psychologically about twice as powerful as the pleasure of gaining; "losses loom larger than gains" (Kahneman and Tversky 1979). According to the expectation of loss aversion, the "stick" may be the stronger motivator than the "carrot" (Imas et al. 2017). The recent EU pilot projects with catch quotas and a fully documented fishery have used the "carrot" (van Helmond et al. 2015; Needle et al. 2015; Mortensen et al. 2017). On the other hand, the EU cod plan (EU 2008) used the "stick" of effort reductions to motivate (groups of) fishers to take up cod avoidance measures, but this was not well received by the fishing sector (Kraak et al. 2013). Since 2018, Germany offers a mobile app to small-scale fishers in the Western Baltic with which they can document where they are fishing. Only when they can prove that they are fishing in water less than 20 m deep are the small-scale fishers allowed to fish in a seasonally closed area (BLE 2018). The provision is formally phrased as a "carrot" (when using the app, they gain the right to fish in the otherwise closed area during the cod spawning season). However, some fishers perceive it as a "stick" (losing the opportunity to fish in the area unless using the

app) because the area closure had never been sufficiently well enforced in previous years.

Kraak et al. (2016) suggested that a "stick" approach may induce a negative emotional response to management regulations, which may in turn induce reduced compliance (see also Imas et al. 2017). Accordingly, a potential trade-off might exist between the higher management effectiveness of a "stick" approach and reduced compliance with regulation. In order to find out what kind of framing would lead to highest uptake of the desired behaviour (e.g. a fully documented fishery and compliance), directed research on the response of fishers to "sticks" and "carrots" is needed.

11.4 Conclusions

In conclusion, much of the past implementation error has been caused because the complex top-down control and lack of trust have undermined potential intrinsic motivation to fish sustainably. Compliance is not necessarily a function of the economic benefits and costs of rule violation: compliance may be more or less, depending on intrinsic motivations and the economic circumstances of the fisher. An increased level of self-decision may lead to more buy-in to sustainable fishing practices and voluntary compliance to catch limits and the Landing Obligation. All else being equal, people in small and self-selected groups are inherently more likely to behave in a "prosocial" manner (Ostrom 1990, 2001). However, transition to a system allowing for more freedom from top-down regulation, with more self-governance and self-regulation, may be difficult. Some key recommendations are given below, several of which can be characterised as "nudges" as defined by Thaler and Sunstein (2009):

- Increase regulators' trust of fishers through a fully documented fishery.
- Increase fishers' trust of regulators by designing simpler legislation, with non-contradictory rules, which do not lead to perverse incentives.
- Increase fishers' trust of scientists and scientists' trust of fishers by setting up industry-science partnerships and collaborative research.
- Increase fishers' mutual trust and their intrinsic motivations to fish sustainably by facilitating and encouraging fishers to organise themselves in small groups with common interests.
- Allow for several levels of organisation to choose from, and allow for self-selection of group membership.
- Incentivise the organisation of fishers into groups through the provision of, for example, extra quota and relative freedom from top-down regulation or through financial instruments.
- Allow small groups of fishers to make their own decisions, where their own rules and sanctions do not necessarily have to be coded in law.

- Incentivise uptake of a fully documented fishery, but carefully consider whether a "stick" or a "carrot" should be used.
- Allow groups of fishers to decide themselves on the methods of implementation of a fully documented fishery.
- Do not only rely on monetary incentives and monetary penalties; these may "crowd out" intrinsic motivations.
- For groups of fishers who know each other, publish examples of cooperative behaviour in the local press. Publishing the good behaviour of named fishers may be a nonmonetary incentive because it fosters pride of being a sustainable (good) fisher.
- Establish QR codes (Quick Response codes) that link a product to an individual fisher to foster a sense of being watched as well as ownership and pride in being a sustainable fisher.
- Use moral reminders in the e-log software, such as a requirement to sign a statement of accurate reporting at the start of their e-log session (instead of at the end), and pictures of watching eyes on the screen.

Acknowledgements The ideas brought forward in this chapter are largely based on the BehavFish workshop funded by the ICES Science Fund and the Fisheries Society of the British Isles (Kraak et al. 2015), and we therefore thank all participants of this workshop: Chris Anderson, Dorothy Dankel, Matteo Galizzi, Mark Gibson, Diogo Gonçalves, Katell Hamon, Ciaran Kelly, Jean-Jacques Maguire, Jeroen Nieboer, Ingrid van Putten, Andrew Reeson, Dave Reid, Andries Richter and Alessandro Tavoni.

References

Ariely, D. (2012). *The (honest) truth about dishonesty.* New York: Harper Collins.

Aronson, E., & Aronson, J. (2012). *The social animal*, 11th edn. London: Macmillan.

BBC (2007). http://news.bbc.co.uk/1/hi/uk/7102241.stm. Accessed 12 June 2018.

Becker, G.S. (1974). Crime and punishment: an economic approach. *Journal of Political Economy, 76*(2), 169–217.

Behavioural Insights Team (2018). https://www.behaviouralinsights.co.uk/. Accessed 26 June 2018.

BLE (Bundesanstalt für Landwirtschaft und Ernährung) (2018). Bekanntmachung zur Fischerei auf Dorsch im Jahr 2018 unter der Ausnahmemöglichkeit innerhalb der Schonzeiten nach der Verordnung (EU) 2017/1970. *Bundesanzeiger* vom 8. Januar 2018.

Booker, C. (2007). Fishing quotas are an ecological catastrophe. https://www.telegraph.co.uk/news/uknews/1570439/Fishing-quotas-are-an-ecological-catastrophe.html. Accessed 21 June 2018.

Borges, L., & Penas-Lado, E. (this volume). Discards in the common fisheries policy: The evolution of the policy. In S.S. Uhlmann, C. Ulrich, S.J. Kennelly (Eds.), *The European Landing Obligation – Reducing discards in complex multi-species and multi-jurisdictional fisheries.* Cham: Springer.

Bowles, S. (2008). Policies designed for self-interested citizens may undermine "the moral sentiments": evidence from economic experiments. *Science, 320*(5883), 1605–1609.

Brekke, K.A., Hauge, K.E., Lind, J.T., Nyborg, K. (2011). Playing with the good guys. A public good game with endogenous group formation. *Journal of Public Economics, 95*(9), 1111–1118.

Cardenas, J.C., Stranlund, J., Willis, C. (2000). Local environmental control and institutional crowding-out. *World Development, 28*, 1719–1733.

Carpenter, J.P., & Seki, E. (2011). Do social preferences increase productivity? Field experimental evidence from fishermen in Toyama Bay. *Economic Inquiry, 49*(2), 612–630.

Chuenpagdee, R. (Ed.). (2011). *World small-scale fisheries. Contemporary visions*. Delft: Eburon.

Congdon, W.J., & Shankar, M. (2018). The role of behavioural economics in evidence-based policymaking. *The ANNALS of the American Academy of Political and Social Science, 678* (1), 81–92.

Couper, A., Smith, H.D., Ciceri, B. (2015). *Fishers and plunderers. Theft, slavery and violence at sea*. London: Pluto Press.

Daw, T., & Gray, T. (2005). Fisheries science and sustainability in international policy: a study of failure in the European Union's Common Fisheries Policy. *Marine Policy, 29*(3), 189–197.

Dhami, S. (2016). *The foundations of behavioral economic analysis*. Oxford: Oxford University Press.

Drupp, M.A., Khadjavi, M., Quaas, M.F. (2016). *Truth-telling and the regulator. Evidence from a field experiment with commercial fishermen* (Kiel Working Paper 2063). https://www.ifw-kiel.de/datenmigration/publikationen/document-store/truth-telling-and-the-regulator-evidence-from-a-field-experiment-with-commercial-fishermen-7902/. Accessed 19 November 2018.

Eliasen, S.Q., & Bichel, N. (2016). Fishers sharing real-time information about "bad" fishing locations. A tool for quota optimisation under a regime of landing obligations. *Marine Policy, 64*, 16–23.

European Union. (2008). Council Regulation (EC) No 1342/2008 of 18 December 2008 establishing a long-term plan for cod stocks and the fisheries exploiting those stocks and repealing Regulation (EC) No 423/2004. *Official Journal of the European Union L, 348*, 20–33.

European Union. (2013). Regulation (EU) No 1380/201308 of the European Parliament and of the Council of 11 December 2013 on the Common Fisheries Policy, amending Council Regulations (EC) No 1954/2003 and (EC) No 1224/2009 and repealing Council Regulations (EC) No 2371/2002 and (EC) No 639/2004 and Council Decision 2004/585/EC. *Official Journal of the European Union. L, 354*, 22–61.

Falk, A., & Kosfeld, M. (2006). The hidden costs of control. *American Economic Review, 96*(5), 1611–1630.

Fishing into the Future (2018). https://www.fishingintothefuture.co.uk/. Accessed 31 May 2018.

Fitzpatrick, M., Frangoudes, K., Fauconnet, L., Quetglas, A. (this volume). Fishing industry perspectives on the EU Landing Obligation. In S.S. Uhlmann, C. Ulrich, S.J. Kennelly (Eds.), *The European Landing Obligation – Reducing discards in complex multi-species and multi-jurisdictional fisheries*. Cham: Springer.

Frey, B.S., & Torgler, B. (2007). Tax morale and conditional cooperation. *Journal of Comparative Economics, 35*(1), 136–159.

Gächter, S., & Schulz, J.F. (2016). Intrinsic honesty and the prevalence of rule violations across societies. *Nature, 531*, 496–499.

Gino, F., Ayal, S., Ariely, D. (2009). Contagion and differentiation in unethical behavior: the effect of one bad apple on the barrel. *Psychological Science, 20*(3), 393–398.

Gürerk, Ö., Irlenbusch, B., Rockenbach, B. (2006). The competitive advantage of sanctioning institutions. *Science, 312*(5770), 108–111.

Hart, P.J.B. (1997). Controlling illegal fishing in closed areas: the case of mackerel off Norway. In: D.A. Hancock, D.C. Smith, A. Grant, et al. (Eds.), *Developing and sustaining world fisheries resources: the state of science and management. 2nd world fisheries congress proceedings*. Brisbane, Australia, pp. 411–414.

Harth, S.N., Leach, C.W., Kessler, T. (2013). Guilt, anger, and pride about in-group environmental behaviour: different emotions predict distinct intentions. *Journal of Environmental Psychology, 34*, 18–26.

Hatcher, A., Jaffry, S., Bennett, E. (2000). Normative and social influences affecting compliance with fishery regulations. *Land Economics, 76*(3), 448–461.

Henrich, J. (2004). Cultural group selection, coevolutionary processes and large-scale cooperation. *Journal of Economic Behavior and Organization, 53*, 3–35.

Hermann, B., Thöni, C., Gächter, S. (2008). Antisocial punishment across societies. *Science, 219* (5868), 1362–1367.

Hetherington, S.J. (2014). MB0125 & C6122: common fisheries policy reform implementation: aligning zero quota species and improving fisheries management – a spurdog case study, Defra Funded Research Programme (2013–2015).

Holland, D.S., & Jannot, J.E. (2012). Bycatch risk pools for the US West Coast Groundfish Fishery. *Ecological Economics, 78*, 132–147.

Imas, A., Sadoff, S., & Samek, A. (2017). Do people anticipate loss aversion? *Management Science, 63*(5), 1271–1284.

Jacobsen, R.B., Wilson, D.C.K., Ramirez-Monsalve, P. (2012). Empowerment and regulation – dilemmas in participatory fisheries science. *Fish and Fisheries, 13*, 291–302.

James, K.M., Campbell, N., Viðarsson, J.R., Vilas, C., Plet-Hansen, K.S., Borges, L., et al. (this volume). Tools and technologies for the monitoring, control and surveillance of unwanted catches. In S.S. Uhlmann, C. Ulrich, S.J. Kennelly (Eds.), *The European Landing Obligation – Reducing discards in complex, multi-species and multi-jurisdictional fisheries*. Cham: Springer.

Jentoft, S. (1998). Social theory and fisheries co-management. *Marine Policy, 22*, 423–436.

Kahneman, D., & Tversky, A. (1979). Prospect theory: an analysis of decision under risk. *Econometrica, 47*(2), 263–291.

Kelleher, K. (2005). *Discards in the world's marine fisheries. An update*. Rome: FAO. No. 470, 131 pp.

Kraak, S.B.M. (2011). Exploring the 'Public Goods Game' model to overcome the tragedy of the commons in fisheries management. *Fish and Fisheries, 12*(1), 18–33.

Kraak, S.B.M., Bailey, N., Cardinale, M., Darby, C., José, A.A.D.O, Eero, M., et al. (2013). Lessons for fisheries management from the EU cod recovery plan. *Marine Policy, 37*, 200–213.

Kraak, S.B.M., Kelly, C., Anderson, C., Dankel, D., Galizzi, M., Gibson, M., et al. (2015). *ICES science fund 2014 project report: insights from behavioral economics to improve fisheries management*. Report of the Workshop jointly funded by ICES and FSBI held 21–23 October 2014 at ICES HQ, Copenhagen, Denmark, 46 pp. http://www.ices.dk/community/icessciencefund/Documents/06_Report%20of%20the%20BehavFish%20Workshop_final.pdf. Accessed 29 June 2018.

Kraak, S.B.M., Drupp, M.A., Quaas, M. (2016). Carrot or stick? Experimental tests on a potential trade-off between effectiveness and compliance of alternative management approaches, ICES CM 2016/B:293. http://www.ices.dk/sites/pub/ASCExtended2016/Shared%20Documents/B%20-%20Predictably%20Irrational%20%E2%80%93%20a%20new%20scientific%20research%20field%20for%20the%20science%20underpinning%20marine-resource%20management/abstract%20Kraak%20Drupp%20Quaas%20-%20carrot%20stick%20and%20prosocial%20behaviour_final.pdf. Accessed 19 Nov 2018.

Kubanek, J., Snyder, L.H., Abrams, R.A. (2015). Reward and punishment act as distinct factors in guiding behaviour. *Cognition, 139*, 154–167.

Kuperan, K., & Sutinen, J.G. (1998). Blue water crime: deterrence, legitimacy, and compliance in fisheries. *Law & Society Review, 32*(2), 309–337.

Le Gallic, B., & Cox, A. (2006). An economic analysis of illegal, unreported and unregulated (IUU) fishing: Key drivers and possible solutions. *Marine Policy, 30*, 689–695.

Mangi, S.C., Dolder, P.J., Catchpole, T.L., Rodmell, D., de Rozarieux, N. (2013). Approaches to fully documented fisheries: Practical issues and stakeholder perceptions. *Fish and Fisheries, 16* (3), 426–452.

Mangi, S.C., Kupschus, S., Mackinson, S., Rodmell, D., Lee, A., Bourke, E., et al. (2018). Progress in designing and delivering effective fishing industry-science data collection in the UK. *Fish and Fisheries*. https://doi.org/10.1111/faf.12279.

Mann, H., Garcia-Rada, X., Houser, D., Ariely, D. (2014). Everybody else is doing it: Exploring social transmission of lying behavior. *PLoS ONE, 9*(10), e109591. https://doi.org/10.1371/journal.pone.0109591.

Mazar, N., Amir, O., Ariely, D. (2008). The dishonesty of honest people: A theory of self-concept maintenance. *Journal of Marketing Research, 45*(6), 633–644.

Mead, N.L., Baumeister, R.F., Gino, F., Schweitzer, M.E., Ariely, D. (2009). Too tired to tell the truth: Self-control resource depletion and dishonesty. *Journal of Experimental Social Psychology, 45*(3), 594–597.

Milinski, M., & Rockenbach, B. (2007). Spying on others evolves. *Science, 317*(5837), 464.

Mortensen, L.O., Ulrich, C., Eliasen, S., Olesen, H.J. (2017). Reducing discards without reducing profit: Free gear choice in a Danish result-based management trial. *ICES Journal of Marine Science, 74*(5), 1469–1479.

Needle, C.L., Dinsdale, R., Buch, T.B., Catarino, R.M.D., Drewery, J., Butler, N. (2015). Scottish science applications of remote electronic monitoring. *ICES Journal of Marine Science, 72*(4), 1214–1229.

Nielsen, J.R. (2003). An analytical framework for studying: Compliance and legitimacy in fisheries management. *Marine Policy, 27*(5), 425–432.

Nielsen, J.R., & Mathiesen, C. (2003). Important factors influencing rule compliance in fisheries lessons from Denmark. *Marine Policy, 27*(5), 409–416.

Norton, M.I., Mochon, D., Ariely, D. (2012). The IKEA effect: When labor leads to love. *Journal of Consumer Psychology, 22*, 453-460.

O'Keefe, C.E., & DeCelles, G.R. (2013). Forming a partnership to avoid bycatch. *Fisheries, 38*(10), 434–444.

Ostrom, E. (1990). *Governing the commons: The evolution of institutions for collective action.* New York: Cambridge University Press.

Ostrom, E. (1998). A behavioral approach to the rational choice theory of collective action: Presidential address, American Political Science Association, 1997 *American Political Science Review, 92*(1), 1–22.

Ostrom, E. (2001). Reformulating the commons. In J. Burger, E. Ostrom, R.B. Norgaard, D. Policansky and B.D. Golsdstein (Eds.), *Protecting the commons. a framework for resource management in the Americas* (pp. 17–41). Washington, DC: Island Press.

Ostrom, E. (2009). A general framework for analyzing sustainability of social-ecological systems. *Science, 325*, 419–422.

Ouwerkerk, J.W., Kerr, N.L., Gallucci, M., van Lange, P.A.M. (2005) Avoiding the social death penalty: Ostracism and cooperation in social dilemmas. In K.D. Williams, J.P. Forgas, W. von Hippel (Eds.), *The social outcast: Ostracism, social exclusion, rejection and bullying*. New York: Psychology Press.

Panagopoulos, C. (2010). Affect, social pressure and prosocial motivation: Field experimental evidence of the mobilizing effects of pride, shame and publicizing voting behavior. *Political Behavior, 32*, 369–386.

Prellezo, R., Kraak, S.B.M., Ulrich, C., et al. (2016). *Research for PECH Committee – the discard ban and its impact on the maximum sustainable yield objective on fisheries: Workshop.* Brussels: European Union.

Richter, A.P., & van Soest, D.P. (2012). Global environmental problems, voluntary action and government intervention. In E. Brousseau, T. Dedeurwaerdere, P.-A. Jouvet, et al. (Eds.), *Global environmental commons: Analytical and political challenges in building governance mechanisms* (pp. 223–248). Oxford: Oxford University Press.

Schou, M. (2011). Introducing transferable fishing concessions – A check list for operative considerations, (15.08.2011), unpublished.

Schou, M. (2015). BalticSea2020, (16.06.2015), unpublished.

SeafoodSource (2018). First QR code sushi on UK menu. https://www.seafoodsource.com/news/environment-sustainability/first-qr-code-sushi-on-uk-menu. Accessed 31 May 2018.

Sumaila, U.R., Alder, J., Keith, G. (2006). Global scope and economics of illegal fishing. *Marine Policy, 30*, 696–703.

Thaler, R.H., & Sunstein, C.R. (2009). *Nudge, improving decisions about health, wealth and happiness*. London: Penguin Books.

This Fish. (2018). http://thisfish.info. Accessed 31 May 2018.

Toonen, H.M., & Bush, S.R. (2018). The digital frontiers of fisheries governance: Fish attraction devices, drones and satellites. *Journal of Environmental Policy & Planning*. https://doi.org/10.1080/1523908X.2018.1461084.

Tversky, A., & Kahneman, D. (1974). Judgment under uncertainty: Heuristics and biases. *Science, 185*(4157), 1124–1131.

Tversky, A., & Kahneman, D. (1991). Loss aversion in riskless choice: A reference-dependent model. *The Quarterly Journal of Economics, 106*(4), 1039–1061.

Tyler, T.R. (2006). *Why people obey the law*. Princeton: Princeton University Press.

van Helmond, A.T.M., Chen, C., Poos, J.J. (2015). How effective is electronic monitoring in mixed bottom-trawl fisheries? *ICES Journal of Marine Science, 72*, 1192–1200.

Vohs, K.D., Mead, N.L., Goode, M.R. (2006). The psychological consequences of money. *Science, 314*, 1154–1156.

Vyrastekova, J., & van Soest, D. (2003). Centralized common-pool management and local community participation. *Land Economics, 79*(4), 500–514.

Wageningen University Research. (2018). Over Kenniskringen. https://www.wur.nl/nl/Onderzoek-Resultaten/Projecten/Kenniskring-Visserij/Over-Kenniskringen.htm. Accessed 31 May 2018.

Wilson, D.C., Nielsen, J.R., Degnbol, P. (Eds.). (2003). *The fisheries co-management experience. Accomplishments, challenges and prospects*. Dordrecht: Kluwer Academic Publishers.

Part IV
Tactical and Technological Options for Reducing Unwanted Catches

Chapter 12
A Marine Spatial Planning Approach to Minimize Discards: Challenges and Opportunities of the Landing Obligation in European Waters

José M. Bellido, Iosu Paradinas, Raúl Vilela, Guillermo Bas, and Maria Grazia Pennino

Abstract A sensible approach to minimize discards is to avoid areas or seasons where unwanted catch may be present. The implementation of a Marine Spatial Planning (MSP) approach to discard management requires the understanding of marine biological processes, as well as fishing conditions at a defined spatial scale. Mathematical models that analyze the spatio-temporal conditions of selected fishing areas allow the definition of different scenarios where discards are minimized by avoiding fishing for unwanted species and/or illegal specimens. Here we show some examples of how particular spatial models can be used for advice on MSP for discards. We introduce a geoserver GIS platform developed to produce maps of discard probability by using a Fishing Suitable Index. We also give an example of simulating virtual fishing closures. The inclusion of a Marine Spatial Planning approach to implement the Landing Obligation will bring some new challenges and opportunities. Finally, we will discuss and suggest some recommendations for its effective and successful implementation.

Keywords Discards GIS platform · Landing obligation · Maps of discard probability · Marine spatial planning · Simulating fishing closures

J. M. Bellido (✉)
Instituto Español de Oceanografía, Centro Oceanográfico de Murcia, Murcia, Spain

Statistical Modeling Ecology Group (SMEG), Departament d'Estadística i Investigació Operativa, Universitat de València, Valencia, Spain
e-mail: josem.bellido@ieo.es

I. Paradinas
Asociación Ipar Perspective, Sopela, Basque Country, Spain

R. Vilela · G. Bas · M. G. Pennino
Instituto Español de Oceanografía, Centro Oceanográfico de Murcia, Murcia, Spain

S. S. Uhlmann et al. (eds.), *The European Landing Obligation*,
https://doi.org/10.1007/978-3-030-03308-8_12

12.1 Introduction

The FAO Fisheries Glossary describes discards as

> *"the proportion of the total organic material of animal origin in the catch, which is thrown away or dumped at sea, for whatever reason. It does not include plant material and post-harvest waste such as offal".*

Additionally, FAO also defines bycatch as

> "the part of a catch of a fishing unit taken incidentally in addition to the target species towards which fishing effort is directed. Some or all of it may be returned to the sea as discards."

The discards may be dead or alive, depending on the selectivity of the fishing gear and the injuries and stress suffered by fishing. Although some species have high survival chances, many of the discards are dead or dying when rejected.

The "discards problem" is a key point in fisheries management around the world (Karp et al., this volume). It is not an easy issue, as it occurs at the core of fishing operations, both from economic, legal and biological points of view. It is basically a decision-making process, i.e. the decision to reject or retain a fish. However, there is usually a common perception from all sides (the public, NGOs, fishing sectors, policymakers, scientists, etc.) that discards are generally negative and that a better solution should be found.

Discarding unwanted catches has many assumed negative environmental and economic effects, especially since very few discarded fish will actually survive. Some of these effects are summarized below:

- Discarding juveniles means lower future catch opportunities and reduced spawning biomass.
- Discarding mature individuals weakens the stock's productivity both in the short and long term.
- Discarding fish, crustaceans, sea birds, sea mammals and non-targeted species undermines the balance of the marine ecosystem.
- Some vulnerable species can become severely depleted even in the absence of any directed fishery (e.g., certain sharks and rays).
- For fishers, discarding is a waste of time and effort in the present, as well as a serious potential loss of future income.

The European Union has recently reformed the Common Fishery Policy (CFP; EU 2013). One of the most important changes in the new CFP is the focus on what is caught rather than what is landed, as well as the introduction of a Landing Obligation, which has been progressively implemented. It is expected that full enforcement of the Landing Obligation will have a direct impact on discard reduction through more responsible and selective fishing.

On the other hand, Marine Spatial Planning (MSP) is also related to the implementation of the Landing Obligation. Fisheries management needs to consider the spatial dynamics where the natural stocks and fleet interact. Life always occurs in a defined space and time and fishing exploits marine living resources. For instance, a sensible approach to minimize discards is to simply avoid areas or seasons where unwanted catch is more likely (O'Keefe et al. 2013; Vilela et al. 2015; Paradinas et al. 2016).

The implementation of a Marine Spatial Planning approach to discard management requires the understanding of marine biological processes, as well as fishing conditions at a defined spatial scale. However, quantifying the fishery importance in an area is a challenging task because of a lack of information due to the inherent constrains of sampling at sea. Mathematical models that analyze the spatio-temporal conditions of considered fishing areas allow the definition of different scenarios of the fishing activity where discards are minimized by avoiding fishing undesirable species and/or illegal specimens (Hobday and Hartmann 2006; Pennino et al. 2014).

Here we show some examples of how particular spatial models can be used for advice on MSP for discards. We introduce a geoserver GIS platform developed to produce maps of discard probability by using a Fishing Suitable Index. This platform is designed to help fishers locate areas where they can maximize yield while minimizing unwanted catch. We also provide an example of simulating virtual fishing closures. This was done to test different possibilities of spatial planning for a purse seine fishery off the Southern Spanish Mediterranean coast.

The inclusion of a Marine Spatial Planning approach to implement the Landing Obligation will bring some new challenges and opportunities. Finally, we discuss and suggest some recommendations for its effective and successful implementation. We conclude that the Landing Obligation should be accompanied by other measures such as improvements in controlling fishing effort, better fishing selectivity, spatio-temporal fishing restrictions for vulnerable sizes and/or areas, effective enforcement and finally the agreement, commitment and support from the fishing sector to comply with the rules and regulations.

12.2 Marine Spatial Planning Approach to Minimize Discards

MSP aims to manage the different and shared uses in a marine area. This is typically referred to as "Ecosystem services", defined according to Wikipedia as *"the many and varied benefits that humans freely gain from the natural environment and from properly-functioning ecosystems"* (https://en.wikipedia.org/wiki/Ecosystem_services.)

Marine Spatial Planning is a public process of analyzing and allocating the spatial and temporal distribution of human activities in marine areas to achieve ecological, economic, and social objectives that usually have been specified through a political process (http://msp.ioc-unesco.org/about/marine-spatial-planning/). It is important to note that MSP is not only conservation planning, although it considers environmental protection and the sustainability of the marine environment for future generations as one of its main objectives. MSP seeks to balance economic development and environmental conservation, and not focus only on the goals of conservation or protection. Additionally, characteristics of Marine Spatial Planning also include ecosystem-based, area-based, integrated, adaptive, strategic, and participatory procedures.

The concept of multi-uses MSP is a relatively new approach to marine management. Actually, the first MSP workshop was held by UNESCO in 2006 in Paris, which makes this a recent movement considering the slow inertia and complexity to develop, establish and implement such a management system. However, since the beginning of its formulation there has been worldwide agreement on the need for such an approach. There are thus already a number of MSP applications in various regions (Maxwell et al. 2015; Pennino et al. 2017).

Although it varies greatly between countries, we may find many different uses of marine areas in Europe. This includes:

1. Extraction of non-renewable and renewable resources, including aquaculture.
2. Transit of people and merchandise, which are ruled by both national and international law.
3. Coastal zones are areas of residence and enjoyment.
4. Areas subject to new uses in the future, such as renewable energies, new methods for extraction of minerals, oils, new fishing grounds, etc.

This shared use can cause conflicts between the different users of ecosystem services. Particularly, in fisheries, there are many conflicts over the use of space:

1. Conflicts between different fishing gears: There are restrictions on some areas to certain fishing gears, for instance depth limits for trawlers or purse seiners.
2. Marine Protected Areas: There are many different types of protection, the most severe of which restrict all fishing, some others restrict specific fishing gears or limit the number of fishing boats, and others offer only seasonal protection.
3. Vulnerable species: Some species may have high levels of protection as they are considered sensitive and vulnerable. They can be scarce or confined to a particular habitat, threatened or their populations are at low levels.
4. Essential Fish Habitats: Some particular areas are protected due to the presence of juvenile fish, which can result in high levels of fishing discards and unwanted catch.

A Marine Spatial Planning approach can provide further insights in fishery management, considering the spatial information on where natural populations and fishing occur. It is well-known that spatial patterns, local movements, migration patterns, and more general geographical scenarios are thought to play an important role in the dynamics of fisheries resources (Warren 1998; Fogarty and Botsford 2007).

However, to get reasonable estimations and predictions of abundance, including a description of variability, models must describe the relevant species interactions, effects of environmental conditions and fishing effort by gear at appropriate spatial and temporal scales (Sims et al. 2008; Elith and Leathwick 2009; Vilela and Bellido 2015). There are obvious relationships between fishing effort, habitat properties, catchability and fishing mortality, and all these features have to be considered in order to enhance fisheries management in the framework of a MSP approach.

A key point to reduce discards and achieve more environmentally responsible fishing is to minimize fishing operations in inappropriate areas. It is quite obvious that any management of unwanted catches should consider measures to avoid areas or periods with high abundances of bycatch and discards. Bellido et al. (2011) numbered the three main pillars to reduce discards and unwanted catch:

1. Avoid fishing in areas of higher discards, related to improving fishing strategies by an adequate selection of the fishing grounds.
2. Better selectivity of fishing gears, allowing avoidance and escape of unwanted catch to increase survivability.
3. Valorization of the unwanted catch, with a progressive positive input of fish processing technology and adding value. This is complimented by research to provide new fish products when needed (burgers, nuggets, etc.) as well to use fish material for other uses (pharmaceutical, cosmetic, aquaculture feed producers, biomedical, etc.).

MSP is totally related to the first pillar. With that aim, sensitive areas for particular species or early life stages should be identified, as well as the derivation of areas of special protection for mobile gears and the classification of more suitable fishing areas for "mobile" and "non-mobile" fishing gears according to habitat and fish community characteristics. This research may allow a better understanding of spatial and temporal distributions and abundances of discards, together with an online handling and updating of information, to assess the health of the ecosystem on a spatial basis.

Although scientific advice is indeed essential for an appropriate MSP implementation, MSP scientists should show awareness, support and openness to advice. They should support stakeholders and policymakers by providing knowledge on every specific spatial scale and provide tools to transfer research to advisory products for better spatial planning. For example, a quite accessible and useful outcome is to provide maps and indicators that help managers and users with better decision-making when considering a space-time scenario.

MSP scientific advice is mainly based on the application of mapping tools together with a set of different statistical models and other computational techniques, all on a spatio-temporal basis (Fig. 12.1). Unfortunately, these tools have often been ignored in fisheries management, mostly due to both a lack of appropriated knowledge of fisheries researchers to apply these tools and a lack of involvement of spatial researchers to provide technical advice for the quantitative and qualitative spatio-temporal analysis.

Here we provide two examples of how to apply these techniques to provide scientific advice for MSP, particularly related to Landing Obligation issues. The first one is related to the establishment of an online GIS platform to inform fishers of areas where they can maximize yield by minimizing discards. The second one is a simulation of spatial fishing closures to test different MSP scenarios for a purse seine fishery in South East Spain. The findings and applications of these two case studies can help fishery managers make decisions to mitigate discards in European waters.

This work was developed in the framework of the iSEAS LIFE+ project, "Knowledge-Based Innovative Solutions to Enhance Adding-Value Mechanisms

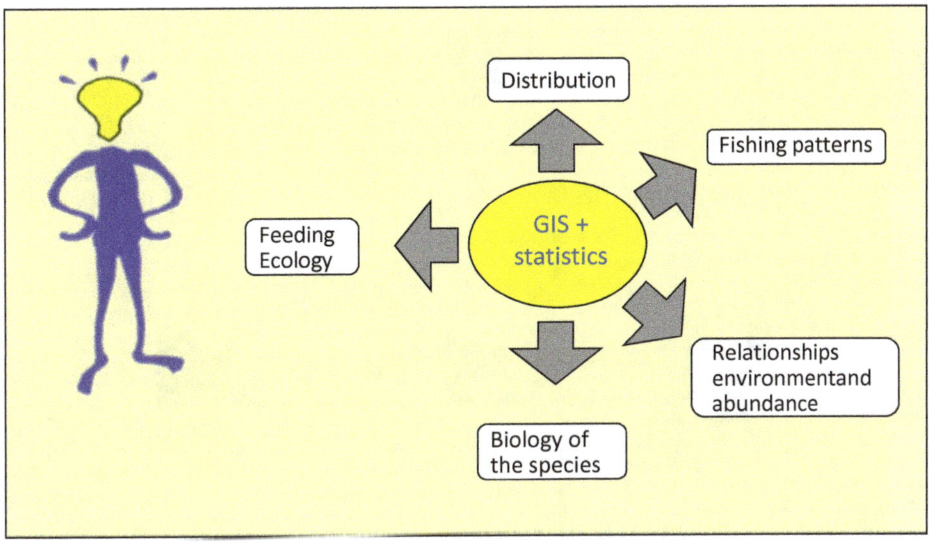

Fig. 12.1 Possibilities of spatial techniques in MSP, how to integrate different statistical models and other computational techniques, all on a spatio-temporal basis

towards Healthy and Sustainable EU Fisheries". iSEAS WP3 was totally devoted to applying spatial models to reduce discards and increase efficiency and these case studies are an example of that approach.

12.2.1 An Online GIS Platform to Mitigate Discards

Over the past decades, a branch of IT (Information Technologies) has gradually evolved, specifically dedicated to mapping and spatial analysis. This technology is usually referred to as Geographical Information Systems (GIS), though it has also been called geo-data systems, spatial information systems, digital mapping systems and land information systems. A GIS comprises a collection of integrated computer hardware and software which together is used for inputting, storing, manipulating, analyzing and presenting a variety of geographical data. Applications of GIS to marine fishery resources, or indeed to any marine applications, have been very limited, being mostly confined to peripheral areas such as coastal zone management, pollution modeling and controls, mariculture and shoreline mapping (Meaden 2000).

One of the objectives of our work was to develop a fully-operative GIS tool that integrates predictive models of suitable fishing areas for Northeast Atlantic fisheries (Fig. 12.2). This is a real time modeling technique, which could help fishers avoid areas or periods with high probability of unwanted bycatch/discards (Vilela and Bellido 2015). Such a model aims to minimize discards/bycatch throughout the study area through flexible real-time modeling of recent catches and discards, which produces maps as final results, i.e., outputs that are easy to interpret by fishers (see Fig. 12.3). The final aim is to provide additional info to skippers to carry out their fishing strategies taking into account which species are in the area and in which

Fig. 12.2 Schematic of the automatic procedure for data acquisition and main page of the GIS platform. (http://iseas.cesga.es/)

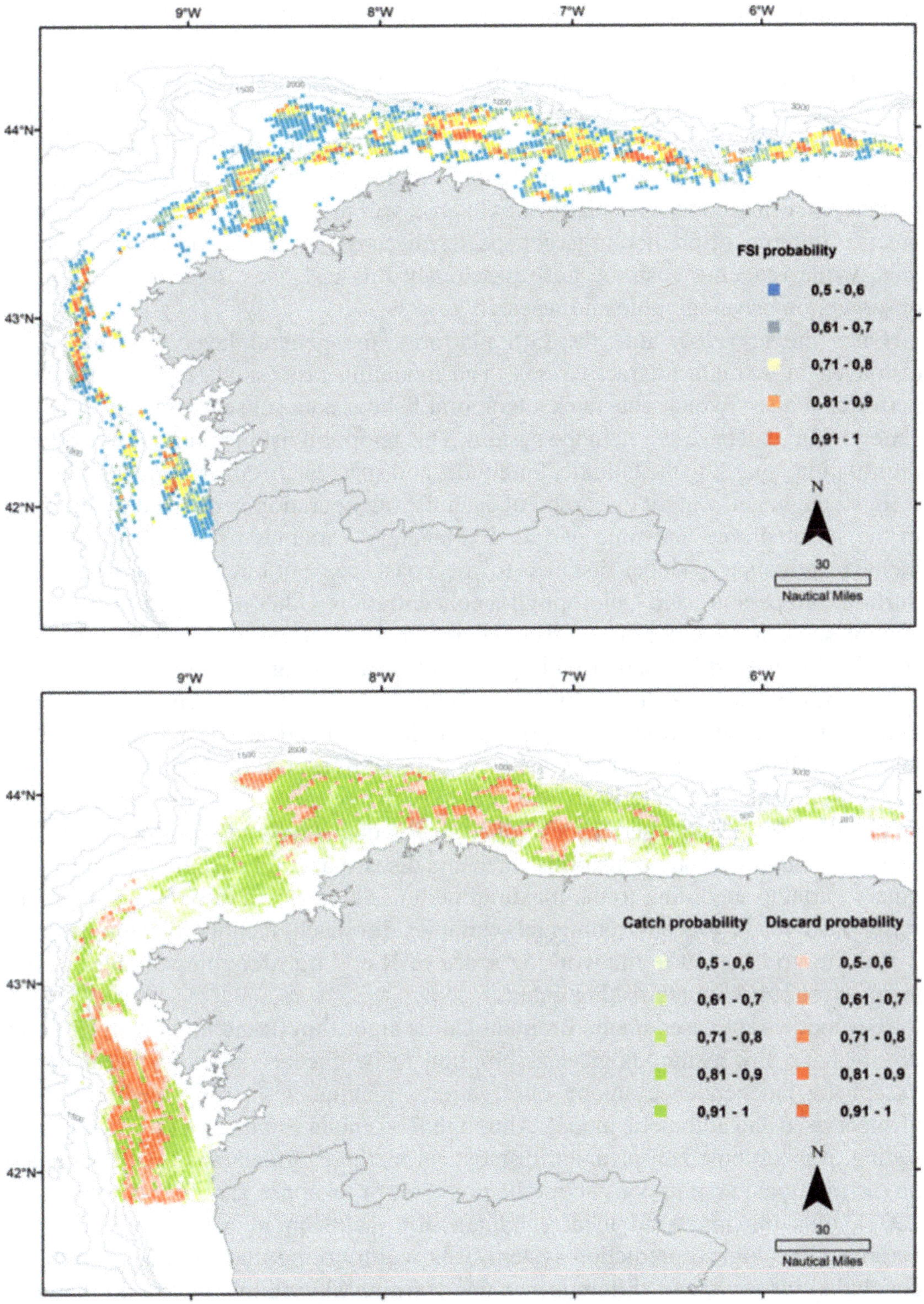

Fig. 12.3 Model results for *Merluccius merluccius* in area ICES VIII-west and area ICES IXa: FSI annual probability (upper image), with the best fishing areas in green; catch probability (bottom left), with higher probabilities of positive catch (i.e. distribution area of the species) in dark green; discard probability (bottom right), with higher probabilities of discard rates of more than 20% in orange

catch composition, and as a consequence, the kind of discards/bycatch to be expected.

Fishery data are provided for the fishing vessels in an automatic way (Fig. 12.2). The participation of fishing vessels is voluntary, but only fishing vessels providing data can benefit from the online results. As a result, fishing vessels will act as online sensors that continuously provide real data from the daily activity of fleets both to the models, as well as to the GIS platform. This fact will make models more accurate and precise (in terms of predicting target species/discards) for fleets working in a given area, while reducing costs because previously this task was done by specialized observers on oceanographic and research vessels.

Once the user logs into the GIS platform, the general layer information is displayed in a small, interactive box. The modeling process starts when the user performs a query which generates a temporal fishing data subset extracted from the general SQL database stored in the system. This temporal fishing data subset collects information regarding the latitude, longitude, and species-specific information on the weight caught and weight discarded of each fishing operation performed by the fleet in the selected area and time period. Explanatory variables used for the analysis include bathymetry, slope, distance to the coast, sea bottom characteristics, Sea Surface Temperature and Chlorophyll-a concentration. Other important explanatory variables, such as the season, fleet characteristics, etc., are intrinsic and self-contained in the fishing data used to perform the modelling.

The model estimates the Fishing Suitability Index (FSI) for a given operational unit (métier), time period and marine area. It is defined as the probability of a location to have low discards, i.e. below a defined threshold discard rate, based on its environmental characteristics and previous fishing activity information (Vilela and Bellido 2015). The script, developed in R open statistical software, reads the temporal data subset and transforms catch and discard data for each record into a FSI binary variable according to the maximum admissible discard rate (threshold). The script also reads the environmental variables for each fishing location from a 2×2 km grid loaded in the working space of R and transfers the variables to the temporal subset with the fishing data.

The models are based on the Breiman-Curtis algorithm (Random Forest; RF) and do not have any methodological assumption to be checked prior to the analysis, except the independence among observations, making it a strong method for an unsupervised and automatic model. Although RF is not a suitable tool for hypothesis testing, it is a robust non-parametric statistical method for data analysis that makes no distributional assumptions about the predictor or response variables (Cutler et al. 2007), thus making it an ideal candidate for inclusion in a flexible and fully automated ecological prediction system. It is worth noting that Crisci et al. (2012) concluded his review of different machine learning algorithms over rocky benthic communities by highlighting the properties of RF as, "*one of the most efficient learning algorithms in terms of prediction accuracy*".

Model results are projected for the whole area using the 2×2 km grid, and resulting maps are stored in a vectorial GeoJSON format in a local folder of the server. The maps and a text file with information regarding the performance and information about the fishing data processed (i.e. number of records with catch, number of records with discards, percentage discarded, average FSI, mean catch, and mean discards) are available.

In the last step, the geoportal shows the GeoJSON vectorial image in a web-mapping service and shows the user a measure of the model performance, according to predefined probability values (FSI index):

- Between 0 and 0.5: Red zone, less suitable areas for fishing (high discard proportion).
- Between 0.5 and 0.8: Orange zone, intermediate areas for fishing (medium discard proportion).
- Between 0.8 and 1: Green area, the most suitable areas for fishing (low discard proportion).

Results vary between species, areas and métiers, achieving good results with less than 50 records for target species, while poor predictions are obtained for those bycatch species less influenced by environmental factors. Some examples of the map tool generated for hake (*Merluccius merluccius*) are presented in Fig. 12.3.

These prediction tools help fishers avoid areas or time periods with high probabilities of obtaining unwanted bycatch/discards since they produce maps as final results, i.e., outputs that are easy to interpret by fishers (Vilela and Bellido, 2015). Also they are able to perform both short term and long term predictions, adapt to any species, area and fishing operational unit and they are quick, i.e., results can be visualized by the user in seconds. These prediction models provide a good predictive accuracy and offer an assessment (or goodness of the fit) about the reliability of the prediction visualized to the user.

This GIS platform is accessible to the public from http://iseas.cesga.es/ where some of these maps and results are displayed.

12.2.2 MSP to Reduce Discards in a Small Pelagic Fishery off South East Spain

Along the Spanish Mediterranean coast, the purse seine fishery mainly harvests small pelagic fish (Fig. 12.4). The most common type of fishery operates during the night and mainly targets anchovy (*Engraulis engrasicolus*) and sardine (*Sardina pilchardus*). Purse seine fisheries are characterized by actively searching for fish using echo acoustics. This makes it a reasonably selective fishery with generally lower discard rates than trawlers, gillnetters and longliners, thus attracting fewer studies related to discards.

However, despite relatively low mean discard rates in purse seiners, a huge variability per haul is present ranging from 0% to 100% of the catch being discarded (Borges et al. 2001). This high variability is influenced by the electronic equipment used to identify and target fish schools, failing to correctly determine the species composition or size of a school and can lead to the whole volume of the haul being discarded.

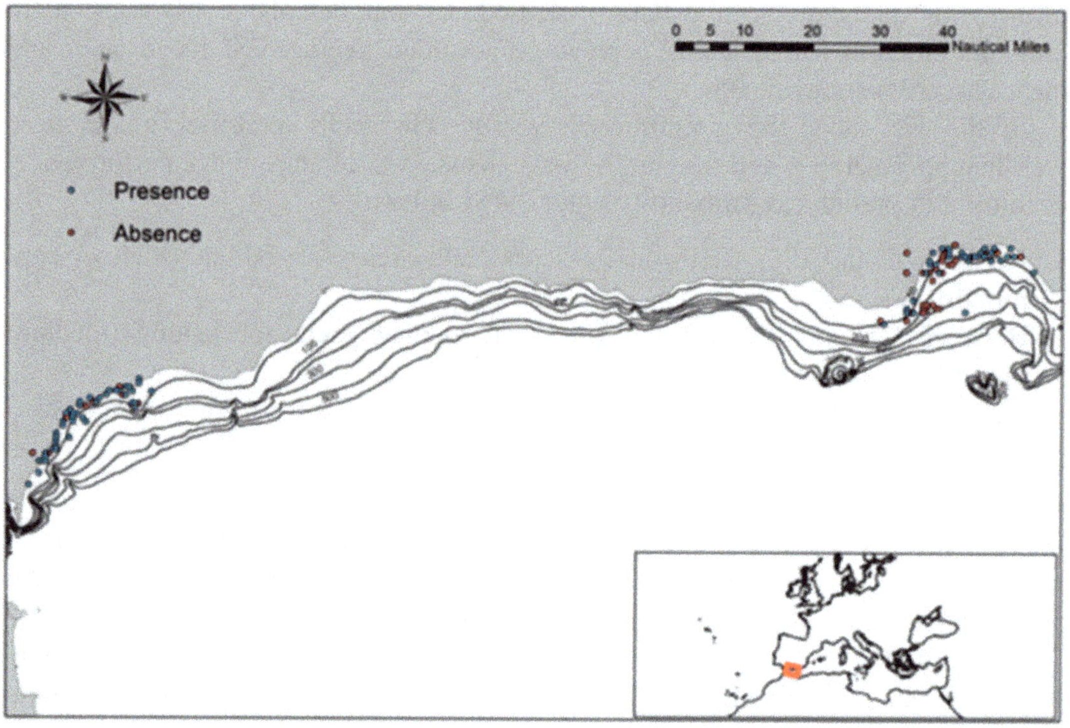

Fig. 12.4 Depiction of fishing locations in the purse seine fishery in South Spanish Mediterranean. Dots represent hauls in the two main fishing grounds, Bay of Málaga (western area) and Bay of Almería (eastern area)

We explored the Spanish Mediterranean purse seine reference fleet with respect to discards to identify potential discard driving patterns. We simulated different fishing closures and explored how they can affect catch compositions and discard patterns. This study is based on data collected by the Spanish Mediterranean on-board observers program, for which the purse seine reference fleet is located in the Northern Alborán Sea (GSA01). The Alborán Sea has 11 ports, although the data set covers the two main sub-areas of the Northern Alborán Sea (Almería and Málaga). The data set comprises 108 fishing trips and a total of 173 fishing operations from 7 different vessels during a 3 years period (2009–2011). It also contains information on the location, time, depth, and moon phase of each haul, as well as characteristics of the vessel such as gross registered tonnage (GRT), length and power (HP).

As usual in fisheries, the abundance of discards in purse seiners is characterized by a relatively high number of zeros. In these cases, data modeling has to take into account both zero and non-zero observations. Models able to do so are known as zero-inflated models. However, most of these models have been developed for discrete data and cannot be implemented for continuous data, such as discard volumes. Data with this characteristic are known as semi-continuous data. In the present analysis we have applied a 2-stage model to simulate the semi-continuous behavior that characterizes discards of the Spanish Mediterranean purse seiners. We first fitted a Bayesian generalized linear mixed model (GLMM, Muñoz et al. 2012) to model the occurrence (presence/absence) of discards according to technical and environmental factors that characterize discard occurrence. Then, conditional to presence, we modeled the abundance of the discards using Bayesian geostatistics. In this case, log-transformed discards were used to down-weight extreme values and ensure a better fit of the models.

Regression models such as generalized linear models (GLM-GLMM) and generalized additive models (GAM) were selected as the main candidate approach to uncover the relationships between the amount of discards, expressed in kilos, and some independent covariates. A stepwise approach, based on the estimation of the Akaike Information Criteria (AIC) and the Deviance Information Criteria (DIC) value for each model, was applied to select the final model. This study did not aim to estimate the differences among vessels, so we introduced vessel as a random noise effect in every model. This way, we expected it to absorb the variability that encompasses different purse seine discarding behaviors for each vessel, which can occur due to various technical, social and/or economic reasons.

In addition to discards, a couple of complementary models were fit. First, a Poisson GLM model was designed to fit the relationship between the number of species caught and bathymetry. Additionally, two linear regression models were fit; one for the retained fraction of the catch and the other for the discarded fraction.

Fisheries management has used a wide range of different fishing closures. Some are used as general regulatory principles, such as minimum depth and proximity to land, and others are usually applied on a more regional scale, such as spatial, temporal, and spatio-temporal closures. Here we present a fast, simulation-based approach to easily quantify the impact of applying any such fishing closures in terms of efficiency, in this case lowering discards. Assuming that the available data is representative of the local purse seine fishery, we first simulated a fishing closure and removed all observed hauls that fell into that fishing closure window. Then, we resampled the resulting hauls with replacement and calculated the mean. We then repeated the process N times and compared the results with the whole population. This process was applied to both Almería Bay (simulating 6 fishing closures) and Málaga Bay (simulating 7 fishing closures).

All sampled hauls summed up to 249 tonnes of retained catch and 25.1 tonnes of discarded fish, of which 15.8 tonnes were non-regulated species under the Landing Obligation. The mean discarding ratio of the fishery is 12.3%, with a bootstrap confidence interval between 9.6% and 15.2%. Discard ratios were higher in Málaga than in Almería for both the whole discarded fraction and the non-regulated fraction (Fig. 12.5). Discarding behavior was very variable; we found differences within Málaga and Almería that showed evidence of heterogeneity of discarding at this regional mesoscale.

Results showed that the most significant variable explaining discards was depth for both areas. The model found a negative relationship between depth and the occurrence probability of discards, i.e., a higher occurrence in shallower waters (< 50 m) and a lower occurrence in deeper waters. Also, the number of species caught and discarded decreased with depth (Fig. 12.6), and retained weight per haul increased with depth, up to approximately 130 m. Discarding behavior was also shown to vary significantly between vessels, suggesting individual skipper's fishing preferences may also influence the catch composition of each haul.

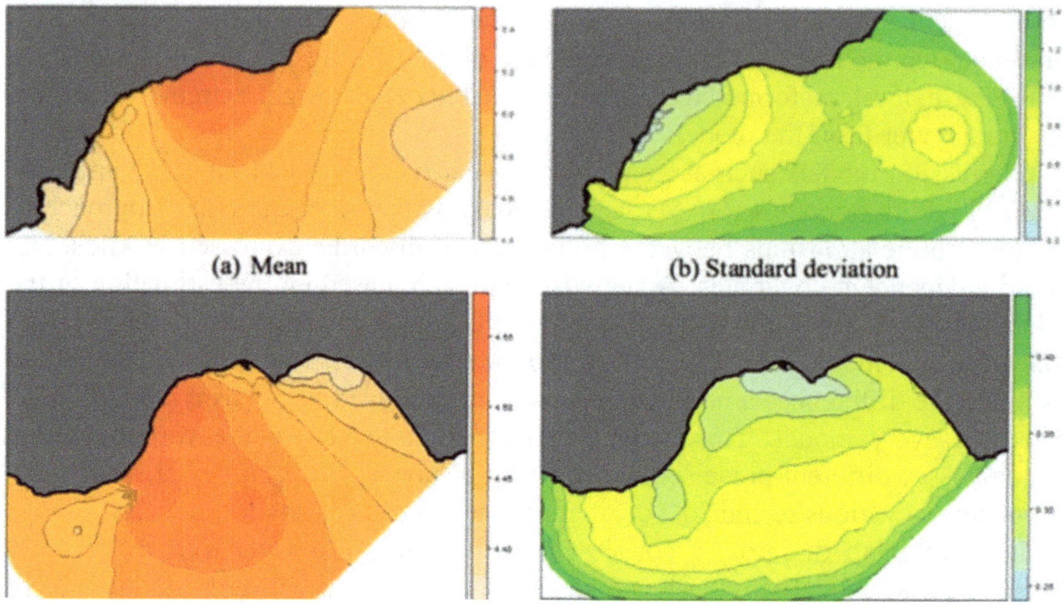

Fig. 12.5 Predicted log-kg of discards in the Bay of Málaga (upper) and Bay of Almería (bottom). Maps show mean (left) and the standard deviation (right) of the predictions

Regarding simulation of fishing closures, both bays show the same pattern and we found no significant differences among fishing closures both in Málaga and Almería (Fig. 12.7). This can be explained by the stable pattern of fish occurrence in both bays as well as a similar discard behavior along the bays. The removal of hauls in every virtual fishing closure is rapidly compensated with hauls in the non-protected areas, generating approximately the same discard level. Hence, the most important spatial effect is therefore bathymetry, and that should be the criteria for any MSP approach to this fishery. As a summary for this fishery, we recommend better enforcement of the purse seine fishery ban in shallow waters (< 35 m), which is currently ignored by many skippers, as the best strategy to minimize purse seine discards in this area.

12.3 Challenges and Opportunities of MSP

Fisheries management is a challenging discipline where biological, social and economic factors converge to form a complex web of interactions, all occurring in a defined geographical scenario. Hence, Marine Spatial Planning presents an intuitive, complementary and natural approach to enrich fisheries management towards more sensible and effective strategies.

MSP requires a profound knowledge of fishing grounds, and not just focused on target species. The monitoring of fishing operations faces formidable technical challenges due to the combination of a target that varies in space and time and a mobile exploitation activity. In many exploited fisheries, the large number of species involved, the high number of fishing gears employed, and the widely dispersed

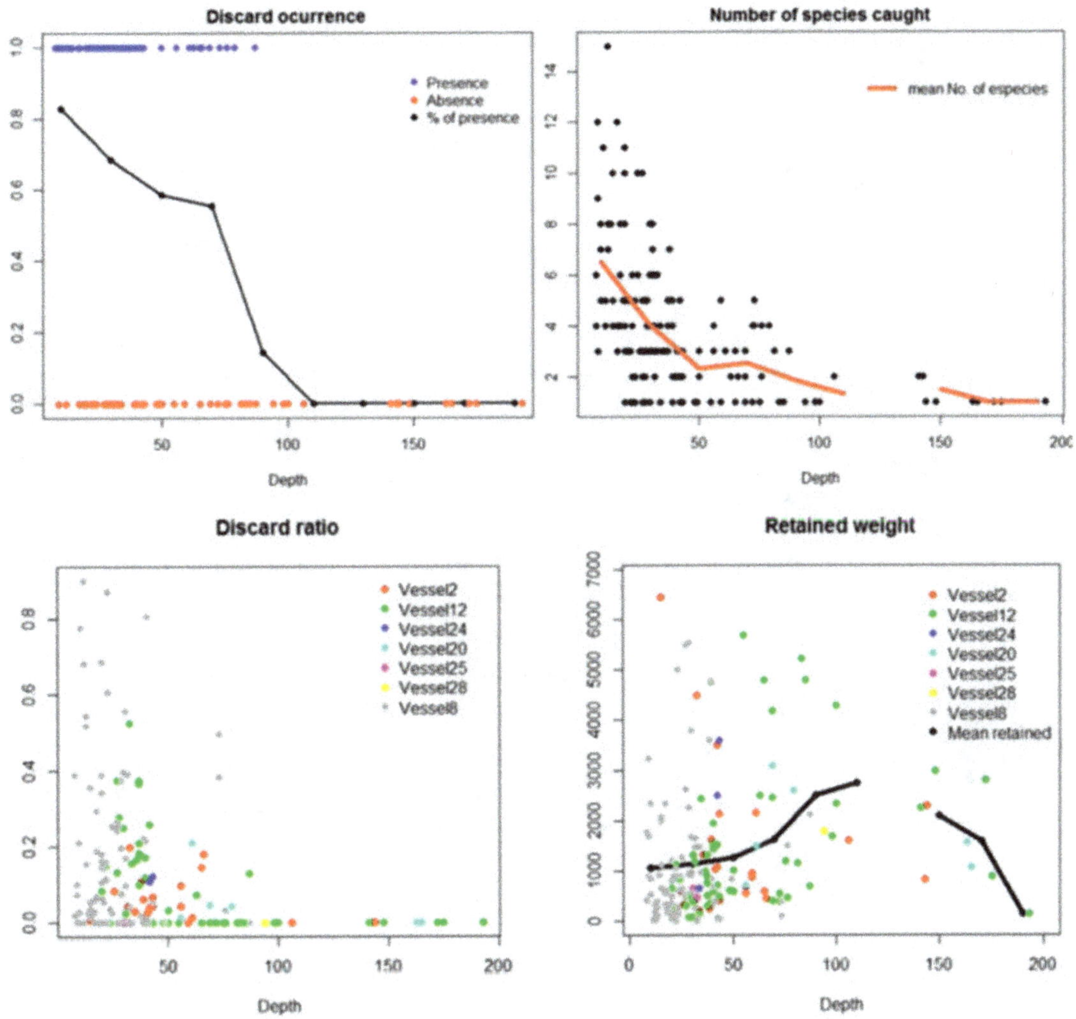

Fig. 12.6 Upper left: presence/absence of discards against depth; in black, the percentage of hauls that present discards every 20 m depth stratum. Upper right: heterogeneity of hauls versus depth; in red, the mean number of species per haul every 20 m depth stratum. Bottom left: discard ratios of each haul versus depth at the location of the haul, differentiated by vessel. Bottom right: mean total retained weight per haul versus depth at the location of the haul, differentiated by vessel. Also shown is the mean retained weight per 20 m depth stratum (i.e. 0–20 m, 21–40 m...), missing data for stratum 120–140 m

landing ports make monitoring, enforcement and compliance measures extremely difficult (Williams and Corral 1999). As a consequence, huge costs and workload are entailed on already overburdened institutions, not always with the required or expected results. In other words, monitoring actions are not as exhaustive as required, in the sense that they are not implemented in every vessel all the time.

Technology could help to implement MSP. To monitor, control, and document full catch and discard information, several alternative data sources are available: inspectors, Electronic Monitoring with video, monitoring via GPS and sensors, naval and air patrols, reference fleets, landing controls, satellite tracking with VMSs (Vessel Monitoring Systems), and the fishers' self-reporting (including

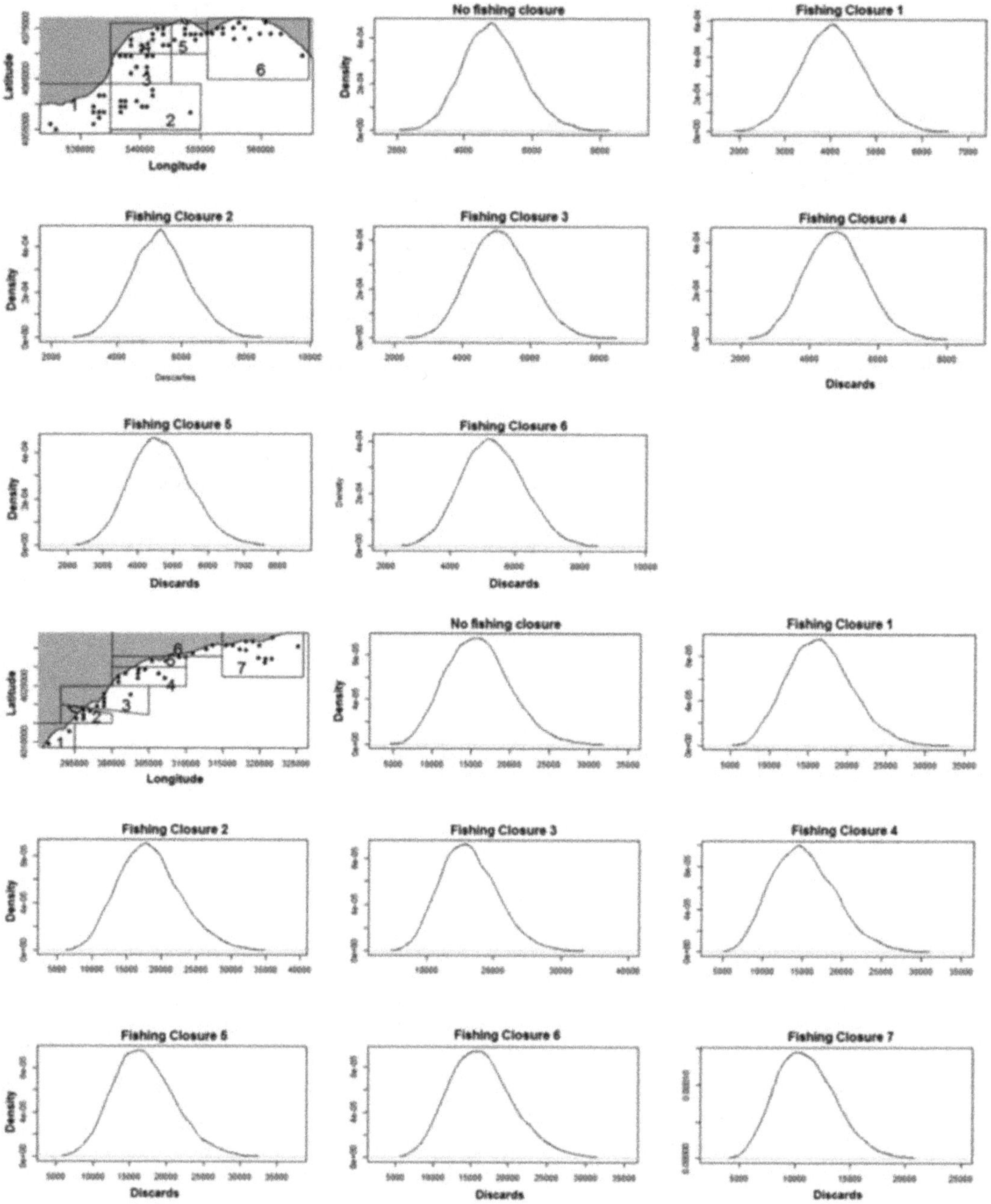

Fig. 12.7 Estimated discards (kg) for every virtual fishing closure for Bay of Almería (top panels) and Bay of Málaga (bottom panels)

logbooks, landing declarations, and sales notes). It must be noted that novel technologies for fisheries monitoring do not replace traditional control and surveillance methods, such as inspections onboard vessels or on shore. However, used correctly, the new technologies should help to better target actions and therefore cut costs and increase effectiveness (James et al., this volume). By crosschecking data collected using the different systems, fisheries authorities can apply risk based control strategies and detect illegal activities that could otherwise go unnoticed.

When fully implemented the main benefits of this system will include:

1. *Quasi* real-time monitoring (depending on when catch information is transmitted) of catches by all fishing vessels operating in EU waters and of fish quotas.
2. Protection of valuable commercial fish stocks by appropriate stock assessments based on real catch data.
3. Efficient and effective data interchange between agencies engaged in fisheries monitoring and control across the European Union.
4. Reduced requirements for manual entry of logbook data into central databases.

The defined area, time scale, and review period may not be the same for different legal obligations, policy and management goals, and operational objectives. The validity of the goals and objectives should be assessed by SMART criteria (Specific, Measurable, Achievable, Realistic and Time-bound).

Policy approaches can be top-down (imposed by government), bottom-up (meeting popular demands from end users), or a combination of the two. The balance between these policy approaches will give an indication of how likely end-users will be to follow enforcement laws.

A real challenge is to avoid duplication of effort by different public agencies and levels of government in MSP activities. It is necessary to provide a rational basis for setting priorities, as well as to direct resources to where and when they are needed most. If resources are limited, then a prioritization exercise could be undertaken to consider the relative importance of ecological, social, economic, and other operational objectives.

A quite useful product to design and start a MSP approach is the UNESCO Step-by-Step guide (Ehler et al. 2009). This guide is directed towards professional marine managers, and it provides a complete view of MSP and describes a logical sequence of steps that are required for successful MSP implementation.

Additional initiatives aimed at progressing towards MSP are underway. MESMA (www.mesma.org) was an EU-FP7 project on monitoring and evaluation of Spatially Managed Marine Areas. One of the main outcomes of this project was the elaboration of a roadmap to implement MSP. Since MESMA has finished, the MSP movement has been gathering throughout the European MSP Platform (http://msp-platform.eu), which provides information and tools for MSP in different European sea basins.

12.4 Summary and Policy Recommendations

The best discard mitigation measure occurs at sea; not catching unwanted bycatch. The key aspect should be better fishing practices to avoid unwanted catch.

A better marine spatial planning approach is needed for fishery management. These types of measures can be more flexible and dynamic in response to spatial and seasonal restrictions related to fishing. Depending on the population, the measures can differ, for instance we suggest spatio-temporal fishing restrictions for vulnerable sizes and/or areas. Regarding spawners, they could be protected through planning for temporal (seasonal) closures. These temporal closures should be applied to all métiers at the same time.

In the cases studied here, depth fishing restrictions appear to be one of the best management measures, and they are totally based in MSP. There are several prohibitions that are essential to our knowledge and should be maintained and, in some cases, better enforced. These are the prohibition on purse seine fishing shallower than 35 m or over sensitive habitats, the prohibition on trawl fishing shallower than 50 m or over sensitive habitats, and the prohibition on all fishing beyond 1000 m.

The Landing Obligation should be accompanied by other measures for its successful implementation. Some of these measures are improvements in fishing effort controls, effective enforcement, and finally an agreement from the fishing sectors to comply with rules and regulations.

Acknowledgments This work was funded by the Project iSEAS, Ref. LIFE13 ENV/ES/000131, "Knowledge-Based Innovative Solutions to Enhance Adding-Value Mechanisms towards Healthy and Sustainable EU Fisheries", cofounded under the LIFE+Environmental Program of the European Union. We are deeply grateful to all the partnerships on this project and particularly to the observers who gather the fishing data within the IEO discarding observer program. The collection of fishery data has been co-funded by the EU through the European Maritime and Fisheries Fund (EMFF) within the National Program of collection, management and use of data in the fisheries sector and support for scientific advice regarding the Common Fisheries Policy.

References

Bellido, J.M., Santos, M.B., Pennino, M.G., Valeiras, X., Pierce, G.J. (2011). Fishery discards and bycatch: Solutions for an ecosystem approach to fisheries management? *Hydrobiologia*, *670*(1), 317–333.

Borges, T.C., Erzini, K., Bentes, L., Costa, M.E., Gonçalves, J.M.S., Lino, P.G., et al. (2001). By-catch and discarding practices in five Algarve (southern Portugal) métiers. *Journal of Applied Ichthyology, 17*, 104–114.

Crisci, C., Ghattas, B., Perera, G. (2012). A review of supervised machine learning algorithms and their applications to ecological data. *Ecological Modelling*, *240*, 113–122. doi:https://doi.org/10.1016/j.ecolmodel.2012.03.001.

Cutler, D.R., Edwards, T.C., Beard, K.H., Cutler, A., Hess, K.T., Gibson, J., Lawler, J.J. (2007). Random forests for classification in ecology. *Ecology*, *88*, 2783–2792. https://doi.org/10.1890/07-0539.1. PMID:18051647.

Ehler, C., & Douvere, F. (2009). Marine spatial planning: A step-by-step approach toward ecosystem-based management. Intergovernmental oceanographic commission and man and the biosphere programme. IOC manual and guides no. 53, ICAM dossier no. 6. Paris: UNESCO. 2009 (English).

Elith, J., & Leathwick, J.R. (2009). Species distribution models: Ecological explanation and prediction across space and time. *Annual Review of Ecology, Evolution, and Systematics*, *40* (1), 677–697. https://doi.org/10.1146/annurev.ecolsys.110308.120159.

EU. (2013). Regulation (EU) No 1380/2013 of the European Parliament and of the Council on 11 December 2013 on the Common Fisheries Policy, amending Council Regulations (EC) No 1954/2003 and (EC) No 1224/2009 and repealing Council Regulations (EC) No 2371/2002 and (EC) No 639/2004 and Council Decision 2004/585/EC.

Fogarty, M.J., & Botsford, L.W. (2007). Population connectivity and spatial management of marine fisheries. *Oceanography*, *20*, 112–123.

Hobday, A. J., & Hartmann, K. (2006). Near real-time spatial management based on habitat predictions for a longline bycatch species. *Fisheries Management and Ecology, 13*, 365–380.

Karp, W.A., Breen, M., Borges, L., Fitzpatrick, M., Kennelly, S.J., Kolding, J., et al. (this volume). Strategies used throughout the world to manage fisheries discards – Lessons for implementation of the EU Landing Obligation. In S.S. Uhlmann, C. Ulrich, S.J. Kennelly (Eds.), *The European Landing Obligation – Reducing discards in complex, multi-species and multi-jurisdictional fisheries*. Cham: Springer.

Maxwell, S. M., Hazen, E. L., Lewison, R. L., Dunn, D. C., Bailey, H., Bograd, S. J., Briscoe, D. K., Fossette, S., Hobday, A. J., Bennett, M., Benson, S., Caldwell, M. R., Costa, D. P., Dewar, H., Eguchi, T., Hazen, L., Kohin, S., Sippel, T., Crowder, L. B., (2015). Dynamic ocean management: defining and conseptualizing real-time management of the ocean. *Marine Policy, 58*, 42–50.

Meaden, G.J. (2000). Application of GIS to fisheries management. In D.J. Wright & D.J. Bartlett (Eds.), Marine and coastal geographical information systems (pp 205–226). London: Taylor & Francis.

Muñoz, F., Pennino, M.G., Conesa, D; López-Quílez, A., Bellido, J.M. (2012). Estimation and prediction of the spatial occurrence of fish species using Bayesian latent Gaussian models. *Stochastic Environmental Research and Risk Assessment, 27*, 1171–1180. https://doi.org/10.1007/s00477-012-0652-3.

O'Keefe, C. E., Cadrin, S. X., Stokesbury, K. D. E. (2013). Evaluating effectiveness of time/area closures, quotas/caps and fleet communications to reduce fisheries bycatch. *ICES Journal of Marine Science, 71*, 1286–1297.

Paradinas, I., Martín, M., Pennino, M.G., López-Quílez, A., Conesa, D., Barreda, D., Gonzalez, M., Bellido, J.M. (2016). Identifying the best fishing-suitable areas under the new European discard ban. *ICES Journal of Marine Science*, 2479– 2487. https://doi.org/10.1093/icesjms/fsw114.

Pennino, M.G., Muñoz, F., Conesa, D., López-Quílez, A., Bellido, J.M. (2014). Bayesian spatio-temporal discard model in a demersal trawl fishery. *Journal of Sea Research, 90*, 44–53.

Pennino, M.G., Vilela, R., Valeiras, J., Bellido, J.M. (2017). Discard management: A spatial multi-criteria approach. *Marine Policy, 77*, 144–151.

Sims, M., Cox, T., Lewison, R. (2008). Modeling spatial patterns in fisheries bycatch: Improving bycatch maps to aid fisheries management. *Ecological Applications 18*, 649–661. https://doi.org/10.1890/07-0685.1.

Vilela, R., & Bellido, J.M. (2015). Fishing suitability maps: Helping fishermen reduce discards. *Canadian Journal of Fisheries and Aquatic Sciences, 72*, 1191–1201.

Warren, W.G. (1998). Spatial analysis for marine populations: Factors to be considered. *Proceedings of the North Pacific Symposium on Invertebrate Stock Assessment and Management, Canadian special publication of fisheries and aquatic sciences, 125*, 2128.

Williams, M.J., & Corral, V.P. (1999). Fisheries monitoring: Management models, compliance and technical solutions. In C.P. Nolan (Ed.), Proceedings of the international conference on integrated fisheries monitoring. Sydney, Australia, 1–5 February 1999 (pp. 37–50). Rome: FAO. 378 p.

Chapter 13
The Best Way to Reduce Discards Is by Not Catching Them!

David G. Reid, Julia Calderwood, Pedro Afonso, Pierre Bourdaud, Laurence Fauconnet, José Manuel González-Irusta, Lars O. Mortensen, Francesc Ordines, Sigrid Lehuta, Lionel Pawlowski, Kristian S. Plet-Hansen, Zachary Radford, Marianne Robert, Marie-Joelle Rochet, Lucía Rueda, Clara Ulrich, and Youen Vermard

Abstract Under the Landing Obligation (LO) fishers will need to reduce or land fish that were previously discarded. In this chapter we look at how they might be able to do that by summarising a number of studies conducted in various European regions. We start by describing a series of "challenge" trials where fishers tried to reduce their discards by whatever (legal) means they thought best. In some cases, they were able to reduce unwanted catches, in others they were less successful. We also interviewed fishers not involved in the trials to ask them what they thought they could do. We explore their approaches which generally fell into three categories:

D. G. Reid (✉) · J. Calderwood
Marine Institute, Oranmore, County Galway, Ireland
e-mail: David.Reid@Marine.ie

P. Afonso · L. Fauconnet · J. M. González-Irusta
Instituto do Mar (IMAR), Department of Oceanography and Fisheries, Marine and Environmental Sciences Centre (MARE), and OKEANOS Research Unit, Universidade dos Açores, Horta, Portugal

P. Bourdaud · S. Lehuta · Y. Vermard
Ifremer, Fisheries Ecol & Modelling Unit, Nantes, France

L. O. Mortensen · K. S. Plet-Hansen · C. Ulrich
National Institute of Aquatic Resources, DTU Aqua, Technical University of Denmark, Kgs. Lyngby, Denmark

F. Ordines · L. Rueda
Centro Oceanográfico de Baleares, Instituto Español de Oceanografía, Palma, Spain

L. Pawlowski · M. Robert
Fishery Technology and Biology Laboratory, IFREMER, Lorient, France

Z. Radford
Centre for Environment, Fisheries & Aquaculture Science (Cefas), Lowestoft, UK

M.-J. Rochet
IFREMER, Centre Atlantique, Nantes, France

S. S. Uhlmann et al. (eds.), *The European Landing Obligation*,
https://doi.org/10.1007/978-3-030-03308-8_13

more selective gear; tactical and strategic changes; and management changes. Scientific data (surveys, landings, and observers data) can also be valuable to help fishers to decide where and when to fish to best avoid unwanted catches and maximise opportunities to catch their quotas. We provide some examples of this type of approach, and also how these can be adapted for use as interactive online apps that fishers can use in planning or whilst at sea.

Keywords Challenge trials · Decision support tools · Discard avoidance · Fine scale mapping · Fish distribution · Fishers · Fishing strategies · Hot-spot maps

13.1 Introduction

Under the Landing Obligation (LO) fishers will need to reduce or land fish that were previously discarded. In this context, understanding how fisheries operate is central to understand how to manage them (Hilborn 2007; Eliasen et al. 2014). An obvious way by which fishers can reduce discards is via improved gear selectivity (O'Neill et al., this volume). Beyond that, the tactical choices made by fishers on "where, when and how to fish" can play a central role in reducing discards (Rijnsdorp et al. 2012; Dunn et al. 2011). This can be implemented in terms of top down control (e.g. closed areas). However, the need for management to provide bottom-up incentives to reduce discards is also well established (Rochet et al. 2014; Condie et al. 2014; Little et al. 2015; Pascoe et al. 2010).

In parallel, the ongoing improvements in data availability open for new and more precise knowledge. Analysis of discard observers' information (e.g. Anon 2011; Viana et al. 2011) provides a better understanding of spatio-temporal patterns of discarding. Catch locations and landings per unit of effort can be determined at fine spatial scales from Vessel Monitoring Systems (VMS) and logbook data (e.g. Gerritsen and Lordan 2011), and increasingly from Electronic Monitoring (EM) data (Plet-Hansen et al. 2017). Bottom trawl surveys can be used to map the locations of species (Fraser et al. 2008), spawning aggregations (Nash et al. 2012) and size structure (Shephard et al. 2011). This information can help fishers to decide where and when to fish to avoid having to catch unwanted fish.

In this chapter we look at how fishers themselves may be able to change the way they operate in order to reduce discards, based on a series of recent studies performed in several European fisheries in the frame of the EU research project DiscardLess (www.discardless.eu). We start by describing a series of 'challenge' trials where several individual fishers tried to reduce their discards by whatever (legal) means they thought best. We also interviewed other fishers not involved in the trials to ask them what they thought they could do. Their approaches generally fell into three categories: more selective gear; tactical and strategic changes; and management changes (Reid 2017). After a description of the trials and their results, we look at other tools to help fishers decide where and when to fish to best avoid unwanted catches, and maximise their opportunities to catch their quotas. At the time of writing this summary, a number of the individual studies presented here were still ongoing and/or unpublished, but a more detailed description of the methods used and preliminary results has been reported in Reid and Fauconnet (2018).

Management changes are beyond the scope of this chapter but are addressed in other chapters of this book.

13.2 What Can Fishers Themselves Do to Reduce Their Discards?

In a series of "challenge experiments", individual vessels and crew were challenged to reduce their discards by whatever legal means available. Intuitively, this could be by (for example) changing the fishing gear, or by changing their fishing tactics, perhaps by shifting areas or seasons. Each vessel fished first with their normal approach (control) and then with the modified approach (test) with the aim of minimising discards over a predetermined period (challenge trial). They reported the adjustments they made and why. Skippers were asked to set themselves a target for discard reduction between the test and the control trips, and this was the core of the "challenge". The targets could have been in terms of reducing discards of TAC species in general, or of those that represent the major 'choke' species in their fishery, i.e. the species for which the available quota is exhausted (long) before the quotas are exhausted of (some of) the other species that are caught together in a (mixed) fishery (Zimmermann et al. 2015). Scientists were sometimes placed on-board to collect catch data, and also to train crews in self-sampling. The catch data were then analysed by scientists to determine the degree of success at reaching these targets.

Challenge trials were done in three different countries and across a number of fisheries. The approach was slightly different in the three countries:

- Ireland – one demersal trawl vessel targeting whitefish (cod, haddock and whiting) and one targeting Norway lobster (*Nephrops norvegicus*) with additional catches of the same fish species (Calderwood et al. 2016).
- Denmark – 12 demersal trawl vessels mainly fishing cod and saithe, with three vessels targeting Norway lobster. The vessels towed a mix of single and twin trawl rigs, and were distributed between the North Sea, the Skagerrak, and the Baltic Sea. (Mortensen et al. 2017).
- France – three vessels targeting a mix of species including cod, whiting, squid, cuttlefish and some pelagic species. The vessels were all demersal trawlers, two < 18 m, and one > 18 m in length (Balazuc et al. 2016).

In Denmark, the main option explored by fishers was gear modification, and the data were mostly collected by the fishers themselves, supplemented with Fully Documented Fishery (FDF) methods (including Electronic Monitoring (EM) with cameras). In Ireland and France, the approaches included both gear and tactical modifications, with full observer coverage.

13.2.1 Gear Based Changes Used in the "Challenge Trials"

Changes to the fishing gear figured strongly in fisher's choices in all challenge trials. This was the main thrust of the Danish study, where the fishers used a variety of different gear modifications. These included:

- Changing mesh size in the codend of the net, usually to a larger mesh size, but in the Baltic Sea some vessels trialled reduced mesh sizes
- Inserting escape panels or separator panels into the net (with two codends for fish going above or below the panel)
- Topless trawl or modified mesh in an escape panel

In the French trials gear changes consisted of:

- The inclusion of a larger mesh cylinder in the extension (CMC)
- Separator panels with two codends
- Increased mesh size in the codend and extension, and T90 mesh

The only gear change in the Irish example was that one of the Irish vessels (the *Nephrops* targeting vessel) used a quad rig *Nephrops* net, with large mesh square mesh panels (SMP) in all four extensions.

The outcomes of these trials were somewhat mixed. In the Danish trials, nine vessels were able to reduce the discard ratio (Discards/Discards + Landings by species by weight) using the tested modifications (three in the North Sea, three in Skagerrak and three in the Baltic Sea), while two vessels (from the North Sea) actually increased their discard ratio and one North Sea vessel showed no difference. The improvements ranged from less than 2% for four of the vessels, 2–7% for four others, and, in one case, a 17.6% improvement (Fig. 13.1).

In the French trials, there was insufficient time after making the gear changes to collect sufficient data to analyse their performance. However, the vessel using the mesh cylinder (CMC) approach reported little loss of commercial catch volume, and in some cases reductions in discard volume. The separator panel with two codends could not be evaluated, but the skipper was still very positive and felt it had value. In general, the fishers did not feel that the changes in codend meshes achieved the results they had hoped for small fish, and there were concomitant losses in commercial sized fish (Balazuc et al. 2016).

In the Irish trials the use of the SMP in the quad rig allowed the vessel to keep fishing significantly longer before choking on the cod that was the main choke species during the control phase of the study. The results are shown in Fig. 13.2, and the reduction in over quota cod is clear, although there was an increase in over quota *Nephrops*.

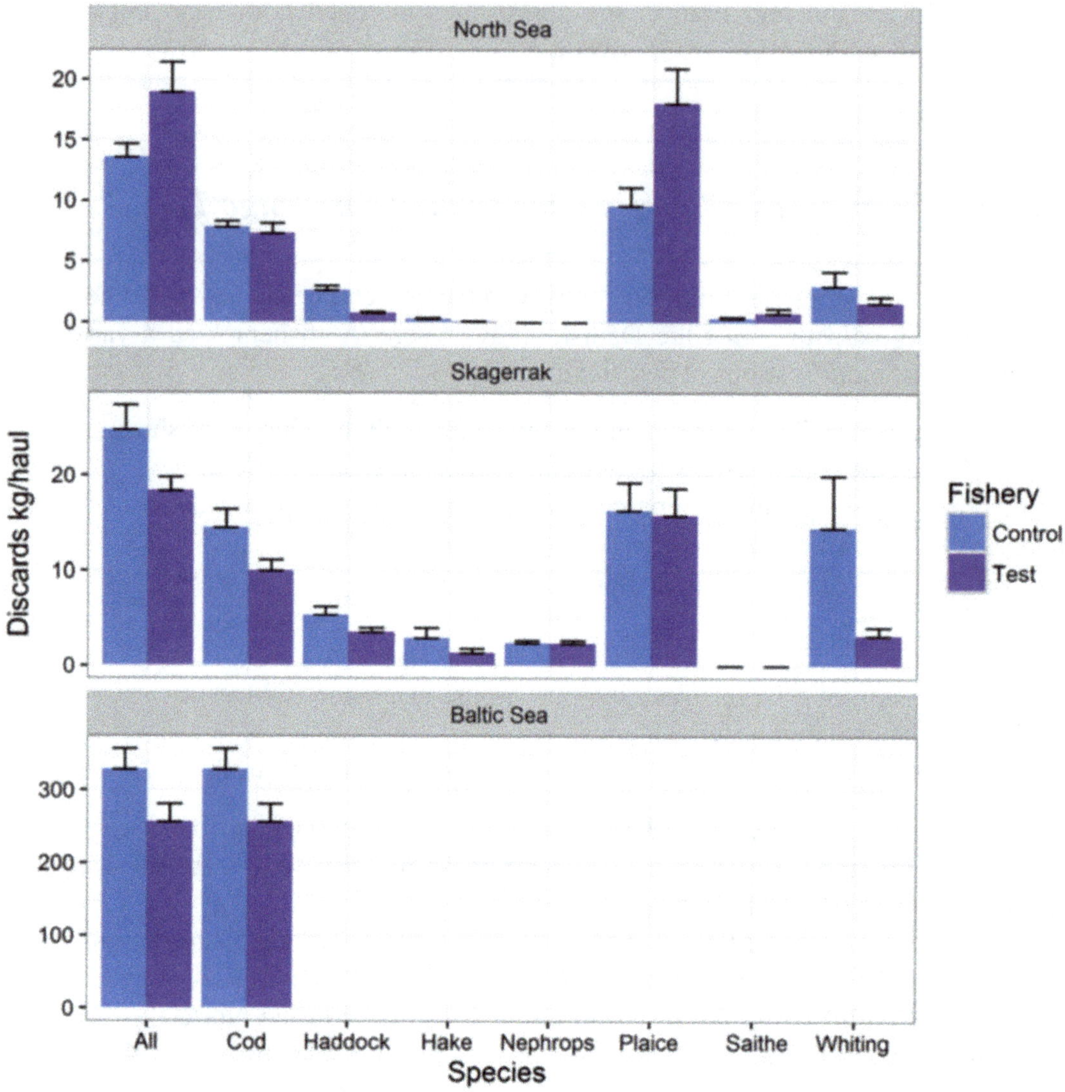

Fig. 13.1 Bar chart showing the average overall discards per haul from each area and the average discards per haul of individual species in each area. Error-bars signify standard error. Note that y-axes differ between areas. (From Mortensen et al. 2017)

13.2.2 Tactical and Strategic Changes Used in the Challenge Trials

Tactical and strategic changes to fishing to reduce discards were mainly tested in the Irish and French Challenge trials. In the Irish trials, the whitefish targeting vessel used changes in both the time of day and also in the depths of fishing. The vessel also tried to move between management areas to maximise the time fishing for the month. The main issue for this vessel in the control period was a very early choke on cod and haddock in all management areas. The combination of area and behavioural changes allowed a small change in choke time across all areas from 4 to 9 days. There was some evidence that the skipper was actually trying to avoid discards during the

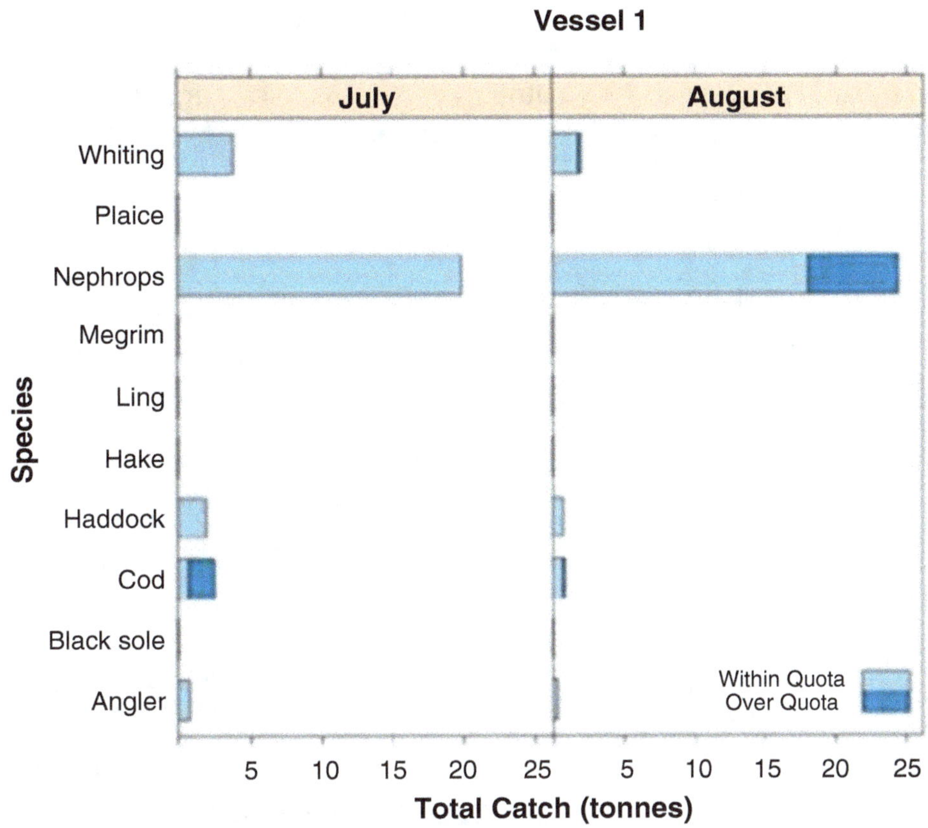

Fig. 13.2 Total catch of quota species for the Nephrops vessel during the 2 months of the trial, with a distinction between within quota landings (light blue) and over quota/< MCRS (over Minimum Conservation Reference Size) landings (dark blue). (From Calderwood et al. 2016)

control periods as well – he had somewhat higher discards in the months prior to the trials than in the control month during the trials. This may have impacted on the outcomes from the changes he made. The *Nephrops* vessel, while focused on the gear changes outlined above, also used movement between management areas to successfully reduce the choke problem (Fig. 13.3).

The strategic changes made by the French vessels were mainly focused on the potential for avoiding "sensitive" areas based on traditional ecological knowledge, characterised by high catch rates of quota species under MCRS. The outcomes suggested that the large vessel already did this in its normal practice, and that scope to do any more was limited. For the smaller vessels, their main operating area with high discards was within the three mile zone along the Channel coast, where almost 70% of their catch was usually discarded (Fig. 13.4). Avoiding this area would clearly help with their landing obligation (LO) requirement. The key issue was that, while discards are high in this zone, it is also their main area of operation. These are small, artisanal vessels, and this area is both close to their home ports and also sheltered from bad weather. As a consequence, the skippers were reluctant to avoid this area during the trials. However, it remains a potential valuable tool for discard mitigation under the LO, and ways to encourage the avoidance of this area should be explored.

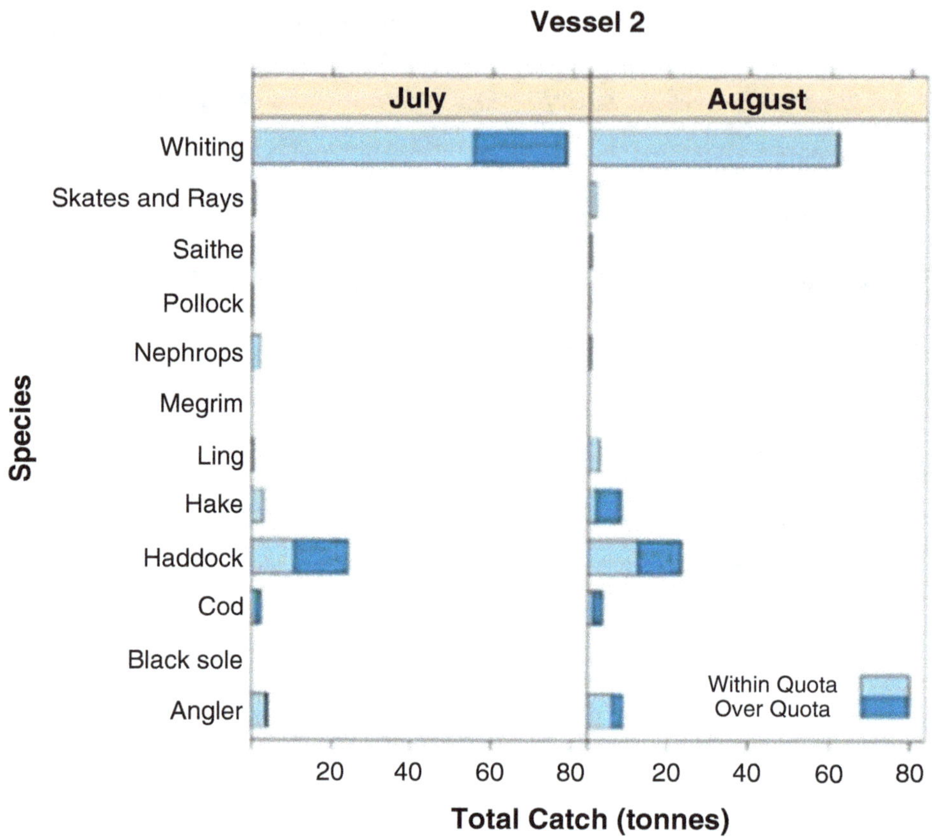

Fig. 13.3 Total catch of quota species for the white fish vessel during the 2 months of the trial, with a distinction between within quota landings (light blue) and over quota/below Minimum Conservation Reference Size (< MCRS) landings (dark blue). The vessel was able to reduce his over quota whiting catch, but could make little change in his over quota cod or haddock catches. (From Calderwood et al. 2016)

13.2.3 Conclusion

The use of modified gears to improve selectivity and reduce the scale of discarding showed some promise during the challenge trials. In all three cases, the use of added panels, changes in codend mesh size and configuration, modifications to the extension, and the use of separator panels with twin codends showed some improvements. However, it should be noted that these improvements were often quite small and would probably not solve all the problems fishers would face under a full implementation of the LO. Additionally, these were the fishers' own trials, and could not always be fully substantiated in a scientific context. One positive approach that could be taken, would be to enhance the institutional paths for a "fast-tracking" of such bottom-up initiatives (O'Neill et al., this volume).

The challenge trials showed that there was some scope for the use of both more selective gear and changes in behaviour, both locally, and in moving between management units, to reduce discards, and mitigate the impacts of the LO on fishing viability. Fishers in all the trials did believe that these changes could make some

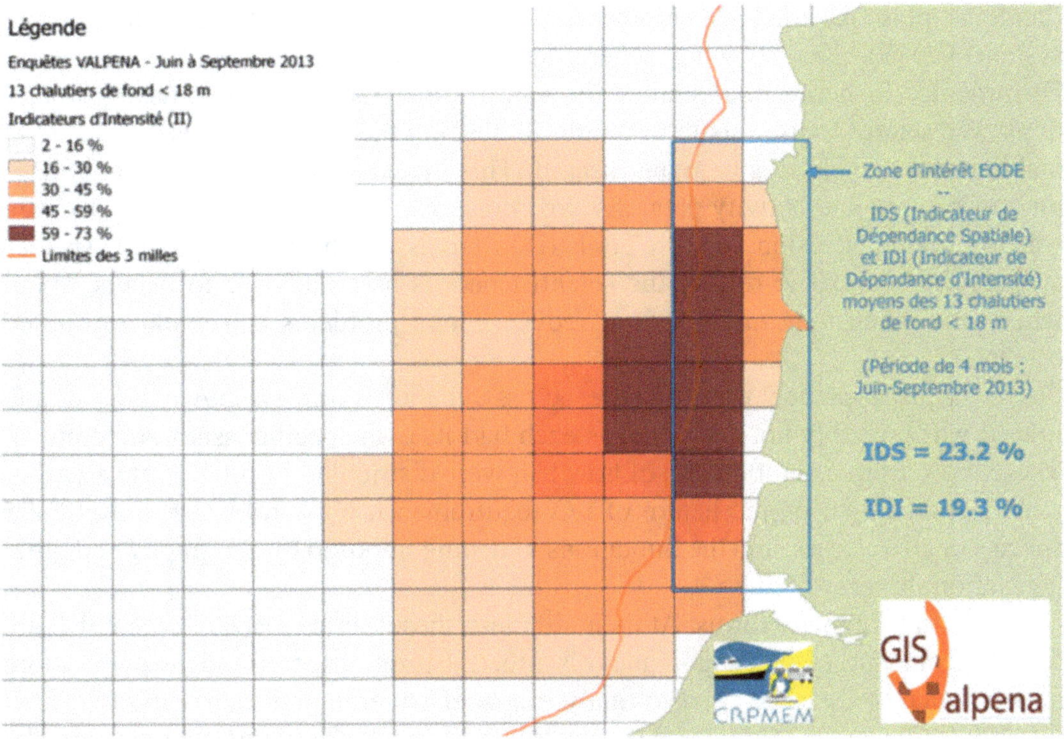

Fig. 13.4 Map showing Intensity Indices (II) of 13 trawlers under 18 meters over the period June–September 2013. (Source: CRPMEM NPdC-P and Gis Valpena)

difference, even if they did not work as well as expected in the limited context of the challenges. It should be noted though, that even when the trials were able to reduce discarding or the impact of choke species, the improvements were generally quite small. So, while such changes may help fishers comply with the LO by reducing discards, they are still not sufficient to avoid significant impacts on economic viability. Notwithstanding this, we consider it desirable to continue working with fishers on both gear and behavioural based responses to the challenges implicit in the LO. The trials were all successful in terms of the level of collaboration and in some of the outcomes, so such approaches should continue.

13.3 Where and When to Fish to Avoid Unwanted Catches – How the Scientists Can Help

Based on the challenge trials and interviews with fishers (Reid 2017), it was clear that tactical changes could help avoid unwanted catches, and we believe that more information would help fishers achieve this. We then looked for ways to provide the detailed knowledge that can come from using scientific data to illustrate the spatial and temporal distributions of the fish, catches and discards.

Fisheries institutions have access to a range of data. These include research vessel surveys showing abundance distributions, observer data showing detailed catch (landings plus discards) by commercial vessels, landings and vessel monitoring system (VMS) data showing where and when catches are made, and Fully Documented Fisheries pilot studies showing full details of complete fishing operations. We set out to use this information to develop the potential to assist fishers in making strategic choices to avoid discard. This included fine-scale, real-time mapping of catches and activity data, discards hotspots, juvenile surveys, etc. One aim was to provide Decision Support Tools (DST) to assess the role of "choke" species at the local scale. The role of the scientist here is as an advisor to fishers, about where and when they might fish to reduce choke problems and avoid unwanted catches.

No single approach was possible across all the examples shown below, and indeed was probably not desirable, as each had its own specific issues and context. These arose from a combination of how fish were distributed i.e. in discrete areas, or widely spread, and on the nature of LO requirements, e.g. avoidance of particular species or size classes, and the limitations in fishing imposed by geography and other legislation drivers.

DST can take many forms. At their simplest, these can be maps of where fish are found (from surveys), caught and discarded (from observers). However, more detailed analyses can be used to analyse spatial patterns and their variation, how discards and catches of numbers of species co-occur in space and time, or not. The information can also be represented in an interactive form using web-based apps. But the DST process can also simply be the provision of understanding discarding and its drivers, e.g. quota management rules, or about the interaction of economic profitability with discarding – is it economically better not to discard? We present examples of all these types of Decision Support information. These cover case studies from the North Sea, through North East Atlantic (European western waters) to the Mediterranean Sea. They cover many different metiers and fleets, from single to multi-species, using a wide variety of fishing gears.

13.3.1 Decision Support Tools Using Survey Data

Fisheries surveys are carried out across the EU and provide valuable data. The use of survey data in helping fishers to decide where and when to fish is illustrated with an example from the Balearic Islands (in the western Mediterranean Sea). The surveys were used to model the spatial patterns of species abundance for the main commercial species. The results from this were a series of maps of species distributions above and below MCRS, species overlaps, fishing grounds, discard hotspots etc. An example showing the density and persistence of thornback ray *Raja clavata* is presented in Fig. 13.5. In a second example, the degree of species overlap is presented in Fig. 13.6, illustrating where more than one species is likely to be caught together. Other data products from this study also made use of observer data to supplement the surveys.

In the Azores, habitat suitability models for 10 species of deep-water sharks and rays were developed based on survey data (Fauconnet et al. 2018). Deep-sea sharks,

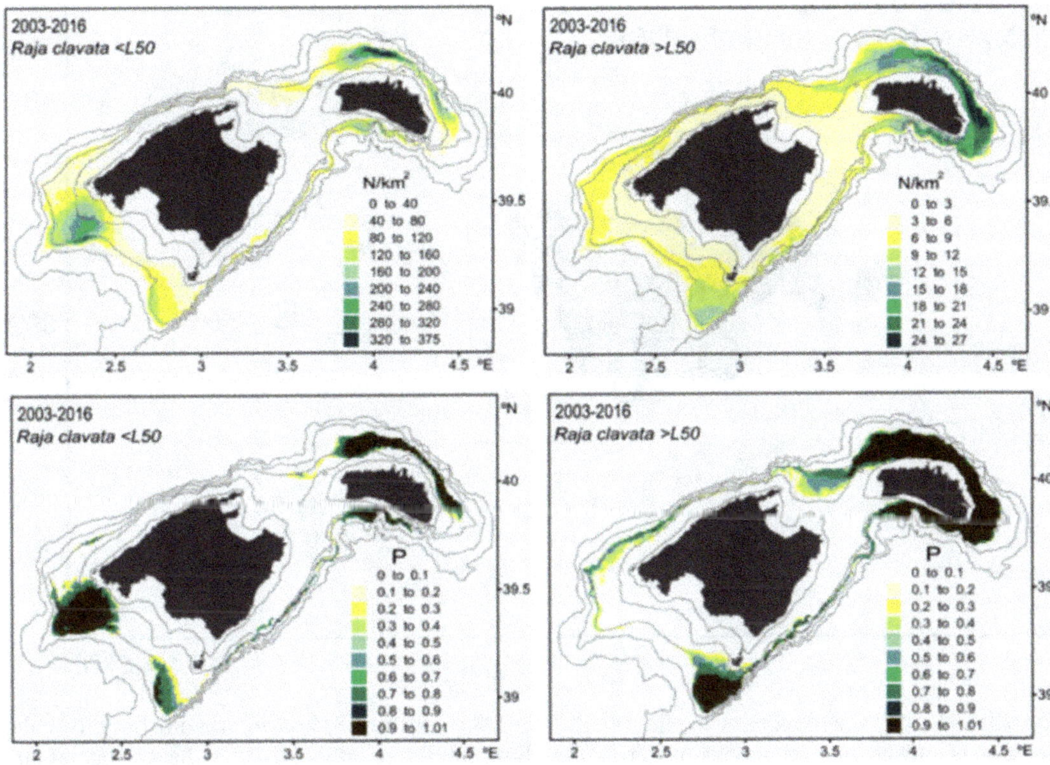

Fig. 13.5 Maps of density (N individuals/km^2; above) and persistence (P, fraction of years; below) of thornback ray individuals from the Balearic Islands under (< L50) and over (L > 50) the size at first maturity (73 cm)

even if only occasionally taken as bycatch of the deep-water longline fisheries in the Azores, could rapidly choke the fisheries of this Portuguese outermost region, as many of those species are currently managed under a zero TAC. Maps predicting occurence by species and combined occurence of all species at the range of the whole Azores EEZ were developed using data from demersal bottom longline surveys carried out from 1996 to 2017 (Fauconnet et al. unpublished data), to help fishers identify areas they should avoid to limit the risks of catching those species (Fig. 13.7). Composite maps combining the distribution of the main shark species caught by the bottom longliners, and by the deep-water drifting longliners were also created to better highlight the main areas to be avoided for those two groups of fishers. This information was completed using fine-scale information on deep-water shark spatial and vertical movements derived from acoustic telemetry data from 2 species: kitefin shark (*Dalatias licha*) (Fig. 13.8) and bluntnose sixgill shark (*Hexanchus griseus*). Telemetry data helped identify potential essential habitats for those species. The study highlights that areas to avoid fishing and limits in fishing depths at some time of the day could be promising mitigation measures for fishers to implement to avoid some species of deep-water elasmobranchs – but not for all.

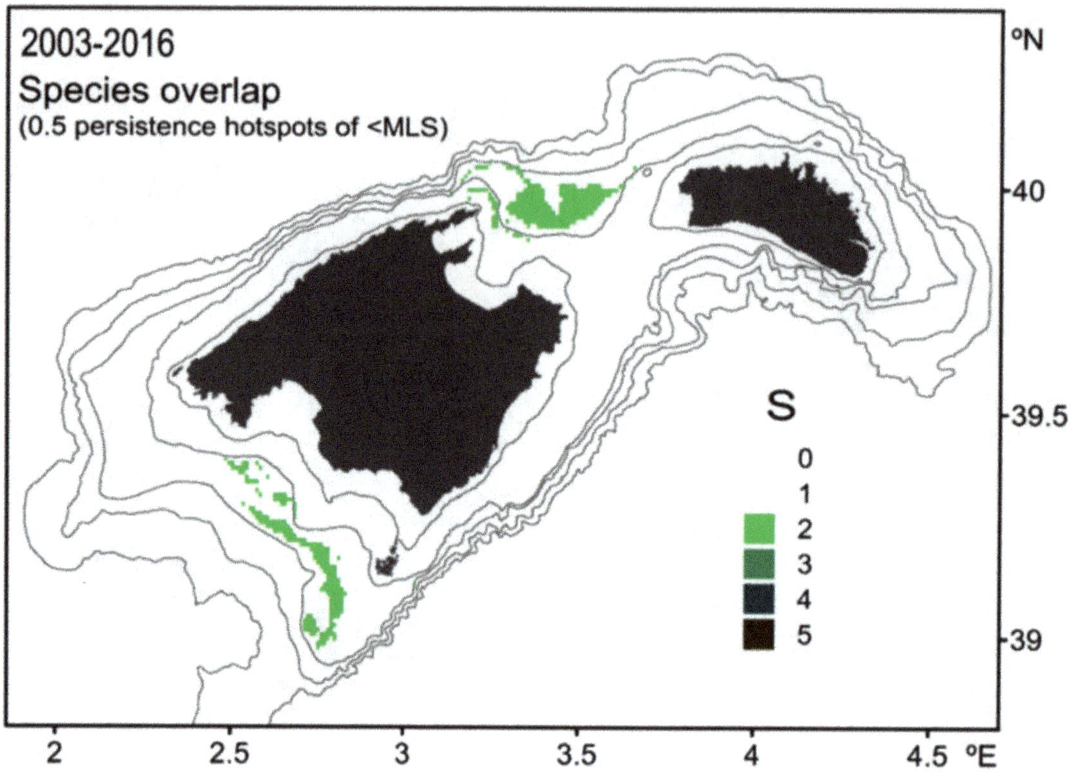

Fig. 13.6 Map of the number of species overlapping (S) in the mixed bottom trawl fishery from the Balearic Islands; S was obtained considering the Minimum Landing Size (MLS) of each species and a persistence level of 0.5

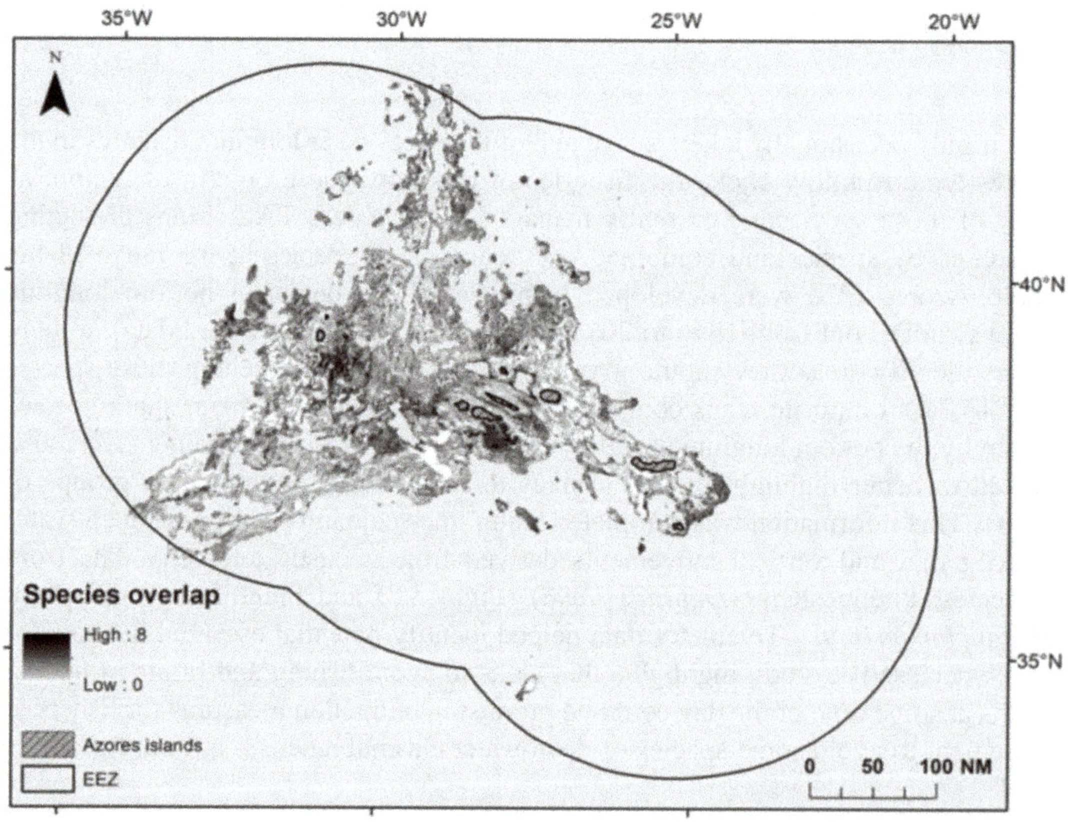

Fig. 13.7 Deep-water elasmobranchs hotspots distribution overlap based on presence/absence distribution of the 10 selected species. Uniform light grey represents areas with no data. (Fauconnet et al. 2018; Fontes et al. 2015)

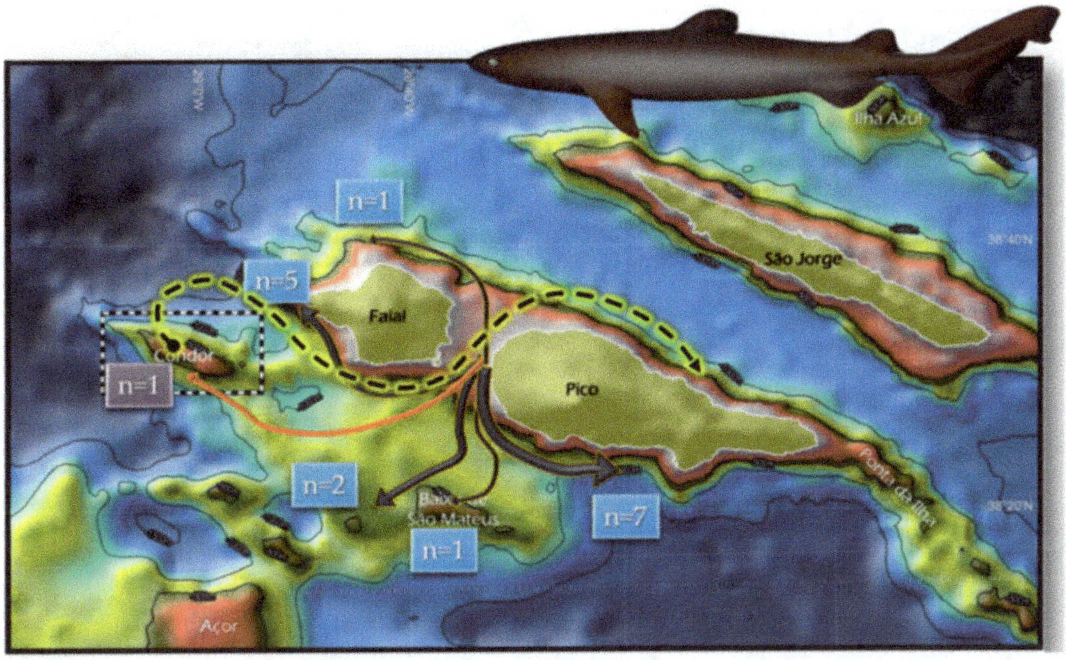

Fig. 13.8 Graphical representation of the movement ranges of individual kitefin sharks tagged with acoustic transmitters at the south of the Faial-Pico channel and monitored using deepwater acoustic receivers in the islands' slopes and neighbouring seamounts; boxes represent the number of sharks undertaking a particular movement. (Fauconnet et al. 2018)

13.3.2 Decision Support Tools Using Observer Data

Observer data come from on-board observers on commercial fishing vessels, and, like surveys, similar coverage is carried out across the EU. Their primary task is to record discards, but they also record fish that go to landings. Thus they represent very detailed information on catches, landed and discarded. It is only possible to deploy observers on a small proportion of all fishing trips, but we were able to combine observer data from France, Ireland and the UK for the Celtic Sea to provide a larger dataset to work on, and some of the results are shown here. Two different approaches are presented as examples of what information can be produced.

13.3.2.1 Where Are Discards Clustered Together?

This study is the first multispecies, fine scale, spatial analysis of landings and discards in mixed fisheries across a multinational context. The core aim was to use observer data to identify where commercial fish were landed and discarded and with what other species. Multivariate analysis (Principal Component Analysis PCA and hierarchical clustering) on combined observer data from Ireland and France between 2010 and 2014 grouped cells of space characterized by homogeneous species profiles in terms of discards (or landings). Each cluster was then plotted on a map with a

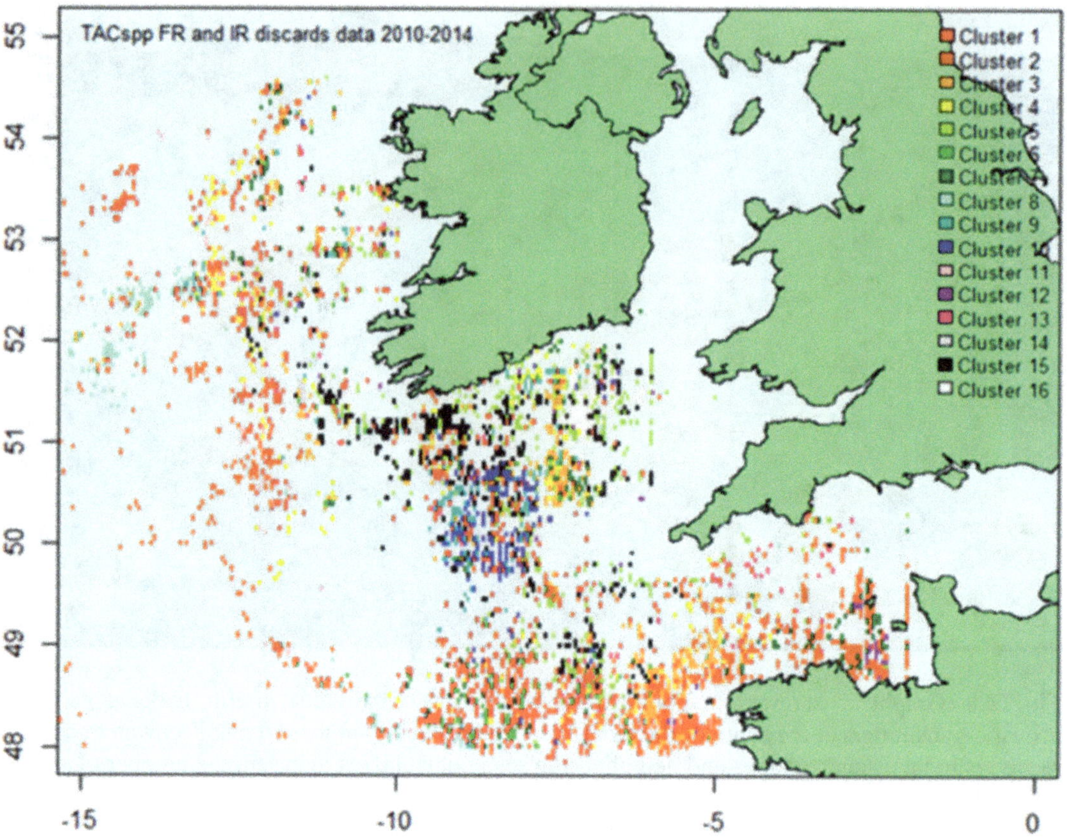

Fig. 13.9 Cluster maps of French and Irish discards. The same colour code was assigned to each $3'*3'$ square belonging to the same cluster. This analysis was for TAC species only

colour code. It provides a global overview of discards and landings locations by species in the central region of the Celtic Sea. What was found was a highly structured fishing ground, with some of the clusters only found in a smaller part of the whole Celtic Sea. For instance, in the map shown in Fig. 13.9, there is a notable patch of the dark blue cluster 10 in the middle of the area. This cluster mainly represents observed discards of Norway lobster and spurdog (*Squalus acanthias*). But the more widely spread red cluster 13, was mainly composed of mackerel (*Scomber scombrus*) and sprat (*Sprattus sprattus*). While some discard clusters corresponded well to landings clusters spatially, this was less common than cases where no obvious common pattern was found. This result suggests that in the central Celtic Sea, landings profiles in terms of species may not predict discards species composition.

13.3.2.2 Mapping Catch Hot Spots to Avoid Unwanted Catches

A valuable support tool for fishers would be to have access to maps showing species abundance hotspots – that is, areas where there would be a high probability of catching a given species, above or below MCRS. This was carried out using a

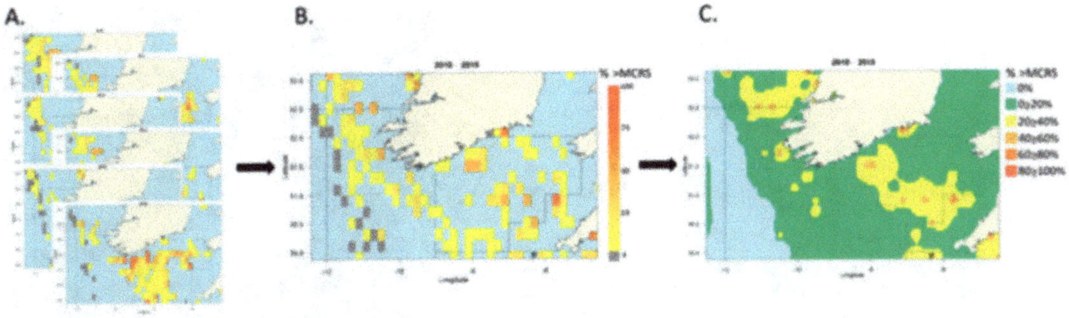

Fig. 13.10 Diagram showing the steps in the map production process. (A. Individual binned maps created for each year; B. Amalgamated map for all years identifying grid cells within consistent binned categories over multiple years; C. Final interpolated map (using inverse distance weighted interpolation)

detailed analysis of observer data from Ireland, France and the UK used as an indicator of the catches taken in the Celtic Sea. The analysis focused on mapping hot spots of CPUE and catch proportion for three key species; cod, haddock and whiting, and over and under MCRS. The analysis can be extended to any species, both commercial and non-commercial. The maps were based on consistent observations of particular catch rates, so only those locations where one would consistently (over 5 years) see high or low levels for these categories were used. The data were then interpolated to provide regional coverage (Fig. 13.10). The maps were then drawn together into a web-based app (discussed below).

13.3.2.3 Detailed Haul-by-Haul Mapping Using Electronic Monitoring Data

Another option to monitor catches is the use of Electronic Monitoring (EM) systems with video on board vessels that can monitor the haul-by-haul catch remotely. Such a system has been trialled in Denmark since 2008 (Bergsson et al. 2017; Ulrich et al. 2015). The analysed footage provides detailed information on discards, covering more trips and hauls compared to observer trips, for vessels carrying the system. An example is shown in Fig. 13.11.

Fine-scale information on landings and discards by haul can be combined with the landing price for the trip to map trade-offs between high-value and high-discard fishing spots, which can potentially complement the fisher's implicit knowledge on the best fishing locations (Plet-Hansen and Ulrich 2018). An example for a single vessel for which such a detailed information is available is presented in Fig. 13.12.

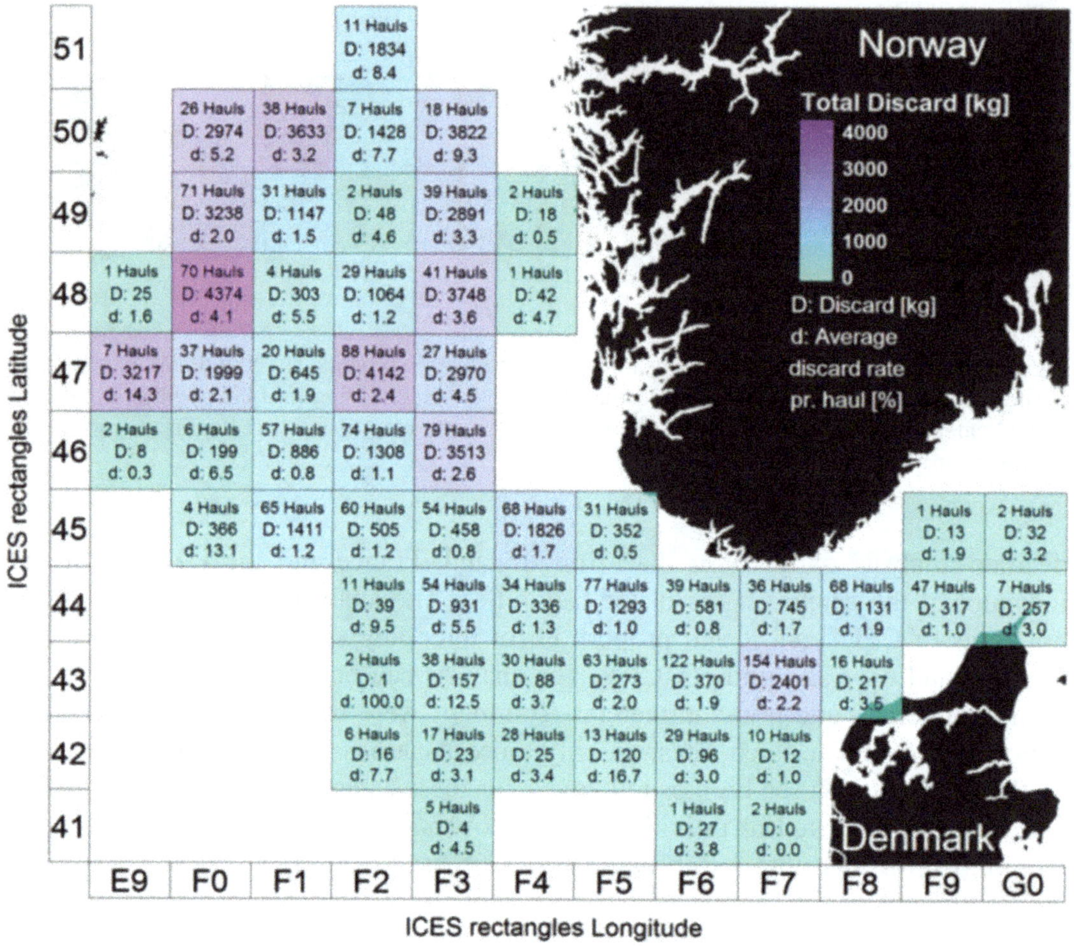

Fig. 13.11 Map of 2016 EM discard data for cod, hake, saithe, whiting and haddock from 12 Danish vessels (sampled hauls with video footage reviewed representing 29% of all their hauls that year). Cyan areas have low total discards in kg, purple areas have higher total discards. Each grid cell is an ICES rectangle. The x-axis shows the ICES rectangles' longitudinal ID, the y-axis shows the ICES rectangles' latitudinal ID. The number of sampled hauls conducted in each ICES rectangle by the 12 vessels is written together with the discard in kg (D), and the average discard rate (Discards/Catch) for the 5 species per haul (in %). (From Bergsson et al. 2017)

13.3.2.4 Combining Surveys and Commercial Catch Data to Provide Year-Round Abundance Distributions

Data from scientific surveys are not available for all times of the year but provide consistent yearly and spatially resolved abundance indices. On-board commercial data cover the whole year, but generally provide a biased perception of stock abundance. The combination of scientific and commercial catches per unit of effort (CPUEs), standardized using statistical methods (in this example a delta-generalized linear model), allows the description of the spatial and temporal (monthly) dynamics of fish distributions in the Eastern English Channel (Bourdaud et al. 2017). Using the scientific survey as a baseline, the degree of reliability of commercial CPUEs was assessed with survey-based distributions using the local overlap between

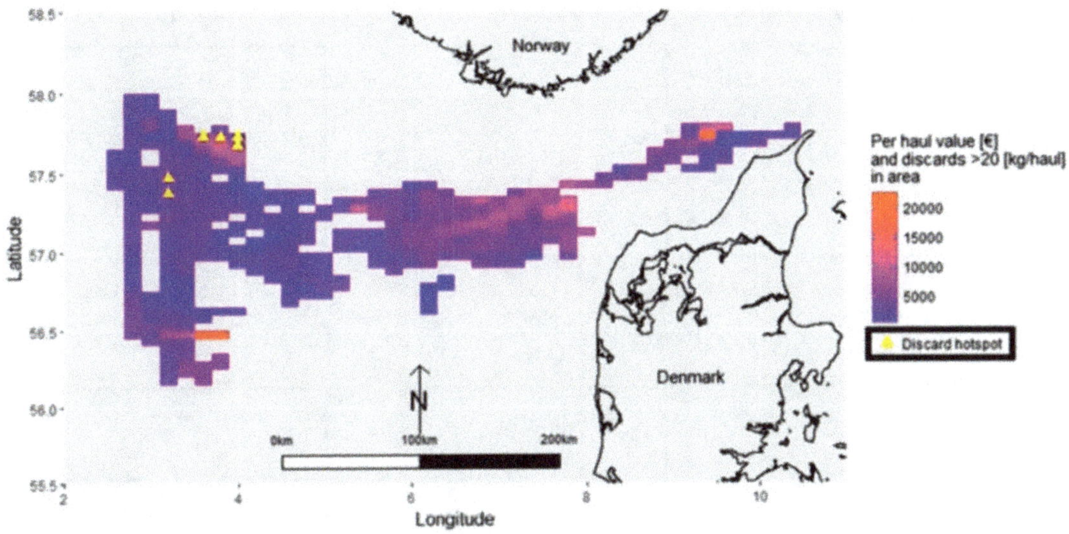

Fig. 13.12 Gridded map of the landing value per haul for a single Danish trawler. The red colour represents the greatest value per haul. The yellow triangles represent the areas of high discard volumes. The discard hotspot areas often coincide with high value hauls but importantly, there are also other high value areas without such high discard levels. (From Plet-Hansen and Ulrich 2018)

distributions. The broader spatio-temporal estimated distribution of the species agreed with qualitative information from the literature, especially for cuttlefish, and this is illustrated in Fig. 13.13. Fine scale consistency (using cells of 0.3°*0.3°) between survey and commercial data was significant for half of the 19 tested species (e.g. whiting, cod). For the other species (e.g. plaice, thornback ray), the results were inconclusive. The approach allowed a more representative mapping of the abundance distribution across the year, that can then be used in both targeting, and avoidance in the context of the LO.

13.4 Web-Based Apps to Help Fishers Plan Where and When to Fish to Avoid Unwanted Catches

In many of the analyses in this chapter, the scientists concerned have been able to produce information, usually in map form, that has the potential to help fishers target their activity to avoid unwanted catches. To make this practically useful, and useable, scientists have started developing a range of web-based apps both to present the information, but critically, to allow the fishers to work with it in their own way. In three of the examples given in the above descriptions, such apps have been developed and are, or will be, refined with fishers to make them as useful as possible.

One example developed in the Balearic Islands is presented in Fig. 13.14. The app allows fishers to choose the species and fishing ground of interest. They can then see

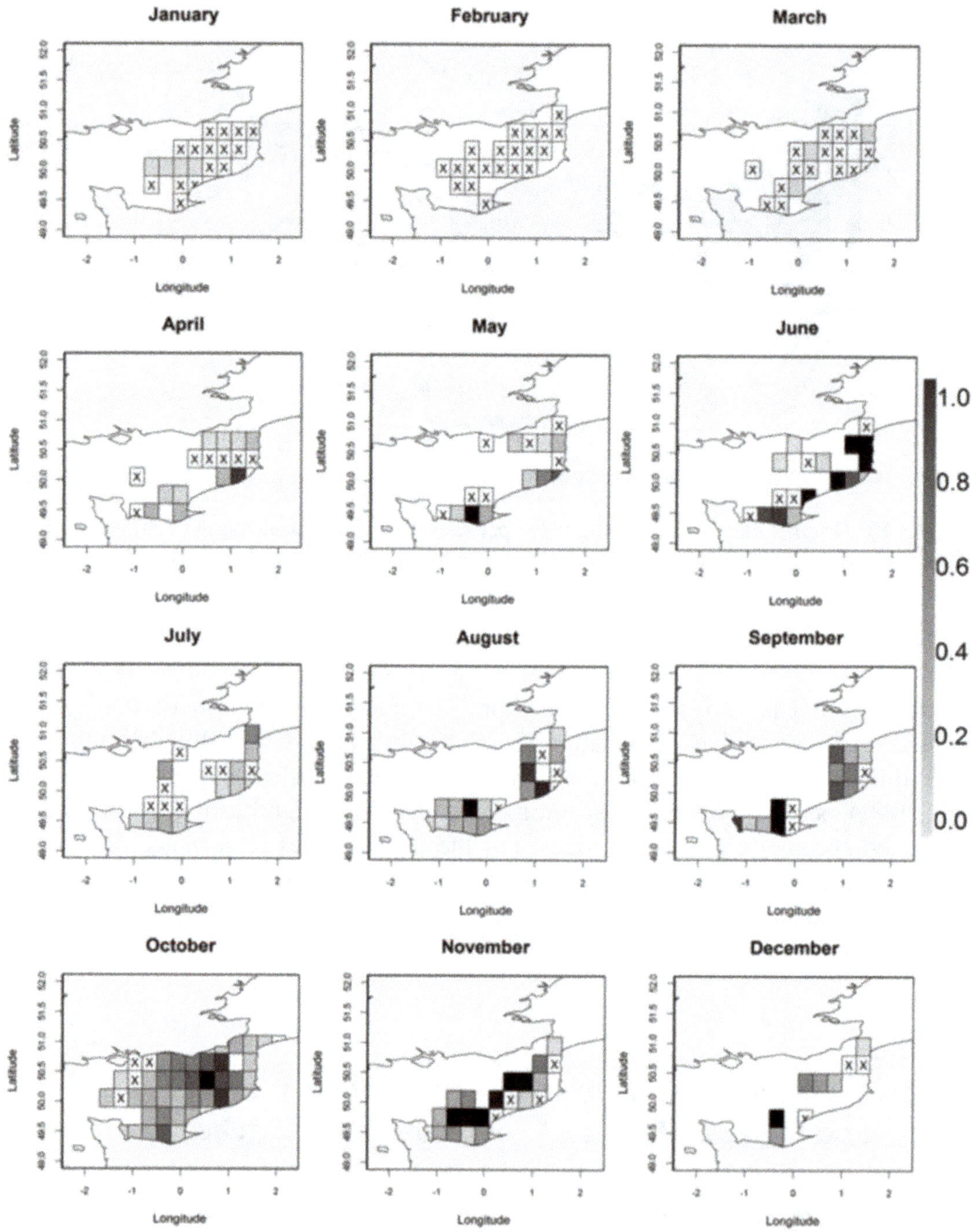

Fig. 13.13 Monthly spatial abundance distribution estimated from on-board commercial observations (OBSMER program in France) and scientific surveys (CGFS, Channel Ground Fish Survey) for cuttlefish. The crosses in the white squares (X) represent areas where no cuttlefish was ever fished during a month. (Figure from Bourdaud et al. 2017)

observer or survey data, as well as discard information, length and maturity data, and other information on the species and the fisheries.

A second example, developed for the Celtic Sea fisheries, is presented in Fig. 13.15, where the fishers can choose the species (or size class) of interest, and

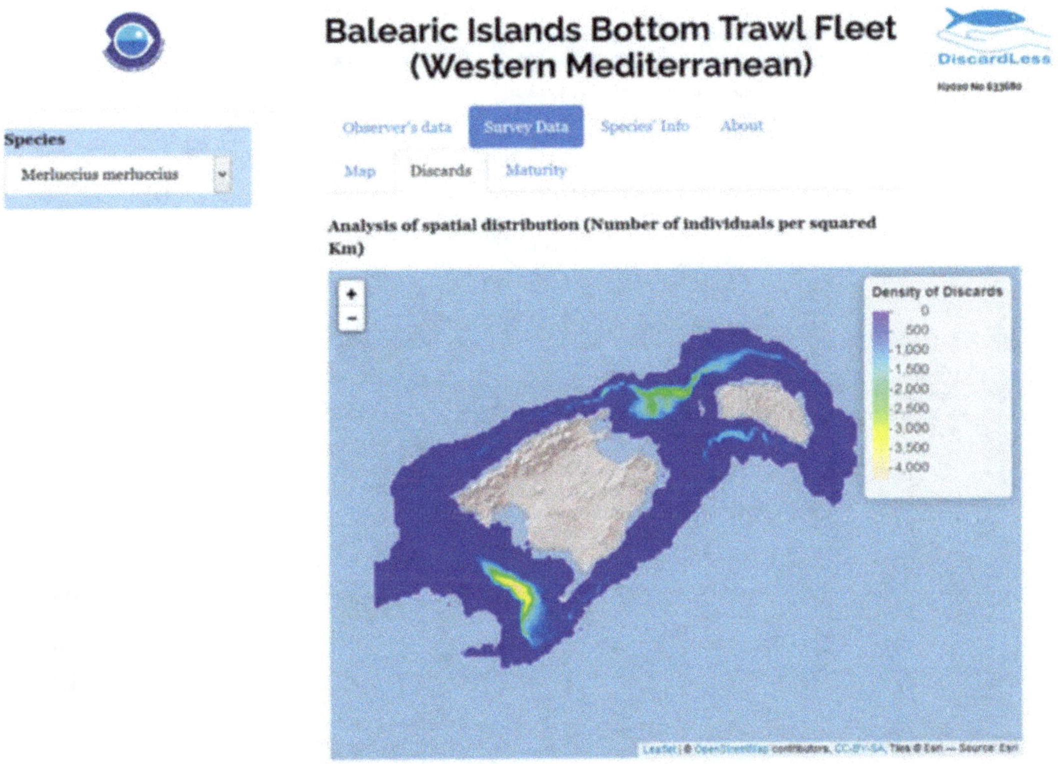

Fig. 13.14 Snapshot of the Shiny App produced as a decision support tool to assist fishers in the Balearic Islands to make choices of fishing location to avoid discards. Density map (number of individuals per km2) of hake discards (individuals under the MLS = 20 cm) is shown

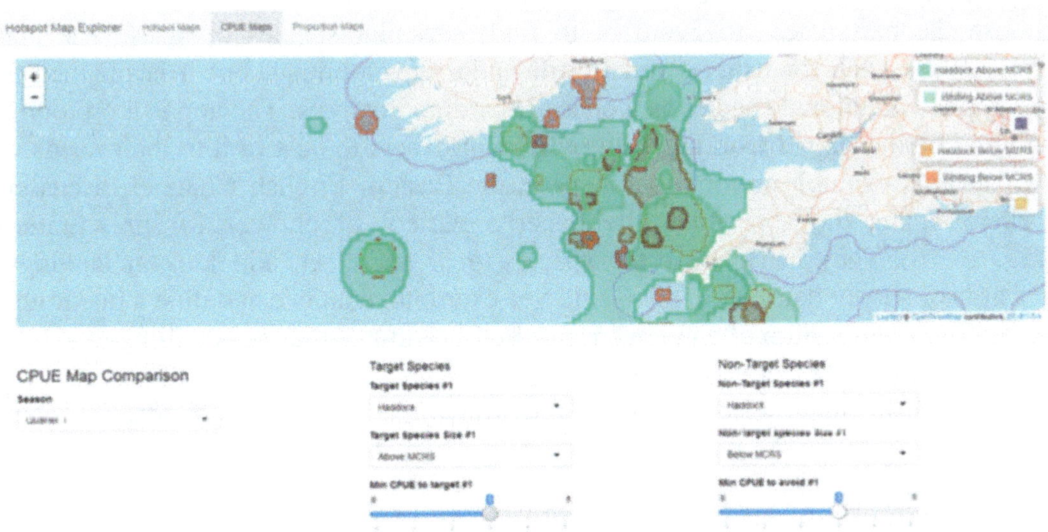

Fig. 13.15 A screenshot of the Shiny App developed to allow stakeholders to select the size, species and quantity of fish they would like to target and/or avoid during different seasons. The resultant map displays layers allowing fishers to balance trade-offs of target and non-target to optimise catch composition.

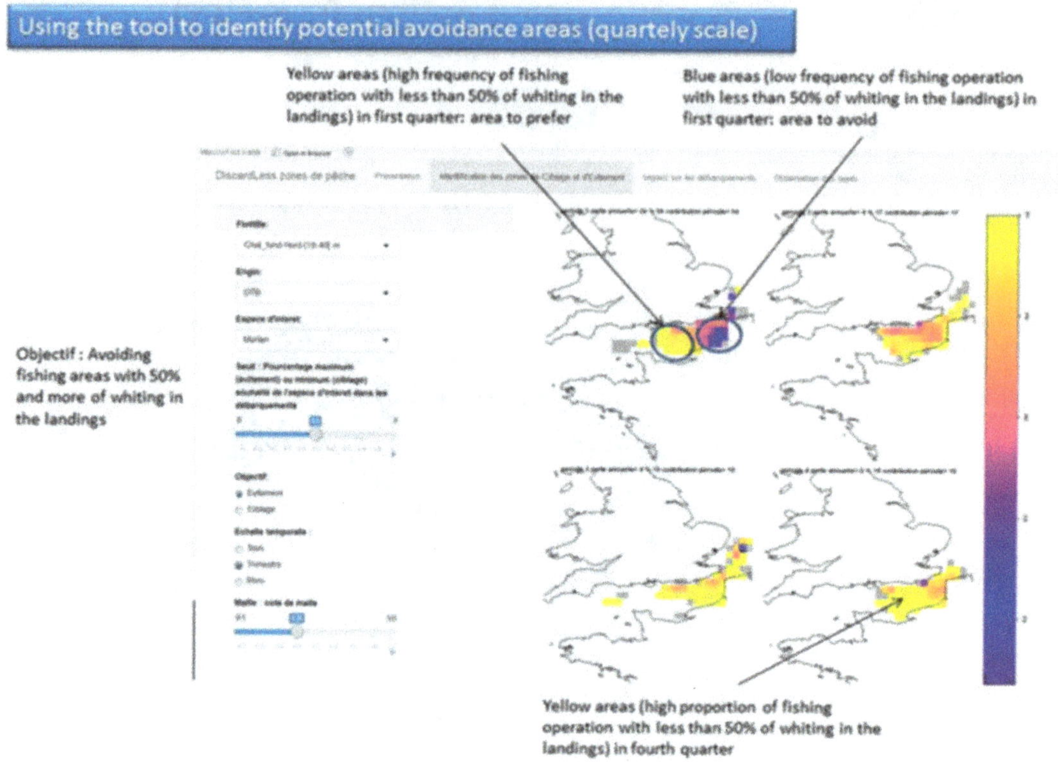

Fig. 13.16 An example of the web-based app for the English Channel showing a variety of ways to present information to avoid discards

then map CPUE or catch proportion at the selected level of intensity. They can also map a number of species or sizes together on one map, and change the levels, to help choose the best places to avoid or to find particular species or sizes. The app represents a DST for fishers, and is still under development, but it is planned to incorporate additional species as well as discarding hotspots. The app is a prototype and will be developed working with individual fishers to best fit it to their needs.

The final example is for the fishery in the Eastern English Channel discussed earlier. Again, the app offers the fisher a variety of choices for presentation (Fig. 13.16). They can choose the species, gear and fleet, and look at landings, discards or survey data. As in the Celtic Sea example, fishers can define a maximum or minimum proportion of a given species in the landings that he is willing to accept depending on the objective i.e. avoiding or targeting the species. The time scale over which the maps are presented can also be controlled, either for a year, quarter or month.

13.5 Conclusions

The original idea we stated in the title was "The best way to reduce discards is by not catching them!" The work presented in this chapter shows some of the ways that this could be achieved.

Gear based changes in selectivity remain the most common, default, way to do this, and we present here some of the broad range of such approaches available. Many of these have been developed by gear technologists, but many also by working fishers or netmakers. In many of our studies, we found that fishers remain innovative and willing to explore the use of different gears to reduce unwanted catches.

Behavioural changes by fishers, i.e. tactical changes in where and when to fish, are a second route to avoiding unwanted catches that have attracted less attention than gear-based approaches. In the challenge trials described here, fishers attempted both gear and behavioural changes in their fishing practices. In some cases, these changes reduced unwanted catches but not in all. One of the reasons for this advanced by several fishers involved in the work was that they lacked the information needed to help them choose where and when to fish to minimise the unwanted component.

It is beyond doubt that fishers know their own fishing activity far better than any scientists could. They are, after all, observing it on a daily basis over many years. But, equally, scientists have information that fills in the wider picture on distributions and abundances of fish, both wanted and unwanted. Taken together, fishers' and scientists' "knowhow" can give the working fisher the best chance to reduce, or possibly eliminate, unwanted catches. We have shown here how surveys, observers' information, landings data, etc. can provide useful information on where the fishers are likely to encounter a given species or size class of that species, as well as those fish commonly encountered together. It needs to be emphasised that this information should be seen as providing a probability of encounter or not, rather than a certainty. The take-home message from this is that there is more chance of approaching the objectives of the landing obligation by combining fishers' and scientists' knowledge than by working apart.

Another key message is that we can identify several different approaches that could help reduce discarding, but they all tend to be specific to local conditions. It should be possible to export the approaches to other fisheries, but only in broad terms. Essentially, the causes of discarding are common, but the solutions tend to be local and specific.

Acknowledgments This work has received funding from the Horizon 2020 Programme under grant agreement DiscardLess number 633680. This support is gratefully acknowledged.

References

Anon. (2011). Atlas of demersal discarding, scientific observations and potential solutions. Marine Institute, Bord Iascaigh Mhara, September 2011. ISBN:978-1-902895-50-5. 82 pp.

Balazuc, A., Goffier, E., Soulet, E., Rochet, M.J., Leleu, K. (2016). EODE – Expérimentation de l'Obligation de DEbarquement à bord de chalutiers de fond artisans de Manche Est et mer du Nord, et essais de valorisation des captures non désirées sous quotas communautaires, Comité Regional de Pêches Maritimes et des Élevages Marins de la Réunion. 136 + 53 pp.

Bergsson, H., Plet-Hansen, K.S., Jessen, L.N., Jensen, P., Bahlke, S.Ø. (2017). Final report on development and usage of REM systems along with electronic data transfer as a measure to monitor compliance with the Landing Obligation – 2016. Danish AgriFish Agency, Ministry of Food, Agriculture and Fisheries. https://doi.org/10.13140/RG.2.2.23628.00645. 61 pp.

Bourdaud, P., Travers-Trolet, M., Vermard, Y., Cormon, X., Marchal, P., et al. (2017). Inferring the annual, seasonal, and spatial distributions of marine species from complementary research and commercial vessels' catch rates. *ICES Journal of Marine Science, 74*(9), 2415–2426.

Calderwood, J., Cosgrove, R., Moore, S.-J., Hehir, I., Curtin, R., Reid, D.G., et al. (2016). Assessment of the impacts of the Landing Obligation on Irish Vessels, BIM Technical Report. http://www.bim.ie/media/bim/content/publications/Lo,report,2016_final.pdf. 22 pp.

Condie, H.M., Grant, A., Catchpole, T.L. (2014). Incentivising selective fishing under a policy to ban discards; lessons from European and global fisheries. *Marine Policy, 45*(0), 287–292.

Dunn, D.C., Boustany, A.M., Halpin, P.N. (2011). Spatio-temporal management of fisheries to reduce by-catch and increase fishing selectivity. *Fish and Fisheries, 12*(1), 110–119.

Eliasen, S.Q., Papadopoulou, K.N., Vassilopoulou, V., Catchpole, T.L. (2014). Socio-economic and institutional incentives influencing fishers' behaviour in relation to fishing practices and discard. *ICES Journal of Marine Science, 71*(5), 1298–1307.

Fauconnet, L., Das, D., González, J.M., Catarino, D., Giacomello, E., Fontes, J., et al. (2018). Can deep-water sharks be avoided? Spatial mitigation measures in the deep-sea longline fisheries in the Azores. In D. Reid and L. Fauconnet (Eds.), Decision support tool for fishers incorporating information from tasks 4.1, 4.2 and information on unwanted catches derived from scientific data. DiscardLess deliverable D4.3.

Fontes, J., Rosa, A., Graça, G., Giacomelo, E., Cardenal, R., Afonso, P., et al. (2015). Can Deep Sea Sharks respond to short term protection? Looking into the deep (2015). European Elasmobranch Association meeting 2015, Peniche, Portugal.

Fraser, H.M., Greenstreet, S.P.R., Fryer, R.J., Piet, G.J. (2008). Mapping spatial variation in demersal fish species diversity and composition in the North Sea: accounting for species- and size-related catchability in survey trawls. *ICES Journal of Marine Science, 65*(4), 531–538.

Gerritsen, H., & Lordan, C. (2011). Integrating vessel monitoring systems (VMS) data with daily catch data from logbooks to explore the spatial distribution of catch and effort at high resolution. *ICES Journal of Marine Science, 68*(1), 245–252.

Hilborn, R. (2007). Managing fisheries is managing people: what has been learned? *Fish and Fisheries, 8*(4), 285–296

Little, A.S., Needle, C.L., Hilborn, R., Holland, D.S., Marshall, C.T. (2015). Real-time spatial management approaches to reduce bycatch and discards: experiences from Europe and the United States. *Fish and Fisheries, 16*(4), 576–602.

Mortensen, L.O., Ulrich, C., Eliasen, S., Olesen, H.J. (2017). Reducing discards without reducing profit: free gear choice in a Danish result-based management trial. *ICES Journal of Marine Science, 74*(5), 1469–1479.

Nash, R.D.M., Wright, P.J., Matejusova, I., Dimitrov, S.P., O'Sullivan, M., Augley, J. (2012). Spawning location of Norway pout (Trisopterus esmarkii Nilsson) in the North Sea. *ICES Journal of Marine Science, 69*(8), 1338–1346.

O'Neill, F.G., Feekings, J., Fryer, R.J., Fauconnet, L., Afonso, P. (this volume). Discard avoidance by improving fishing gear selectivity: Helping the fishing industry help itself. In S.S. Uhlmann, C. Ulrich, S.J. Kennelly (Eds.), *The European Landing Obligation – Reducing discards in complex, multi-species and multi-jurisdictional fisheries*. Cham: Springer.

Pascoe, S., Innes, J., Holland, D., Fina, M., Thébaud, O., Townsend, R., et al. (2010). Use of incentive-based management systems to limit bycatch and discarding. *International Review of Environmental and Resource Economics, 4*(2), 123–161

Plet-Hansen, K.S., Eliasen, S.Q., Mortensen, L.O., Bergsson, H., Olesen, H.J., Ulrich, C. (2017). Remote electronic monitoring and the landing obligation – some insights into fishers' and fishery inspectors' opinions. *Marine Policy, 76*, 98–106.

Plet-Hansen, K.S., Ulrich, C. (2018). Trade-off between value of landings and discard impact. Presentation at IIFET 2018, Seattle, United States.

Reid, D. (2017). Initial avoidance manuals by case study including tactical, strategic and gear based approaches agreed by scientists and fishers. DiscardLess Deliverable D4.1. http://www.discardless.eu/deliverables/entry/initial-avoidance-manuals-by-case-study. 40 pp. Accessed 30 Aug 2018.

Reid, D., & Fauconnet, L. (2018). Decision support tool for fishers incorporating information from tasks 4.1, 4.2 and information on unwanted catches derived from scientific data. DiscardLess Deliverable D4.3.

Rijnsdorp, A.D., van Overzee, H.M.J., Poos, J.J. (2012). Ecological and economic trade-offs in the management of mixed fisheries: a case study of spawning closures in flatfish fisheries. *Marine Ecology Progress Series, 447*, 179–194.

Rochet, M.-J., Catchpole, T., Cadrin, S. (2014). Bycatch and discards: from improved knowledge to mitigation programmes. *ICES Journal of Marine Science, 71*(5), 1216–1218.

Shephard, S., Gerritsen, H.D., Kaiser, M.J., Truszkowska, H.S., Reid, D.G. (2011). Fishing and environment drive spatial heterogeneity in Celtic Sea fish community size structure. *ICES Journal of Marine Science, 68*(10), 2106–2113.

Ulrich, C., Olesen, H.J., Bergsson, H., Egekvist, J., Håkansson, K.B., Dalskov, J., et al. (2015). Discarding of cod in the Danish Fully Documented Fisheries trials. *ICES Journal of Marine Science, 72*(6), 1848–1860.

Viana, M., Graham, N., Wilson, J.G., Jackson, A.L. (2011). Fishery discards in the Irish Sea exhibit temporal oscillations and trends reflecting underlying processes at an annual scale. *ICES Journal of Marine Science, 68*(1), 221–227.

Zimmermann, C., Kraak, S., Krumme, U., Santos, J., Stotera, S., Nordheim, L. (2015). Research for PECH Committee – Options of handling choke species in the view of the EU landing obligation – the Baltic plaice example. European Parliament. 100 pp. https://doi.org/10.2861/808965.

Chapter 14
Discard Avoidance by Improving Fishing Gear Selectivity: Helping the Fishing Industry Help Itself

Finbarr G. O'Neill, Jordan Feekings, Robert J. Fryer, Laurence Fauconnet, and Pedro Afonso

Abstract To address the challenges of the Landing Obligation, fishers need to be able to adjust the selective performance of each fishing operation in response to what they observe on the fishing grounds and to what they bring on board. This will include strategies on where and when to fish but also on how to fish, which we examine here. In particular, we focus on ways to encourage and support fishers to design, develop and test selective gears that will avoid unwanted catches in the first place. To this end, we highlight the necessity to increase awareness of existing solutions, the importance of understanding the capture process and how fish react to fishing gears, and the need to evaluate the economic implications of new gears. We examine the success of science-industry collaborations and emphasise the benefits of a flexible regulatory environment. Looking ahead, the fishing industry needs to keep up-to-date with new technologies that can be used to observe the interaction of fish and their gears and with new approaches to modifying selectivity.

Keywords Catch avoidance · Discard reduction · Fish reactions to fishing gear · Industry participation · Selective fishing gears

F. G. O'Neill (✉) · J. Feekings
Technical University of Denmark, National Institute of Aquatic Resources, Hirtshals, Denmark
e-mail: barone@aqua.dtu.dk

R. J. Fryer
Marine Laboratory, Marine Scotland Science, Aberdeen, Scotland, UK

L. Fauconnet · P. Afonso
Marine and Environmental Sciences Centre and OKEANOS Research Unit, Universidade dos Açores, Horta, Portugal

S. S. Uhlmann et al. (eds.), *The European Landing Obligation*,
https://doi.org/10.1007/978-3-030-03308-8_14

14.1 Introduction

Fishing with selective gears has long been used as a management measure to promote the sustainable exploitation of commercial fisheries. In the European Union and many other jurisdictions, there are measures based on the technical specifications of the gears deployed. For example, minimum mesh size, maximum twine thickness and the use of devices such as square mesh panels and sorting grids have been prescribed in demersal trawl fisheries (Broadhurst 2000; Graham 2010); beam size and number of dredges in beam trawl and dredge fisheries; net height, length and hanging ratio in gill net fisheries (He and Pol 2010); hook size, shape and type of bait in longline fisheries (Løkkeborg et al. 2010); and escape gap dimensions and number of traps in creel and pot fisheries (Thomsen et al. 2010).

The concern of these gear based technical conservation measures has generally been the selectivity or capture of just one or a small number of species. This was usually the main target species but in some instances was a protected species or a marine bird or mammal. The advent of the EU Landing Obligation (LO) brings with it a need to consider the selective performance of a fishing gear in relation to all of the species in the catch to which the Landing Obligation applies.

To develop the range of gears that will be capable of dealing with the specific challenges posed by the Landing Obligation, will require input from fishers, net and gear makers, fishing gear technologists and fish behaviourists. Fishers, in particular, have developed many fishing techniques and fishing gears throughout the world. They can be broadly classified as either active gears, which are towed or are encompassing (Fig. 14.1a, b and c), or static gears, which trap, entangle or hook (Fig. 14.1d, e and f). The design and deployment of these gears are very diverse and depend on fish behaviour, the fishing grounds and the resources available to fishers.

Here, rather than review the many different ways the selective performance of these gears can be improved, we set out a loose framework of initiatives and measures that we believe would support the fishing industry to develop their own selective gears. To facilitate these efforts, we highlight the need for fishers and gear manufacturers to be made aware of existing solutions, to have an understanding of the capture process and an appreciation of how fish react to their gears. We also consider how to encourage the successful development and implementation of more selective gears, the measures that need to be put in place to maximise the likelihood of success of science-industry collaborations, and the benefit of having a flexible regulatory environment. Furthermore, the fishing industry needs to keep up-to-date with technological advances that can be used to observe the interaction of fish and their gears and with new approaches to modifying selectivity.

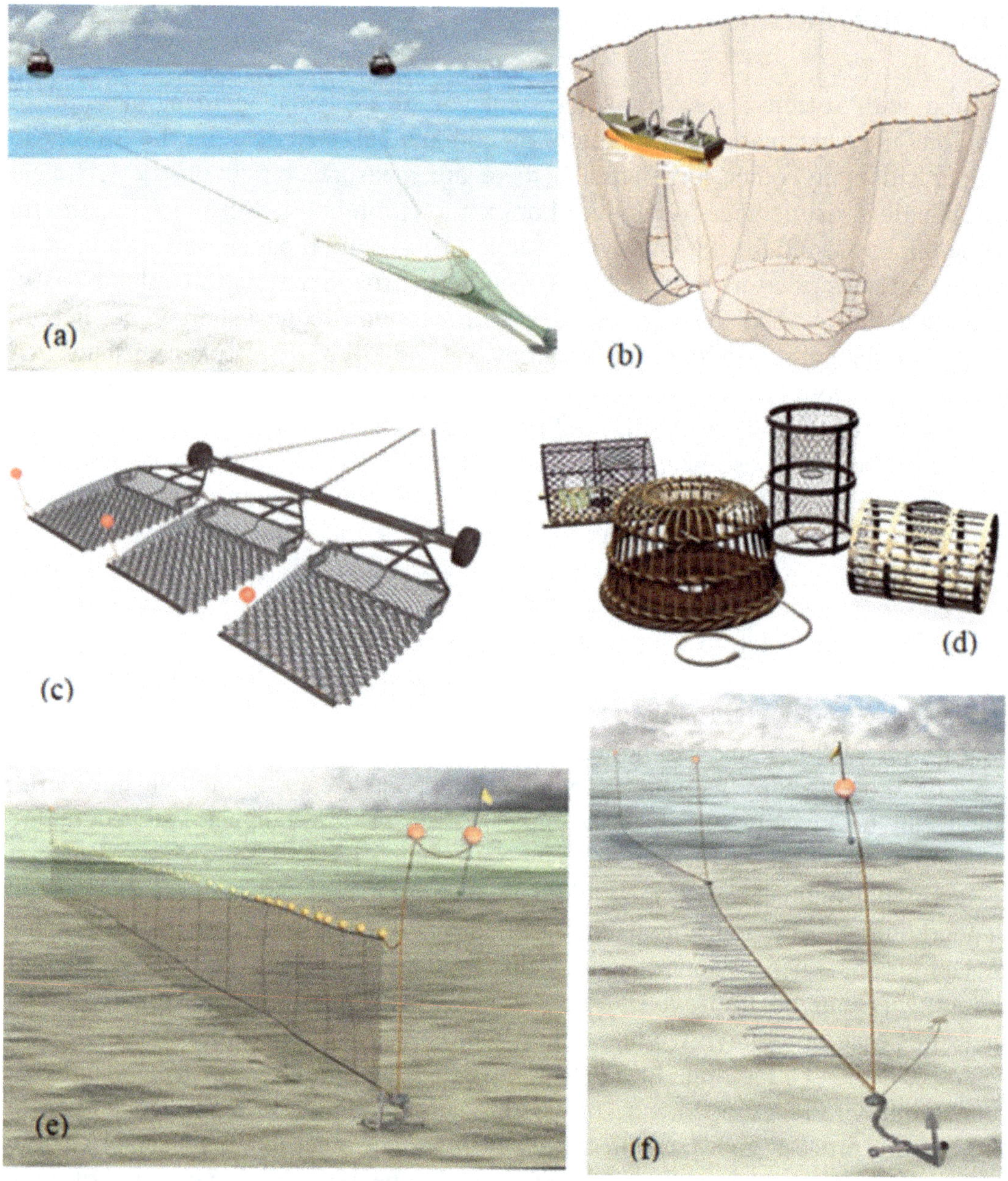

Fig. 14.1 (**a**) A pair trawl team towing a single demersal trawl. (**b**) A purse seine net. (**c**) Three scallop dredges being towed on a single beam. (**d**) Different types of pots and creels. (**e**) A gillnet. (**f**) A longline (Taken from Galbraith et al. 2004)

14.2 Making Best Use of Existing Information

There have been many studies of how design changes to fishing gears can alter the size and species profile of the catch. Although these studies may only focus on a limited number of species and the gears tested may only have a local or regional relevance, very often the principles applied have a general significance. Accordingly, there is a lot to be learned from the results of previous trials, and an obvious starting point in the development of a selective gear is a thorough review of trials that have already taken place.

The reporting, however, of these studies is varied. Some go unreported, some are reported locally or regionally, while some may be reported internationally (in scientific conferences and peer-reviewed publications). There is also great variation in the level of detail presented: in some cases, it may be only the conceptual idea or a brief design description along with overall results, while in others detailed gear specifications and full haul-by-haul catch data are supplied. To make best use of this information, it needs to be available in an easily accessible way, and, where appropriate, the results should be consolidated and synthesised.

Initiatives such as the DiscardLess factsheets (O'Neill and Mutch 2017, http://www.discardless.eu/) (Fig. 14.2) and the Gearing up project (https://gearingup.eu/) describe and summarise many of the catch-comparison and selectivity trials that have taken place in the North East Atlantic and adjacent seas and present the results in a searchable format. These initiatives provide an invaluable repository to help fishers, gear manufacturers and fishing gear technologists find practical solutions to the problems they face. However, there is a need to continue to build on them, extend their geographic range and broaden their scope to more gear types (to date they are almost entirely for towed demersal gears). In the first instance, this could be done by linking them to the factsheets that have been produced in some of the regional gear development studies, which have taken place in many countries. There is also a need to ensure the longer-term support and survival of these databases and to secure commitments from relevant bodies (such as ICES, International Council for the Exploration of the Sea or the FAO, United Nations' Food and Agriculture Organisation) to resource them beyond the life of the projects within which they have been developed.

Where sufficient data exists, there may be scope to combine the results from a number of trials to explore a broad range of selective gear options for use in a fishery. Madsen (2007) carried out such an analysis for cod (*Gadus morhua*), Perez Comas and Pikitch (1994) for a range of gadoid species and ICES (2007) for Norway lobster (*Nephrops norvegicus*). Another study estimated the proportion of several species that enter a trawl above or below a given height, which will be very useful in designing species selective gears such as low headline or raised footrope trawls (Fryer et al. 2017) (Fig. 14.3). A meta-analysis of target and bycatch species in longline fisheries shows that the performance of circle hooks versus J-hooks is species dependent with higher catch rates of some species with circle hooks and higher rates of other species with J-hooks (Reinhardt et al. 2018), and another assessed the performance of a wide range of bycatch reduction devices in relation to the capture of sharks and rays and the target species (Favaro and Côté 2015).

These types of analyses can be very powerful because they incorporate results from trials where typically only one or two parameters are tested to produce empirical models that predict selection across a wide range of gear parameters, leading to a better understanding of the relative influence of all of these parameters. A meta-analysis of haddock (*Melanogrammus aeglefinus*) selectivity describes the selective performance of trawl codends in terms of the mesh size, the number of

changing the groundgear
to reduce the capture of small flatfish in a Nephrops trawl

TARGET SPECIES

Nephrops **and mixed round and flatfish**

AREA, VESSEL

24 twin trawl catch comparison hauls were carried out in the North Sea on board the Zenith BF106 (671 HP)

GEAR MODIFICATION

The catching performance of a low headline 'Letterbox' trawl with 200mm spherical bobbins in the 15m centre section of the groungear is compared with a similar gear with 200mm rockhopper discs in the centre section.

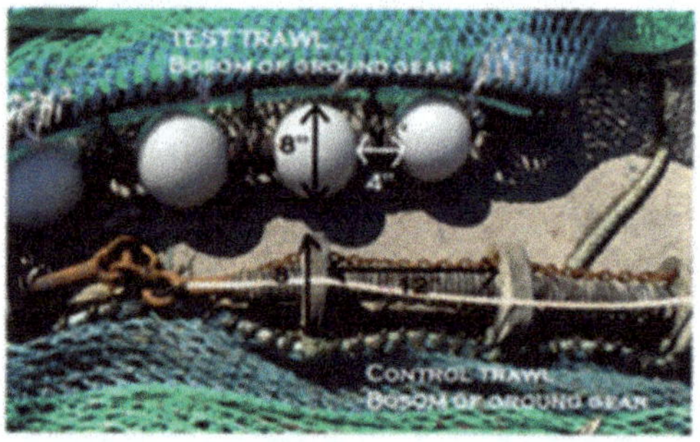

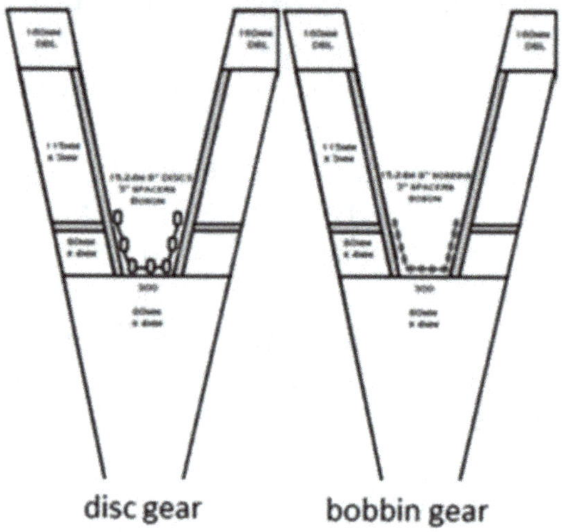

	discs	bobbins	
	Total Weight (kg)		% diff
Lemon Sole	144	148	2.9
Plaice	261	257	-1.4
Witch	108	96	-11
Megrim	173	132	-24
Comon dab	59	50	-16
Long rough dab	178	81	-54

RESULTS

using the spherical bobbins reduced the catches of flatfish species.

this was length dependent and smaller flatfish were less likely to be retained than larger ones.

For plaice and lemon sole there were greater catches of the larger individuals.

marine scotland science

The Scottish Government

FURTHER INFORMATION

Matt Kinghorn

mathew.kinghorn@gov.scot

SCOTTISH FISHERMEN'S FEDERATION

DiscardLess

Fig. 14.2 Two examples of the DiscardLess factsheets summarising gear trials in Scotland and Turkey (Taken from O'Neill and Mutch 2017)

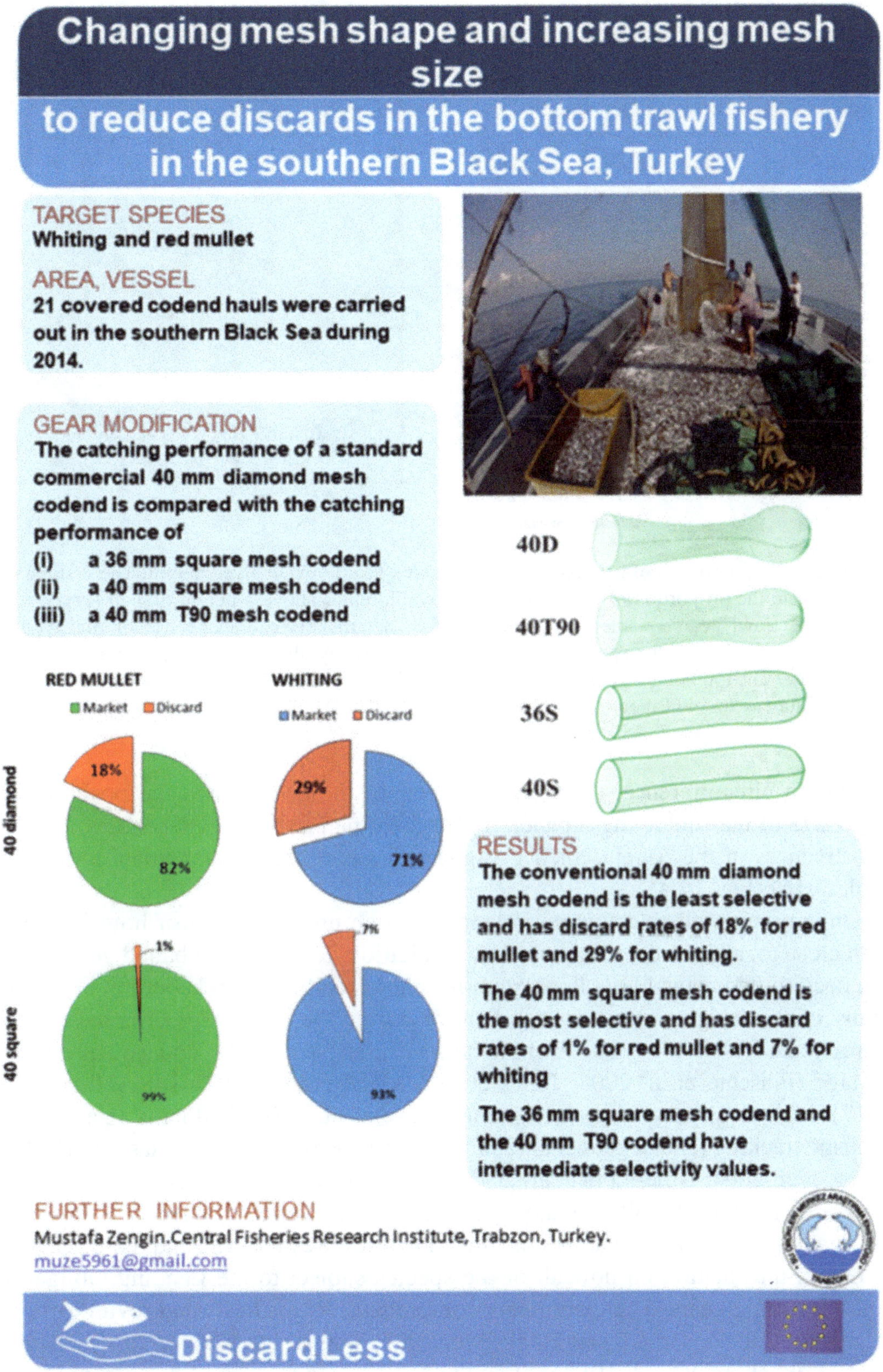

Fig. 14.2 (continued)

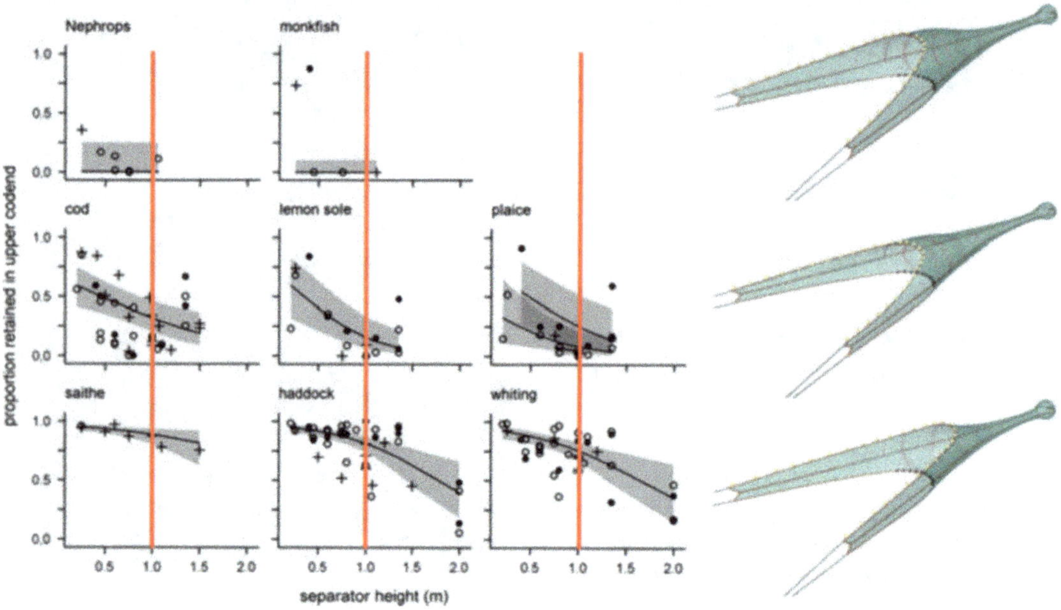

Fig. 14.3 The proportion of fish that will enter a trawl gear above a given height. The vertical red lines indicate the proportion of each species that would enter above a height of 1 m (Fryer et al. 2017). The trawl gears on the right illustrate how net makers can make use of this type of information to influence the species profile entering a gear by altering the height and position of the headline. The top net is a standard trawl, the middle one is a low headline trawl and the bottom one is a cutaway trawl (Taken from O'Neill and Mutch 2017)

meshes in circumference, the twine diameter of the codend and the position and mesh size of the square mesh panel. It also identified a seasonal dependence on the effectiveness of the panel, which it suggested was related to fish condition (Fryer et al. 2016) (Fig. 14.4).

Underpinning all selective improvements is an understanding of how fish react both close to, and inside, a fishing gear. In addition, for static and baited gears, there is a need to understand how fish search for and catch food (Løkkeborg et al. 2014). Many of the insights into how fish behave during the capture process have come from visual observations by divers (Main and Sangster 1981), underwater camera footage (Piasente et al. 2004; Jones et al. 2008; Bryan et al. 2014; Anders et al. 2017), high frequency acoustic cameras (Rose et al. 2005; Williams et al. 2013), acoustic tracking (Handegard et al. 2003), laboratory experiments (Glass et al. 1993; Breen et al. 2004; Utne-Palm et al. 2018) and from experimental fishing trials at sea (Engås et al. 1998; Ingólfsson and Jørgensen 2006; Ryer et al. 2010).

Again, much of what is known is limited to a few key species, and there is a need to obtain information on how all of the species subject to the Landing Obligation react to fishing gears. There are many comprehensive synthesis and review articles (e.g. Wardle 1993; Broadhurst 2000; Ryer 2008; Winger et al. 2010; Thomsen et al. 2010, Løkkeborg et al. 2010) but often these are aimed at a scientific audience. This information needs to be distilled and presented in a way that is accessible to nonspecialists. O'Neill and Mutch (2017) produce a brief summary of the different

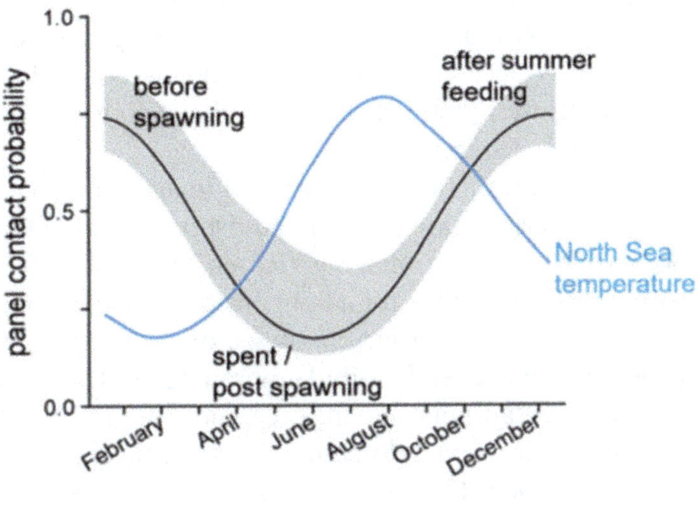

Fig. 14.4 One of the outputs of a meta-analysis of haddock codend selectivity data (Fryer et al. 2016). The panel contact probability, which characterises the effectiveness of the square mesh panel, was shown to vary seasonally. This was out of sync with the average North Sea water temperature but corresponded well with periods of high and low fish condition, suggesting that the selective performance of square mesh panels is related to fish condition

stages of the fish capture process of towed demersal gears, highlighting how different parts of the gear may influence selection and identifying possible design changes which may alter the selectivity of the gear. Their intention was to increase awareness of the possible modifications that can be made to gears so that fishers and net makers can design and develop gears with a selective performance suitable for their particular fishery.

14.3 Obtaining New Insights and Enhancing the Capacity to Make Real-Time Decisions

As camera technologies have improved and become miniaturised and less expensive, they are more frequently being used by researchers and fishers to obtain footage of fish reactions to their gears (Struthers et al. 2015). The cheapest and most easy-to-use systems record footage which can be viewed subsequently when the gear is retrieved. Some fishers have made these investments themselves, but in some areas, there are schemes where cameras, specifically designed to be attached to fishing gears, can be borrowed or leased (http://www.fast-track.dk/). The ability to view their own gears, to observe how fish react to them and to see the effects of design changes will enable fishers to find tailored solutions to the specific catch and quota restrictions they are subject to under the LO.

There are also systems which provide real-time footage, but these require transmission cables to the surface which can be difficult to handle and are more expensive. Nevertheless, such an ability would allow fishers to make real-time decisions regarding their fishing operation. These could be as simple as deciding to continue or stop fishing, based on observations of what fish are on the ground or entering their gear; or they could be used in conjunction with remotely controllable instruments that, for example, open/close a codend or operate flaps/doors that direct fish into different compartments of a fishing gear.

Current developments in camera technology and image processing will improve the ability to make direct observations of fish and fishing gears. 3D camera systems employing methods such as stereo imaging (Rosen and Holst 2013) or ‘time of flight’ (which measures the time taken for a light pulse to reach the object and return) are being improved and developed (https://www.utofia.eu/) and may soon permit position and size measurement, even in turbid environments. These systems allied to advances in image analysis, artificial intelligence and machine learning will allow real-time species identification and automatic analysis of acquired images, permitting the skipper or a control system to make real-time decisions (http://smartfishh2020.eu/; https://www.deepvision.no/).

Acoustic systems have been used in pelagic fisheries, from estimating the size and density of fish schools to tracking individuals, and more recently, to differentiate between and within species (Trenkel et al. 2016). Such developments are likely to be particularly useful for catch identification during the early hauling stages of purse seine fisheries. At present direct methods such as hand-lining and dip-netting are used to determine the species and size profile of the catch, but these can often only be used during the latter stages of hauling, when overcrowding may have occurred and the survival of released catches is likely to be low (Marçalo et al., this volume). Sampling methods, which can be used during the early stages of a haul, such as shooting a ‘mini-trawl’ into a purse seine (Isaksen 2013) or deploying ROVs fitted with some of the camera systems mentioned above are currently being developed.

Another way of obtaining new insights into gear technology is through the analysis and modelling of data collected in trials. Mixed models are well suited for analysing these data because they estimate the effects of practical importance while accounting for the different sources of variation in the data. The past decade has seen many advances in the statistical methods and software available for fitting mixed models, and they are now routinely used to analyse standard gear trials, such as estimating the selection of a trawl from a covered codend or paired tow experiment, or to compare the catch of one gear with that of another (Millar et al. 2004; Holst and Revill 2009). They also offer exciting possibilities for analysing the data from nonstandard trials, and recently, Browne et al. (2017) used a multinomial mixed model to analyse a quad-rig catch-comparison trial where four test codends were fished simultaneously. The main purpose of the trials was to assess the catch performance of the quad-rig, which is increasingly used in Irish Nephrops fisheries. However, the trials, and the methods for analysing them, suggest how more efficient experiments might be designed in the future, with multiple codends being fished in each haul. Mixed models are also a standard approach to synthesising the results of multiple trials and were used, for example, in the meta-analysis of haddock described in the previous section. A challenge moving forward is to make better use of sparse data, particularly for choke species. In single trials, simplifying assumptions are often needed to get models to converge and the power to detect effects can be low. One possibility might be to combine data across multiple trials and to exploit or assume correlations in selection between a data-sparse species and data-rich species that have similar behavioural or morphological characteristics.

14.4 Successful Development and Implementation

Innovative selective gears are much more likely to be taken up by the fishing industry if fishers are involved at every stage of the development process, from the conceptual design and experimental refinement through to commercial trials (Kennelly and Broadhurst 2002). In recent years, most industry-science collaborations in Europe have taken such a bottom-up approach (Armstrong et al. 2013; Mortensen et al. 2017), and a number of gears such as the netting grids in the Scottish *Nephrops* fishery (Kynoch et al. 2012; Drewery et al. 2012) have been successfully introduced. This type of approach creates a sense of ownership and control over the gears developed and often results in a range of technical solutions, all of which may achieve the same objective (Feekings et al. submitted). In contrast, when gears are imposed and introduced into legislation in a top-down approach, with little or no involvement of the fishing industry, there can be a reluctance to use the gear effectively, and indeed, additional alterations may even be made to compromise the selective improvement of the new gear (Krag et al. 2016).

In some cases, incentives have been offered to encourage participation in industry-science collaborations. These can be in the form of financial support for gear construction, or additional effort or quota to offset losses that are incurred during trials or commercial operations (O'Neill et al. 2014). There have also been nonmonetary incentives where maybe a prize is offered, such as the WWF Smart Gear competition (https://www.worldwildlife.org/), or where the publicity or recognition for being involved is considered sufficient reward.

Fishing is an economic activity and uncertainty surrounding the costs and benefits of gear modifications may make vessel owners reluctant to make gear changes due to potential losses in time and revenue both during trial periods and when the gear is being used commercially. Hence, it is very important that fishers can evaluate the financial implications of developing and using more selective gears. To help vessel operators make these assessments, Seafish (a UK Non-Departmental Public Body) has developed a Best Practice Guidance and a Financial Assessment Spreadsheet for industry-led gear trials (http://www.seafish.org/). A fisher can input his own data (on fuel and crew costs, etc.), the cost of the gear modifications/changes and catch data from the corresponding gear trials to assess the financial consequences to his/her business, of testing and introducing new and modified gears.

To develop and introduce selective gears, fishers need to be able to operate in a much more flexible environment. The regulatory regime, in many jurisdictions, is prescriptive, rigid and difficult to change. This can be a serious impediment to gear development as often new or modified gears may not comply with legislation, although they may be more selective than what is currently prescribed. Furthermore, to fully address the challenges brought about by the Landing Obligation, fishers will have to be able to react to what they see on the fishing grounds and to modify the selective performance of their gears accordingly. An optimal regulatory approach would be one which is output rather than input driven, that is, one that focusses on what is landed on deck rather than on how it is caught. This would give fishers flexibility to develop and use gears that best suit their specific circumstances (Eliasen et al. 2019).

14.5 Alternative Technologies to Improve Species and Size Selectivity

While a lot can be done to develop more selective gears with existing technologies and knowledge, it is also important to consider alternative approaches and new developments. The selective performance of a fishing gear depends on design parameters such as mesh and hook size, and on the response of the species under consideration to the various optical, acoustic, magnetic, electric, hydrodynamic and/or chemical stimuli the gear generates (Popper and Carlson 1998; Jordan et al. 2013; Løkkeborg et al. 2014). In recent years, due to technological developments which can generate and/or modify these stimuli, and an improved understanding of how fish react to them, there has been an increasing focus on harnessing such stimuli to modify fishing gear selectivity.

Light has long been used by fishers to capture squid and pelagic species (Arimoto et al. 2010; Ben-Yami and Pichovich 1988) and, with the onset of robust low-powered LED light sources, it is being considered again in many contexts. Bryhn et al. (2014) increased the catch efficiency of larger cod (*Gadus morhua*) in pots by using green lights, while Nguyen et al. (2017) improved the catchability of snow crab (*Chionoecetes opilio*) by using LED lights in their traps. In trials on the west coast of the USA, Hannah et al. (2015) were able to reduce the capture of some fish species by up to 90% with no loss of ocean shrimp (*Pandalus jordani*) by placing LED lights on the fishing line of their shrimp trawls. There have also been successful trials with luminous netting materials, fibre optic cables and lasers to direct fish into or within a trawl (Karlsen et al. 2018; O'Neill et al. 2018; Hreinsson et al. 2018). To fully exploit the potential of light to improve the selective performance of commercial fishing gears, more research needs to be done on how parameters such as the wavelength, intensity, polarisation and strobing of light can be used to modify the behavioural reaction of fish (Ben-Yami 1976; Marchesan et al. 2005; Arimoto et al. 2010; Königson et al. 2002).

Sound has been used to guide, ranch, condition and aggregate fish in fishing operations (Yan et al. 2010). It has also been used to deter and repel fish and marine mammals from water intake pipes and fishing gears. In gillnet fisheries, active devices such as pingers, which emit an acoustic signal, have been attached to gears to reduce the bycatch of cetaceans by alerting them to the presence of the gear (Rihan 2010). There have been many advances of parametric sound technology where a 'beam' of sound is transmitted directionally and focused at high intensity on to a relatively small area (Gan et al. 2012). These developments open up the possibility of creating 'walls' or surfaces of sound which could be used to direct and herd fish and marine mammals in a more focussed way. There are also passive approaches where the acoustic reflectivity of the gear is enhanced by treating the netting material or attaching acoustic reflectors to the gear so that they are more easily detected by echo-locating species (He and Pol 2010).

In longline fisheries there have been attempts to take advantage of elasmobranchs' ability to detect weak electromagnetic fields to reduce their capture by using electropositive metals and magnets (Robbins et al. 2011; Kaimmer and Stoner 2008; O'Connell et al. 2014). While success has been limited thus far, and there are issues related to manufacturing costs, deterioration in water and large-scale deployment (Favaro and Côté 2015), there is the possibility that alternative metals and compounds will offer cheaper and cost-effective solutions (O'Connell et al. 2014). In trawl fisheries, electricity has been used to increase catchability by stimulating benthic species from the seafloor, to direct and aggregate fish so that they can be caught more easily by conventional means and to improve the performance of selective devices by exploiting species and size differences in their behavioural responses (Polet 2010). In the southern North Sea flatfish fishery, electrodes produce an electric field which induces a cramp response that bends fish in a U-shape, making it easier for the ground gear to get underneath them so they enter the trawl (van Marlen et al. 2014; Depestele et al. 2018). Other examples of using electricity in trawl fisheries include the Belgian and Chinese shrimp fisheries (Polet et al. 2005; Yu et al. 2007) and the razor clam (*Ensis* spp.) fishery in the West of Scotland (Murray et al. 2016).

There are a number of examples where the hydrodynamics of towed gears have been exploited to improve selectivity. Veil nets in shrimp fisheries, rising panels in codend extensions and the flex deflector modify the flow in the gear to direct fish and crustaceans onto or closer to grids and square mesh panels (Graham 2003; Santos et al. 2016). The Hydrodredge deflects a water flow on to the seabed to raise great scallops (*Pecten maximus*) from the seabed (Shephard et al. 2009), and Jordan et al. (2013) suggest that water jets directed downwards, ahead of a trawl gear could elicit an early response from elasmobranchs, allowing them to avoid capture. There is also potential to create regions of low flow behind screens and bluff bodies and turbulent regions which, if the associated vortices are an appropriate strength and size, can be used to encourage fish to hold station and perhaps increase their probability of contact with a selectivity device (Liao 2007; Laird et al. 2016).

The gustatory and olfactory senses are of particular importance in baited gears and Løkkeborg et al. (2010) and Thomsen et al. (2010) highlighted the potential of artificial baits, longer-lasting baits and a better understanding of species-specific differences in bait performance to improve the selective performance of longline and pot fisheries. Gilman et al. (2008) have shown that using fish instead of squid for bait reduced shark bycatch in pelagic longlines, while Stroud et al. (2014) have shown

Table 14.1 Technologies that may be useful in elasmobranch bycatch mitigation by gear and sensory modality

Sensory modality	Baited hook and line (longline)	Gill net	Trawl	Purse seine
Olfaction	Surfactants, semiochemicals	Surfactants, semiochemicals		Remote attraction/ bait stations
	Bait type			
	Dead sharks			
Hearing	Not recommended			
Vision	Light sticks: wavelength and flicker	Net illumination	Flashing lights	
	Bait colour	Net colour		
	Leader type/colour	Predator models		
	Dead sharks			
Mechanosensory lateral line/pit organs			Water jets	
Electrosensory	Magnets, lanthanide metals, battery-powered electric devices	Powered electric field 'barrier' Magnetic field 'barrier'	Electric pulse generators	
Other		Pre-net fence (tactile)		

Taken from Jordan et al. (2013)

that a necromone produced from putrefied shark tissue was 100% repellent to competitively feeding Caribbean reef sharks (*Carcharhinus perezi*) and blacknose sharks (*Carcharhinus acronotus*).

It is evident that there is great scope to make better use of all the senses that target and bycatch species have. Again, it is very important that this information is relayed to fishers, gear manufacturers, gear technologists and fisheries managers in a readily accessible way. Jordan et al. (2013) provide a very useful summary table which identifies new and existing technologies that should undergo further testing for use in elasmobranch bycatch mitigation. They classify potential solutions for a range of gear types in terms of the sensory modality that the fish will use and which we reproduced here (Table 14.1) as an example of what could be very usefully extended to other species.

14.6 In Summary

The specific challenges posed by the Landing Obligation will depend on the catching performance of the fishing gears, the total allowable catches and the size profile and spatial distribution of the stocks. Hence, not only will the issues that arise vary from year to year and from fishery to fishery, they will most likely vary from vessel to vessel and from trip to trip. As a result, the most practicable way forwards is one where each fisher is in a position to adjust the selective performance of each fishing operation. This will include strategies on where and when to fish, as addressed by Reid et al. (this volume), but also, on how to fish, as is set out here in this chapter. In this regard, fishers need to be able to modify the selective performance of their fishing gear in response to what they observe on the fishing grounds and to what they bring on board. They need to be made aware of existing solutions and have an appreciation of how their gears catch. This includes understanding the mechanistic aspects of their gear's performance (and its dependence on parameters such mesh or hook size, etc.) and an awareness of the sensory stimuli generated by fishing gears and the corresponding responses of the species they catch. They need to operate in a regime which encourages their participation in a meaningful way and in a regulatory environment that permits them to develop and use new and modified gears.

Ultimately, fishers are best placed to identify the challenges brought about by the landing obligation, and in collaboration with gear makers, fishing gear technologists and fish behaviourists, are those most likely to find solutions that are both acceptable and effective.

Acknowledgements Part of this work has received funding from the Horizon 2020 Programme under grant agreement DiscardLess number 633680. This support is gratefully acknowledged.

References

Anders, N., Fernö, A., Humborstad, O.-B., Løkkeborg, S., Rieucau, G., Utne-Palm, A.C. (2017). Size-dependent social attraction and repulsion explains the decision of Atlantic cod Gadus morhua to enter baited pots. *Journal of Fish Biology, 91*(6), 1569–1581.

Arimoto, T., Glass, C.W., Zhang, X. (2010). Fish vision and its role in fish capture. In P. He (Ed.), Behavior of marine fishes: Capture processes and conservation challenges. Oxford: Wiley-Blackwell.

Armstrong, M.J., Payne, A.I.L., Deas, B., Catchpole, T.L. (2013). Involving stakeholders in the commissioning and implementation of fishery science projects: Experiences from the UK Fisheries Science Partnership. *Journal of Fish Biology, 83*, 974–996.

Ben-Yami, M. (1976). *Fishing with light. FAO fishing manuals*. Fishing News Books Limited, Surrey.

Ben-Yami, M., Pichovich, A. (1988). *Attracting fish with light* (FAO Training Series No. 14). Rome: FAO.

Breen, M., Dyson, J., O'Neill, F.G., Jones, E., Haigh, M. (2004). The swimming performance of haddock at prolonged and sustained swimming speeds, and its role in their capture by towed fishing gears. *ICES Journal of Marine Science, 61*, 1071–1079.

Broadhurst, M.K. (2000). Modifications to reduce bycatch in prawn trawls: A review and framework for development. *Reviews in Fish Biology and Fisheries, 10*(1), 27–60.

Browne, D., Minto, C., Cosgrove, R., Burke, B., McDonald, D., Officer, R., Keatinge, M. (2017). A general catch comparison method for multi-gear trials: Application to a quad-rig trawling fishery for Nephrops. *ICES Journal of Marine Science, 74*, 1458–1468.

Bryan, D.R., Bosley, K.L., Hicks, A.C., Haltuch, M.A., Wakefield, W.W. (2014). Quantitative video analysis of flatfish herding behavior and impact on effective area swept of a survey trawl. *Fisheries Research, 154*, 120–126.

Bryhn, A.C., Konigson, S.J., Lunneryd, S.-G., Bergenius, M.A.J. (2014). Green lamps as visual stimuli affect the catch efficiency of floating cod (Gadus morhua) pots in the Baltic Sea. *Fisheries Research, 157*, 187–192.

Depestele, J., Degrendele, K., Esmaeili, M., Ivanović, A., Kröger, S., O'Neill, F.G., Parker, R., Polet, H., Roche, M., Teal, L.R., Vanelslander, B., Rijnsdorp, A.D. (2018). Comparison of mechanical disturbance in soft sediments due to tickler-chain SumWing trawl versus electro-fitted PulseWing trawl. *ICES Journal of Marine Science*. https://doi.org/10.1093/icesjms/fsy124.

Drewery, J., Watt, M., Kynoch, R.J., Edridge, A., Mair, J., O'Neill, F.G. (2012). *Catch comparison trials of the Flip Flap netting grid trawl*. Marine Scotland Science Report 08/12.

Eliasen, S.Q., Feekings, J., Krag, L., Veiga Malta, T., Mortensen, L.O., Ulrich, C. (2019). The landing obligation calls for a more flexible technical gear regulation in EU waters – Greater industry involvement could support development of gear modifications. *Marine Policy*. 173–180.

Engås, A., Jørgensen, T., West, C.W. (1998). A species-selective trawl for demersal gadoid fisheries. *ICES Journal of Marine Science, 55*, 835–845.

Favaro, B., Côté, I.M. (2015). Do by-catch reduction devices in longline fisheries reduce capture of sharks and rays? A global meta-analysis. *Fish and Fisheries, 16*, 300–309.

Feekings, J.P., O'Neill, F.G., Krag, L.A., Ulrich, C., Veiga-Malta, T. (submitted). An evaluation of European initiatives established to encourage industry-led development of selective fishing gears. *Fisheries Management and Ecology*.

Fryer, R.J., O'Neill, F.G., Edridge, A. (2016). A meta-analysis of haddock size-selection data. *Fish and Fisheries, 17*, 358–374. https://doi.org/10.1111/faf.12107

Fryer, R.J., Summerbell, K., O'Neill, F.G. (2017). A meta-analysis of vertical stratification in demersal trawl gears. *Canadian Journal of Fisheries and Aquatic Sciences, 74*, 1243–1250.

Galbraith, R.D. and Rice, A. after Strange, E.S. (2004). *An introduction to commercial fishing gear and methods used in Scotland*. Scottish Fisheries Information Pamphlet. No. 25, 2004.

Gan, W.S., Yang, J., Kamakura, T. (2012). A review of parametric acoustic array in air. *Applied Acoustics, 73*, 1211–1219.

Gilman, E., Clarke, S., Brothers, N., Alfaro-Shigueto, J., Mandelman, J., Mangel, J., Petersen, S., Piovano, S., Thomson, N., Dalzell, P., Donoso, M., Goren, M., Werner, T. (2008). Shark interactions in pelagic longline fisheries. *Marine Policy, 32*, 1–18.

Glass, C.W., Wardle, C.S., Gosden, S.J. (1993). Behavioural studies of the principles underlying mesh penetration by fish. *ICES Marine Science Symposia, 196*, 92–97.

Graham, N. (2003). By-catch reduction in the brown shrimp, Crangon crangon, fisheries using a rigid separation Nordmore grid (grate). *Fisheries Research, 59*, 393–407.

Graham, N. (2010). Technical measures to reduce bycatch and discards in trawl fisheries. In P. He (Ed.), *Behavior of marine fishes: Capture processes and conservation challenges*. Oxford: Wiley-Blackwell.

Handegard, N.O., Michalsen, K., Tjøstheim, D. (2003). Avoidance behaviour in cod (Gadus morhua) to a bottom-trawling vessel. *Aquatic Living Resources, 16*, 265–270.

Hannah, R.W., Lomeli, M.J.M., Jones, S.A. (2015). Tests of artificial light for bycatch reduction in an ocean shrimp (Pandalus jordani) trawl: Strong but opposite effects at the footrope and near the bycatch reduction device, *Fisheries Research, 170*, 60–67.

He, P., & Pol, M. (2010). Fish behavior near gillnets: Capture processes and influencing factors. In P. He (Ed.), *Behavior of marine fishes: Capture processes and conservation challenges*. Oxford: Wiley-Blackwell.

Holst, R., Revill, A., (2009). A simple statistical method for catch comparison studies. *Fisheries Research, 95*, 254–259.

Hreinsson, E., Karlsson, H., Gudmundsson, G., Jonsdottir, H., Thorhallsson, T. (2018). *Catching Northern Prawn without benthic contact*. ICES WGFTFB 2018 REPORT. ICES CM 2018/EOSG:12.

ICES. (2007). Report of the Workshop on Nephrops Selection (WKNEPHSEL). ICES CM 2007/FTC 1, 49pp.

Ingólfsson, Ó.A., & Jørgensen, T. (2006). Escapement of gadoid fish beneath a commercial bottom trawl: Relevance to the overall trawl selectivity. *Fisheries Research, 79*, 303–312.

Isaksen, B. (2013). *Fish sampling by shooting a mini trawl into the purse-seine net*. Norwegian Institute of Marine Research, Havforskningsnytt nr. 2–2013.

Jones, E.G., Summerbell, K., O'Neill, F.G. (2008). The influence of towing speed and fish density on the behaviour of Haddock in a trawl cod-end. *Fisheries Research, 94*, 166–174.

Jordan, L.K. Mandelman, J.W. McComb, D.M. Fordham, S.V., Carlson, J.K., Werner, T.B. (2013). Linking sensory biology and fisheries bycatch reduction in elasmobranch fishes: A review with new directions for research. *Conservation Physiology, 1*(1).

Kaimmer, S., Stoner, A.W., (2008). Field investigation of rare-earth metal as a deterrent to spiny dogfish in the Pacific halibut fishery. *Fisheries Research, 94*, 43–47.

Karlsen, J.D., Melli, V., Krag, L.A. (2018). *The luminous net VISIONET – A guiding swimway to the exit or a stressor?* ICES WGFTFB 2018 REPORT. ICES CM 2018/EOSG:12

Kennelly, S.J., & Broadhurst, M.K. (2002). By-catch begone: Changes in the philosophy of fishing technology. *Fish and Fisheries, 3*(4), 340–355.

Königson, S., Fjälling, A. & Lunneryd, SG., (2002). Reactions in individual fish to strobe light. Field and aquarium experiments performed on whitefish (*Coregonus lavaretus*). *Hydrobiologia, 483*, 39–44.

Krag, L.A., Herrmann, B., Feekings, J.P., Karlsen, J.D. (2016). Escape panels in trawls – A consistent management tool? *Aquatic Living Resources, 29*, 306.

Kynoch, R.J., Edridge, A., O'Neill, F.G. (2012). Catch comparison trials with the Faithlie Cod Avoidance Panel (FCAP). *Scottish Marine and Freshwater Science, 3*(8).

Laird, A., Cahill, J., Liddell, B. (2016). *Kons covered fisheyes BRD trial*. Report Northern Prawn Fishery. http://www.afma.gov.au/wp-content/uploads/2017/05/Kons-Covered-Fisheyes-BRD-Trial-Report-Northern-Prawn-Fishery-2016_FINAL.pdf.

Liao, J.C. (2007). A review of fish swimming mechanics and behaviour in altered flows. *Philosophical Transactions of the Royal Society, 362*, 1973–1993.

Løkkeborg, S., Fernö, A., Humborstad, O.-B. (2010). Fish behavior in relation to longlines. In P. He (Ed.), *Behavior of marine fishes: Capture processes and conservation challenges*. Oxford: Wiley-Blackwell.

Løkkeborg, S., Siikavuopio, S.I., Humborstad, O.-B., Utne-Palm, A.C., Ferter, K. (2014). Towards more efficient longline fisheries: Fish feeding behaviour, bait characteristics and development of alternative baits. *Reviews in Fish Biology and Fisheries, 24*, 985–1003.

Madsen, N. (2007). Selectivity of fishing gears used in the Baltic Sea cod fishery. *Reviews in Fish Biology and Fisheries, 17*, 517-544.

Main, J., & Sangster, G.I. (1981). *A study of the fish capture process in a bottom trawl by direct observations from a towed underwater vehicle* (Scottish fisheries research report No. 23). Aberdeen: Department of Agriculture and Fisheries for Scotland, 23 pp.

Marçalo, A., Breen, M., Tenningen, M., Onandia, I., Arregi, L., Gonçalves, J.M.S. (this volume). Mitigating slipping related mortality from purse seine fisheries for small pelagic fish: Case studies from European Atlantic waters. In S.S. Uhlmann, C. Ulrich, S.J. Kennelly (Eds.), *The European Landing Obligation – Reducing discards in complex multi-species and multi-jurisdictional fisheries*. Cham: Springer.

Marchesan, M., Spoto, M., Verginella, M., Ferrero, E.A. (2005). Behaviour effects of atificial light on fish species of commercial interest. *Fisheries Research, 73*, 171–185.

Millar, R.B., Broadhurst, M.K., MacBeth, W.G. (2004). Modelling between-haul variability in the size selectivity of trawls. *Fisheries Research, 67* 171–181.

Mortensen, L.O., Ulrich, C., Qvist Eliasen, S., Olesen, H.J. (2017). Reducing discards without reducing profit: Free gear choice in a Danish result-based management trial. *ICES Journal of Marine Science, 74*(5), 1469–1479.

Murray, F., Copland, P., Boulcott, P., Robertson, M., Bailey, N. (2016). Impacts of electrofishing for razor clams (Ensis spp.) on benthic fauna. *Fisheries Research, 174*, 40–46.

Nguyen, K.Q., Winger, P.D., Morris, C., Grant, S.M. (2017). Artificial lights improve the catchability of snow crab (Chionoecetes opilio) traps. *Aquaculture and Fisheries, 2*(3), 124–133.

O'Connell, C.P., Stroud, E.M., He, P. (2014). The emerging field of electrosensory and semiochemical shark repellents: Mechanisms of detection, overview of past studies, and future directions. *Ocean & Coastal Management, 97*, 2–11.

O'Neill, F.G., & Mutch, K. (2017). Selectivity in trawl fishing gears Scottish Marine and Freshwater. *Science, 8*(1).

O'Neill, F.G., Lines, E.K., Kynoch, R.J., Fryer, R.J., Maguire, S. (2014). A short-term economic assessment of incentivised selective gears. *Fisheries Research, 157*, 13–23.

O'Neill, F.G., Summerbell, K., Barros, L. (2018). *Some recent trials with illuminated grids*. ICES WGFTFB 2018 REPORT. ICES CM 2018/EOSG:12.

Perez Comas, J.A., & Pikitch, E.K. (1994). The predictive power of empirical relationships describing size selectivity, with application to gadoid fish. *Fisheries Research, 20*, 151–164.

Piasente, M., Knuckey, I.A., Eayrs, S., McShane, P.E. (2004). In situ examination of the behaviour of fish in response to demersal trawl nets in an Australian trawl fishery. *Marine and Freshwater Research, 55*, 825–835.

Polet, H. (2010). Electric senses of fish and their application in marine fisheries. In P. He (Ed.), Behavior of marine fishes: Capture processes and conservation challenges. Oxford: Wiley-Blackwell.

Polet, H., Delanghe, F., Verschoore, R. (2005). On electrical fishing for brown shrimp (Crangon crangon) II. Sea trials *Fisheries Research, 72*, 13–27.

Popper, A.N., & Carlson, T.J. (1998). Application of sound and other stimuli to control fish behavior. *Transactions of the American Fisheries Society, 127*, 673–707.

Reid, D.G., Calderwood, J., Afonso, P., Fauconnet, L., Pawlowski, L., Plet-Hansen, K.S., Radford, Z., Robert, M., Rochet, M.-J., Ordines, F., Rueda, L., Mortensen, L., Ulrich, C., Vermard, Y. (this volume). The best way to reduce discards is never to catch them in the first place! In S.S. Uhlmann, C. Ulrich, S.J. Kennelly (Eds.), *The European Landing Obligation – Reducing discards in complex multi-species and multi-jurisdictional fisheries*. Cham: Springer.

Reinhardt, J.F., Weaver, J., Latham, P.J., Dell'Apa, A., Serafy, J.E., Browder, J.A., Christman, M., Foster, D.G., Blankinship, D.R. (2018). Catch rate and at-vessel mortality of circle hooks versus J-hooks in pelagic longline fisheries: A global meta-analysis. *Fish and Fisheries, 19*, 413–430.

Rihan, D. (2010). Measures to reduce interactions of marine megafauna with fishing operations. In P. He (Ed.), *Behavior of marine fishes: Capture processes and conservation challenges*. Oxford: Wiley-Blackwell.

Robbins, W.D., Peddemors, V.M., Kennelly, S.J. (2011). Assessment of permanent magnets and electropositive metals to reduce the line-based capture of Galapagos sharks, Carcharhinus galapagensis. *Fisheries Research, 109*, 100–106.

Rose, C., Stoner, A.W., Matteson, K. (2005). Use of high-frequency imaging sonar to observe fish behaviour near baited fishing gear. *Fisheries Research, 76*, 291–304.

Rosen, S., & Holst, J.C. (2013). DeepVision in-trawl imaging: Sampling the water column in four dimensions. *Fisheries Research, 148*, 64–73.

Ryer, C.H. (2008). A review of flatfish behavior relative to trawls. *Fisheries Research, 90*(2008), 138–146.

Ryer, C.H., Rose, C.S., & Iseri, P.J. (2010). Flatfish herding behaviour in response to trawl sweeps: A comparison of diel responses to conventional sweeps and elevated sweeps. *Fishery Bulletin, 108*, 145–154.

Santos, J., Herrmann, B., Mieske, B., Stepputtis, D., Krumme, U., Nilsson, H., (2016). Reducing flatfish bycatch in roundfish fisheries. *Fisheries Research 184*, 64–73.

Shephard, S., Goudey, C.A., Read, A., Kaiser, M.J. (2009). Hydrodredge: Reducing the negative impacts of scallop dredging. *Fisheries Research, 95*, 206–209.

Stroud, E.M., O'Connell, C.P., Rice, P.H., Snow, N.H., Barnes, B.B., Elshaer, M.R., Hanson, J.E. (2014). Chemical shark repellent: Myth or fact? The effect of a shark necromone on shark feeding behavior. *Ocean & Coastal Management, 97*, 50–57.

Struthers, D.P., Danylchuk, A.J, Wilson, A.D.M., Cooke, S.J., (2015). Action cameras: Bringing aquatic and fisheries research into view. *Fisheries, 40*(10), 502–512.

Thomsen, B., Humborstad, O.-B., Furevik, D.M. (2010). Fish pots: Fish behavior, capture processes, and conservation issues. In P. He (Ed.), *Behavior of marine fishes: Capture processes and conservation challenges*. Oxford: Wiley-Blackwell.

Trenkel, V.M., Handegard, N.O., Weber, T.C. (2016). Observing the ocean interior in support of integrated management. *ICES Journal of Marine Science, 73*, 1947–1954.

Utne-Palm, A.C., Breen, M., Løkkeborg, S., Humborstad, O.-B. (2018). Behavioural responses of krill and cod to artificial light in laboratory experiments. *PLoS ONE, 13*(1), e0190918. https://doi.org/10.1371/journal.pone.0190918

van Marlen, B., Wiegerinck, J.A.M., van Os-Koomen, E., van Barneveld, E. (2014). Catch comparison of pulse trawls and a tickler chain beam trawl. *Fisheries Research, 151*, 57–69.

Wardle, C.S. (1993). Fish behaviour and fishing gear. In T.J. Pitcher (Ed.), *Behaviour of teleost fishes* (pp. 609–643). London: Chapman & Hall.

Williams, K., Wilson, C.D., Horne, J.K. (2013). Walleye pollock (Theragra chalcogramma) behavior in midwater trawls. *Fisheries Research, 143*, 109–118.

Winger, P.D., Eayrs, S., Glass, C.W. (2010). Fish behavior near bottom trawls. In P. He (Ed.), *Behavior of marine fishes: Capture processes and conservation challenges*. Oxford: Wiley-Blackwell.

Yan, H.Y., Anraku, K., Babaran, R.P. (2010). Hearing in marine fish and its application in fisheries. In P. He (Ed.), *Behavior of marine fishes: Capture processes and conservation challenges*. Oxford: Wiley-Blackwell.

Yu, C., Chen, Z., Chen, L., He, P. (2007). The rise and fall of electrical beam trawling for shrimp in the East China Sea: Technology, fishery, and conservation implications. *ICES Journal of Marine Science, 64*, 1592–1597.

Chapter 15
Mitigating Slipping-Related Mortality from Purse Seine Fisheries for Small Pelagic Fish: Case Studies from European Atlantic Waters

Ana Marçalo, Mike Breen, Maria Tenningen, Iñigo Onandia, Luis Arregi, and Jorge M. S. Gonçalves

Abstract The release of unwanted catches (UWC) from purse seines, while the catch is still in the water, is known as "slipping". Once thought to be a benign process, compared to discarding UWC overboard from the fishing vessel, it is now recognised that "slipping" can lead to significant mortality in the released fish if done inappropriately. In this chapter, we examine purse seining and slipping operations, and discuss what drives slipping and potential mitigation measures to reduce slipping mortality. We use three examples of purse seine fisheries for small pelagic species in the North-east Atlantic; from Norway, Portugal and Spain. The ideal solution (identifying and avoiding UWC before the net is set) requires the development of tools to enable fishers to better characterise target schools in terms of key selection criteria, e.g., with respect to species, individual size and catch biomass. Such tools are being developed, based primarily on hydro-acoustic technology. However, some UWC in purse seine catches are inevitable, and operational improvements in slipping practices have been shown to significantly reduce stress and mortality in the released UWC. We conclude with a discussion on the challenges currently facing the implementation of the European Union (EU) Landing Obligation with regards to minimising slipping related mortality.

Keywords Bycatch and slipping mitigation · Delayed mortality · Pelagic fish · Purse seine

---

A. Marçalo (✉) · J. M. S. Gonçalves
Centro de Ciências do Mar-CCMAR, Universidade do Algarve, Faro, Portugal
e-mail: amarcalo@ualg.pt

M. Breen (✉) · M. Tenningen
Institute of Marine Research (IMR), Bergen, Norway
e-mail: michael.breen@hi.no

I. Onandia · L. Arregi
AZTI, Txatxarramendi ugartea z/g, Sukarrieta, Bizkaia, Spain

S. S. Uhlmann et al. (eds.), *The European Landing Obligation*,
https://doi.org/10.1007/978-3-030-03308-8_15

15.1 Introduction

Purse seining is a fishing method for targeting large (e.g. tunas) and small species (e.g. mackerels, sardines and anchovies) that school or aggregate close to the surface. It has been the most productive fishing method throughout the world for the past six decades, accounting for approximately one third of the global catch by weight (Watson et al. 2006). Incidental catches of dolphins first raised awareness of bycatch issues in tuna purse seine fisheries; although unwanted catches (UWC) of some species and sizes of teleosts, including undersized tunas and some species of elasmobranchs, are also common for these fisheries (Hall et al. 2000; Kelleher 2005; Megalofonou et al. 2005; Hall and Roman 2013). Purse seine fisheries target catches of low species and size diversity, which contributes to a sporadic occurrence of UWC (Broadhurst et al. 2006; Borges et al. 2008), but reported rates of UWC are usually low, and spatially and temporally variable (i.e., 3.5% for tuna fisheries, Gilman et al. 2017; and 1.6–27% for small pelagic fisheries, Kelleher 2005; Borges et al. 2001).

Purse seines, particularly for small species, are generally considered to be a non-selective fishing gear once a target school has been encircled, primarily because of the small mesh sizes used in the main body of the net, typically < 20 mm. Therefore, the release of UWC generally happens in one of two ways: firstly, by "slipping" all or part of the UWC out of the net while it is still in the water; or secondly, by "discarding", when the catch is taken aboard and any unwanted components are removed and returned to the sea alive or dead. As with other fisheries, "discarding" is associated with an array of potentially fatal stressors for the affected animals, and the likelihood of survival is generally assumed to be low (e.g. Davis 2002; Breen et al. in review).

Conversely, because the catch never leaves the water, "slipping" was once assumed to be a benign method of releasing UWC from the net, without harming it. However, experiments have demonstrated that "slipping" of small pelagic species like mackerel (*Scomber scombrus*) (Lockwood et al. 1983; Huse and Vold 2010), herring (*Clupea harengus*) (Tenningen et al. 2012), sardine (*Sardina pilchardus*) (Marçalo et al. 2006; Marçalo et al. 2010) and sardinops (*Sardinops sagax*) (Mitchell et al. 2002) may result in unacceptably high rates of mortality. Some of this research has shown that during the final phase of the capture process, the catch can become highly crowded, with densities >250 kg.m^{-3} (Tenningen et al. 2012), and can be exposed to potentially fatal stressors, including hypoxia, exhaustion and physical injury from contact with the net and catch. This established that the mortality of slipped fish is directly related to their treatment within the net, with mortality increasing with increasing crowding densities and crowding time (Lockwood et al. 1983; Tenningen et al. 2012; Marçalo et al. 2010). This concern about slipping-related mortality has led to recent regulations in some European fisheries that ban the practice of slipping, unless the released fish can survive [e.g. EU Landing Obligation (EU 2013); Norwegian Seawater Fisheries Regulations (NSFR 2014].

In this chapter we review purse seine fishing targeting small pelagic species in the North East Atlantic, with reference to mitigating slipping-related mortality. We will use three different purse seine fisheries (Norwegian coastal and offshore; Spanish Cantabrian Sea and North-western waters; and Portuguese sardine fishery) as case studies to explore the following topics:

- Case study overviews including fishing operations and slipping practices;
- Mitigation measures to minimise slipping related mortality; and
- Challenges currently facing the implementation of EU Landing Obligation with regards to minimising slipping related mortality.

15.2 Purse Seine Fishing Targeting Small Pelagic Species in the Atlantic

15.2.1 Overviews of Case Studies

Three case studies of purse seine fisheries for small pelagic species in the NE Atlantic are briefly reviewed here.

15.2.1.1 Norwegian Inshore and Offshore Fisheries

Close to 400 purse seine vessels operate in Norway (Table 15.1), which account for about 30% of Norway's total annual landings (Norwegian Directorate of Fisheries 2018). Most of the vessels are relatively small coastal purse seiners (15–55 m length), but about 65% of the catch is taken by 73 large ocean-going purse seiners (Table 15.1). Mackerel, herring and capelin (*Mallotus villosus*) are the main target species, and in 2017 about 800,000 tonnes were captured (of which 63,000 t taken in the EU Zone) with a landed sale value of €490 M.

15.2.1.2 Spanish Cantabrian Sea and North-West Fisheries

Purse seine landings in this fishery are estimated at 170,450 tonnes, representing 37% of the total landings by the fleet operating in EU waters (STECF 2017), with a value of more than €140 M. Purse seining employs 6276 fishers out of a total of 28,078 employed in fisheries in Spanish waters (MAPAMA 2017). Five main Spanish purse seine fisheries can be identified in the fishing area, here we focus on the purse seine fisheries in the Cantabrian Sea and North-west targeting small pelagics (Table 15.1).

Table 15.1 Purse seine fleet and gear characteristics from case studies

Country – fleet	No of vessels	Hold capacity (tonnes)	Power (HP) mean (range)	LOA (m) mean (range)	GRT (tonnes) mean (range)	Net length (m)	Net depth (m)	Min mesh size (mm)	Refs
Norway – offshore	73	≤ 1500	4660	63 (45–80)	~2500 (1000–4000)	~400–850	~100–265	15–18	a, b
Norway – inshore	~300	≤ 500	~500	~35 (15–55)	~300 (~100–500)	~400–700	~100–150	15–18	a, b
Spain – Cantabrian Sea and NW waters	262	≤ 100	330 (210–780)	23 (12–36)	83 (52–247)	≤ 600	120	14	c, d
Portugal – mainland	180	≤ 20	265 (15–600)	17 (6–27)	30 (5–95)	≤ 800	≤ 120	18	e, f

Sources: a – Norwegian Fisheries Directorate DataBase 2018; b – NSFR 2014; c – MAPAMA 2017; d – Spanish regulation for purse seine gears, 2015; e – DGRM 2017; f – Marçalo 2009

15.2.1.3 Portuguese Sardine Fishery

The Portuguese sardine fishery accounts for about 50% by weight of catches landed in the country's mainland ports (DGRM. DATAPESCAS 2017). The fishery is coastal (operating within the continental shelf), and the fleet is comprised of 180 vessels (Table 15.1). Sardine is the target species and accounts for more than 90% of total catch weight and value (Silva et al. 2015). Horse mackerel (*Trachurus trachurus*), chub mackerel (*Scomber colias*) and anchovy (*Engraulis encrasicolus*) account for a smaller part of the landings (Stratoudakis and Marçalo 2002; Feijó et al. 2018). Landings in 2017 were 45,488 tonnes with a value of more than €51 M (DGRM. DATAPESCAS 2017).

15.2.2 Purse Seine Fishing Operations

Once a potential target school has been located, typically using hydro-acoustic fish detection technology (e.g. sonar and echo-sounders), it is quickly surrounded by a wall of netting. The net is supported at the surface with a float-line, while at the bottom, a heavy, "leaded" rope and a series of heavy metal rings ("purse rings") ensures the net sinks quickly around the catch (Fig. 15.1a). The purse seine is typically set rapidly (~10 mins) by a single vessel, although in Portugal a small auxiliary vessel ("skiff"; 6–7 meters) assists in the operation. When the net has been fully deployed, or "set", a wire rope, or "purse line", that passes through the purse rings is then pulled into the vessel, closing the net beneath the school to prevent the fish from diving and escaping (Fig. 15.1b). The net is then gradually hauled in using hydraulic winches (e.g. a Triplex Power block), which progressively reduces the volume contained by the net, and crowds the catch into the bunt end (Fig. 15.1a) until the density becomes sufficiently high to inspect what is in the catch. In Portugal and Spain this final stage of hauling, as the fish are crowded in the bunt, is often done manually (i.e., with the crew hauling the net aboard the vessel by hand). After the catch is inspected, if the decision is made to harvest it, it is then transferred into the boat; either by brailing (i.e., using large dip nets called brails in Portugal and Spain) or using fish pumps (in Spain and Norway), while the rest of the catch is left in the water.

There are substantial differences between fleets with respect to individual catch sizes, as well as vessel and gear dimensions (Table 15.1). Catches in the Spanish and Portuguese fisheries are typically around 3–5 tonnes, but can exceed 10 tonnes, while in the Norwegian fisheries they are typically much larger (50–500 tonnes) and can exceed 1000 tonnes. The most striking difference between the fleets is the size range and hold storage capacity of the vessels, with the Norwegian offshore fleet dwarfing those of the other fleets. In contrast, differences in net dimensions among fleets are comparatively small.

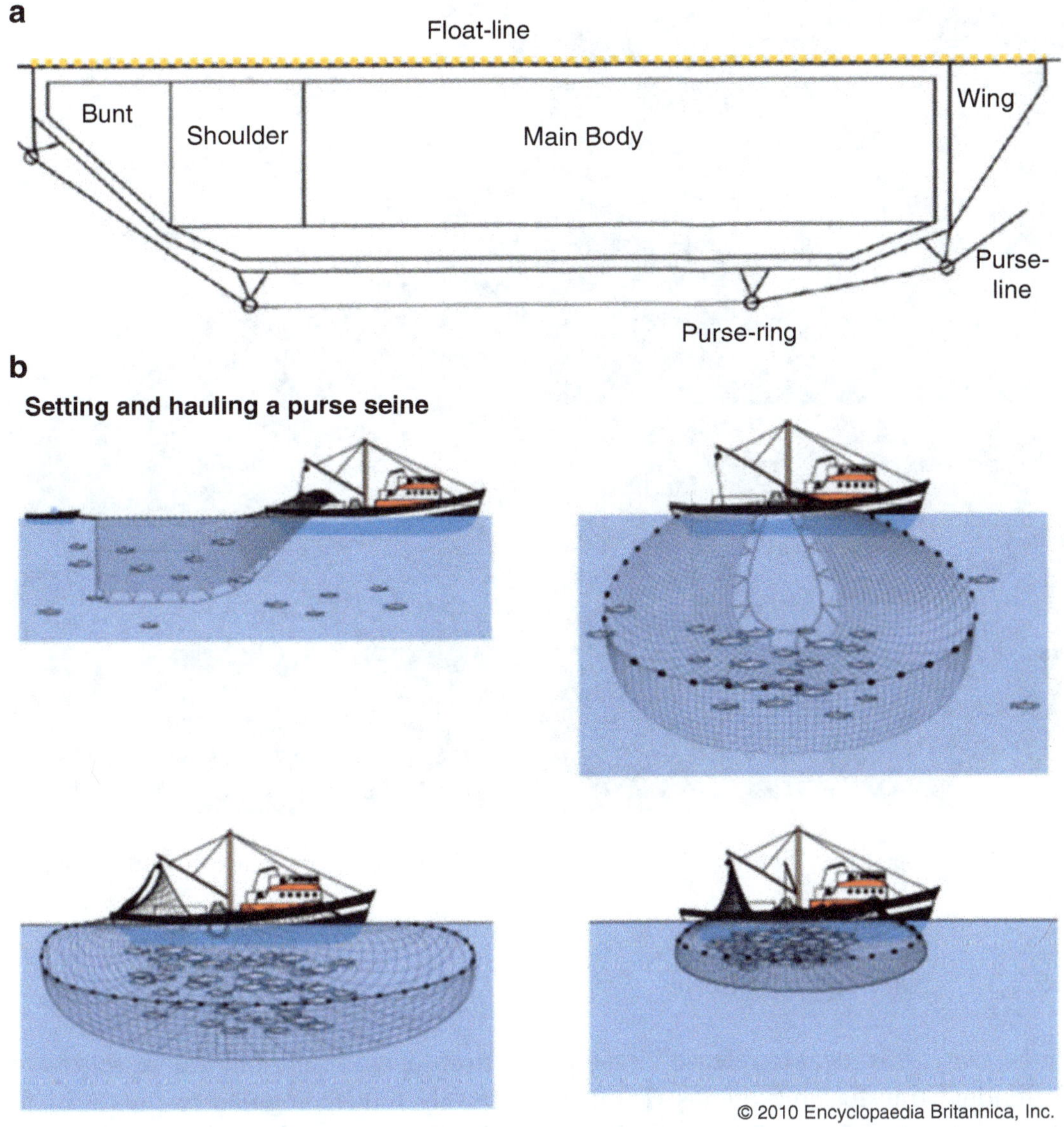

Fig. 15.1 Diagram of a purse seine net: (**a**) principal components) and (**b**) purse seining operations: setting the net and hauling

15.2.3 Slipping Practices – Drivers and Methods

As with many fisheries, UWC in purse seine fisheries can result from a variety of economic (i.e., catch quality, market price/demand) and regulatory (i.e., quotas, sizes, protected species) drivers that may result in the catch being slipped. In addition, due to the dimensions of the net and the nature of the target species to form large and sometimes dense schools, purse seining can take catches far in excess of the holding capacity of the fishing vessel. This issue is particularly relevant for smaller vessels, which may only have the capacity to retain a proportion of the catch in the net (Table 15.1).

Fig. 15.2 Slipping operations in different purse seine fisheries: (**a**) Norwegian; (**b**) Portuguese; (**c**) Spanish

15.2.3.1 Norwegian Fisheries

In Norway, it is legal to release viable UWC from purse seine, but the vessels have individual transferrable quotas (ITQ) to purposely reduce regulatory pressures to discard (Gullestad et al. 2015). Furthermore, all pelagic fish sales are controlled by a single authority ("*Norges Sildesalgslag*"), so landings (including bycatch) from individual boats are closely monitored and controlled. While economic drivers may also influence slipping practices, it is generally felt that at present the main driver for slipping in the Norwegian fleet is catch size, particularly amongst the coastal fleet. However, there are currently no reliable estimates of the magnitude or frequency of these slipping events.

With regard to slipping practices, fishers' behaviour is now, in principle, driven by regulations governing such practices in Norwegian and EU waters. Before the introduction of these regulations, unwanted or excessive catches were released late in the capture process, when the catch was very crowded, by slipping over the float-line, or by partially opening the bunt-end as the net was hauled aboard (Fig. 15.2a). In 2014, regulations were introduced to manage slipping practices in mackerel fisheries in Norwegian waters. The regulations prescribe that the seine must be prepared for release of UWC (i.e. bunt-end opened) before 7/8 of the net length is hauled (marked by a visible float), and UWC must be released through an opening sufficiently large to permit the fish to swim out freely (NSFR 2014; Rule §48a). Following the introduction of the EU Landing Obligation (LO), slipping practices in EU waters were regulated from 2015 by Commission Delegated Regulations (CDRs), for both North-Western Waters and the North Sea (EU 2014a, b). These enact High Survival Exemptions (HSE) to the LO (see Chap. 3 for more details; Rihan et al., this volume) for herring and mackerel fisheries, provided the catch is released before a certain proportion of the net is hauled, referred to as the "point of retrieval", (also marked by a visible float); where the limits are 80% of the net for mackerel; and 90% for herring. Furthermore, the vessel and purse seine should be equipped with electronic instruments to document when, where and the extent to which the purse seine has been hauled.

15.2.3.2 Spanish Fisheries

In the Cantabrian and NW Spanish waters, anchovy is slipped due to the low market price of small-sized individuals and fish under the Minimum Conservation Reference Size (MCRS), while mackerel and horse mackerel are slipped when their quotas are exhausted. On the other hand, sardine (ICES divisions 8a, b and d) and Atlantic chub mackerel have no quota limitations, but they may be discarded due to low market prices. From observer data in these fisheries, the frequency of slipping events was estimated to be 8.3% (17 slipping events in 204 sets) (Arregi et al. unpublished data).

The South Western Waters' purse seine fisheries (Portuguese, Spanish and French) are permitted to slip several species (mackerel, horse mackerel, and anchovy) under a High Survival Exemption (HSE) in the EU Commission Delegated Regulation (CDR) for South Western Waters (EU 2014c; Rihan et al., this volume). Furthermore, the conditions of this HSE recognise the small-scale nature of purse seine operations in these fleets and is far less prescriptive in its conditions for slipping practices, stating: "[. . .] catches may be released, provided the net is not fully taken aboard". The slipping method differs from other fisheries, where fishers will roll the fish over the headline. Here, the bunt and the first 5 to 10 pursing rings are detached and the catch is released before it becomes too crowded (Fig. 15.2c). Catches from this fleet are used fresh, for human consumption, so quality is an important factor. This influences how the fishers handle the catch, where crowding density is kept under about 80 kg of fish per cubic metre to avoid abrasion and crushing of the catch.

15.2.3.3 Portuguese Sardine Fishery

The Portuguese sardine fishery is currently affected by a historically low spawning stock biomass for the southern-Iberian stock, which is below safe biological limits (Silva et al. 2015). This led to strict management measures (e.g. seasonal ban and daily quotas per vessel), which have been applied by the Portuguese government since 2012 (Silva et al. 2015) and ICES recommending a zero TAC since 2018 (ICES 2018). As with Norway, all landings and sales are handled through a single national authority (DocaPesca). So, during the sardine ban, fishers are targeting other pelagic species, including Atlantic chub mackerel and horse mackerel; and slipping sardine, if caught. Conversely, during the open sardine season there is an increased slipping of other species (due to the high market prices for sardine), but also excess sardine, because daily sardine quotas are very small (usually around 1.5–2.0 tonnes per vessel). However, neither sardine nor chub mackerel are currently listed amongst the species in the HSE for SW waters (EU 2014c). Other drivers for slipping in this fishery may be: vessel capacity the presence of non-commercial species, undersized fish, or a mixture of species, which will devalue the catch at auction (Stratoudakis and Marçalo 2002; Marçalo 2009; Feijó et al. 2018). The slipping practice typically occurs at the very end of the fishing operation and involves rolling the fish over the float-line (Fig. 15.2b).

15.3 Mitigation Measures to Reduce Slipping Related Mortality

To effectively reduce slipping-related mortality, it is necessary to release any unwanted catch as early in the capture process as possible, before the fish become fatally stressed. To do this the fisher requires tools and methods to: (a) characterise the potential catch so that the decision to take or release it can be made as early in the capture process as possible, ideally before the net has even been set; and, (b) where slipping is unavoidable, release any unwanted catch in a controlled way with minimal stress to the fish (Breen et al. 2012; Breen et al. in review; CRISP 2018). Here, we describe research in three main areas of development that aims to provide such tools: (1) pre-catch identification of fish schools (with respect to species, quantity and fish size) using hydro-acoustic methods to prevent catching unwanted fish; (2) monitoring the catch and net during the haul to provide information on the catch density, fish size and quality at an early stage in the haul, while slipping is still acceptable; and (3) modifications to purse seine design and practices to promote the survival of slipped fish (Fig. 15.3).

15.3.1 Pre-Catch Identification – Minimize the Need for Slipping

Skippers use experience and knowledge about the behaviours of different species to evaluate school size and species based on received echoes on their sonar and echo-sounder screens. However, having accurate quantitative estimates of school characteristics will further improve catch estimation and reduce UWC. Furthermore,

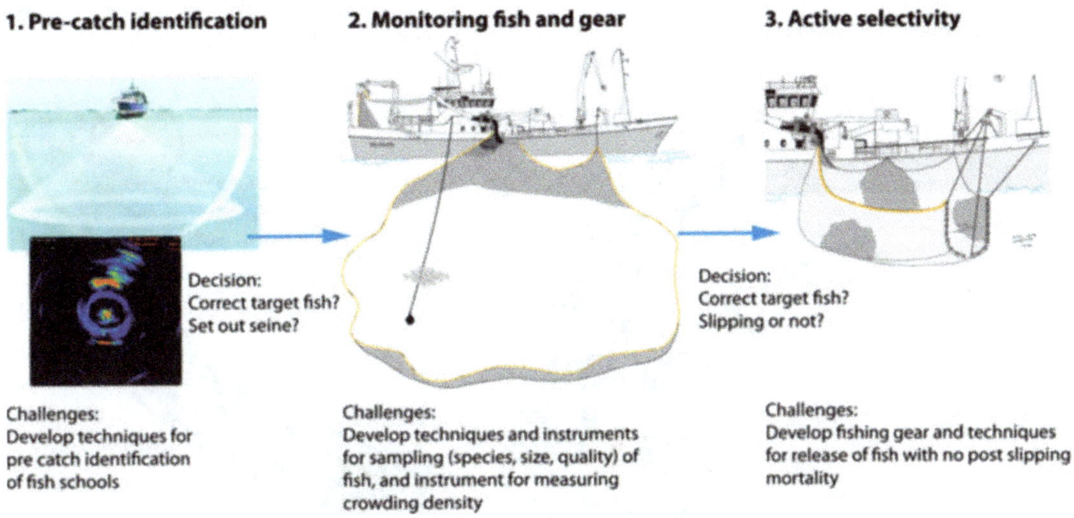

Fig. 15.3 The three-stage strategy to provide purse seine fishers with the tools and methods necessary for avoiding unwanted catches and reducing slipping-related mortality. (From Breen et al. 2012)

avoiding UWC can have significant economic benefits for fishers, through reduced fuel costs and improved catch quality and prices (Larsen and Dreyer 2013). Information about the species in a school, school morphology and geographical distribution can, to some degree, be estimated using multi-frequency echo-sounders (Horne 2000; Korneliussen et al. 2009). The echo strengths at different frequencies are species-specific, due to variation in fish morphology (e.g. presence or absence of a swimbladder) and the relative frequency response $r(f)$, i.e. the ratio of the backscattered energy at frequency f to that at 38 kHz, can be used to distinguish between some species. Individual fish size within a school can also be estimated using a high resolution broadband echo-sounder, if individual targets can be detected (CRISP 2018). In recent years, significant progress has also been made in using multi-beam sonars to quantify fish school sizes (Nishimori et al. 2009; Vatnehol et al. 2017) and behaviour (Gerlotto and Paramo 2003; Holmin et al. 2012). In Norway, research and development in hydro-acoustic pre-catch identification is a well-functioning cooperation between research institutes, the fishing industry and companies delivering fisheries instrumentation (e.g. CRISP 2018; LSSS 2018; DABGRAF 2018; SEAT 2018).

15.3.2 *Early-Catch Monitoring*

Nevertheless pre-catch identification is not always accurate, especially when schools are large and dense. So, it is also necessary to have tools to monitor and characterise the catch early in the capture process before the fish become too crowded in the net. However, monitoring a school inside the net is challenging, even using acoustic

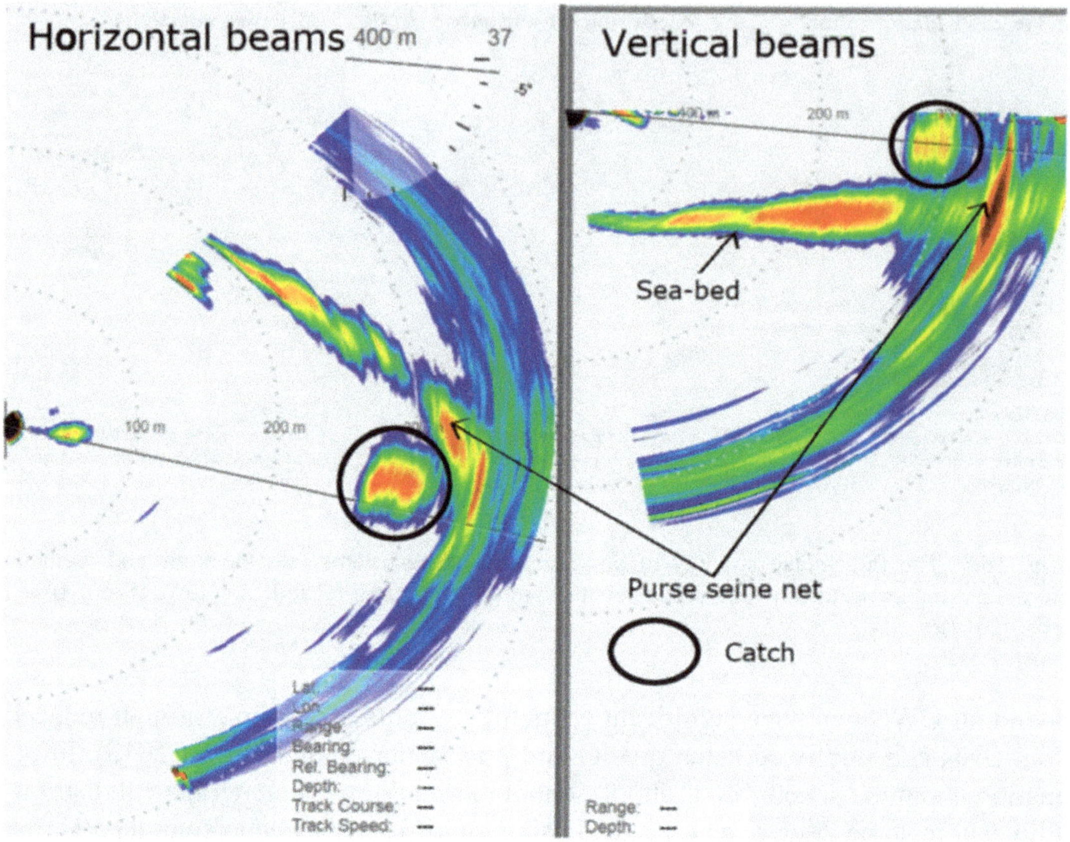

Fig. 15.4 Image from the Simrad SN90 sonar (Kongsberg Maritime AS) of a school of North Sea herring in a purse seine, with the wall of the net clearly visible. Left panel – horizontal view; right panel – vertical view

technologies. Omnidirectional sonars are usually retracted into the hull during purse seining to avoid damage, making them unsuitable for monitoring schools during capture. But multi-beam sonar, mounted on a research vessel, has been used to monitor and describe the behaviour of schools captured by purse seine (Tenningen et al. 2017). Multi-beam sonars on fishing vessels, with side-looking transducers are now commercially available (e.g. Kongsberg Maritime SN90) so work is in progress to obtain a better understanding of fish behaviour, densities and school biomass inside the seine (Fig. 15.4).

Multi-beam sonar has also been used to describe purse seine shape and volume during seine hauling (Tenningen et al. 2015). This work has provided a better understanding of how the volume available for captured fish schools varies under different fishing conditions and the impact that may have on the survival of slipped fish.

In addition to acoustic methods, efforts have been made to develop tools for obtaining sub-samples of catches, monitoring the catch visually and collecting data on environmental conditions in the net. However, this is technically challenging because the seine is large and dynamic, making it difficult to attach any monitoring instruments to it directly. Examples of promising methods include: a small sampling

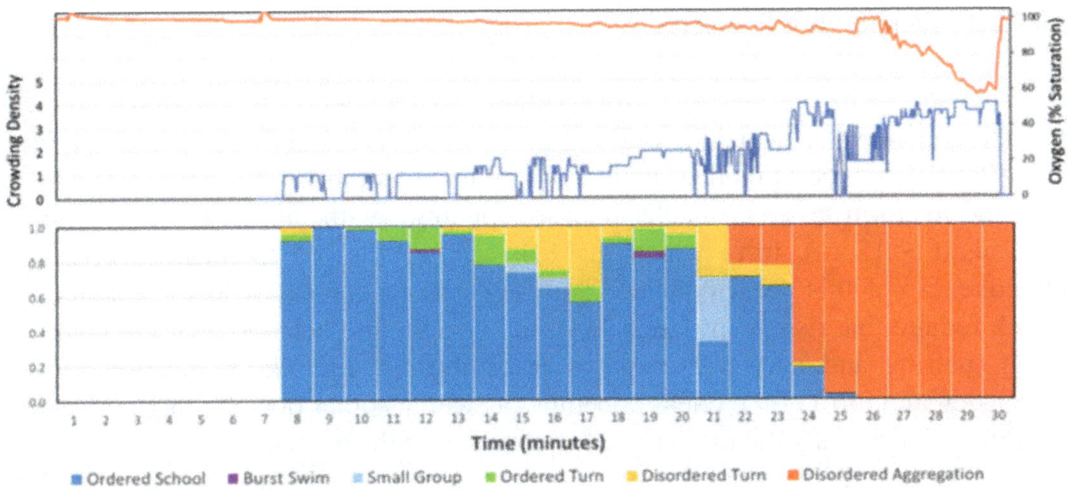

Fig. 15.5 An example from a single commercial purse seine cast showing that crowding density (blue line; ordinal score) and dissolved oxygen concentration (red line; % saturation) (top), and behaviour (below), changed over time (Behaviour summarised in 1 min bins). (From Breen et al. in prep)

trawl, deployed using a pneumatic canon (Isaksen 2013; Peña et al. 2018); a monitoring probe deployed in the same way and equipped with cameras and instruments for measuring oxygen, temperature and depth (Breen et al. in prep); and measuring the size frequency distribution inside the catch with a stereo-camera (see SINTEF FKiN Project 2018).

15.3.3 Late Capture and Release

Research has demonstrated that slipping unwanted catches from the purse seine can be a responsible catch control practice, if done before the captive fish become too crowded and in a way that maintains their ordered behaviour. For example, in controlled experiments, mackerel could tolerate moderate crowding densities (~88 kg m^{-3} for up to 1 h) and relatively low oxygen concentrations (~40% saturation) without significant mortality (as observed up to 8 days after stressor treatment) (Handegard et al. 2018). Furthermore, preliminary observations suggest that such conditions do not develop in commercial catches until late in the haul-back phase, particularly in slipped catches (Breen et al. in prep; Fig. 15.5). In addition, changes in behaviour have been observed at sub-lethal and potentially lethal crowding densities that could be used as potential indicators of stress in the catch, but practical challenges for effectively monitoring such indicators remain (Breen et al. in review).

Amongst our case studies, there are several examples of research demonstrating the effectiveness of good slipping practices in reducing stress and promoting survival in released catches.

15.3.3.1 Norwegian Fisheries

A practical code of "best practice" for conducting slipping operations was developed in collaboration with the fishing industry and Norwegian Fisheries Directorate (Vold et al. 2017). This "best practice" includes recommendations for using the bunt-end of the net to form a controllable release opening (with minimum dimensions, i.e. length > 18 m), from which the fish can be allowed to swim freely. The effectiveness of this "best practice" was assessed in context with the regulation, which require that fish swim freely from the net. So, the behaviour of the released catch (herring and mackerel) was observed during slipping, in relation to the dimension and form of the release opening, as well as other operational parameters. It showed that initially the fish were reluctant to leave the net but eventually, once the catch began to swim out of the net, they typically retained an ordered schooling behaviour. However, some disordered behaviour was observed as well, and this typically occurred later in the slipping process and was more likely to be seen in larger catches (Vold et al. 2017).

15.3.3.2 Spanish Cantabrian and NW Fisheries

To assess whether different components of UWC could be released without a significant slipping mortality, the survival of several species (mackerel, horse mackerel, anchovy, sardine and Atlantic chub mackerel) was assessed after exposing them to different crowding periods (0–50 min) during commercial purse seine operations. Several experiments were done (in 2013/14) aboard vessels that alternated purse seine with pole and line targeting tunas during summer. These vessels were equipped with several large water tanks (~10m^3, water replacement rate 400 l.min^{-1}) specifically designed to maintain live bait (caught with the vessel's purse seine gear) and thus provide a useful facility for this research. Sub-samples of the catch, using a fish pump, were taken at different time intervals after the catch had been crowded in the bunt, and then behaviour and survival of the fish was monitored in the "live-bait" tanks for between 2 and 6 days. The results generally showed high survival rates (horse mackerel: 89.7–100%; anchovy: 54.2–97.8%; sardine: 83.9–100%; and Atlantic chub mackerel: 100%); but for mackerel a large range: 3–100%. These results were presented as supporting evidence for a successful application for a HSE (STECF 2014; EU 2014b). However, there was considerable variability, particularly with respect to species, crowding time, crowding density and catch size (Fig. 15.6). Furthermore, following advice from the ICES Workshop on Methods for Estimating Discard Survival (WKMEDS; ICES 2016) work is ongoing to substantiate these results with experiments using longer monitoring periods (up to 20 days), to avoid underestimating mortality (ICES 2018; in press; see also Rihan et al., this volume).

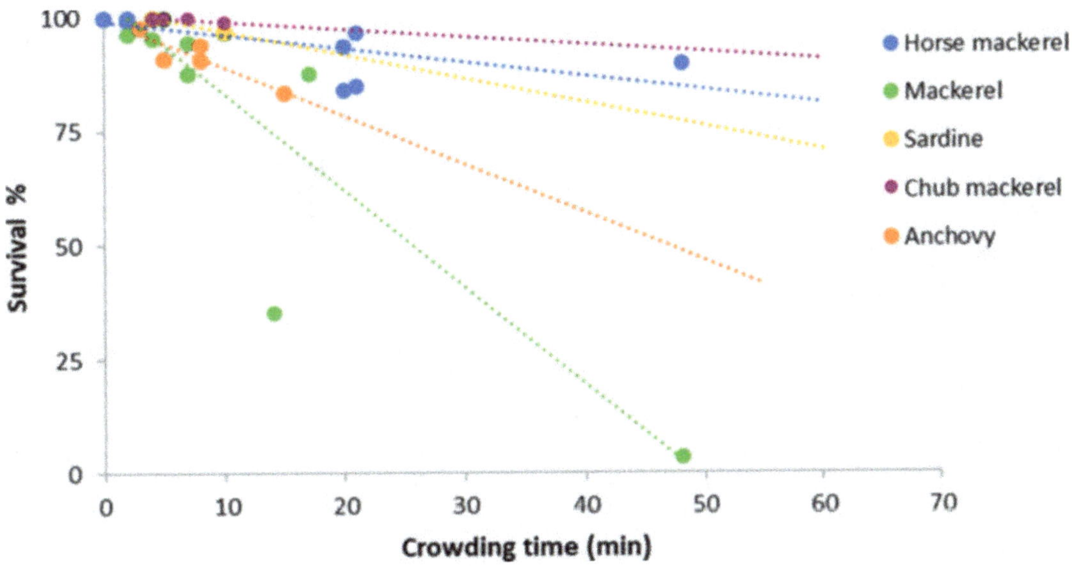

Fig. 15.6 Post-crowding survival of five small pelagic species (mackerel, horse mackerel, chub mackerel, anchovy and sardine) subsampled from Spanish purse seines after different time intervals (0–50 min). (From Arregi et al. unpublished data)

15.3.3.3 Portuguese Sardine Fishery

Previous research in this fishery showed that survival of released sardines was likely to be greater during earlier stages of the hauling phase, when crowding was less (Marçalo et al. 2010). During meetings with fishers to discuss practical methods to mitigate the slipping problem, it was suggested that during the closed season, sardines could be released from the remainder of the catch through an opening created by putting weights over the float-line. This utilised differences in the behaviour of different species in the catch to selectively release the sardines. That is, sardines when in a mixed catch with other small pelagic species, usually swim close to the surface, while other species (e.g., chub mackerel) swim down in the net. Experiments were done to assess the effectiveness of this method in promoting the survival of slipped sardines, compared to the standard method of rolling the fish over the float-line and a control (non-slipped and non-crowded sardines) (Fig. 15.7; Marçalo et al. 2018). After transferring samples of released and control fish to onshore aquaria, survival rates were monitored in captivity for 28 days. Survival (at asymptote) of sardines in the three replicates from the standard slipping was low (12.8%; 8.9–15.2 at 95% CI), however the modified slipping procedure did significantly improve survival (survival at asymptote of 44.7%; 39.3–50.1% at 95% CI), which was comparable to the control fish (survival at asymptote of 43.6%; 38-0–49.3 at 95% CI).

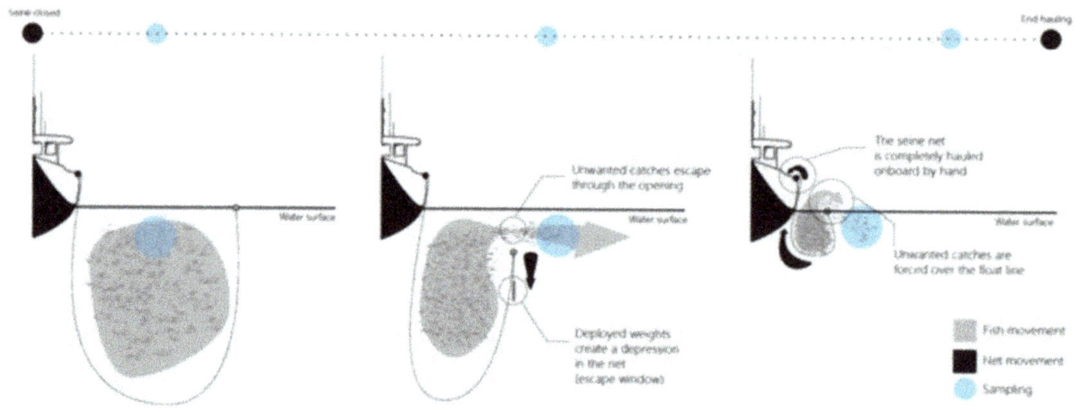

Fig. 15.7 Diagram of sampling at sea for the control, modified and standard slipping techniques in the Portuguese purse seine MINOUW case study (in Marçalo et al. 2018)

15.4 Challenges for the Landing Obligation in Regulating Slipping in Purse Seine Fisheries

In this chapter, we have reviewed slipping in three NE Atlantic purse seine fleets operating in small pelagic fisheries. This has shown that there are several issues driving slipping practices, including economic (e.g., market value) and regulatory (e.g., quotas, MCRS) pressures. For purse seine fishing, there is an additional driver, i.e., the capacity for vessel to hold excessively large catches. Furthermore, there are substantial differences across the described fleets with respect to individual catch composition and sizes, vessel power and capacity. The Norwegian offshore fleet is characterised as having relatively new and large vessels and has invested considerably in the latest fish finding (hydro-acoustic), gear handling and catch storage technologies, compared to the smaller vessels in the Norwegian (coastal), Portuguese and Spanish fleets. This diversity in fishing practices, resources and investment, as well as regional economic and social difficulties, will differentially affect slipping practices and has been cited as a major challenge for introducing the Landing Obligation in some EU Member States (Veiga et al. 2016; Maynou et al. 2018).

Various operational and technological solutions are described in this chapter (some still being developed) which have the potential to promote the survival of unwanted catches released from purse seine fisheries. As with many fisheries, the ideal solution for dealing with UWC is to avoid catching it in the first instance (see O'Neill et al., this volume; Reid et al., this volume). In purse seining, this means providing the fisher with the tools to characterise the catch (in terms of species composition, quality/size range and catch volume) before setting the seine, or at least early in the capture process. For this, hydro-acoustic technologies are being examined as the most promising technological solution, with the potential for describing species composition, size frequency distribution and catch biomass. However, we also identified several novel methods (e.g., the canon-deployed sampling trawl) that could provide "low tech" and more affordable solutions for at least partly characterising the catch before it is fully crowded next to the boat.

In most fisheries at present it is not until the catch is inside the net, and at least partially crowded, that the fisher has enough information to be able to decide to bring the catch on board or not. If the fisher is legally obliged to take the catch onboard, even if it entirely or partly consists of UWC, this can present him/her with several challenges, particularly if the vessel is small. Firstly, assuming the vessel has the capacity to take the catch onboard, storage space may be limited making it difficult to keep the UWC separate from the marketable catch, as required by the LO (Villasante et al. 2016a, b). Furthermore, there is currently no on-shore infrastructure for accommodating and processing the UWC – at least in most southern European Member States (Veiga et al. 2016; Maynou et al. 2018). More critically, if a vessel does not have the capacity to take all of the catch on board, the fisher is presented with a serious dilemma that could threaten the safety of his/her vessel and crew. There are examples of fisheries where the catch in such cases is shared between nearby vessels (e.g., in Portugal and Spain; Feijó et al. 2018). However, the delay associated with transferring the catch to other vessels has been linked with a substantial reduction in catch quality, and hence price (Digre et al. 2016). Thus, this presents the manager with the challenge of how to regulate/incentivise such practices to ensure the vessel receiving the excess catch is suitably compensated.

Several studies in this chapter have shown that, provided that the catch does not become too crowded, it can be released in a viable state with a high likelihood of survival. Each of these methods relies on providing a suitably large opening in the net to allow the fish to swim out before they become too crowded and for too long (Fig. 15.8). Furthermore, in the case of the Portuguese fishery, it was also shown that sardine could be released selectively, while retaining other components of the catch, because of behavioural differences between species when captive in the net (Marçalo et al. 2018). If the whole catch is to be released, this is a relatively simple operation, as demonstrated in the Spanish fishery, whereby the bunt end/wing is opened and several purse rings are released. However, if only a proportion of the catch is to be released, more control over the opening and release process is required. Furthermore, there is likely to be a critical point during the slipping process, after which the fish have become too stressed and their survival will be compromised. How this release process can be effectively and safely monitored and controlled to ensure that only viable fish are released remains a considerable challenge.

The EU Landing Obligation (LO) recognises that regional and fisheries specific differences in UWC will require tailored solutions, which can be facilitated through Commission Delegated Regulations (CDRs), and include exemptions like the High Survival Exemption (HSE) (Rihan et al., this volume). Particularly relevant to the slipping problem is the HSE, because if a fishery can demonstrate to the EU Commission that any released UWC has a high likelihood of survival, it may be exempted from the LO. Comparable regulations also apply in Norway (Karp et al., this volume). Unfortunately, there are disparities in some of these regulations that present fishers with some considerable challenges. For example, in the EU CDR for pelagic fisheries in SW waters (EU 2014a, b, c), a HSE has been granted to release

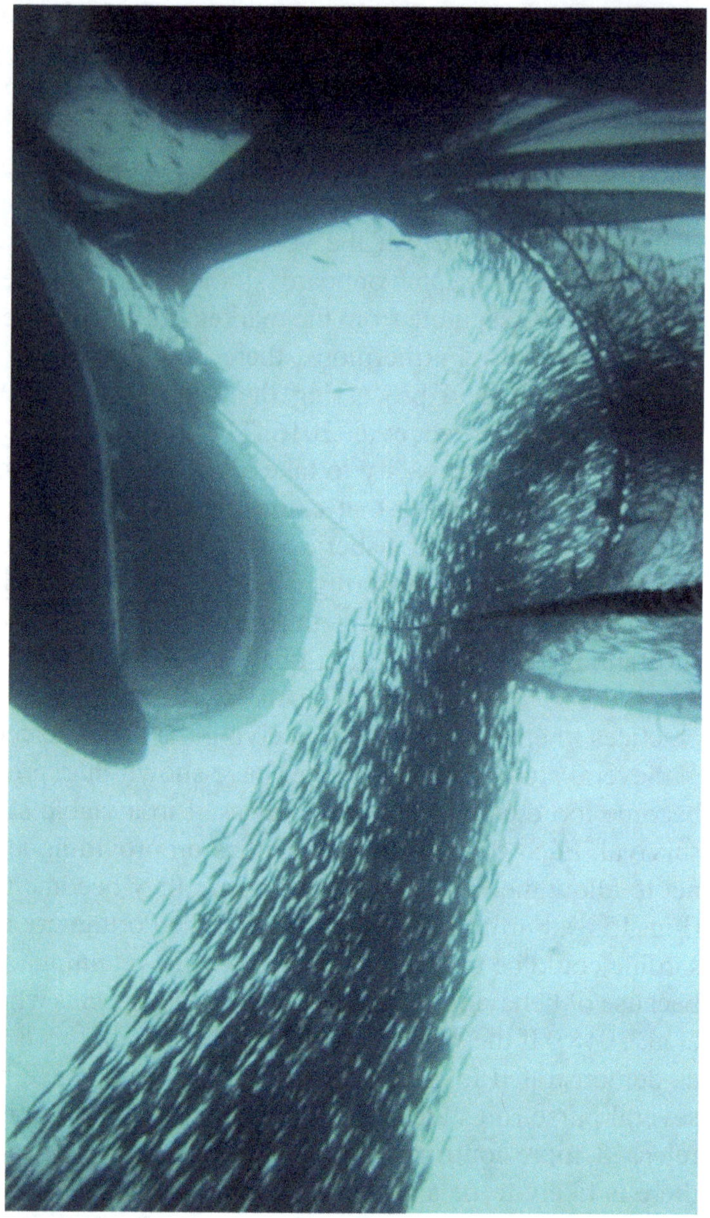

Fig. 15.8 View from beneath the vessel as a school of herring swims out of a purse seine during slipping. [Source: IMR]

unwanted catches of anchovy, mackerel, horse mackerel and jack mackerel. Excluded from this exemption is sardine, despite data being available when the HSE was first proposed, which demonstrated high survival (83.9–100% up to 6 days post-treatment; Arregi et al. 2014). The spawning stock biomass for sardine in Cantabrian Sea and Atlantic Iberian waters (ICES divisions 8.c and 9.a) has been below safe biological limits since 2009, with ICES currently advising a zero TAC (in 2018) and Portuguese fisheries having at least a partial (seasonal) ban on catches. Therefore, although it would clearly be beneficial for this stock, fishers who responsibly release catches of sardines would be in breach of the LO. In another example, the EU CDR for pelagic fisheries in NW waters and the North Sea (EU 2014a, b) permits the slipping of mackerel and herring provided the release is completed before 80%/90%, respectively, of the net has been hauled (“the point of retrieval”). This contrasts with the Norwegian regulations for the mackerel fishery, which (following consultation with fishers and researchers) stipulate that the release opening must have been prepared before a comparable “point of release” (87.5% [7/8th] of the net) but that the release may continue beyond this point; enabling the fishers to assess their catch and control the slipping operation. In addition, the Norwegian regulations stipulate how the catch should be released, i.e. through an opening sufficiently large to allow the released fish to swim out freely (NSFR 2014; Rule §48a). These are practices that are already used in the Norwegian inshore purse seine fishery for transferring catches into holding cages (Breen et al. 2012) and that have been shown in work reviewed in this chapter to promote survival.

Another major challenge, shared by many fisheries, is monitoring fishing practices and ensuring compliance with the regulations (discussed in detail by Nuevo et al. (this volume)). Reliable estimates of slipping rates and any associated mortality would enable fisheries managers to account for this additional fishing mortality in stock assessments and any resultant advice on catch limits (e.g. Breen and Cook 2002; Mesnil 1996), as well as monitor the effects of slipping regulations on fishing practices. However, there are currently no known monitoring programmes targeting slipping practices. Furthermore, slipping practices will prove particularly challenging to monitor, because the catch is not taken aboard the vessel before it is released. Effective monitoring is likely to require on-board observers and/or electronic monitoring (EM). Even then methods for reliably characterising the status and species composition of the released catch are still to be developed. With regards to EM, the EU CDRs for pelagic fisheries in NW waters and the North Sea (EU 2014a, b) stipulate that all slipping operations should be monitored with an electronic recording system documenting when, where and the extent to which the net has been hauled. Interestingly, no commercially available technology currently exists to monitor purse seine hauling, effectively prohibiting all slipping operations in NW waters and the North Sea. Most concerning of all with regards to compliance, is that many fishers are yet to fully appreciate the implications of the LO (Maynou et al. 2018), although many do voluntarily take steps to avoid unwanted catches (Marçalo et al. 2018).

15.5 Conclusion

Management strategies and regulations currently attempting to address the slipping problem are, in reality, still in early development. Therefore, as we gain more knowledge in this area, management strategies are likely to require modifications to better suit the fisheries they are regulating. Furthermore, it is recognised that the successful implementation of fishing regulations is best done in close consultation with all stakeholders to ensure that what is prescribed is practical, safe, economically viable, effective and something that the fishers will actually implement (Karp et al., this volume). In terms of policy and research, priority should be given to avoiding

UWC; avoidance is the most effective way of reducing slipping mortality and is likely to increase profitability for the fisher and therefore implementation. However, some level of UWC in purse seine fisheries for small pelagic species is inevitable. Methods for releasing UWC that promote high survival for the slipped catch must continue to be developed in collaboration with fishers to ensure they are practical, effective and implemented.

References

Arregi, L., Onandia, I., Ferarios, J.M., Ruiz J., Basurko, O.C. (2014). Assessing fish survival from slipping in purse seine fisheries of European southern waters. AZTI-Tecnalia, Sukarrieta (p. 44). [Report Presented to STECF Plenary 14-02].

Borges, T.C., Erzini, K., Bentes, L., Costa, M.E., Gonçalves, J.M.S., Lino, P.G., et al. (2001). By-catch and discarding practices in five Algarve (southern Portugal) métiers. *Journal of Applied Ichthyology, 17,* 104–114.

Breen, M., Isaksen, B., Ona, E., Pedersen, A.O., Pedersen, G., Saltskår, J., et al. (2012). A review of possible mitigation measures for reducing mortality caused by slipping from purse-seine fisheries. *ICES CM 2012*, C:12.

Breen, M., & Cook, R. (2002). Inclusion of Discard and Escape Mortality Estimates in Stock Assessment Models and its likely impact on Fisheries Management. *ICES CM 2002/V, 27,* 15pp.

Breen, M., Anders, N., Humborstad, O.-B., Nilsson, J., Tenningen, M., Vold, A. (In review). Catch welfare in commercial fisheries. In F. Kristiansen, & Van de Vis Pavlidis (Eds.), *Fish Welfare*. Springer.

Breen, M., Saltskår, J., Anders, N., Totland, B., Øvredal, J.T., Tenningen, M., Handegard, N.O., Peña, H. (In review). A novel method for monitoring the behaviour of mackerel (Scomber scombrus) in relation to crowding and oxygen concentrations in commercial purse seine catches. Submitted to *PLoS One*.

Broadhurst, M.K., Suuronen, P., Hulme, A. (2006). Estimating collateral mortality from a towed fishing *gear. Fish and Fisheries, 7*, 180–218.

CRISP. (2018). CRISP: The Centre for Research-based Innovation in Sustainable fish capture and Processing technology. http://crisp.imr.no/en/projects/crisp/

DABGRAF. (2018). http://cmr.no/projects/10397/dabgraf/

Davis, M.W. (2002). Key principles for understanding fish bycatch discard mortality. *Canadian Journal of Fisheries and Aquatic Sciences, 59*, 1834–1843.

DGRM. (2017). *DATAPESCAS. N° 115/Janeiro-Dezembro 2017.* Direção Geral dos Recursos Naturais, Segurança e Serviços Marítimos, Lisboa (p. 11).

Digre, H., Tveit, G.M., Solvang-Garten, T., Eilertsen, A., Aursand, I.G. (2016). Pumping of mackerel (Scomber scombrus) onboard purse seiners, the effect on mortality, catch damage and fillet quality. *Fisheries Research, 176*, 65–75.

EU. (2013). REGULATION (EU) No 1380/2013 OF THE EUROPEAN PARLIAMENT AND OF THE COUNCIL of 11 December 2013 on the Common Fisheries Policy, amending Council Regulations (EC) No 1954/2003 and (EC) No 1224/2009 and repealing Council Regulations (EC) No 2371/2002 and (EC) No 639/2004 and Council Decision 2004/585/EC.

EU. (2014a). Discard Plan for certain pelagic fisheries in north-western waters. Commission Delegated Regulation (EU), No 1393/2014.

EU. (2014b). Discard Plan for certain pelagic fisheries in the North Sea. Commission Delegated Regulation (EU) No 1395/2014.

EU. (2014c). Discard Plan for certain pelagic fisheries in south-western waters, Commission Delegated Regulation (EU) 1394/2014.

Feijó, D., Marçalo, A., Bento, T., Barra, J., Marujo, D., Correia, M., Silva, A. (2018). Trends in the activity pattern, fishing yields, catch and landing composition between 2009 and 2013 from onboard observations in the Portuguese purse seine fleet. *Regional Studies in Marine Science*. Available online 1 January 2018, https://doi.org/10.1016/j.rsma.2017.12.007.

Gerlotto, F., & Paramo, J. (2003). The three-dimensional morphology and internal structure of clupeid schools as observed using vertical scanning multibeam sonar. *Aquatic Living Resources, 16*, 113–122.

Gilman, E., Suuronen, P., Chaloupka, M. (2017). Discards in global tuna fisheries. *Marine Ecology Progress Series, 582*, 231–252. https://doi.org/10.3354/meps12340.

Gullestad, P., Blom, G., Bakke, G., Bogstad, B. (2015). The "Discard Ban Package": Experiences in efforts to improve the exploitation patterns in Norwegian fisheries. *Marine Policy, 54,* 1–9. https://doi.org/10.1016/j.marpol.2014.09.025.

Hall, M.A., Alverson, D.L., and Metuzals K.I. (2000). By-catch: Problems and solutions. *Marine Pollution Bulletin, 41,* 204–216.

Hall, M., & Roman, M. (2013). Bycatch and non-tuna catch in the tropical tuna purse seine fisheries of the world. *FAO Fisheries and Aquaculture Technical Paper No. 568*. Rome, FAO (p. 249).

Handegard, N.O., Tenningen, M., Howarth, K., Anders, N., Rieucau, G., Breen, M. (2017, December). The loss of schooling function in mackerel in response to crowding but not hypoxia. *PLoS ONE, 12*(12), e0190259. https://doi.org/10.1371/journal.pone.0190259.

Holmin, A.J., Handegard, N.O., Korneliussen, R.J., Tjostheim, D. (2012). Simulations of multi-beam sonar echos from schooling individual fish in a quiet environment. *Journal of the Acoustical Society of America, 132,* 3720–3734.

Horne, J.K. (2000). Acoustic approaches to remote species identification: An introductory review. *Fisheries Oceanography, 9,* 356–371.

Huse, I., and Vold A. (2010). Mortality of mackerel (*Scomber scombrus L.)* after pursing and slipping from a purse-seine. *Fisheries Research, 106*, 54–59.

ICES. (2016). Report of the Workshop on Methods for Estimating Discard Survival 5 (WKMEDS 5), 23–27 May 2016, Lorient, France. *ICES CM 2016/ACOM:56* (p. 51).

ICES. (2018). Advice on fishing opportunities, catch, and effort. *Bay of Biscay and the Iberian Coast Ecoregion* pil.27.8c9a. https://doi.org/10.17895/ices.pub.4495.

ICES. (2018; In press). ICES WKMEDS Guidance on Method for Estimating Discard Survival. Breen, M. & Catchpole, T. (Eds.). ICES Cooperative Research Report. In press.

Isaksen, B. (2013). Fish sampling by shooting a mini-trawl into the purse seine. IMR Marine Research News, 2-2013. https://www.hi.no/filarkiv/2013/07/hi_nytt_2_2013_til_web.pdf/en

Karp, W.A., Breen, M., Borges, L., Fitzpatrick, M., Kennelly, S.J., Kolding, J., et al. (this volume). Strategies used throughout the world to manage fisheries discards – Lessons for implementation of the EU Landing Obligation. In S.S. Uhlmann, C. Ulrich, S.J. Kennelly (Eds.), *The European Landing Obligation – Reducing discards in complex, multi-species and multi-jurisdictional fisheries*. Cham: Springer.

Kelleher, K. (2005). Discards in the world's marine fisheries. An update. *FAO Fisheries Tech Paper, No. 470*. Rome, FAO (p. 131).

Korneliussen, R.J., Heggelund, Y., Eliassen, I.K., and Johansen, G.O. (2009). Acoustic species identification of schooling fish. *ICES Journal of Marine Science, 66,* 1111–1118.

Larsen, T.A., & Dreyer, B. (2013). Ringnot – Struktur og lønnsomhet. Tromsø: Nofima (ISBN 978-82-8296-111-0) 25, p. Nofima rapportserie (34/2013).

Lockwood, S.J., Pawson, M.G., and Eaton, D.R. (1983). The effects of crowding on mackerel (*Scomber scombrus* L.) physical condition and mortality. *Fisheries Research, 2,* 129–147.

LSSS. (2018). http://cmr.no/projects/10396/lsss/. Accessed 7 July 2018.

MAPAMA. (2017). Ministerio de Agricultura y Pesca, Alimentación y Medio Ambiente (MAPAMA), Estadísticas Pesqueras Noviembre 2017. Recovered 1/04/2018: http://www.mapama.gob.es/es/estadistica/temas/estadisticas-pesqueras/

Marçalo, A., Mateus, L, Correia, J.H.D., Serra, P., Fryer, R., Stratoudakis, Y. (2006). Sardine (*Sardina pilchardus*) stress reactions to purse-seine fishing. *Marine Biology, 149,* 1509–1518.

Marçalo, A. (2009). Sardine (*Sardina pilchardus*) delayed mortality associated with purse-seine slipping: contributing stressors and responses. Tese de doutoramento. Faculdade de Ciências e Tecnologia. Universidade do Algarve. 189pp.

Marçalo, A., Marques, T.A., Araújo, J., Pousão-Ferreira, P., Erzini, K., Stratoudakis, Y. (2010). Fishing simulation experiments for predicting effects of purse-seine capture on sardine (*Sardina pilchardus*). *ICES Journal of Marine Science, 67*, 334–344.

Marçalo, A., Guerreiro, P.M., Bentes, L., Rangel, M., Monteiro, P., Oliveira, F., Afonso, C., Pousão-Ferreira, P., Benoît, H., Breen, M., Erzini, K., Gonçalves, J.M.S. (2018). Effects of different slipping methods on the mortality of sardine, *Sardina pilchardus*, after purse-seine capture off the Portuguese Southern coast (Algarve). *PLoS One, 13*(5), e0195433. https://doi.org/10.1371/journal.pone.0195433.

Maynou, F., Gil, M.D.M., Vitale, S., Giusto, G.B., Foutsi, A., Rangel, M., Rainha, R., Erzini, K., Gonçalves, J.M.S., Bentes, L., Viva, C., Sartor, P., Carlo, F., Rossetti, I., Christou, M., Stergiou, K., Maravelias C.D., Damalas, D. (2018). Fishers' perceptions of the European Union discards ban: Perspective from south European fisheries. *Marine Policy, 89,* 147–153. https://doi.org/10.1016/j.marpol.2017.12.019.

Megalofonou, P., Yannopoulos, C., Damalas, D., De Metrio, G., Deflorio, M., de la Serna, J.M., Macias, D.. (2005). Incidental catch and estimated discards of pelagic sharks from the swordfish and tuna fisheries in the Mediterranean Sea. *Fisheries Bulletin, 103,* 620–634.

Mesnil, B. (1996). When discards survive: Accounting for survival of discards in fisheries assessments. *Aquatic Living Resources, 9,* 209–215.

Mitchell, R.W., Blight, S.J., Gaughan, D.J., Wright, I.W. (2002). Does the mortality of released *Sardinops sagax* increase if rolled over the headline of a purse-seine net? *Fisheries Research, 57,* 279–285.

Nishimori, Y., Iida, K., Furusawa, M., Tang, Y., Tokuyama, K., Nagai, S., Nishiyama, Y. (2009). The development and evaluation of a three-dimensional, echo-integration method for estimating fish-school abundance. *ICES Journal of Marine Science, 66,* 1037–1042.

NSFR. (2014). Regulations relating to sea-water fisheries. Norwegian Ministry of Fisheries and Coastal Affairs. Amended 7th April 2014. https://www.fiskeridir.no/.../20140407-regulations-relating-to-sea-water-fisheries.pdf

Norwegian Fisheries Directorate DataBase. (2018). Fishers, fishing vessels and licenses. https://www.fiskeridir.no/English/Fisheries/Statistics/Fishers-fishing-vessels-and-licenses. Accessed 7 July 2018.

O'Neill, F.G., Feekings, J., Fryer, R.J., Fauconnet, L., Afonso, P. (this volume). Discard avoidance by improving fishing gear selectivity: Helping the fishing industry help itself. In S.S. Uhlmann, C. Ulrich, S.J. Kennelly (Eds.), *The European Landing Obligation – Reducing discards in complex, multi-species and multi-jurisdictional fisheries*. Cham: Springer.

Peña, H., Saltskår, J., Totland, B., Vold, A., Breen, M., Ingolfsson, Ò.A., Tenningen, M. Øvredal, J.T. (2018). Purse Seine Catch Control. Final Project Report. Rapport fra Havforskningen 3-2018. (In Norwegian). https://www.hi.no/filarkiv/2017/12/3-2018-pscc_sluttrapp_av.pdf/nb-no

Reid, D.G., Calderwood, J., Afonso, P., Fauconnet, L., Pawlowski, L., Plet-Hansen, K.S., et al. (this volume). The best way to reduce discards is by not catching them! In S.S. Uhlmann, C. Ulrich, S.J. Kennelly (Eds.), *The European Landing Obligation – Reducing discards in complex, multi-species and multi-jurisdictional fisheries*. Cham: Springer.

Rihan, D., Uhlmann, S.S., Ulrich, C., Breen, M., Catchpole, T. (this volume). Requirements for documentation, data collection and scientific evaluations. In S.S. Uhlmann, C. Ulrich, S.J. Kennelly (Eds.), *The European Landing Obligation – Reducing discards in complex, multi-species and multi-jurisdictional fisheries*. Cham: Springer.

SEAT. (2018). http://cmr.no/projects/10414/seat/. Accessed 7 July 2018.

STECF. (2014). Scientific, Technical and Economic Committee for Fisheries (STECF) – 46th Plenary Meeting Report (PLEN-14-02). 2014. Publications Office of the European Union, Luxembourg, EUR 26810 EN, JRC 91540, 117 pp.

STECF. (2017). Scientific, Technical and Economic Committee for Fisheries – *The 2017 Annual Economic Report on the EU Fishing Fleet (STECF-17-12)*. Publications Office of the European Union, Luxembourg, 2017, ISBN 978-92-79-73426-7. https://doi.org/10.2760/36154, PUBSY No. JRC107883.

Silva, A., Moreno, A., Riveiro, I., Santos, B., Pita, C., Rodrigues, J.G. et al. (2015). Research for Pech Committee- Sardine Fisheries. Resource Assessment and Social and Economic Situation. Directorate- General for Internal Policies IP/B/PECH/IC/2015_133.

SINTEF FKiN Project. (2018). Catch control in purse seine fishing. https://www.sintef.no/fangstkontroll

Stratoudakis, Y., & Marçalo, A. (2002). Sardine slipping during purse-seining off northern Portugal. ICES Journal of Marine Science, 59, 1256–1262.

Tenningen, M., Vold, A., Olsen, R.E. (2012). The response of herring to high crowding densities in purse-seines: Survival and stress reaction. *ICES Journal of Marine Science, 69,* 1523–1531.

Tenningen, M., Macaulay, G.J., Rieucau, G., Pena, H., Korneliussen, R.J. (2017). Behaviours of Atlantic herring and mackerel in a purse-seine net, observed using multibeam sonar. *ICES Journal of Marine Science, 74,* 359–368.

Tenningen, M., Pena, H., Macaulay, G.J. (2015). Estimates of net volume available for fish shoals during commercial mackerel (*Scomber scombrus*) purse seining. *Fisheries Research, 161,* 244–251.

Vatnehol, S., Pena, H., Ona, E. (2017). Estimating the volumes of fish schools from observations with multi-beam sonars. *ICES Journal of Marine Science, 74,* 813–821.

Veiga, P., Pita, C., Rangel, M., Gonçalves, J.M.S., Campos, A., Fernandes, P.G., et al. (2016). The EU landing obligation and European small-scale fisheries: What are the odds for success? *Marine Policy, 64,* 64–71.

Villasante, S., Pierce, G.J., Pita, C., Guimeráns, C.P., Rodrigues J.G., Antelo, M., et al. (2016a). Fishers' perceptions about the EU discards policy and its economic impact on small-scale fisheries in Galicia (North West Spain). *Ecological Economics, 130,* 130—138.

Villasante, S., Pita, C., Pierce, G.J., Guimeráns, C.P., Rodrigues, J.G., Antelo, M., et al. (2016b). To land or not to land: How do stakeholders perceive the zero discard policy in European small-scale fisheries? *Marine Policy, 71,* 166–174.

Vold, A., Anders, N., Breen, M., Saltskår, J., Totland B. og Øvredal J.T. (2017). Beste praksis for slipping fra not. Utvikling av standard slippemetode for makrell og sild i fiske med not. Faglig sluttrapport for FHF-prosjekt 900999. Rapport fra Havforskningen nr. 6 2017. ISSN 1893–4536 (online).

Watson, R., Revenga, C. & Kura, Y. (2006). Fishing gear associated with global marine catches. I Database development. Fisheries Research, 79(1–2), 97–102.

Chapter 16
Onboard and Vessel Layout Modifications

Jónas R. Viðarsson, Marvin I. Einarsson, Erling P. Larsen, Julio Valeiras, and Sigurður Örn Ragnarsson

Abstract The purpose of this chapter is to discuss challenges that the EU Landing Obligation presents to the onboard handling of unwanted catches and how vessel layout modifications can be applied to meet these challenges. The key challenge the industry is facing is having to bring ashore catches of little or no value, which requires significant effort to handle and takes up valuable space that is, in many cases, not available. Considering that 85% of EU fishing vessels are under 12 metres long and 97% are under 24 metres, it is evident that the majority of the EU fleet has limited options when it comes to handling and stowage of catches that would have been discarded prior to the implementation of the Landing Obligation. The Landing Obligation only applies to species subject to catch limits, which means that the current set-up on vessels can, for the most part, accommodate the fish of legal size that needs to be landed. The main challenge is catches of undersized fish that are not permitted to be used for direct human consumption. For those catches, the simplest approach is to handle them as targeted catch, which will allow them to be used for higher-value products such as pet food, pharmaceuticals, food supplements, etc. This is, however, not applicable for the majority of the fleet, due to a lack of space and the labour effort required. Solutions such as bulk storage and simple silage preservation are alternatives that are being explored for smaller vessels. The larger vessels have more options, such as full silage production, fish protein hydrolysate and fish meal production.

Keywords Automated classification · Bulk landing · Fish protein · Fish silage · Hydrolysate · MCRS · Onboard handling · Vessel modifications

---

J. R. Viðarsson (✉) · M. I. Einarsson · S. Ö. Ragnarsson
Matís ohf. Icelandic food and biotech R&D, Reykjavík, Iceland
e-mail: jonas@matis.is

E. P. Larsen
DTU Aqua, National Institute for Aquatic Resources, Technical University of Denmark, Kongens Lyngby, Denmark

J. Valeiras
Instituto Español de Oceanografía (IEO), Centro Oceanográfico de Vigo, Vigo, Spain

S. S. Uhlmann et al. (eds.), *The European Landing Obligation*,
https://doi.org/10.1007/978-3-030-03308-8_16

16.1 Introduction

Discards have been a part of fishing practices in most fisheries around the world since fisheries began. Fishers have selected what fish to keep and what to release or throw back into the sea long before quotas and catch limits were invented. The introduction of catch limits has, however, created new incentives for discarding, as fishers try to maximise the value of their catches under quota regimes. Unwanted catches (UWC), such as low-value bycatches, undersized fish, catches exceeding quotas and catches of target species that are unlikely to attain premium prices are thrown back, and much of these catches are dead or dying. This has been the practice in European fisheries under the Common Fisheries Policy (CFP) of the European Union (EU). European fishers have annually discarded more than 1.5 million tons of fish in order to maximise the value of their catch and to meet with regulations (EC 2011). This practice has been the subject of increasing levels of debate (Borges and Penas Lado, this volume). As a result, the European Commission has introduced a Landing Obligation as a part of the 2013 reform of the CFP (EC 2013). This means that all catches of species subject to catch limits, i.e. where total allowable catches (TAC) have been set and in the Mediterranean catches of species subject to minimum sizes, will have to be landed and will be counted against quotas. The obligation is gradually implemented. The first fisheries were subject to this Landing Obligation in the beginning of 2015, and by 2019, all EU fisheries are required to land the entire catch of all species subject to catch limits.

The Landing Obligation presents a number of challenges for the European seafood sector. Fishing strategies of individual fishers will have to be enhanced; selectivity of fishing gear will need to be improved; onboard handling, sorting, storing and monitoring of compliance will need to be reconsidered; land-based processing will have to adjust to different supplies; and markets will be affected. The purpose of this chapter is to discuss challenges that the Landing Obligation presents to the onboard handling of unwanted catches and how vessel layout modifications can be applied to meet with these challenges.

16.2 Challenges

In recent years, the main focus of the EU authorities, researchers and the seafood industry working on the implementation of the Landing Obligation has been on how to avoid unwanted catches (see Reid et al., this volume; O'Neill et al., this volume) and how to facilitate efficient monitoring, control and surveillance (MCS) of unavoidable unwanted catches (see also James et al., this volume; Nuevo et al., this volume). There has, however, been much less attention given to what to do with the unwanted catches which, prior to the implementation of the Landing Obligation, would have been discarded after being caught. Exceptions to the Landing Obligation and limits on permitted uses of the unwanted catches do have to be taken carefully into consideration when contemplating onboard handling and stowage. Species not covered bycatch limits, species where high survivability can be demonstrated and catches falling under the de minimis exceptions can still be discarded under the Landing Obligation; but everything else will need to be landed. In addition, catches under minimum conservation reference size (MCRS) need to be landed but cannot be used for direct human consumption. With all this in mind, it is clear that space for classification, proper handling and stowage will become an issue for much of the EU fishing fleet when the Landing Obligation is fully implemented. The available alternatives for addressing that challenge are scarce and are generally only applicable for larger vessels – but 85% of the EU fleet are under 12 metres long and 97% are under 24 metres (EU 2016), which severely reduces available solutions.

Catches falling under the Landing Obligation can be broken into two basic groups, i.e. catches under MCRS that cannot be used for direct human consumption and catches above MCRS that can be used for human consumption. Bycatch of species not subject to catch limits (or minimum size in the Mediterranean) can still be discarded as before. The main challenges for onboard handling are connected to catches under MCRS, as the larger fish destined for human consumption can, for the most part, simply be diverted to the traditional onboard handling processes that are already available. However, stowage of low-value species with catch limits can have an effect on the duration of fishing trips, as stowage space is limited. The MCRS catches have to be recorded and stowed separately from the catches intended for human consumption (EU 2015/812), and vessels greater than 12 metres in length overall also have to place their catches in boxes, compartments or containers separately for each stock, just as with any other catches (EC 1224/2009). This basically means that any mixing of species during stowage onboard is prohibited. Vessels will therefore need to have two separated compartments for stowage, i.e. one for < MCRS catches and another for human consumption catches; vessels greater than 12 metres in length will need to sort everything into boxes, tubs or other such compartments. It is therefore evident that significantly more space will be required for onboard classification, handling and stowage, in addition to added labour.

There are, however, a number of available alternatives for adapting onboard handling to the Landing Obligation, and some of them require modifications of the vessels and their equipment. These are highly dependent on each fleet type. The main challenges and available alternatives for each fleet type are discussed below.

16.2.1 Small Coastal Vessels

About 85% of the EU fishing fleet consists of vessels that are under 12 metres in length (EU 2016). Unlike larger vessels, they do not have the necessary space onboard to handle and stow their catch (Viðarsson et al. 2016). The catches of these vessels are however generally quite limited, as they are most often counted in kilogrammes or hundreds of kilogrammes per fishing trip. These catches are almost solely landed on the day of capture, which is why the lack of proper bleeding, cleaning, chilling, sorting and storing is often not of major concern. These vessels are allowed to stow their catches without sorting each species into boxes or other such compartments, which makes it much easier to fit everything onboard (EC 1224/2009). They are, however, required to record and stow < MCRS catches separately from other catches (EU 2015/812). The Landing Obligation as such should therefore not create major challenges for this fleet sector in regard to onboard handling, as all < MCRS can simply be stored in bulk in boxes or larger containers. The problem is that the < MCRS catches being landed after each fishing trip are so small that they do not create enough incentives for buyers to source them (Viðarsson et al., 2017). Special solutions will therefore have to be implemented to aggregate these catches so that they become large enough to attract the attention of potential buyers. This is a subject discussed in another chapter of this book (Iñarra et al., this volume).

16.2.2 Small- and Intermediate-Sized Vessels

About 12% of the EU fleet are vessels between 12 and 24 metres in length (EU 2016). This is a highly diverse fleet targeting most commercial species in European waters, such as crustaceans, molluscs, groundfish and pelagics. They use a range of fishing gears, including dredges, bottom trawls, pots, gill nets, longline, handline, Danish seine and purse seine. The space onboard these vessels is limited when it comes to onboard handling of unwanted catches (Viðarsson et al. 2016). As these vessels are generally at sea for several days, it is important that all catches are properly handled, i.e. bled, gutted, cleaned, chilled, sorted and stowed, in accordance to the need of each species. These vessels are required to sort catches according to species into boxes and to stow catches < MCRS separately from catches intended for human consumption. The challenge for these vessels is therefore twofold, i.e. to ensure proper onboard handling of all catches, including sorting by species and intended utilisation (< MCRS or human consumption), and to appropriately stow all species. Due to the limited space onboard vessels of this size, both on the processing deck and in the hold, this can present major challenges. It is therefore likely that in some cases, the Landing Obligation will result in the need for investment in new equipment on the processing deck, e.g. storage boxes for separating between species and sizes, increased throughput capabilities of bleeding and cleaning tanks and increased numbers of stowage boxes, which should preferably be of different colours depending on intended usage (< MCRS or human consumption). Separation panels in the hold so that storage of < MCRS catches are clearly separated from other catches are also advised. Such separation panels could be adjustable so that the space required for each type of catches would not have to be fixed. Finally, it is likely that the Landing Obligation will require increased numbers of crew or longer working time because of the additional handling requirements and that the fishing trips will be shortened due to the lack of stowage space.

There are limited alternatives for small vessels to handle unwanted catches beyond what has been described above (Viðarsson et al. 2017). There are, however, options available which may need special permission from authorities to implement, or even changes in regulations. One such option is bulk storage of < MCRS catches, i.e. to store all catches below MCRS mixed in large boxes or compartments. The sorting would then have to be done after landing. These options are already being explored, for example, AZTI Tecnalia in Spain has been working on the development of an automatic system for the quantification and classification of catches landed in bulk (Melado et al. 2018), and the life iSEAS project has been looking into similar solutions (iSEAS 2018). Such systems, if approved by the authorities, would allow for bulk landing of < MCRS catches, or even entire catches. This solution could be a major contributor to solving the main challenges associated with a lack of space and human capital onboard this sector of the fleet. It may also reduce the cost that the fishers have to pay for renting boxes. It is however not permitted at the moment according to current EU regulations (EC 1224/2009).

Another option is to produce silage from < MCRS catches onboard the vessels (Viðarsson et al. 2017). Simple and relatively compact equipment can be fitted onboard vessels of this size that produces basic silage. What is needed is a powerful mincer, acid dispenser and a tank for storage. The fish are minced and mixed with organic acid in a storage tank. Around 2–3% of 85% formic acid is most commonly used, i.e. 20–30 kg of acid per ton of raw material. The acid lowers the pH of the silage, which gives it an extremely long shelf-life. Silage of this type is, however, not very valuable. But in any case, this option currently contradicts EU legislation that requires catches to be placed in boxes, compartments or containers separately for each stock (EC 1224/2009). It may also prove difficult to validate what is actually in the silage, as fishermen may claim that catches not subjected to the Landing Obligation have been used as raw material for the silage. For similar reasons, it may also contradict the EU regulation that requires < MCRS catches to be stowed separately from other catches (EU 2015/812).

There are, however, examples where onboard silage production has been permitted within CFP fisheries. For example, some Danish fishing vessels have been fitted with silage equipment, and the Danish authorities have given them an exemption from the regulations, with the condition that what goes into the silage is recorded via camera (FiskerForum 2008; Fiskeritidende 2016; FiskerForum 2017).

In 2016, the Danish fishing trawler Juli-Ane RI-568, from the fishing port of Hvide Sande, was renovated and extended to 23.95 metres overall length (Fiskeritidende 2016; FiskerForum 2017). In connection with the renovation, it was decided to install a silage system with grinder, automatic dosing of formic acid and storage tanks with the capacity of 16 tons – all done in acid proof steel. CCTV was installed to monitor and document the catch, according to specifications from the authorities. The system has been tested but has never reached the level of continuous day-to-day use, mainly because of management problems, and the hardware has been exposed to more strain than expected (Larsen 2018).

Previous to the trials onboard Juli-Ane RI-568, the 40 metre Danish fishing trawler Tobis HG-306, from the harbour of Hanstholm, ran similar experiments in 2008. The vessel was equipped with a silage system and 20 ton storage tanks, which was supposed to be a practical full load for a lorry (FiskerForum 2008). The system was tested in the Baltic and North Sea, but the results were not as positive as hoped for, especially because the quality of the silage was poorer than expected and the transportation cost was higher than anticipated. As a result, the system was uneconomic and was therefore taken out of use.

The figure below shows upper and lower deck of a Danish fishing boat that is around 25 metres in length and has been fitted with a basic silage system. The handling on the upper deck ensures proper bleeding and cleaning, as well as sorting what goes in to silage preservation. On the lower deck is the target catch sorted and iced into boxes. There is also an option to stow unwanted catches in bulk storage in differently coloured tubs.

The steel tank on the lower deck is an ice slurry machine that produces slurry to use during cleaning on the upper deck and can also be used in the boxes on the lower deck. The machine next to it produces flake ice that is used on the target catches stowed in boxes. The different coloured tubs are intended for bulk storage of unwanted catches, and the intermediate bulk container (IBC) stores the silage. It is then simple to replace tubs with IBC, or vice versa, if necessary.

The silage tanks used in the experiments onboard Juli-Ane and Tobis were able to carry 16 and 20 tons of silage and only had to be emptied or replaced when they had been filled, which could take several fishing trips. The tanks were fitted on the upper deck of the vessels, and the weight therefore had an effect on the stability of the vessels, in addition to taking up considerable space. Silage tanks of that size can be fitted below deck where they have less impact on stability, e.g. instead of oil or water tanks, but they must then be made of stainless steel and be easily emptied and cleaned. In the figure above (Fig. 16.1), the silage is stored in 1000 litre IBC in the hold of the vessel, which can be easily replaced after each fishing trip. The official discard rates of the Danish bottom trawl fleet operating in the North Sea prior to the implementation of the Landing Obligation was 0.9% (STECF 2015) which indicates that one IBC tank should easily be enough to stow the silage produced in a single fishing trip. The average discard rates in Skagerrak and Kattegat are, however, much higher, which is primarily explained by high discard rates in the Nephrops fishery.

The Danish silage trials have shown that the quality of the silage is the dominant factor that determines if this solution is economically viable or not.

The experiences from these two trials have not been very promising. The solutions have met with opposition from crew members for taking up too much space, and the value of the silage has not been as expected. The conclusion of one of the skippers on these trial vessels was that modification of an older fishing vessel to accommodate silage production is a significant challenge and that a vessel that is purpose-built for silage production from the start would be more likely to be successful (FiskerForum 2017).

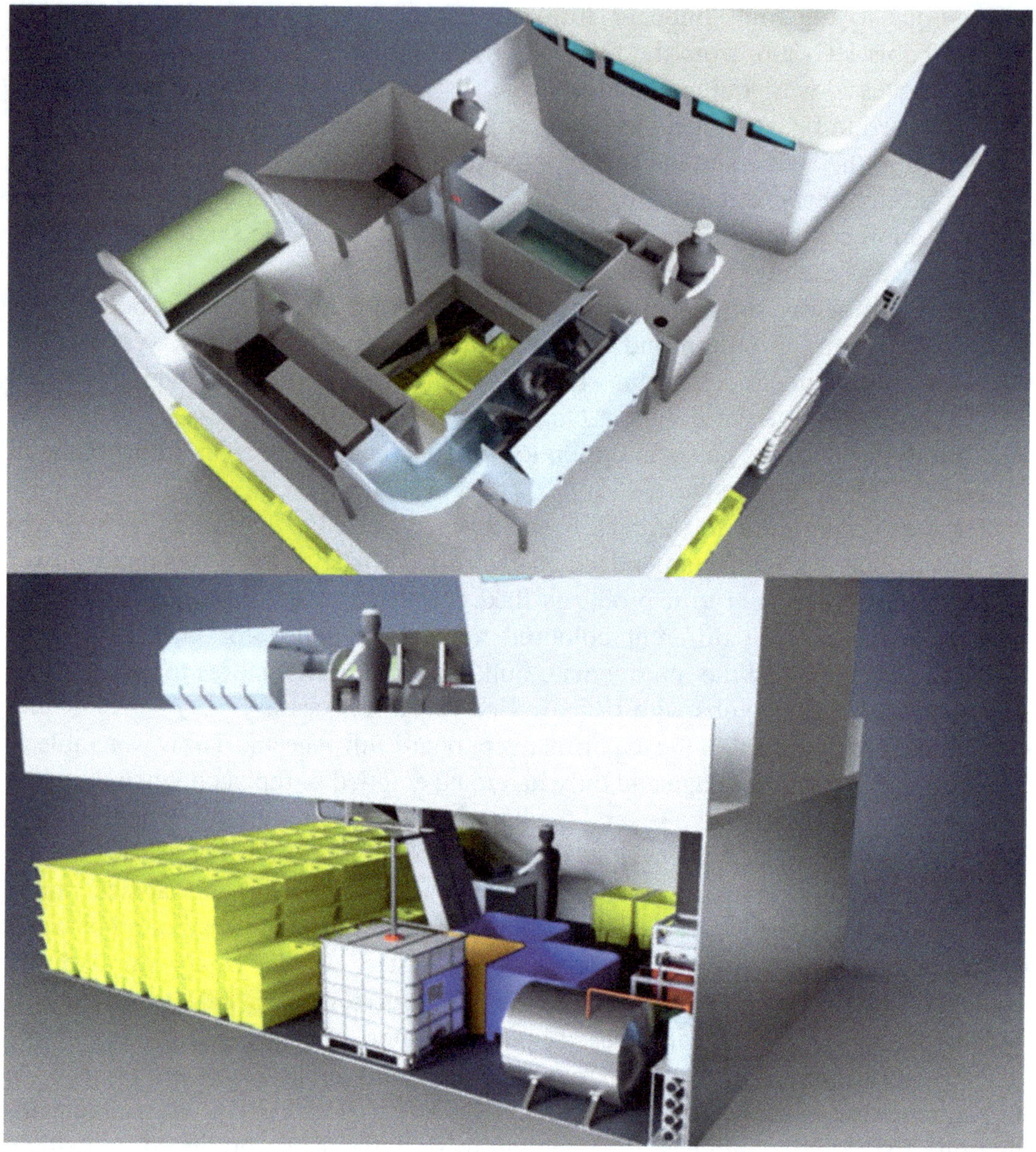

Fig. 16.1 Upper and lower deck of a medium-sized Danish seiner. The handling on the upper deck ensures proper bleeding and cleaning, as well as sorting what goes in to silage preservation. On the lower deck is the target catch sorted and iced into boxes. There is also an option to stow unwanted catches in bulk storage in differently coloured tubs

The main potential buyers of fish silage are either the fishmeal and fish oil sector or feed producers, e.g. mink feed or pig feed (Iñarra et al., this volume). The fishmeal and fish oil producers do not have a positive view of silage because their equipment is, in most cases, not made from acid proof steel. They need to add a base to the silage to bring the pH to 7.0. This is an additional cost in the production both in manpower and chemicals. Feed producers for mink and pig farms prefer fresh or frozen fish over silage, especially when the silage is of the variable quality shown in the Danish trials.

16.2.3 Larger Fresh Fish and Factory Vessels

Options for onboard handling of unwanted catches increase with increasing size of the vessels (Viðarsson et al. 2016). The bleeding, gutting, cleaning, chilling, sorting and stowing of unwanted catches require significant space and labour effort if it is to be done properly; and the same applies if additional onboard handling is applied such as silage or fishmeal production. Investment in onboard solutions and the ability to add crewmembers also become more applicable as space and throughput increase. Solutions that increase automation also usually require significant throughput in order to be economically sensible. Larger vessels with high throughput can be equipped with automatic species and size grading equipment (Skaginn 3X 2018a). These solutions are already on the market and use computer vision for identifying species and sizes. Flowline graders are also used for size grading onboard some fishing vessels (Marel 2018). Using such equipment can obviously increase throughput, reduce labour costs and make grading more accurate. This is already being done on bottom trawlers in Iceland (Skaginn 3X 2018b).

One option for larger vessels is to equip them with silage production units (Viðarsson et al. 2017). These units in their simplest form consist of a mincer, acid dispenser, two primary silage tanks and secondary silage tanks. The mincer shreds the material apart, which is then pumped into the primary silage tanks. In these tanks, commonly referred to as day tanks, formic acid is mixed in proportions with the raw material and heated up to 25–30 °C to speed up the digestion and to create a more uniform product. Each tank has a pump for constant circulation of the material to prevent settling of bones and other particles. When the material has been kept under these conditions for approximately 24 hours, it is pumped into secondary storage tank(s), which can be located at any place where space can be found onboard the vessel, as long as they can be easily emptied and cleaned. This type of silage has a high water content, which requires considerable space, and the value per cubic metre is rather low. By modifying the equipment, it is possible to separate the oil from the rest of the silage and then remove some (and in some cases most) of the water by evaporation, reducing the need for storage space and increasing the value of the final product(s). An example of such a mechanism on a 40 m bottom trawler is shown in Fig. 16.2.

Silage tanks full of silage are heavy and will therefore have an effect on the stability of the vessel. Locating the storage tanks as far below deck as possible will help in reducing those effects, but it is very important that safety issues concerning stability are taken into consideration when developing silage units for fishing vessels.

Production of fish protein hydrolysis (FPH) is another option, where hydrolysis is used to separate the bones from the rest of the fish (Viðarsson et al. 2017). The bones are then filtered out and what's left is a "soup" to which enzymes are added. The addition of those enzymes allows oils, proteins and amino acids to be extracted, which can be of high value as ingredients to animal feed or human food supplements. However, this requires complicated and expensive machinery, and it is therefore unlikely it will be a suitable solution for many vessels. The Norwegian freezer trawler Molnes M-69-G was recently equipped with such FPH system. This was

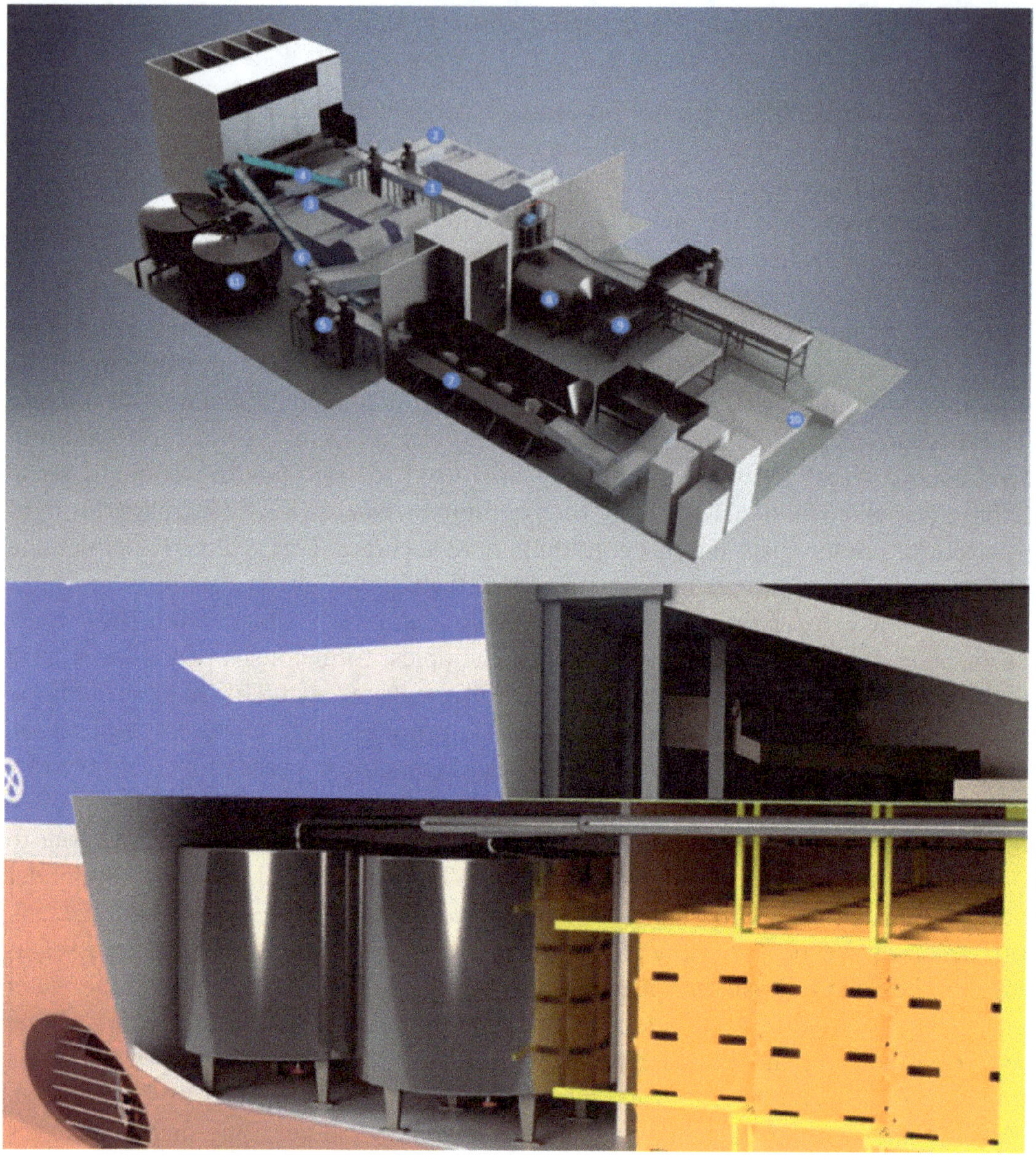

Fig. 16.2 Overview of a concept prospecting deck of a 40 metre bottom trawler with silage production. No.1 – Target catch (only bled) / not bled. No.2 – Target catch (bled in seawater / not to be gutted). No.3 – Target catch (bled in seawater / to be gutted). No.4 – Path for UWC/MCRS. No.5 – Gutting. No.6 – Path for viscera. No.7 – Rotary cooling / cleaning tank. No.8 – Slurry ice buffer tank. No.9 – Fish buffer tank with slurry ice. No.10 – Elevator down to hold. No.11 – Silage unit. The primary (day tanks) are located on the processing deck, but the evaporators and the storage tanks can be located elsewhere

part of a major renovation that was done on this 66-metre-long and 14-metre-wide bottom trawler. The results on the long-term economic viability of the investment for Molnes are still unknown.

The largest fishing vessels, particularly factory vessels, can be equipped with compact onboard fishmeal or fish protein plants (Amof-fjell 2018; Haarslev 2018; Héðinn 2018; Viðarsson et al. 2017). This is a solution that has been used onboard factory vessels for decades and has been proven to be a practical and cost-efficient alternative. The plants that are available today are relatively compact and require little manpower. The products do, however, require significant storage space, and the investment cost can be substantial. With regard to the Landing Obligation, the issue of documenting what is in the fishmeal might also be an issue, as with silage production.

16.3 Discussion

Very little progress has been achieved with regard to vessel modifications to meet the requirements of the Landing Obligation. The available solutions are generally not applicable for the EU fleet, the potential products are low value and require significant manpower, and the space onboard vessels to accommodate these low-value catches is scarce or not available. Vessel owners are reluctant to invest in technology that is unlikely to be economically viable, and enforcement by authorities has been undertaken in such a manner that places little pressure on the vessel owners to react.

At the moment, it seems that the most applicable short-term solution for most vessels is to either get permission to land unwanted catches in bulk, where the classification will then take place at official weighing stations on land; for larger vessels, and where there is an economic benefit, silage systems or fishmeal plants may be appropriate. More complicated solutions need to be investigated in the long term, including FPH systems (Iñarra et al., this volume).

Many consider the production of pet food, cosmetics, food supplements and even pharmaceuticals, when discussing potential products derived from unwanted catches. And there are, in fact, opportunities in such products, but they depend on the landed catch being of the highest quality and with the fish being processed onshore. The raw materials used for such products are generally specific parts of the fish, and not the whole fish. For such products, the only onboard modifications necessary are therefore to make sure that all catches are properly handled, i.e. bled, cleaned, chilled, sorted and stored. This has been the process in other countries with long experiences of discard bans, such as Iceland, Norway and the Faroe Islands, where the emphasis has been on landing all catches at sufficient quality so that the land-based processing can make as much value from them as possible (Karp et al., this volume). To begin with, the economic returns for the fishermen were insignificant, but as the volume increased, processes improved and markets established, the returns to the fishermen have grown.

By the end of 2017, around EUR 30 million of the European Maritime and Fisheries Fund (EMFF) had been committed to projects related to the Landing Obligation across ten member states (EC 2018). Most of the funding had been allocated to projects focusing on gear selectivity and MCS, as well as investment in ports and processing. Very little had been allocated to fleet investment, but countries such as Denmark, Spain, the Netherlands and Italy have allocated some funds to such projects for 2018 and onwards. In the coming years, it is therefore likely that more focus will be on onboard handling and vessel modifications to meet the requirements of the Landing Obligation.

Acknowledgements The content of this chapter is largely based on work carried out in the DISCARDLESS project and the Life iSEAS project. The DISCARDLESS project has received funding from the European Union's Horizon 2020 research and innovation programme under grant agreement DiscardLess No 633680. The Life iSEAS project has been co-funded under the LIFE +Environment programme of the European Union (LIFE13 ENV/ES/000131). This support is gratefully acknowledged.

References

Amoffjell. (2018, April 20). Compact fish meal plants. Retrieved from www.amof-fjell.com: http://www.amof-fjell.com/plant-design/compact-fish-meal-plants/

Borges, L., & Penas Lado, E. (this volume). Discards in the common fisheries policy: The evolution of the policy. In S.S. Uhlmann, C. Ulrich, S.J. Kennelly (Eds.), *The European Landing Obligation – Reducing discards in complex, multi-species and multi-jurisdictional fisheries*. Cham: Springer.

Reid, D.G., Calderwood, C., Afonso, P., Fauconnet, L., Pawlowski, L., Plet-Hansen, K.S., et al. (this volume). The best way to reduce discards is by not catching them! In S.S. Uhlmann, C. Ulrich, S.J. Kennelly (Eds.), *The European Landing Obligation – Reducing discards in complex, multi-species and multi-jurisdictional fisheries*. Cham: Springer.

EC 1224/2009. (2018). Council regulation (EC) No 1224/2009. Retrieved from www.eur-lex.europa.eu: https://eur-lex.europa.eu/legal-content/EN/TXT/PDF/?uri=CELEX:32009R1224&from=EN

EC. (2011). Commission staff working paper SEC (2011) 891: Impact assessment. Brussels: European Commission. Retrieved from europarl.europa.eu: http://www.europarl.europa.eu/registre/docs_autres_institutions/commission_europeenne/sec/2011/0891/COM_SEC(2011)0891_EN.pdf.

EC. (2013). Regulation (EU) 1380/2013. Brussels: European Commission. Retrieved from http://eur-lex.europa.eu/LexUriServ/LexUriServ.do?uri=OJ:L:2013:354:0022:0061:EN:PDF

EC. (2018). Commission staff working document accompanying the document communication from the commission on the State of Play of the Common Fisheries Policy and Consultation on the Fishing Opportunities for 2019, COM (2018) 452 final. Brussels, 11.6.2018 SWD (2018) 329 final. Retrieved from https://eur-lex.europa.eu/legal-content/EN/TXT/PDF/?uri=CELEX:52018SC0329&from=EN

EU 2015/812. (2018). Regulation (EU) 2015/812 of the European parliament and of the council. Retrieved from www.http://eur-lex.europa.eu: http://eur-lex.europa.eu/legal-content/EN/TXT/PDF/?uri=CELEX:32015R0812&from=EN

EU. (2016). Facts and figures on the common fisheries policy. Brussels: EU. Retrieved from www.ec.europa.eu: https://ec.europa.eu/fisheries/sites/fisheries/files/docs/body/pcp_en.pdf

FiskerForum (2008, December 26). Ensilering af fiskeindvolde til søs en success. Retrieved from http://www.fiskerforum.dk/erhvervsnyt/2008/261208_ensilering.asp

FiskerForum. (2017, July 13). Ensilering kan gøre bifangsten værdifuld. Retrieved from www.fiskerforum.dk: http://www.fiskerforum.dk/erhvervsnyt/a/ensilering-kan-goere-bifangsten-vaerdifuld-13072017

Fiskeritidende. (2016, May 3). Ensilagetanke skal sikre en bedre udnyttelse af råvaren. Retrieved from www.fiskeritidende.dk: http://fiskeritidende.dk/ensilagetanke-skal-sikre-bedre-udnyttelse-raavaren/

Haarslev. (2018, April 20). Compact fish meal plant. Retrieved from www.haarslev.com: https://www.haarslev.com/products/compact-fish-meal-plant/

Héðinn. (2018, April 20). The Hedinn protein plant at sea. Retrieved from www.hedinn.com: https://hedinn.weebly.com/at-sea.html

Iñarra, B., Bald, C., Cebrián, M., Antelo, L.T., Franco-Uría, A., Vázquez, J.A., et al. (this volume). What to do with unwanted catches: Valorisation options and selection strategies. In S.S. Uhlmann, C. Ulrich, S.J. Kennelly (Eds.), *The European Landing Obligation – Reducing discards in complex, multi-species multi-jurisdictional fisheries*. Cham: Springer.

iSEAS. (2018). Proyecto LIFE iSEAS: Objetivos propuestos y resultados alcanzados. Retrieved from http://lifeiseas.eu/archivos/?drawer=web*iSEAS_Conference_CETMAR_05042018

James, K.M., Campbell, N., Viðarsson, J.R., Vilas, C., Plet-Hansen, K.S., Borges, L., et.al. (this volume). Tools and technologies for the monitoring, control and surveillance of unwanted catches. In S.S. Uhlmann, C. Ulrich, S.J. Kennelly (Eds.), *The European Landing Obligation – Reducing discards in complex, multi-species and multi-jurisdictional fisheries*. Cham: Springer.

Karp, W.A., Breen, M., Borges, L., Fitzpatrick, M., Kennelly, S.J., Kolding, J., et al. (this volume). Strategies used throughout the world to manage fisheries discards – Lessons for implementation of the EU Landing Obligation. In S.S. Uhlmann, C. Ulrich, S.J. Kennelly (Eds.), *The European Landing Obligation – Reducing discards in complex, multi-species and multi-jurisdictional fisheries*. Cham: Springer.

Larsen, E. (2018). Interview with crew of Juli-Ane on 14.06.2018.

Marel. (2018, April 24). Marine whole fish grader. Retrieved from www.marel.com: https://marel.com/fish-processing/systems-and-equipment/grading%2D%2Dbatching/marine-whole-fish-grader/767?prdct=1&parent=767&pc=3

Melado, A., Olabarrieta, I., de Zarata, A., Pardo, M., & Inarra, B. (2018). Report on the automatic system for by-catches quantification and classification and the battery of specific fluorescent DNA probes. DiscardLess Deliverable 6.3.

Nuevo, M., Morgado, C., Sala A. (this volume). Monitoring the implementation of the Landing Obligation: The last Haul programme. In S.S. Uhlmann, C. Ulrich, S.J. Kennelly (Eds.), *The European Landing Obligation – Reducing discards in complex, multi-species and multi-jurisdictional fisheries*. Cham: Springer.

O'Neill, F.G., Feekings, J., Fryer, R.G., Fauconnet, L., Afonso, P. (this volume). Discard avoidance by improving fishing gear selectivity: Helping the fishing industry help itself. In S.S. Uhlmann, C. Ulrich, S.J. Kennelly (Eds.), *The European Landing Obligation – Reducing discards in complex, multi-species and multi-jurisdictional fisheries*. Cham: Springer.

Skaginn 3X. (2018a, April 20). Vision whole fish grader. Retrieved from www.skaginn3x.com: http://skaginn3x.com/products/vision-whole-fish-grader

Skaginn 3X. (2018b). Fish processing deck. https://www.youtube.com/watch?v=ukrqO0_sNiU

STECF. (2015). STECF data for the North Sea (area 3B2), Kattegat (area 3A) and Skagerrak (area 3B). The scientific, technical and economic Committee for Fisheries. http://datacollection.jrc.ec.europa.eu/dd/effort/graphs, landings and discards.

Viðarsson, J.R., Einarsson, M.I., & Ragnarsson S.Ö. (2016). Report containing identification and recommendations on innovative, applicable and practical solutions for on-board handling of unavoidable unwanted catches. DiscardLess Deliverable 5.2. DOI 10.5281/zenodo.229325.

Viðarsson, J.R., Ragnarsson, S.Ö., Einarsson, M.I, Sævarsson, B., Sævarsdóttir, R., & Szymczak, P. (2017). Report on the 3D drawings and cost-benefit tools developed for Icelandic, North Sea and Bay of Biscay case studies. DiscardLess Deliverable 5.4.

Chapter 17
What to Do with Unwanted Catches: Valorisation Options and Selection Strategies

Bruno Iñarra, Carlos Bald, Marta Cebrián, Luis T. Antelo, Amaya Franco-Uría, José Antonio Vázquez, Ricardo I. Pérez-Martín, and Jaime Zufía

Abstract The European Common Fisheries Policy (CFP) has established a landing obligation (LO) and the need for proper management of bycatches without incentivising their capture. Food use is the priority option but only unwanted catches (UWC) above minimum conservation reference size (MCRS) can be used for direct human consumption. As a result, other options, such as animal feeds, industrial uses or energy, should be considered to valorise landed < MCRS individuals. Two approaches have been developed to help select the best available option for processing UWC. The first methodology is based on a multi-criteria decision analysis (MCDA) using an analytic hierarchy process (AHP) that considers technical, economic and market criteria. As a sample case, we chose the Basque fleet fishing in the Bay of Biscay, developed within the H2020 DiscardLess project. The second approach is based on the simultaneous analysis of both economic and environmental aspects. This was applied to the case of Spanish bottom trawlers operating in ICES sub-Divisions VIIIc and IXa. Finally, various food products and bio compounds from typical UWC biomass were obtained in a pilot food processing plant developed within the LIFE iSEAS project.

B. Iñarra (✉) · C. Bald · M. Cebrián · J. Zufía
AZTI, Efficient and Sustainable Processes area, Parque Tecnológico de Bizkaia, Derio, BI, Spain
e-mail: binarra@azti.es

L. T. Antelo · J. A. Vázquez · R. I. Pérez-Martín
Marine Research Institute IIM-CSIC, Vigo, Spain

A. Franco-Uría
Department of Chemical Engineering, School of Engineering, University of Santiago de Compostela, Santiago de Compostela, Spain

S. S. Uhlmann et al. (eds.), *The European Landing Obligation*,
https://doi.org/10.1007/978-3-030-03308-8_17

Keywords ε-constraint approach · Analytic hierarchy process · Biomolecules · Biorefinery · Bycatches · Discards management · In-land management · Landing obligation · Multi-criteria decision analysis · Unwanted catches · Valorisation

17.1 Introduction

In 2013, the European Commission introduced a Landing Obligation (LO) or 'discard ban' which stated that all catches of species subject to catch quotas and/or minimum conservation reference size (MCRS) must be landed and will be counted against quota. The LO is being gradually implemented from 2015 to 2019 for all regulated species across the EU. In the meantime, various strategies are needed to mitigate any potential negative impacts of the LO on fishing-dependent industries and communities.

Even after implementation of strategies to reduce bycatches a fraction of unwanted catches (UWC) may still be caught and will have to be landed. These UWC may not be used or sold for human consumption, thus other appropriate valorisation options are needed.

When choosing the most suitable use of UWC, one must first consider the reason for discarding these fractions. For those UWC that are >MCRS and of adequate quality for direct marketing, but which were discarded due to lack of fresh market demand, new (transformed) food products must be developed to prevent flooding the fresh fish market. For those UWC > MCRS of insufficient quality for human consumption or for UWC < MCRS, whose use for direct human consumption is forbidden by LO, a wide range of available technological alternatives exist. Not all of them may be equally feasible, however.

The Waste Framework Directive of the EU Parliament (2008) (Fig. 17.1) has established a hierarchy of management options for any food waste or by-product. The preferred choice is always prevention and reduction. In the case of fisheries discards, this is represented by reduction of bycatches via increased gear selectivity and optimisation of fishing strategies (see e.g. O'Neill et al., this volume; Reid et al., this volume). Second, the food by-product should be kept in the food chain as either fresh fish or transformed products (subject to legislation) or by the production of food ingredients. Third, bio-products (i.e., valuable compounds or biomolecules for food, cosmetic or other uses) should be obtained if possible. Fourth is the production of feed for aquaculture, pet-food and other animal feeds. The production of fishmeal is the most common use of fish by-products, used mainly for aquaculture, and is a straightforward option for the treatment of UWC if a fishmeal plant is located nearby.

Other lower-value options can also be imagined and evaluated, such as products for industrial uses, the production of energy, composting or incineration. A final option, putting UWC in a landfill, is not considered a valorisation option.

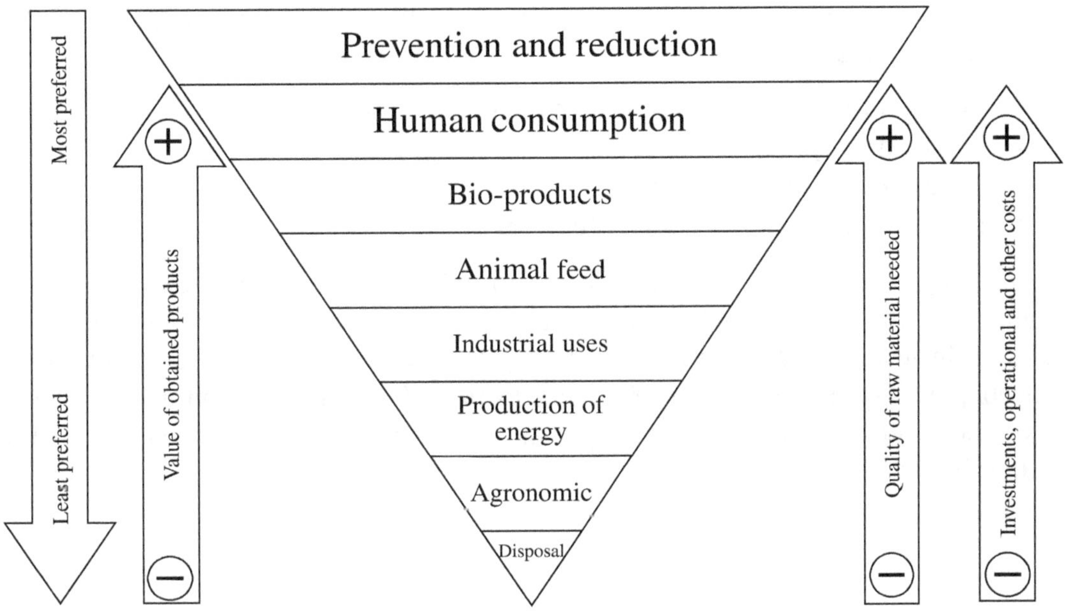

Fig. 17.1 Fish waste management hierarchy adapted from the European waste framework directive and the United States Environmental Protection Agency

To select the most suitable valorisation solution in a specific scenario, many criteria must be considered at the same time. This complicated task requires a suitable systematic methodology. All aspects that may determine the feasibility of a valorisation option can be classified into four main categories:

1. Characteristics of raw material which determine logistics needs and potential end products:
   - Variability, seasonality and geographic dispersion of landings
   - Physicochemical and microbiological characteristics of landed catches
2. Technical paramcters, related to the technical feasibility of a solution, such as:
   - Maturity of the production process
   - Ratio, quality and purity of the product obtained
   - Availability of technology and equipment at an industrial scale
   - Feasibility of modifications on board vessels
   - Availability of shore-based facilities for storage, preservation, logistics, and processing
3. Market aspects that affect product characteristics and their marketability:
   - Compliance with health, environmental and other specific regulations for each use
   - Existence of potential users or market acceptance for new products
   - Existence of a gap in the market or of demand for an existing product
   - Competitors and analogues
   - Quality requirements and volume of available product to satisfy demand

4. Economic aspects are the factors that affect the economic feasibility of the solution such as:

   - Minimum volume of raw material for viable production
   - Final value / price of product
   - Expected cost-benefit ratio
   - Efficient use of existing infrastructures

In many cases, it is almost impossible to collect all the information needed to evaluate all the options, and feasibility studies are quite expensive and time consuming. Instead, we first identify more than 30 products from valorisation options in this chapter. Two methodologies for the selection of valorisation options have been developed with the aim of creating a common, systematic and objective pathway for addressing the decision process.

17.2 Potential Uses of Unwanted Catches

Existing valorisation options and their resulting products were reviewed and compiled. Then, following the waste management hierarchy (Fig. 17.1), these products were classified as follows: food applications, bio-products, animal feed, industrial uses, energy production or agronomic uses (Table 17.1). This prioritisation does not refer to the value of the product, as different markets or uses can strongly affect value.

There is a global trend towards increased demand for fish for human consumption compared to other uses. Fish protein contributes an average 17% of the total animal protein intake globally and in some countries up to 23% (FAO 2017). For UWC > MCRS, there are many solutions in the area of food product innovation. Global consumer trends in industrialised countries show opportunities for new product developments in the categories of processed and ready-to-eat food. Fish pulp, obtained from fish muscle as a basis for making restructured products or surimi derivatives, are intermediate products that can be good options when a critical mass is available and the freshness of the raw material is guaranteed.

When UWC cannot be used for direct human consumption, there are many alternative uses that are gaining in importance and may lead to important revenue streams. A brief description is given below, and more detailed data sheets can be found in Iñarra et al. (2018).

Seafood contains a variety of high-value biomolecules that can be used in food, pharmaceuticals and cosmetics, such as in the animal feed industry (pet food, aquaculture and cattle).

1. Bioactive Peptides: come from the extensive enzymatic hydrolysis of fish protein. They present biological activities, such as antihypertensive, antibacterial, anticoagulant, anti-inflammatory or antioxidants, which make them valuable for food, pharmaceuticals, cosmetics and feed products.

Table 17.1 Main valorisation options of unwanted catches by product category

Category	Valorisation option
Food Applications	New fish products
	Surimi
	Fish pulp
Bio-products	Bioactive peptides
	Polyunsaturated fatty acids
	Enzymes
	Chondroitin sulphate
	Fat-soluble vitamins
	Minerals
	Astaxanthin
	Collagen
	Gelatine
	Sterols
	Insulin
	Protamine
	Hyaluronic acid
	Chitin/chitosan
	Phospholipids
	Peptone
	Squalene
Animal feed	Fishmeal
	Fish oil
	Mink feed
	Marine beef/bait
	Direct pig feed
	Protein concentrate
	Protein hydrolysate
	Silage
	Insects growth medium
Industrial uses	Leather
	Fish oil
	Minerals
	Chitin/chitosan
	Pearl essence
Energy	Biogas
	Biodiesel
Agronomic uses	Fertilisers
	Compost

2. Polyunsaturated fatty acids (PUFAs): come from the purification of fish oil, obtained from viscera or from fatty fishes, and are fats with more than one unsaturation (double bonds) present in their chain. PUFAs include important compounds such as essential fatty acids that are correlated with human cardiovascular health.

3. Proteases and proteolytic enzymes: are extracted from by-products, especially viscera, that contain a substantial proportion of digestive enzymes, with different specific functions. These include collagenases, trypsin, pepsin, chymotrypsin, elastase, and carboxypeptidase. These enzymes extracted from fish are active at low temperature and pH. Proteases play a key role in a wide variety of physiological processes, biotechnology, food processing and other industries.
4. Chondroitin sulphate: is obtained by an enzymatic or chemical hydrolysis process to deproteinise cartilage phases from the skeleton of cartilaginous fish, sharks and rays, followed by successive purification steps. Chondroitin sulphate gives cartilage its mechanical and elastic properties and gives it a large part of its resistance to compression. Chondroitin sulphate is used as a dietary supplement with anti-inflammatory properties, to ease arthritis symptoms.
5. Fat-soluble vitamins: are obtained by solvent extraction of vitamins from fish oil. Vitamins are classified as either fat soluble (vitamins A, D, E and K) or water soluble (vitamins B and C) depending on how they act within the body. Fish liver oil, rich in vitamins A and D, is used in pharmaceutical, cosmetic and food applications.
6. Minerals (Calcium, $CaCO_3$, hydroxyapatite): are obtained from fish bones and shells of bivalve molluscs (mussels, clams, etc.). They can be used as mineral supplements in nutraceutical products (for human or animals) as food ingredient and in technically lower value applications such as soil improvers or mineral fertilisers.
7. Dye/pigments (Astaxanthin): is extracted mainly from crustacean shells. It is used as a pigment in aquaculture, in fish and crustacean feed and as antioxidant in nutraceutical formulations.
8. Collagen: is obtained by an acid or base treatment of spines, scales and skin. The amino acid content of collagen differs from other proteins because of their high content of proline and hydroxyproline. Collagen is widely used in the pharmaceutical and cosmetic industries and as food supplement.
9. Gelatine: is obtained from the partial hydrolysis of collagen. There are two main types of gelatines: Type A is obtained from the acid hydrolysis procedure and Type B from the alkaline hydrolysis procedure. Gelatine is used as gelling agent in food, pharmaceuticals and cosmetics. Fish gelatines are preferred for low temperature gelling needs.
10. Sterols: are obtained from either plants or animals. Phytosterols, which have received considerable attention in recent years due to their cholesterol-reducing properties, can be found in marine organisms whose diet is mainly made up of phytoplankton. An important source of phytosterols is bivalves (filtering organisms). Phytosterols are increasingly demanded as functional ingredients in the food and beverage industry.
11. Insulin: is extracted from the viscera of various fish. Insulin is a peptide hormone produced by beta cells of the pancreatic islets, and by the Brockmann body in some teleost fish. Insulin regulates the amount of glucose (sugar) in the blood and is required for the body to function normally. It is used for treating diabetes.

12. Protamine: is a purified mixture of simple proteins obtained mainly from wild salmon sperm. Protamine is a protein (molecular weight around 4–5 kDa) which works to maintain and protect DNA from being damaged. It is used in pharmaceuticals as a drug that reverses the anticoagulant effects of heparin by binding to it.
13. Hyaluronic acid: is a glycosaminoglycan present in skin, bones and joints. It is obtained by successive extraction and purification steps. Its function is to give elasticity to skin, bones and joints. It is used in regenerative skin cosmetics, in cosmetic surgery injections or in the recovery of joint injuries.
14. Chitin/Chitosan: Chitin is obtained by deproteinisation and discoloration of the exoskeleton of arthropods. Chitosan is obtained by further deacetylation of chitin by chemical-enzymatic processes. Chitosan, in various modified forms and different degrees of purity, can be used in a wide range of applications. It is used in food applications, in edible films or in microencapsulation of ingredients; in pharmaceuticals, in nutritional supplements as fat binder; in aquaculture and ruminant feeding to reduce infections and improve yield; in medicine as material in histocompatible tissues and contact lenses; and in cosmetics in foams. One of its main applications is as a food-grade flocculant in water treatment and in paper manufacturing.
15. Phospholipids (PLs): are extracted from fish oil using different procedures. Marine PLs contain essential omega-3 PUFAs, some of which are only present in marine sources. PLs are used as emulsifiers in the food industry, or as emollients in cosmetics, antibacterials or in drug delivery.
16. Squalene: is extracted mainly from shark liver. A hydrocarbon compound, isoprenoid, is intermediate in the synthesis of cholesterol, hormones and vitamin D. It is used in cosmetics as a lubricant and in pharmacy or dietary supplements as an immunostimulator.
17. Peptones: are produced by controlled enzymatic hydrolysis of proteins. Peptones are a mixture of polypeptides and amino acids formed during the enzymatic degradation of proteins. They are the main source of nitrogen in the organic media for bacterial culture. They are used in the manufacture of culture media for microbiology and biotechnology (industrial fermentations).

The most common use of fish by-products is the production of fishmeal and fish oil that is mainly used for animal feed. This alternative may also be of major interest for the processing of UWC as there are many infrastructures available that are generally close to harbours. However, many other valorisation options focused on animal feed can be considered.

18. Fishmeal: is obtained from any fish or fish by-products. After a thermal process to coagulate the protein and separate the oil, fishmeal is a brown powder rich in protein. The colour is affected by the fish species, particle size, fat and moisture content. Fishmeal is mainly used in animal feed with aquaculture consumption accounting for >60%, pigs 25%, and poultry 8%.
19. Fish oil: is obtained in the same process as fishmeal production. Fish oil is a liquid product composed mainly of fatty acids that are highly unsaturated, with

variable amounts of phospholipids, glycerol ethers and wax esters. Fish oil has different uses that can vary depending on its composition. About 80% of fish oil is used in aquaculture and ~13% is destined for human consumption. When it does not meet feed quality standards it can be used in technical applications as solvent for painting or in biodiesel production.

20. Protein concentrates: Are dehydrated and ground products with a variable protein content, which may or may not taste or smell of fish, depending on the production method used. The aim is to achieve a stable product with a protein concentration higher than that of fish muscle. This type of product can use species that are not appropriate for direct consumption as well as the waste from fish processing industries. These concentrates are used for animal feed but due to their high nutritional value they can also be used for human consumption or as a protein source in the elaboration of different foods.
21. Protein hydrolysates: are prepared from the protein fraction of whole fish, by-products or processing waters. Hydrolysates are produced by chemical or enzymatic hydrolysis and consist of mixtures of amino acids and peptides (fragments of protein) of varying sizes depending on the degree of hydrolysis. Protein hydrolysates provide different technological properties such as flavouring or texturising agents. Their biological activities are also being studied scientifically.
22. Silage: Is a liquid protein hydrolysate made from whole fish or from processed residues. The hydrolysis is carried out by endogenous proteolytic enzymes located in the viscera and in the meat of the fish under acidic conditions. Acid conditions limit the growth of degradative bacteria. It is used mainly as a protein supplement in animal feed (cattle, poultry and aquaculture) and as a base to produce fish sauce.
23. Mink feed: any fish or fish by-product can be used to feed mink for the fur industry.
24. Marine bait: discarded species can be used as effective pot bait in the crab fishery.
25. Insect meal and oil: are obtained after rearing insects on a fish substrate to increase protein content. Insect meal can be used for animal feed.

At the bottom of the valorisation hierarchy are other technical options such as energy production and agronomic uses:

26. Pearl essence: is extracted from fish scales. Guanine is an iridescent substance that is found in the epidermal layer and scales. The suspension of guanine in a solvent is called "essence of pearls". It was formerly used in cosmetics and paints.
27. Fish leather: is the cured and tanned skins of fish. Fish leather can be used to make a wide variety of items such as jewellery, accessories, belts, wallets, bags and in shoes. It can also be used for a wide variety of crafts.
28. Biogas: is produced through the anaerobic digestion of organic matter. This is a complex biological process in which anaerobic bacteria decompose organic matter in environments with little or no oxygen. The process produces biogas (55–65% methane, 35–45% carbon dioxide, and other gases) used as energetic

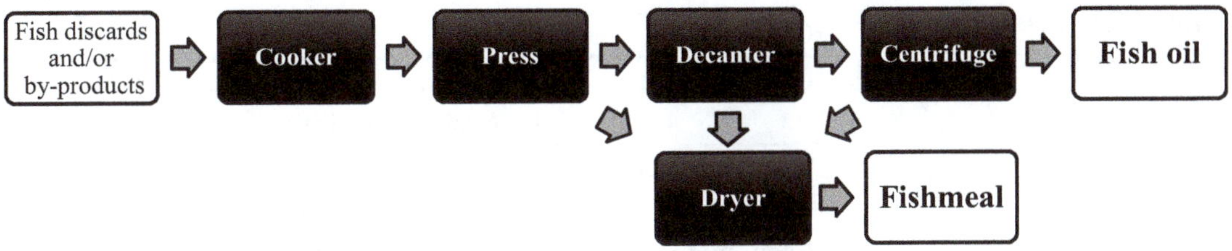

Fig. 17.2 Simplified stepwise schema of a fishmeal and oil production process

source for heating or producing electricity. A digested substrate is produced that can be used as fertilizer.

29. Compost/fertilisers: are obtained by an aerobic decomposition process carried out by the microorganisms of the organic matter. Compost made from fish usually consists of fish waste, sawdust, wood-bark chips and is covered with leaf compost to make a compost pile. The compost is used as a soil amendment or fertiliser. Fish protein hydrolysates can also be used as fertiliser.

Many of these processes of UWC valorisation can be run concurrently in a biorefinery scheme. For instance, when producing a food product, a first biorefinery step takes place when fish meat is separated from viscera, heads and bones. The latter can then be further processed to obtain other valuable products. Biorefinery is the integrated sustainable process that transforms a biological raw material (animal or plant material) into a spectrum of marketable products (e.g. food, feed, materials or chemicals) or energy (e.g. fuels, power and heat).

The simplest biorefinery scheme is to obtain fishmeal and fish oil where, when using a good stickwater recovery system, all the treated raw material provides marketable products. A simplified processing scheme is shown in Fig. 17.2. In brief, the raw material (UWC or fish by-product) is thermally treated and coagulated protein and oil are then separated and recovered.

17.3 Simplified Methodology for the Selection of Potential Uses

The methodology for the selection of the potential uses for UWC in a specific scenario has been developed within the H2020 DiscardLess project (Grant Agreement n° 633,680) and is based on a multi-criteria decision analysis (MCDA) using an analytic hierarchy process (AHP) method. MCDA provides a reliable framework for procedures to rank alternative options and prioritise, based on their assessment across selected criteria, and such methods have been widely and effectively applied in different environmental areas (San Martin et al. 2017).

AHP was introduced as the most appropriate method, because it allows the problem to be partitioned into smaller decision sets that are addressed one at a time. The first step is to define and evaluate the criteria, which must be done case by case, as it is adapted to the subject of the study and stated by consensus.

Table 17.2 Categories and Criteria for the MCDA

Category	Criteria	Units
CS dependent	A: Available raw material	T/year
	B: Available facilities	N° facilities
Technical factors	C: Yield	%
	D: Technology maturity	
Economic factors	E: Value of the product	€/kg
	F: Potential market	Kg/year
	G: Production costs	€/kg
	H: Competing companies	Kg/year

Principles applied to the selection criteria to evaluate the main parameters involved in the process are:

- *systemic principle*: a criteria system should reflect the essential characteristic and the whole performance system
- *measurability principle*: the criteria should be measurable in quantitative values or qualitative criteria should be transformed into numbers
- *comparability principle*: criteria must be either comparable or normalised.

For the final evaluation, the selected criteria have been divided into three groups: Case Study (CS) dependent, technical and economic criteria (Table 17.2).

"Available raw material" is the amount of UWC that can be processed in this way.

The existence of "available facilities" and local infrastructures in the region should be positive.

"Yield" criteria represent the result of the proportion of fish that can enter this valorisation option as well as the yield of the process for the production of the valuable compound or final product.

"Technology maturity" refers to the industrial feasibility and necessary investment cost for implementing the solution. Maturity generally implies technical feasibility (availability of the technology) and lower implementation costs.

"Value of the product" stands for the market value of the product or compound.

"Potential market" is an indicator of demand for the product to be marketed.

"Production costs" account for the different costs involved in the production of the product.

"Competing companies" reflects the quantities produced or the size of the competitor companies.

The methodology has four main steps:

1. Data gathering for each valorisation option (including new options when identified)
2. Evaluate the facilities available for each option
3. Evaluate the amount of UWC available for each option
4. Complete the evaluation and prioritisation table

For the first step – data gathering for each valorisation option – an exhaustive list of valorisation options according to the end-product have been assembled in a single sheet that allows a preliminary evaluation of the proposed solutions to be obtained (Iñarra et al. 2018). Further available options can be added to this list and can be weighted during the evaluation.

Second, the evaluation of existing and available facilities is carefully done in each case study. The selection of an option that is already industrialised has a great advantage and can be a straightforward and short-term solution.

Third, historical discard data or UWC landing data are used to determine and evaluate the amount of raw material available for each option. When using historical discard data, a careful evaluation becomes essential. It cannot be supposed that 100% of discards will be landed due to selectivity gear improvement, better fishing strategies or minimal applications.

After estimating the amount of UWC landed and considering the species and quality, the amount of raw material for each option can also be estimated, keeping in mind that catches under minimum conservation reference size (MCRS) cannot be used for direct human consumption.

A table linking the species with their possible valorisation options has been made (Iñarra et al. 2018). Table 17.3 shows the part of that table used for the Bay of Biscay case study (for details, see Sect. 17.5.1). The selected options are based on the species composition (Froese and Pauly 2018).

The quantitative values obtained in the first step of the methodology for each evaluation criterion and for all the valorisation options are then evaluated jointly. Subsequently, average ranges are established to define different levels for each criterion prioritisation (Table 17.4). Each range is then assigned a score. The more favourable the evaluation, the higher the score (Table 17.5); this is usually 5, 3, 1 and 0. This allows small differences to be highlighted in following calculations. Considerations are that some factors have a negative effect (competing companies and production costs) and therefore have a high qualitative factor if their quantitative factor is low.

A weighting coefficient has been assigned to each prioritisation criterion and each valorisation option will obtain a score (a value between zero and one) based on the following equations:

$$V_{CS} = (x_1 \bullet A + x_2 \bullet B)/(5 \bullet (x_1 + x_2))$$

$$V_{tech} = (x_3 \bullet C + x_4 \bullet D)/(5 \bullet (x_3 + x_4))$$

$$V_{eco} = (x_5 \bullet E + x_6 \bullet F + x_7 \bullet G + x_8 \bullet H)/(5 \bullet (x_5 + x_6 + x_7 + x_8))$$

Where V_{CS} is the score obtained for the case study-dependent criteria, V_{tech} is the score of the technical criteria and V_{eco} is the score of the economic criteria. x_1 to x_8 are the weighting coefficient values assigned to each criterion for the prioritisation and A, B, C, D, E, F, G, the score normalised from 0 to 5 for each criterion as shown in Table 17.5.

Table 17.3 Valorisation options for distinct species based on a case study as part of the H2020 DiscardLess project in the Bay of Biscay

	ANY FISH	Blackbellied angler(*Lophius budegassa*)	European hake (*Merluccius merluccius*)	Horse mackerel (*Trachurus trachurus*)	Atlantic mackerel (*Scomber scombrus*)	Megrim (*Lepidorhombus whiffiagonis*)	Angler (*Lophius piscatorius*)	Blue whiting (*Micromesistius poutassou*)
New fish products		X	X	X	X	X	X	X
Surimi		X	X	X	X	X	X	X
Fish pulp		X	X	X	X	X	X	X
Bioactive peptides				X	X			
Polyunsaturated fatty acids		X	X	X	X	X	X	X
Enzymes								
Chondroitin sulphate								
Fat-soluble vitamins				X	X			
Minerals		X	X	X	X	X	X	X
Astaxanthin								
Collagen			X		X	X		
Gelatine			X		X	X		
Sterols				X				
Insulin		X	X	X	X	X	X	X
Protamine								
Hyaluronic acid								
Chitin/chitosan								
Phospholipids				X	X			

Peptone	X	X	X	X	X	X	X	X
Squalene								
Fishmeal	X	X	X	X	X	X	X	X
Fish oil	X	X	X	X	X	X	X	X
Mink feed		X	X	X	X	X	X	X
Marine beef/bait		X	X	X	X	X	X	X
Direct pig feed		X	X	X	X	X	X	X
Protein concentrate	X	X	X	X	X	X	X	X
Protein hydrolysate	X	X	X	X	X	X	X	X
Silage	X	X	X	X	X	X	X	X
Insect growth		X	X	X	X	X	X	X
Leather								
Fish oil	X	X	X	X	X	X	X	X
Minerals		X	X	X	X	X	X	X
Chitin/chitosan								
Pearl essence			X	X	X			
Biogas	X	X	X	X	X	X	X	X
Compost	X	X	X	X	X	X	X	X
Fertilisers	X	X	X	X	X	X	X	X

Table 17.4 Normalisation of range values of prioritisation criteria

Category	Criteria	Units	High	Medium	Low	Null
Case study dependent	A: Available raw material	t/year	> 2000	1000–2000	500–1000	< 500
	B: Available facilities	N° facilities	> 2	2	1	0
Technical	C: Yield	%	> 50	10–50	< 10	< 0.05
	D: Technology maturity	–	High	Medium	Low	Experimental
Economic	E: Value of the product	€/kg	> 50	5–50	0.5–5	< 0.5
	F: Potential market	t/year	> 1000	100–1000	5–100	< 5
	G: Production costs	€/kg	> 50	5–50	0.5–5	< 0.5
	H: Competing companies	t/year	> 500	100–500	< 100	0

Table 17.5 Assignment of numerical scores to each value range

Category	Criteria	5	3	1	0
Case study dependent	A: Available raw material	High	Medium	Low	Null
	B: Available facilities	Many and/or nearby	Not many and/or far away	Experimental and/or pilot	Null
Technical	C: Yield	High	Medium	Low	Null
	D: Technology maturity	High	Medium	Low	Experimental
Economic	E: Value of the product	High	Medium	Low	Null
	F: Potential market	Big	Medium	Small	Null
	G: Production costs	Very low	Low	Medium	High
	H: Competing companies	Low-none	Medium	High	Saturated

Weighting coefficient values, x_1 to x_8 (Table 17.6) should highlight the importance of each criterion in the final decision and are usually values between 1 and 10:

- Key/critical factor: 10 points
- Very important factor: 7 points
- Factor with some relevance: 3 points
- Factor with small relevance: 1 point

The final score or priority value (V_p) for each solution comes from the product of the technical and economical score.

Table 17.6 Prioritisation evaluation of valorisation options for the Bay of Biscay case study

	CS dependent			Technical parameters			Economical parameters					
Criterion	**A**	**B**	**V_{CS}**	**C**	**D**	**V_{tech}**	**E**	**F**	**G**	**H**	**V_{eco}**	**V_p**
Weighting coefficient	10	7	**10**	7	3	**7**	10	10	7	1	**3**	
New fish products	5	5	**1.00**	5	5	**1.00**	3	5	3	1	**0.73**	**0.96**
Surimi	5	0	**0.59**	3	5	**0.72**	3	3	1	1	**0.49**	**0.62**
Fish pulp	5	5	**1.00**	3	5	**0.72**	1	5	5	5	**0.71**	**0.86**
Bioactive peptides	5	1	**0.67**	3	3	**0.60**	5	3	1	3	**0.64**	**0.64**
Polyunsaturated fatty acids	5	3	**0.84**	3	5	**0.72**	3	5	3	1	**0.73**	**0.78**
Enzymes	0	0	**0.00**	1	1	**0.20**	1	1	1	1	**0.20**	**0.10**
Chondroitin sulphate	0	3	**0.25**	1	3	**0.32**	5	3	1	1	**0.63**	**0.33**
Fat-soluble vitamins	5	0	**0.59**	1	3	**0.32**	3	5	1	1	**0.63**	**0.50**
Minerals	5	1	**0.67**	3	5	**0.72**	1	5	5	3	**0.70**	**0.69**
Astaxanthin	0	0	**0.00**	1	5	**0.44**	5	3	1	3	**0.64**	**0.25**
Collagen	5	3	**0.84**	1	5	**0.44**	3	3	3	1	**0.59**	**0.66**
Gelatine	5	3	**0.84**	1	5	**0.44**	3	3	3	1	**0.59**	**0.66**
Sterols	5	0	**0.59**	1	1	**0.20**	3	3	0	3	**0.45**	**0.43**
Insulin	5	0	**0.59**	3	1	**0.48**	3	1	0	0	**0.29**	**0.50**
Protamine	0	0	**0.00**	1	3	**0.32**	5	1	1	1	**0.49**	**0.18**
Hyaluronic acid	0	0	**0.00**	1	3	**0.32**	5	3	1	1	**0.63**	**0.21**
Chitin/chitosan	0	0	**0.00**	1	5	**0.44**	5	3	3	1	**0.73**	**0.26**
Phospholipids	5	0	**0.59**	1	3	**0.32**	3	3	1	3	**0.50**	**0.48**
Peptone	5	1	**0.67**	3	5	**0.72**	1	1	3	1	**0.30**	**0.63**
Squalene	0	0	**0.00**	1	3	**0.32**	5	3	1	1	**0.63**	**0.21**
Fishmeal	5	5	**1.00**	5	5	**1.00**	1	5	5	1	**0.69**	**0.95**
Fish oil	5	5	**1.00**	5	5	**1.00**	1	5	5	1	**0.69**	**0.95**
Mink feed	5	3	**0.84**	5	3	**0.88**	1	1	5	3	**0.41**	**0.79**
Marine beef/bait	5	1	**0.67**	5	3	**0.88**	1	1	5	3	**0.41**	**0.71**
Direct pig feed	5	1	**0.67**	5	3	**0.88**	1	3	5	3	**0.56**	**0.73**
Protein concentrate	5	3	**0.84**	5	5	**1.00**	3	5	3	3	**0.74**	**0.88**
Protein hydrolysate	5	3	**0.84**	5	5	**1.00**	3	5	3	3	**0.74**	**0.88**
Silage	5	0	**0.59**	5	5	**1.00**	1	3	5	3	**0.56**	**0.73**
Insect growth	5	0	**0.59**	5	1	**0.76**	1	3	5	5	**0.57**	**0.65**
Leather	0	3	**0.25**	1	5	**0.44**	3	1	3	5	**0.47**	**0.35**
Fish oil	5	5	**1.00**	5	5	**1.00**	0	5	5	0	**0.61**	**0.94**
Minerals	5	3	**0.84**	3	5	**0.72**	0	3	5	3	**0.49**	**0.74**
Chitin / chitosan	0	0	**0.00**	3	5	**0.72**	1	3	3	1	**0.44**	**0.32**
Pearl essence	5	0	**0.59**	1	5	**0.44**	1	1	3	3	**0.31**	**0.50**
Biogas	5	1	**0.67**	1	3	**0.32**	1	3	3	1	**0.44**	**0.51**
Compost	5	1	**0.67**	3	3	**0.60**	0	3	5	1	**0.47**	**0.62**
Fertilisers	5	1	**0.67**	3	5	**0.72**	1	3	5	1	**0.54**	**0.67**

A Available raw material, *B* Available facilities, *C* Yield, *D* Technology maturity, *E* Value of the product, *F* Potential market, *G* Production costs, *H* Competing companies, V_{CS}: case study score, V_{tech} technical score, V_{eco} economic score, V_p priority value

$$V_p = (y_1 \cdot V_{CS} + y_2 \cdot V_{tech} + y_3 \cdot V_{eco})/(y_1 + y_2 + y_3)$$

Where y_1 to y_3 are the weighting coefficient values assigned to each category (Table 17.6).

The weighing coefficient value for each prioritisation criterion and category was assigned through consensus within the DiscardLess project expert team. As an example, the availability of raw material (criterion A) was considered a critical factor and was therefore assigned a weighing coefficient value of 10.

The current values were used considering the present situation where the LO is being implemented, and thus the preferred options are those building on existing infrastructures (high V_{cs} weighting coefficient). These coefficients can be changed for a long-term solution by increasing the weighting coefficient of V_{eco}.

This methodology not only allows simultaneous evaluation of all the criteria but also the technical and economical evaluation of each valorisation option separately, and evaluation of the weight of the case study dependent criteria. However, only a preliminary diagnostic is provided which should be discussed further with stakeholders.

The implementation of any selected valorisation option resulting from this methodology will need a more detailed feasibility study.

In Table 17.6, the example of the prioritisation analysis for a specific case study is presented, i.e. the discards of the Bay of Biscay.

17.4 Methodology for the Selection of Inland Management Alternatives Based on Economic and Environmental Impacts

When selecting the optimal processing routes of the different potentially available biomasses in terms of sustainability, both the economic and environmental objectives must be considered simultaneously. A common approach in the scientific literature There is a major tendency in the scientific literature to apply different optimisation strategies to study the trade-off between these two conflicting objectives. The optimisation of processing routes integrating both criteria was for example applied to the case of bio-refineries involving different feedstock types (Martinez-Hernandez et al. 2013; Murillo-Alvarado et al. 2013; Bernardi et al. 2013; Santibanez-Aguilar et al. 2014; Antelo et al. 2015).

In a second case study below, we define an optimisation screening approach adapted to the particularities of marine biomass discarded in trawling fleets of Galicia (NW Spain) operating in ICES areas VIIIc and IXa. These fleets are characterised by highly mixed catches where determined levels of discards are very significant.

To simplify and systematise the selection of optimal valorisation pathways, we have defined a simplified general network (Fig. 17.3) to represent these valorisation alternatives. In this approach, developed in the framework of the LIFE iSEAS project (LIFE13 ENV/ES/000131), each layer of the above-mentioned

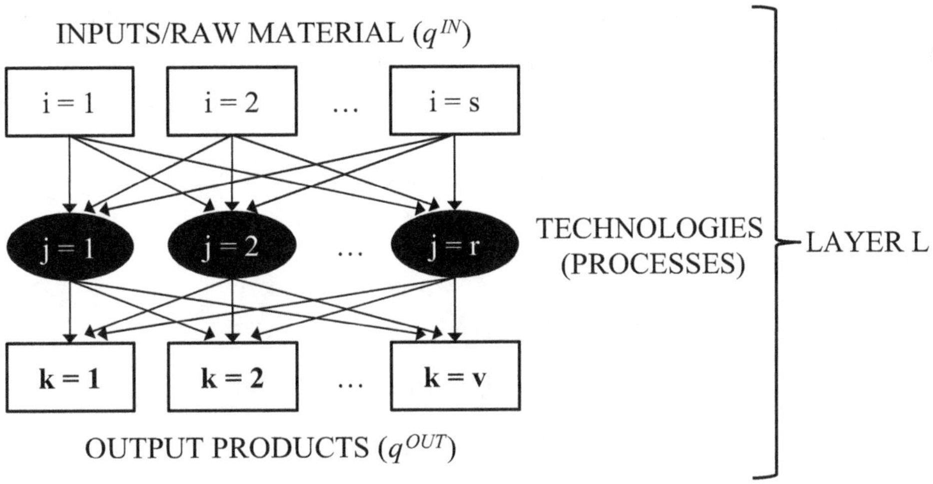

Fig. 17.3 General superstructure of the technology model

superstructure is constituted by s input products (i.e., skins, bones, viscera, muscle, entire specimen, etc. – vector $\boldsymbol{q}^{IN}$) that can be processed by a set r of defined processing technologies to obtain v final or output products (i.e., bioactive peptides, fish protein hydrolysates, chitin/chitosan, gelatine, etc.) This formulation allows easy connection of L layers (when further pre-processing or downstream processing is required) by considering the final products of a given layer l the raw material of the next processing network/layer $l + 1$. We call this the *technology model.*

Our case study describes a situation where 100% of all currently discarded biomass must be landed. Changes in the composition of the considered target fisheries as well as the uncertainties of future legislation might lead to significant variations in the proposed case. In addition, the economic assessment includes costs of the main processing utilities. Water use and CO_2 emissions were considered in the environmental assessment. Installation and personnel costs, as well as solid residues derived from processing (considering that these are sent to fish meal production at zero cost), were not included in the analysis. The reason for these assumptions is the increasing complexity of the network and the uncertainty regarding the available data in the case study.

To evaluate the optimal/best pathways for achieving an integral valorisation of discarded biomass in practice, both economic (J_{eco}) and environmental (J_{en}) objectives are considered and evaluated through a multi-objective approach. The aim is to maximise the economic objectives while minimising environmental impacts.

The *ε-constraint approach* was used to convert multi-objective problems into a set of single objective problems (by incorporating one of the objectives as an inequality constraint) to obtain uniform distributed Pareto fronts – due to their easy implementation and capability. In this case, we considered that ε varies between 0 and 700. Computing the Pareto-optimal set can be challenging due to the highly constrained and nonlinear nature of processing systems. This drawback can be addressed by using suitable global optimisation (GO) methods. Therefore, the optimal solution of the proposed optimisation problem was calculated using scatter search (*eSS*) as implemented in the MEIGO toolbox (Egea et al. 2014).

This optimisation problem can be mathematically represented as:

$$max_X J = J_{eco}(X) = J_{sales}(X) - J_{PC}(X)$$

Subject to:

$$J_{env}(X) = (J_{CO_2}(X) + J_w(X)) \leq \varepsilon$$

$$q^{OUT}(l+1) \leq M_o \quad (Plant\ capacity)$$

$$X^{lower} \leq X \leq X^{upper}$$

where $\boldsymbol{X}$ is the decision vector to be found by means of the optimisation problem (in this case, the set of fractions or percentages of raw material q^{IN} processed by a given technology j) and $\boldsymbol{X_{lower}}$, $\boldsymbol{X_{upper}}$ are the lower and upper bounds for the decision variables, respectively.

The economic part of the objective function represents the profit of the process, which is defined by product sales (J_{sales}) minus production costs (J_{PC}).

The environmental impact of each process was characterised by the ecological footprint (EF). This is an indicator that considers the energy and raw material flow into and out of any particular system, converting them into the spaces of land or water needed by nature to produce and/or assimilate these flows. In this case, environmental criteria for process selection included CO_2 emissions (from electricity and fuel consumption – J_{CO_2}) and water consumption (J_w). The calculation of EF implies the conversion of units for these flows to space units, usually hectares (ha). For that purpose, values of energy intensity as well as natural and/or energy productivity are required.

The main results show that in general the most optimal processing routes correspond to the production of high value-added products (biopeptides, enzymes and chondroitin sulphate), not only due to high sale prices, but also for the lower environmental impact associated with their production processes as compared to other products (fish meal, fish oil, chitin/chitosan or gelatine). However, chondroitin sulphate production should be considered with caution, as the production obtained was much lower than the plant production capacity. In this case, more biomass (from other fishing métiers or fish-processing industries) would be necessary to guarantee the economic feasibility of the valorisation schemes. Fish meal and fish oil, chitin and gelatine were not preferred, mainly due to the high CO_2 emissions and water consumption associated with these processes.

17.5 Case Studies

The selection of potential uses and solutions must be made in each case study. The simplified methodology described is shown for the case studies of Bay of Biscay (BoB), while the methodology based on economic and environmental parameters is applied to the Galician fleet.

17.5.1 Basque Fleet in the Bay of Biscay

The information needed for the application of the simplified methodology for the selection of valorisation options was gathered within the H2020 DiscardLess project. The Bay of Biscay (BoB) is a highly productive fishery zone, in which the Basque fleet operates bottom trawlers (Prellezo et al. 2016).

Data related to main UWC, that represent over the 95%, are presented in Table 17.7.

From these data two problems arise:

1. Mackerel and horse mackerel >MCRS but with low value account for ~ 2200 t/year. Blue whiting, with an important quantity of ~300 t/year, can be included in this group.
2. Hake < MCRS account for ~300 t/year

From the study of the infrastructures in the Basque country, there are different facilities that can be of interest for the valorisation of these products:

- The *Cofradía* (fisher associations) of Bermeo has facilities for fish mince production.
- There are several fish processing industries in the surrounding areas.
- There is a fish by-product valorisation facility that produces fish meal and fish oil.
- In all harbours, freezing facilities are available for conserving UWC.

Table 17.7 Main unwanted catches in the Bay of Biscay bottom-trawl fishery estimation (data 2014)

	Discards (t)			Catch (t)			Landed		
Species	2011	2012	2013	2011	2012	2013	2011	2012	2013
Anglerfish	3	0	5	197	178	254	194	178	249
Black-bellied angler	11	3	10	147	215	415	136	212	405
Blue whiting	117	439	226	250	488	282	133	49	56
Hake	217	309	365	2916	2401	2370	2698	2092	2005
Horse mackerel	3049	2091	1467	3227	2317	1618	178	226	151
Mackerel	3339	1620	990	3728	1693	1035	389	73	45
Megrim	5	1	8	176	210	306	170	209	298

Following the standard methodology for valorisation option selection, discard data were used to obtain scores for criteria A (available raw material) as shown in Table 17.6. Also, the infrastructures available in the Basque Country and surrounding areas were evaluated to complete column B (available facilities).

The valorisation options with the best scores for the main discarded species are:

- Mackerel and horse mackerel study: new food products (96%) or fish pulp (86%) have the higher score. This option can also be taken into consideration for blue whiting (~300 kg/year) if not < MCRS.
- Due to legislation, hake (< MCRS) cannot be directed to food products so the next options are: fishmeal and fish oil, with a score of 95% or fish protein hydrolysate, with a score of 88%. This last option could score higher if hydrolysates were focused on human consumption as a flavouring agent.
- This option is also the main possibility when quantities of other UWC are too low for more specific solutions.

As stated above, these results are preliminary. More detailed evaluation of the solution implementation was performed to corroborate their technical and economic viability. Pilot tests for producing a mackerel hamburger and hake FPH were performed to assess the full value chain and verify its profitability. In both processes, the by-products (e.g. bones) were diverted to the fishmeal processing plant in a biorefinery scheme.

17.5.2 Assays at Pilot Plant Level of Some of the Proposed Alternatives

The pilot plant studies were developed in the framework of the LIFE iSEAS project. Their aim was to establish a so-called *bionode* or *iDVP* (*Integral Discards Valorisation Point*) that applies a biorefinery concept for this important quantity of marine biomass that has to be managed or processed quickly and efficiently to avoid wasting it. The plant is divided into three rooms: (i) Chilled room storage; (ii) Food processing area (iDVP1, Fig. 17.4 and; (iii) Non-food products area (iDVP3, Fig. 17.5). Production lines to be implemented in the iDVP1 are based on the use of the muscle (fish mince) for food purposes, including a line of restructured products (Fig. 17.6). In addition, the iDVP3 (Fig. 17.5) includes a line for fish protein hydrolysates and bioactive peptides as well as several production lines to obtain valuable bio-compounds such as collagen, chondroitin sulphate or hyaluronic acid, among others (Fig. 17.7). iDVP1 fish by-products and undersized specimens of fish species subject to TAC regulation (mainly hake and megrim) are used as raw material. They may be complemented with industrial by-products such as tuna skins, blue shark skin and head, etc. from fish processors nearby.

The production of fish mince using a fish fraction separator (Fig. 17.4), working with initial loads of 50–100 kg of headless and eviscerated fresh fish per batch, was

Fig. 17.4 iDVP1 fish fractions separator

Fig. 17.5 iDVP3 located in the Port of Marín (Spain)

Fig. 17.6 Fish mince blocks obtained in the iDVP1

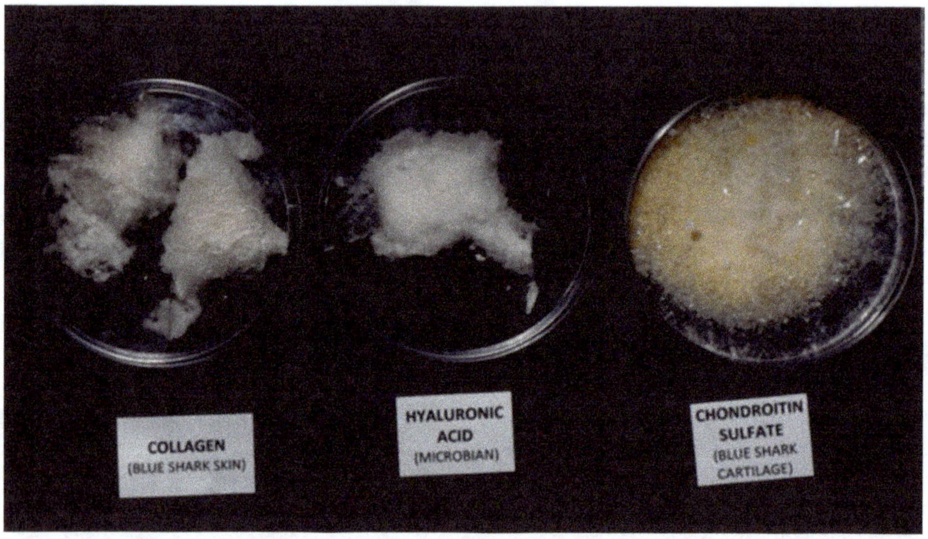

Fig. 17.7 Biocompounds obtained in the iDVP3

successful for all the discards evaluated: pouting, mackerel, sea robins, grenadier, blue whiting, hake, red scorpion fish, etc. Yields and chemical properties of fish mince blocks (Fig. 17.6) were extensively reported (Blanco et al. 2018). The rest of the heads and skin/bones were then processed in iDVP3. Different processed and restructured foods (burgers, nuggets and fingers) were formulated and satisfactorily tested on various tasting days organised for the food industry. The response of diners to the panels of evaluation (scoring the organoleptic, taste and presentation features) was always ‘good’ or ‘excellent’ for all the products prepared.

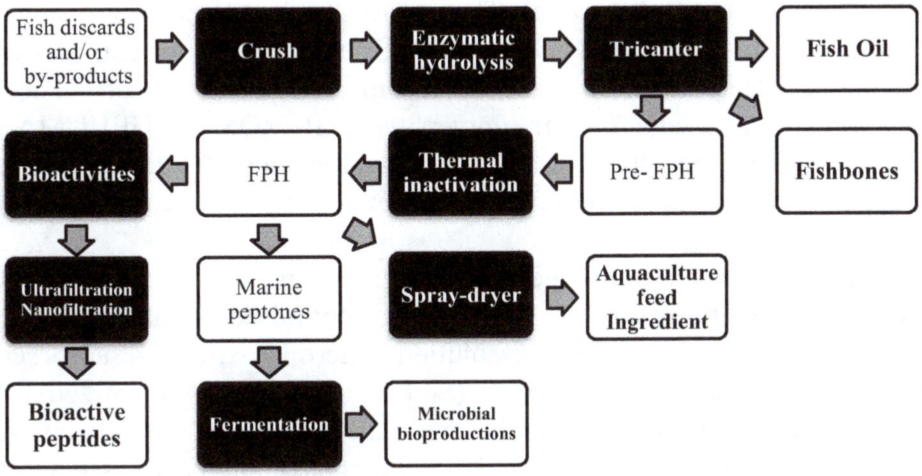

Fig. 17.8 Simplified scheme of fish protein hydrolysates (FPH) process

Another process studied was the production of fish protein hydrolysates (FPH) generated by commercial proteolytic enzymes (alcalase or esperase) using whole fish or by-products from discards (heads and skin/bones). Initially, batch enzyme proteolysis was performed in 5 L vessels and was subsequently validated in 500 L pilot plant reactors (iDVP3). The scheme of FPH-production is shown in Fig. 17.8. Unpublished results indicated that in all substrates hydrolysis (digestion of organic material) was complete. Bioactivities of peptides (anti-oxidant and antihypertensive) were analysed and the potential application of FPH as marine peptones for the formulation of low-cost media for microbial bioproduction is currently being evaluated. In the case of fatty fish species, fish oil was recovered from liquid FPH by centrifugation using a tricanter and the fatty acid profiles were determined. In addition, the chemical characterisation and application of mineral fraction obtained (clean fish bones) are also being studied.

Processes for production of other high-added value biocompounds from fish discards and fishing by-products were optimised at a laboratory (1 to 5 L per batch) and then tested at a pilot plant (iDVP3). The isolation of glycosaminoglycans, such as hyaluronic acid (AH) from fish vitreous humour and chondroitin sulphate (CS) from fish cartilage was done using materials from the eyes of tuna, blue shark and swordfish and cartilage from catshark, blue shark and ray. In both cases, the steps to produce those polysaccharides are generally based on the combination of enzymatic proteolysis, protein electrodeposition, chemical selective precipitation and re-dissolution of carbohydrates, membrane purification and drying (Murado et al. 2012; Blanco et al. 2015; Murado et al. 2010; Vázquez et al. 2016). Despite these efforts, the yields of hyaluronic acid from vitreous humour are lower than those observed by fermentation with *Streptococcus zooepidemicus* bacteria (Vázquez et al. 2010). Condroitin sulphate was obtained from catsharks (*Scyliorhinus canicula*) discarded by Galician fishing fleet. These were processed in iDVP3 (50 kg of cartilaginous material). Chondroitin sulphate from the heads of blue shark (*Prionace glauca*) and head-skeletons of ray (*Raja clavata*) was also successfully produced at pilot plant scale (loads of 40–60 kg). The resulting tailored and well-characterised glycosaminoglycans are in most cases non-existent in the chemical and clinical product market. This successful trial has led to a large number of collaborative projects with Spanish and foreign universities (IBEROS, BLUEHUMAN and CVMar+I) for the formulation of micro and nanodevices for tissue engineering and biosensor applications (Novoa-Carballal et al. 2017; Valcarcel et al. 2017; Ferreira et al. 2018).

Regarding the purification of gelatins, skins of several fish discards (pouting, mackerel, etc.) have been studied. Following an optimised set of stages based on different washes in acid and alkaline solutions, thermal extraction, adsorbent and membrane purification and final drying (Sousa et al. 2017), the extractive yields were low and not very satisfactory. Moreover, the strength of the gelatins was not remarkable. Only the skins of catsharks led to acceptable yield and viscosity values. However, skins obtained from processing fish by-products such as tuna (*Thunnus albacares*) and blue shark proved to be the best materials to produce high strength gelatine, pure collagen and derivatives (Blanco et al. 2017). Productions at pilot plant volume (50–100 kg of substrates) improved upon yields observed at lab scale (1–3 kg). Potential applications for our gelatins in the formulation of different foods and tissue regenerative biomaterials is currently under study.

Finally, the production of chitin and subsequent deacetylation to chitosan was optimised using two types of substrates: crustacean exoskeletons (Vázquez et al. 2017a) and squid gladius (pens) (Vázquez et al. 2017b). In the first case, enzymatic deproteinisation, acid demineralisation, alkaline hydrolysis and thermal/alkaline deacetylation were sequentially performed obtaining chitosan with more than 90% deacetylation degree (DD). The validation of this methodology at the pilot plant was tested on crab (*Polybius henslowii*) discards. In the second approach, endoskeletons by-products from *Illex argentinus* were valorised in a simplified protocol combining deproteinisation using proteases and deacetylation by alkaline solutions at high temperature. High purity chitosans (DD > 93%) with molecular weights (143–339 kDa) were produced. In general, the proposals of valorisation developed in IIM-CSIC were validated using iDPV1 and iDPV3 equipment. A schematic view of the biocompounds generated in iDVP3 plant is displayed in Fig. 17.8. Energy and mass flows calculations, LCA and process integration lines studies will complete these results to optimise our flexible and integral multipurpose pilot plant under the concept of Integral Marine Biorefinery.

17.6 Conclusions

Unwanted catches can be valorised in many different ways, depending on their composition. Not all of those ways are always feasible, however. A well-structured systematic methodology is therefore needed to help choose the best potential valorisation route. A first approach, based on a multi-criteria decision analysis (MCDA) using an analytic hierarchy process (AHP) method, was applied to the case of the Basque fleet in the Bay of Biscay. The best options identified for these discarded species were new fish products based on fish mince; fishmeal and oil; and protein hydrolysates. A second approach, based on the analysis and optimization of economic and environmental parameters, was applied to the highly mixed discards of the Galician fleet. Here, the preferred options were fish mince blocks, protein hydrolysates with bioactive peptides and chondroitin sulphate. For both approaches, pilot trials were performed to demonstrate their feasibility. The good results obtained in both cases indicate that both methodologies can be useful when developing valorisation strategies for UWC in other regions.

Acknowledgments DiscardLess project has received funding from the European Union's Horizon 2020 Framework Programme for Research and Innovation under grant agreement no. 633680. Life iSEAS has been co-funded under the LIFE+Environment Program of the European Union (LIFE13 ENV/ES/000131).

References

Antelo, L.T., De Hijas-Liste, G.M., Franco-Uría, A., Alonso, A.A., Pérez-Martín, R.I. (2015). Optimisation of processing routes for a marine biorefinery. *Journal of Cleaner Production*, *104*, 489–501. https://doi.org/10.1016/j.jclepro.2015.04.105.

Bernardi, A., Giarola, S., Bezzo, F. (2013). Spatially explicit multiobjective optimization for the strategic design of first and second generation biorefineries including carbon and water footprints. *Industrial and Engineering Chemistry Research, 52*(22), 7170–80. https://doi.org/10.1021/ie302442j.

Blanco, M., Fraguas, J., Sotelo, C.G., Pérez-Martín, R.I., Vázquez, J.A. (2015). Production of chondroitin sulphate from head, skeleton and fins of *Scyliorhinus canicula* by-products by combination of enzymatic, chemical precipitation and ultrafiltration methodologies. *Marine Drugs*, *13*, 3287–3308.

Blanco, M., Vázquez, J.A., Sotelo, C.G., Pérez-Martín, R.I. (2017). Hydrolysates of fish skin collagen: An opportunity for valorizing fish industry byproducts. *Marine Drugs*, *15*, 131.

Blanco, M., Domínguez-Timón, F., Pérez-Martín, R.I., Fraguas, J., Ramos-Ariza, P., Vázquez, J. A., Borderías, A.J., Moreno, H.M. (2018). Under evaluation. Valorization of recurrently discarded fish species in trawler fisheries in North-West Spain. *Journal of Food Science and Technology*.

Egea, J.A., Henriques D., Cokelaer, T., Villaverde, A.F., MacNamara, A., Danciu, D.P, Banga, J. R., Saez-Rodriguez, J. (2014). MEIGO: An open-source software suite based on metaheuristics for global optimization in systems biology and bioinformatics. *BMC Bioinformatics, 15*(1), 136. https://doi.org/10.1186/1471-2105-15-136.

European Parliament Council, 2008/98/EC of The European Parliament and of the Council of 19 November 2008 on waste and repealing certain Directives. 2008. Brussels.

FAO. (2017). FAO yearbook. Fishery and Aquaculture Statistics. 2015/FAO annuaire. Statistiques des pêches et de l'aquaculture. 2015/FAO anuario. Estadísticas de pesca y acuicultura. 2015. Rome/Roma, Italy/Italie/Italia.

Ferreira, V.R.A., Azenha, M.A., Mêna, M.T., Moura, C., Pereira, C.M., Pérez-Martín, R.I., Vázquez, J.A., Silva, A.F. (2018). Cationic imprinting of Pb (II) within composite networks based on bovine or fish chondroitin sulfate. *Journal of Molecular Recognition, 31*(3), e2614.

Froese, R., & Pauly, D. (Eds.). (2018). FishBase. www.fishbase.org, version (02/2018).

Iñarra, B., Bald, C., Cebrián, M., Pérez-Villareal, B., Zufía, J. (2018) *Guide for the selection of valorisation options of by-catches*. Derio: AZTI. ISBN: 978-84-944022-4-1

Martinez-Hernandez, E., Campbell, G., Sadhukhan, J. (2013). "Economic value and environmental impact (EVEI) analysis of biorefinery systems." *Chemical Engineering Research and Design, 91*(8), 1418–26. https://doi.org/10.1016/j.cherd.2013.02.025.

Murado, M.A., Fraguas, J., Montemayor, M.I., Vázquez, J.A., González, P. (2010). Preparation of highly purified chondroitin sulphate from skate (*Raja clavata*) cartilage by-products. Process optimization including a new procedure of alkaline hydroalcoholic hydrolysis. *Biochemical Engineering Journal, 49*, 126–132.

Murado, M.A., Montemayor, M.I., Cabo, M.L., Vázquez, J.A., González, M.P. (2012). Optimization of extraction and purification process of hyaluronic acid from fish eyeball. *Food and Bioproducts Processing, 90,* 491–498.

Murillo-Alvarado, P.E., Ponce-Ortega, J.M., Serna-González, M., Castro-Montoya, A.J., El-Halwagi, M.M. (2013). Optimization of pathways for biorefineries involving the selection of feedstocks, products, and processing steps. *Industrial and Engineering Chemistry Research, 52*(14), 5177–90. https://doi.org/10.1021/ie303428v.

Novoa-Carballal, R., Pérez-Martín, R., Blanco, M., Sotelo, C.G., Fassini, D., Nunes, C., Coimbra, M.A., Silva, T.H., Reis, R.L., Vázquez, J.A. (2017). By-products of *Scyliorhinus canicula*, *Prionace glauca* and *Raja clavata*: A valuable source of predominantly 6S sulphated chondroitin sulphate. *Carbohydrate Polymers, 157*, 31–37.

O'Neil, F.G., Feekings, J., Fryer, R.J., Fauconnet, L., Afonso, P. (this volume). Discard avoidance by improving fishing gear selectivity: Helping the industry help themselves. In S.S. Uhlmann, C. Ulrich, S.J. Kennelly (Eds.), *The European discard policy – Reducing unwanted catches in complex multi-species and multi-jurisdictional fisheries*. Cham: Springer.

Prellezo, R., Carmona, I. García, D. (2016). The bad, the good and the very good of the landing obligation implementation in the Bay of Biscay: A case study of Basque trawlers. *Fisheries Research, 181*, 172–185.

Reid, D.G., Calderwood, J., Afonso, P., Bourdeau, P., Fauconnet, L., et al. (this volume). The best way to reduce discards is by not catching them! In S.S. Uhlmann, C. Ulrich, S.J. Kennelly (Eds.), *The European discard policy – Reducing un-wanted catches in complex multi-species and multi-jurisdictional fisheries*. Cham: Springer.

San Martin, D., Orive, M., Martínez, E., Iñarra, B., Ramos, S., González, N., Guinea de Salas, A., Vázquez, L., Zufía, J. (2017). Decision making supporting tool combining AHP method with GIS for implementing food waste valorisation strategies. *Waste and Biomass Valoriztion, 8*, 1555–1567.

Santibañez-Aguilar, J.E., González-Campos, J.B., Ponce-Ortega, J.M., Serna-González, M., El-Halwagi, M.M. 2014. Optimal planning and site selection for distributed multiproduct biorefineries involving economic, environmental and social objectives. *Journal of Cleaner Production, 65*(15), 270–294.

Sousa, S.C., Vázquez, J.A., Pérez-Martín, R.I., Carvalho, A.P., Gomes, A.M. (2017). Valorization of by-products from commercial fish species: Extraction and chemical properties of skin gelatins. *Molecules, 22*, 1545.

Valcarcel, J., Novoa-Carballal, R., Pérez-Martín, R.I., Reis, R.L., Vázquez, J.A. (2017). Glycosaminoglycans from marine sources as therapeutic agents. *Biotechnology Advances, 35*, 711–725.

Vázquez, J.A., Montemayor, M.I., Fraguas, J., Murado, M.A. (2010). Hyaluronic acid production by *Streptococcus zooepidemicus* in marine by-products media from mussel processing wastewaters and tuna peptone viscera. *Microbial Cell Factories, 9*(1), 46.

Vázquez, J.A., Blanco, M., Fraguas, J., Pastrana, L., Pérez-Martín, R.I. (2016). Optimisation of the extraction and purification of chondroitin sulphate from head by-products of *Prionace glauca* by environmental friendly processes. *Food Chemistry, 198*, 28–35.

Vázquez, J.A., Ramos, P., Mirón, J., Valcarcel, J., Sotelo, C.G., Pérez-Martín, R.I. (2017a). Production of chitin from *Penaeus vannamei* by-products to pilot plant scale using a combination of enzymatic and chemical processes and subsequent optimization of the chemical production of chitosan by response surface methodology. *Marine Drugs, 15,* 180.

Vázquez, J.A., Noriega, D., Ramos, P., Valcarcel, J., Novoa-Carballal, R., Pastrana, L., Reis, R.L., Pérez-Martín, R.I. (2017b). Optimization of high purity chitin and chitosan production from *Illex argentinus* pens by a combination of enzymatic and chemical processes. *Carbohydrate Polymers, 174*, 262–272.

Part V
Control, Monitoring and Surveillance

Chapter 18
Tools and Technologies for the Monitoring, Control and Surveillance of Unwanted Catches

Kelly M. James, Neill Campbell, Jónas R. Viðarsson, Carlos Vilas, Kristian S. Plet-Hansen, Lisa Borges, Óscar González, Aloysius T. M. van Helmond, Ricardo I. Pérez-Martín, Luis Taboada Antelo, Jorge Pérez-Bouzada, and Clara Ulrich

Abstract A key requirement for the successful implementation of the Landing Obligation is the need to monitor and regulate unwanted catches at sea. This issue is particularly challenging because of the large number of vessels and trips that need to be monitored and the remoteness of vessels at sea. Several options exist in theory, ranging from patrol vessels to onboard observers and self-sampling. Increasingly though, technology is developing to provide remote Electronic Monitoring (EM) with cameras at lower costs. This chapter first provides an overall synthesis of the pro's and con's of several monitoring tools and technologies. Four EM technologies already trialled in EU fisheries are then summarised. We conclude that it is now possible to conduct reliable and cost-effective monitoring of unwanted catches at sea, especially if various options are used in combination. However,

K. M. James · N. Campbell
Marine Scotland Science, Aberdeen, UK

J. R. Viðarsson
Matís ohf. Icelandic food and biotech R&D, Reykjavík, Iceland

C. Vilas · R. I. Pérez-Martín · L. T. Antelo
Instituto de Investigacións Mariñas (IIM CSIC), Vigo, Spain

K. S. Plet-Hansen · C. Ulrich (✉)
DTU Aqua, Technical University of Denmark, Kgs Lyngby, Denmark
e-mail: clu@aqua.dtu.dk

L. Borges
FishFix, Brussels, Belgium

Ó. González · J. Pérez-Bouzada
Marine Instruments, Nigrán, Spain

A. T. M. van Helmond
Wageningen Marine Research UR, Ijmuiden, CP, The Netherlands

S. S. Uhlmann et al. (eds.), *The European Landing Obligation*,
https://doi.org/10.1007/978-3-030-03308-8_18

effective monitoring is a necessary condition for the successful implementation of the Landing Obligation but insufficient unless it is implemented with a high level of coverage and with the support of the fishing industry.

Keywords Compliance · Electronic Monitoring · Observers · Unreported discards · Video

18.1 Introduction

Discarding in fisheries is driven by a combination of economic and regulatory factors. Fishers may choose to discard fish which are small, damaged, or of low-value to free up space on their vessels for more valuable catches, or they may lack sufficient quota to legally land a species, resulting in them being obliged or incentivised to discard that part of their catch. Global studies have systematically estimated high levels of discard rates in many European fisheries in the North East Atlantic (Kelleher 2005; Zeller et al. 2018). Catchpole et al. (2017) estimated that prior to the establishment of the Landing Obligation, discards of demersal species regulated by quota represented on average 30% of the catch. Of these, small fish (under Minimum Conservation Reference Size) represented only 30–40% of the total discards, highlighting that most discards in Atlantic EU countries are due to quota restrictions and/or are market driven. The scale of the issue demonstrates that there are strong incentives to discard when the practice is unregulated. Thus, a key challenge of the Landing Obligation is adequate enforcement, with two purposes: i) to force selectivity improvements and reduce incentives to discard and ii) to provide reliable catch statistics, since bias in catch estimation has a direct and negative impact on the precision of stock assessments.

A number of approaches are in theory available to conduct MCS (Monitoring, Control and Surveillance) activities. "Monitoring "is defined as the collection of data on catch and fishing effort;" Control "as the regulations and legislation required to stop illegal discarding, and "Surveillance" is defined as the tools available to measure compliance with the Landing Obligation.

This chapter aims first at reviewing the currently available and emerging options for the MCS of discarding of unwanted catches. Many approaches have been applied in a variety of fisheries for decades and their pro's and con's are therefore well-established. Then, this chapter focuses on a more in-depth review and analysis of the recent experiences gained in Europe with Remote Electronic Monitoring (REM –or simply EM) by summarising the main technologies currently available in European fisheries, and their use so far.

18.2 Available and Emerging Measures for MCS

A literature review for the various MCS options was conducted and synthesised. The options are briefly presented sequentially, and their advantages and disadvantages are summarised in Table 18.1.

Table 18.1 The associated advantages and disadvantages of tools used to monitor and control fishing activities

	Advantages	Disadvantages
Aerial and patrol vessel surveillance	High visibility deterrent whilst in sight	Only a short-term deterrent
	Can collect data on fishing effort	Cannot collect data on discards, or any other types of biological data
	If vessels are seen to be partaking in illegal activity, they can be prosecuted	Illegal fishing activity can take place when surveillance vehicle is not in the vicinity
	UAV/USVs have lower operational costs	High costs
	Can observe non-national vessels	Can be adversely affected by weather
		Discarding can still occur illegally
VMS	Offers 100% coverage of fishing vessel movement if installed on all fishing vessels	Only transmits data every 1–2 h so cannot offer detailed information of vessel trips
	Can identify and record non-compliant spatial behaviour	Non-compliant behaviour can still occur between GPS transmissions
	Functions in poor weather and requires no housing of an observer	Cannot collect data on discards, or any other types of biological data
	Can provide data on vessel speed in which fishing effort can be calculated from	Vessel speed may not be an indication of fishing activity taking place
	There is no self-interest of data	Can be switched off or otherwise interfered with (though fines may occur)
		Discarding can still occur illegally
Observers	There is a high confidence in data collected and detailed biological samples can be taken (otolith, gonads, etc. samples).	They can seldom sample the entire trip due to working time restrictions, tiredness, poor weather and illness.
	Observers can play a role in compliance	Data gathered at sea cannot be quality assured directly
	They provide a strong link between fisheries and industry	There can be a safety risk for observers at sea
	Observers are able to detect rare and protected species	Sampling is very costly
		Observer and deployment effects
Self-sampling	A large amount of data can be available at a low cost	Enthusiasm may drop with time
	Data has often been found to be high in quality, and consistent with observers	Data may be biased, or even fabricated and therefore data quality needs to be ensured

(continued)

Table 18.1 (continued)

	Advantages	Disadvantages
	Sense of ownership of data, fishers feels that they are trusted by the authority and scientists. Feel involved in the data collection process	Extensive training may be required
		May not work well for rare or protected species
		Discarding can still occur illegally
Electronic Monitoring with video	Can identify and record non-compliant behaviour and therefore is an effective deterrent	Non-compliant behaviour can still occur around the cameras
	Species identification can be done by shore based analysis.	The technology cannot provide some biological data and reviewers require at least 2 weeks training and auditing process
	Historical videos can be reviewed if a risk of non-compliance is detected	The technology requires significant support to maintain and manage equipment
	Can function in poor weather and requires no housing of an observer	Cameras may not be suitable for monitoring catch in high volume fishing gears (such as trawl and seine)
	Transmits GPS signals every 10 s	There is potential for the technology to be tampered with
	There is no self-interest of data	Can be a considerable investment to get the equipment
	Length data can also be collected	

18.2.1 Aerial and Patrol Vessel Surveillance

The use of aircraft (airplanes, helicopters) and patrol vessels for the MCS of fishing activities is a conventional method used in a variety of modern fisheries (Mangi et al. 2015). There are a number of advantages and disadvantages associated with both aerial and patrol vessel surveillance (European Union 2011; Course 2015). With regards to advantages, both act as a high visibility deterrent for fishing vessels, and it has been observed that discarding is unlikely to occur whilst aerial or vessel patrols are in the vicinity. Also, both are able to observe non-national vessels which may partake in illegal fishing, providing fishers with a "level-playing field "where all boats in the area are under equivalent surveillance. Finally, they are able to monitor fishing on vessels of all sizes, including small ones.

There are however a number of disadvantages. Both aerial and patrol vessels have limitations in the monitoring of discarding, as they only supply data regarding fishing effort and not catch quantities or composition. Even then, data is limited, as coverage is extremely low. In the UK for example, aerial surveillance only monitored 0.026% of fishing effort (hours at sea) in 2013, and patrol vessel surveillance 0.05% of fishing effort (Course 2015). With such a small level of surveillance, this tool may only provide a short-term deterrent as there is no assurance that fishers will continue to comply when vehicles leave the area. A further disadvantage is the high cost associated with aerial and patrol vessel surveillance. It was estimated in 2011 that the Norwegian government spent £86 m a year for the coastguard, which used 70% of their time to enforce its discard ban (European Union 2011). Despite this, patrol vessel surveillance remains at the heart of the control activity deployed by EU Member States together with the European Fisheries Control Agency EFCA (Nuevo et al., this volume).

Unmanned Aerial Vehicles (UAVs, or drones) and unmanned surface vessels (USV) offer a cheaper mechanism for the surveillance of unwanted catches (Miller et al. 2013; Selbe 2014; Linchant et al. 2015). High-resolution optical cameras mounted on drones or USVs enable their operators to visually observe discarding in a range of weather conditions and during both day and night. Drones can also be used to monitor the bycatch of marine mammals. However, lacking a crew, USV's are unable to intervene if an illegal activity is observed, and can be vulnerable to hostile acts from vessels engaged in illegal activities. Furthermore, the legal position of USVs is unclear in some situations and operating a USV inside the national waters of another country could be considered a hostile act (Selbe 2014).

18.2.2 Vessel Transmitted Information and Vessel Detection Systems

Vessel-transmitted information is a general term for all routinely collected control data transmitted from fishing vessels to relevant on-shore authorities. The most common information covered is positional, such as Vessel Monitoring System (VMS) data - a well-established tool used in fisheries management and surveillance globally. The installation of VMS transmitters on fishing vessels is mandatory for all EU vessels above 12 m in length.

VMS data is transmitted via satellite from fishing vessels on a variable timescale, often between 30 min and 2 h intervals. In addition to location, transmissions can provide information regarding a vessel's speed and direction. The data transmitted through VMS can be used to infer spatial distribution of fishing effort (Needle and Catarino 2011) which then has a wide range of scientific and monitoring applications (e.g., Murawski et al. 2005; Lee et al. 2010; Aanes et al. 2011, ICES 2017).

VMS systems are present and active on fishing vessels at all times, and therefore represent a long-term deterrent to illegal fishing in closed areas (Davis 2000; Needle and Catarino 2011; Skaar et al. 2011). Additionally, VMS offers a less expensive alternative to surveillance vehicles, and the data provided are entirely autonomous from the skipper. But their utility remains nevertheless limited, since VMS does not provide information on catch. This is further reduced by infrequent transmissions, resulting in low resolution of spatial data and the potential for illegal fishing activity to take place between transmissions.

Electronic catch reporting (e-log) is another widely applied MCS tool. Catches are entered by the vessel's skipper into an electronic logbook system and transmitted to control authorities on a daily or haul-by-haul basis. This means that accurate catch records can be made available to inspectors in advance of boarding and preliminary figures for catches aboard a vessel in advance of dockside inspection, reducing the potential for misreporting and high-grading. However, a key issue with e-log is the lack of incentive to accurately report discards at sea, if not constrained by other regulatory frameworks auditing them (Ulrich et al. 2015), so discard reporting must be framed in a dedicated self-sampling program (see Sect. 18.2.4 below).

The coupling of both sources of information (VMS and e-log) represents a powerful tool for the fine-scale mapping of catch patterns (e.g. Bastardie et al. 2010; Gerritsen et al. 2012; Hintzen et al. 2012; Ducharme-Barth et al. 2018; Russo et al. 2018), which can thus inform discard reduction strategies in real time. Such an approach was taken by the Scottish "real-time closures" scheme as one part of the North Sea cod recovery program, which aimed to establish a rolling set of closures and effort-penalised areas. These areas were based on the CPUE of cod, calculated from VMS-based effort and electronic catch reports made by fishers, at a spatial resolution of one quarter of an ICES statistical rectangle (Bailey et al. 2010). Non-compliance with this scheme, monitored through VMS, resulted in vessels losing the additional time at sea which they were granted for participating.

A further tool available to monitor fishing vessels is satellite surveillance technology and Vessel Detection Systems (VDS). VDS can detect vessels at sea under most weather conditions and information can be cross-checked with VMS positions to identify the vessel. Fishers are unable to detect VDS whilst at sea, and therefore VDS systems may be a long-term deterrent to fishing in prohibited areas. However, the coverage remains limited because of the high costs associated with satellite imaging. An alternative approach to spatial monitoring is the use of automatic identification system (AIS) data. AIS is an automatic tracking system installed on ships and initially developed by vessel traffic services as a collision avoidance mechanism. Vessels fitted with AIS transceivers can be tracked by base stations located along coast lines and when out of range of terrestrial networks, through a number of satellites fitted with specialised AIS receivers.

An advantage of AIS over other VMS-approaches is that since its primary purpose is navigational, the data are freely available and emitted more frequently. AIS data have been used by academics and NGOs (e.g. Natale et al. 2015; Russo et al. 2016; ICES 2017) to study fishing patterns and demersal impacts. A hindrance in its use as a discard monitoring tool is that it does not monitor catches nor provide information on gear used. Furthermore, as the system is based upon VHF radio transmissions, the range from land over which it is reliable is variable, but often around 60 km. Finally, a disadvantage is that much of the fleet is not required to carry AIS. The International Maritime Organization's International Convention for the Safety of Life at Sea requires AIS to be fitted aboard vessels larger than 300 GT, and all passenger ships regardless of size. Therefore the usefulness of AIS in monitoring compliance with discard regulations is overall limited.

18.2.3 Onboard Observers

Onboard observers are a key part of both MCS and scientific data collection in fisheries globally (Kennelly and Borges 2018; Fernandes et al. 2011). Observers usually remain on a vessel throughout a trip and collect data on the quantities and composition of the catch, discard rates, biological characteristics (such as length, weight and age), fishing effort (Cotter and Pilling 2007; Mangi et al. 2015), and collect tissue samples and otoliths. Observers may also have a role in enforcing fishery regulations, by increasing compliance trough changing fisher's behaviour or by documenting any illegal fishing activities taking place during the trip (Porter 2010). But in many jurisdictions, there is a clear regulatory distinction between scientific and control functions of observers.

Observers are arguably the most valuable source of data on catch and fishing effort, and data collected by observers programs have been used extensively in fisheries management (Benoit and Allard 2009). For example, near real-time management of discarding in Alaskan fisheries is achieved using the high-quality data recorded through a full coverage observers program (Kennelly 2016). Data may also be used to monitor the bycatch of vulnerable species (Piovano and Gilman 2017). Observers can also act as a bridge between science and industry (Mangi et al. 2015), which may contribute to increased compliance with legislation such as the Landing Obligation.

Onboard observers are thus an appropriate tool for the MCS of unwanted catches and the precision of observers data is generally high. However, the main issue is the often-limited coverage of observers programs due to high costs, lack of human resources and/or safety concerns, among others. Observer programs in Scotland and England for example only covered 0.3% of the fishing fleet in 2013 (Course 2015) and observer programs in Fiji only covered 16.7% of the long-line fishery (Piovano and Gilman 2017). A low coverage may not guarantee that the data collected is representative of the whole fleet. Additionally, there is evidence that fishers may exhibit a change in behaviour when observers are onboard (known as observer's effect), leading to bias in the data collected (Liggins et al. 1997), and observed in Europe particularly since the LO came into force (Borges and Dalskov 2018). This can also happen if the data collected by onboard observers is to be used to inform future quota decisions and management. This was documented in the North Pacific Groundfish Observer Program, where fishers avoided areas of high bycatch because data collected by observers are used to extrapolate bycatch rates for quota deduction (Faunce and Barbeaux 2011).

Additionally, with observer coverage on only a sample of the fleet, non random deployment effects may occur (Benoit and Allard 2009; Faunce and Barbeaux 2011). In Europe, efforts are made to avoid this by designing statistically sound sampling programs in the frame of the EU Data Collection Framework (European Union 2016; Rodríguez-Gutierrez et al. 2018).

Whatever the purpose of the observers program, when working under discard reduction measures such as the Landing Obligation, observers should be strongly protected (by having safety training and procedures in case of emergency, adequate regulatory framework, successful prosecutions when interfered, among others), as they inevitably are perceived by fishers to have an enforcement role. In Europe, with the majority of the industry having negative views towards the Landing Obligation (Mangi et al. 2015; Plet-Hansen et al. 2018), there is potential for increased hostility from fishers towards onboard observers (Porter 2010). Ultimately, if the programme coverage is low, an effort should be made to increase the sampling levels. This is not only to guarantee observers safety but to avoid bias in the data collected (Kennelly and Borges 2018).

18.2.4 Self-Sampling

Another solution for the MCS of unwanted catches is the use of data collected or sampled by fishers. Information collected and reported are typically related to catch (total catch and catch composition) and fishing activity (location, duration of fishing activity). Fishers may also be required to take samples, such as tissue samples and otoliths, from the catch (Pennington and Helle 2011). Data may be recorded electronically or on paper and entered into a database upon return. This information can then be processed and incorporated in stock assessments and management purposes, therefore having a role in the control of fishing activity.

Self-sampling and recording by fishers is a technique used for data collection in a variety of fisheries, and may be mandatory through legislation. In principle, EU fishers have been legally required to document discards over 50 kg since 2011, although this measure has largely not been enforced (Ulrich et al. 2015). However, the self-reporting may also be voluntary. In the Norwegian purse-seine fishery, fishers are paid to measure a sub-sample of fish from selected catches as well as collect otolith, stomach and genetic samples (Pennington and Helle 2011).

There are a number of advantages and disadvantages associated with the use of self-sampling for the MCS of unwanted catches (Lordan et al. 2011; Kraan et al. 2013). The major attraction of self-sampling for data collection is that a significant increase in sampling coverage can be achieved at little cost. Many studies have found that fishers welcome being involved with the data collection; although enthusiasm may drop over time (Mangi et al. 2014, 2015). Such engagement with the management process is a key ingredient to success in many fisheries. Additionally, fishers do not need to provide extra accommodation or room on vessels for outside observers. If correctly executed and following unbiased sampling protocols, data collected through self-reporting can be of high quality and used in stock assessment, as in the New Zealand rock lobster potting fishery (Starr 2000).

Though self-sampling is an effective, low-cost method, there are many disadvantages associated with the sampling technique. With negative attitudes towards the Landing Obligation widespread in the fishing industry, non-compliant behaviour may be common and self-reported data may be biased by non-random sampling or even fabricated (Ticheler et al. 1998; Graham et al. 2011; Mangi et al. 2016; Gray and Kennelly 2017). Data precision may also be below the level required for stock assessments. Data collected by fishers must therefore be quality assured and it is unlikely that self-sampling could be used as a stand-alone tool for monitoring compliance with the Landing Obligation.

18.2.5 Electronic Monitoring with Video

Electronic monitoring with video (EM) has been praised by many as a practical, innovative, and applicable solution for MCS in fisheries (Mangi et al. 2015; Course 2015; Mortensen et al. 2017). Through the combination of video cameras (initially analogue and closed circuit (CCTV), now mainly digital), GPS and sensor data, EM can be used to collect information regarding spatial fishing effort and catch data, which can then be used for monitoring and compliance. EM is already used in many fisheries in the world, as a full MCS program in North America and Australia, but with numerous trials also ongoing in South America and the Pacific. In EU, EM has been trialled in a number of fisheries since 2008, mainly associated with the Cod Catch Quota Management with fully documented fisheries (FDF) (Kindt-Larsen et al. 2011; Needle et al. 2015; Ulrich et al. 2015), but also to observe protected species (Kindt-Larsen et al. 2016).

EM with video meets most of the criteria necessary for the MCS of unwanted catches and has important advantages (McElderry 2006; Mangi et al. 2015; Course 2015). EM records from many sensors and at a much higher frequency than VMS or AIS (usually several times a minute). This information provides very rich granularity to distinguish specific vessel behaviours (e.g., gear setting, hauling, haul back, catch stowage, transit, etc.). EM offers thus the opportunity for 100% surveillance of fishing activities. Furthermore, EM has the ability to monitor illegal discarding, with video covering upper deck and lower deck discharge chute(s). Detection of illegal activity could potentially be used in prosecution (McElderry and Turris 2008; Diamond and Beukers-Stewart 2011). With EM systems recording vessel location and behaviour throughout a fishing trip, the technology is considered a plausible long-term deterrent to non-compliant behaviour (Course 2015).

EM is also suitable for monitoring unwanted catches, providing detailed data such as catch composition and length frequencies through video analysis (Needle et al. 2015; Sandeman et al. 2016). Such data can then be used for quota management or for the control of unwanted catches, for example, by closing fishing grounds if catches appear to be comprised of a large percentage of juveniles, vulnerable, or otherwise non-target species. Finally, while initial purchase and installation costs can be significant, running costs are low, and the amortized cost over the life of the equipment is thus very low as compared to human observers.

EM is however not without shortcomings. The main concern is the usually strong reluctance of fishers to accept onboard cameras that can be watched by the authorities. This lack of support from the industry is a major threat against the successful implementation of all MCS tools (Lordan et al. 2011; Kennelly 2016; Plet-Hansen et al. 2018). Incentives have been used to gain support by offering e.g. increased quota, days-at-sea, access to fishing grounds or more flexible gear use. In the cases where EM has been successfully implemented as full MCS programs, EM was first introduced offering incentives, and later made mandatory to all.

An older concern regarding the use of video footage was that data quality could be inconsistent. A meta-analysis by Wallace et al. (2013) found that, in almost all of the 59 EM studies reviewed, data quality was either poor or missing for a proportion of the study. However, modern technology including digital cameras has significantly improved data quality, so these issues are now less of a concern (Bergsson et al. 2017). Nevertheless, monitoring through video may remain challenging on vessels in mixed fisheries catching high volumes and with a diverse species composition (van Helmond et al. 2015). Considerations must thus be made regarding camera type and set up, and changes to conveyor belt layout may be necessary to reduce the volume of fish per video frame.

Another concern is about data quantity. If fisheries were to widely apply EM for the collection of catch data, this would represent a very large volume of data. If inspection is conducted manually by fisheries inspectors, a large onshore team of video viewers would need to be trained and employed to analyse such data. Viewing strategies and technology are therefore required to overcome this. First, viewing time can be reduced by selecting a representative sample of the fishing trips rather than all hauls. Second, video review involving machine learning and artificial intelligence to automatically analyse video footage is advancing (French et al. 2015; Bergsson et al. 2017).

In any case, even if only a portion of video recordings is reviewed, a key element of EM is that the awareness that everything is recorded and can be inspected anytime is expected to have an effective deterrent effect and increase compliance by fishers (Ulrich et al. 2015).

18.3 Overview of the EM Technology Trialled in EU

Several EM trials have been done in several EU countries since 2008, for different purposes and with different technologies. Initial trials used the EM Observe™ technology developed by Archipelago Marine Research (Canada), but new software was later developed within the EU. We briefly summarise the main features and technical characteristics of four systems: The EM Observe™ system (now operated by Marine Instruments since 2017), the Black Box developed by Anchor Lab K/S (Denmark), the Electronic Eye developed by Marine Instruments (Spain), and the iObserver system developed by CSIC (Spain) (Table 18.2).

The EM Observe™ is the first commercial EM system in the world and is used in several national monitoring programs with 100% fleet coverage of fleets comprising

Table 18.2 Overview of four EM systems tested in European countries since 2008

	Black box video system	EM observe	iObserver	Electronic eye
Company	Anchor Lab K/S (Denmark) http://www.anchorlab.dk/EFM.aspx?tab=About	During the EU trials: Archipelago Marine Research Ltd. (Canada). Since 2017, operated by Marine Instruments http://www.archipelago.ca/fisheries-monitoring/electronic-monitoring/	CSIC (Spain) http://lifeiseas.eu/iobserver/	Marine Instruments (Spain) http://www.marineinstruments.es/monitoring-systems/electronic-eye/?lang=en
Applications in EU	Denmark: (i) Danish trial for Catch Quota Management (CQM). 12 Demersal trawlers, Danish seiners and gillnetters (2014–2016) 2.836 hauls audited for 5 gadoids species. (ii) Minimizing discards in Danish fisheries (MINIDISC project). 12 Danish seiners and trawlers (2014–2015). 1.018 hauls audited for 7 species. (iii) The Black Box R2 version of the system, is used for the sensor system required for all vessels fishing for common mussels (*Mytilus edulis*) in Denmark.	Denmark: Danish trial for CQM (2008–2014). 24 demersal trawlers, Danish seiners and gillnetters. Danish trial on documentation of harbour porpoise bycatch by gillnetters (2010–2011). England: Several English CQM trials on otter trawls, gill nets, long liners, beam trawlers, small vessels (2010–2015) Germany: German North Sea CQM Trial (2011–2016) Scotland: Scottish CQM Trial (2008-present). Sweden: Swedish trial on gillnetters bycatch documentation (2008). The Netherlands: Dutch North Sea cod CQM (2011–	Spain: Trials performed onboard Spanish oceanographic vessels, not commercial vessels. Trials on board two oceanographic vessels. 10 surveys in the regions ICES-Spain; ICES-West Ireland; and NAFO, were performed with a total number of 270 days at sea in which the iObserver was used in 780 hauls, taking over 170,000 pictures. Trials on board two commercial vessels. 9 surveys, with a total number of 36 days at sea, were carried out so far in ICES-Spain regions VIIIc and IXa. The iObserver was used in 162 hauls taking around 35,000	Spain: System installed and in operation in more than 20 Spanish tuna purse seiners and supply vessels operating in the Atlantic and Indian Ocean with automatic image capture for fishing monitoring and bycatch control on-board, according to the standards set by the corresponding Regional Fisheries Organizations (ICCAT and IOTC). Scotland: System installed and in operation on 8 Scottish scallop dredge vessels to comply with the regulations and control set by Marine Scotland.

(continued)

Table 18.2 (continued)

	Black box video system	EM observe	iObserver	Electronic eye
		2015). Dutch sole REM trial with beam trawlers (2015)	pictures. 17 species already included in the catalogue.	
Published Scientific references	Bergsson et al. (2017); Mortensen et al. (2017); Plet-Hansen et al. (2018) and van Helmond et al. (n. d.)	French et al. (2015); Kindt-Larsen et al. (2011), (2016); Mangi et al. (2015); Needle et al. (2015); Ulrich et al. (2015) and van Helmond et al. (2016, 2015, 2017)	Vilas et al. (2018a,b)	Ruiz 2013, Ruiz et al. (2014, 2016)

more than 200 vessels. It has been trialled by North Sea countries during various cod catch quota trials in the period 2008–2016 (see references in Table 18.2). Data are recorded on high capacity hard drives which are manually retrieved and replaced when the fishing vessel returns to port. The data are analysed using the EM Interpret software.

The AnchorLab Black Box system was developed to further support the EM trials in Denmark and has been used in a diversity of fisheries. Its main feature is the improvement of video storage and data transmission, where EM data are transmitted to the data receiver via GSM, Wi-Fi, 3G, 4G/LTE, LTE-A or satellite. The analyser software has a number of features facilitating length measurements including grid overlay and measuring line. A low power version to suit small-scale vessels and automated species identification for the system are under development.

The Marine Instruments' Electronic Eye eEYE™ is in operation on a number of Spanish tuna purse seiners mainly to monitor the bycatch of endangered, threatened and protected species. It is also installed on some Scottish scallop dredgers. Data are stored in an internal hard drive and can be downloaded via USB or Wi-Fi. The cameras can also be visualised from the bridge.

The iObserver system has been developed by the scientific institute CSIC in Spain, but is not yet in operation onboard commercial vessels. It is not a full EM system as it does not observe the fishing deck, but is mainly focused on developing algorithms for robust automatic species recognition and size estimation of fish passing on the conveyor belt.

The four systems are quite different in their set up and operation and offer different capabilities. In their current state of development at the time of writing, they are not fully automated and still require human intervention for footage viewing, and their base price (system alone) is in the order of 6000–10,000 EUR per vessel. A direct comparison is however not possible as the systems have not been trialled on the same vessels and for the same purposes.

Additionally, a number of other EM systems are used throughout the world, but have not been trialled in Europe.

18.4 Discussion

18.4.1 Comparison of EM with Other MCS Options

This chapter has reviewed the pro's and con's of available and emerging approaches to the MCS of unwanted catches. All tools have advantages and disadvantages, but the potentials of EM technology seem nevertheless to surpass those of other more conventional tools presently used. With over 25,000 fishing days at sea monitored by EM studies, the conclusion is that such technology can be efficient and a practical method for the MCS of fishing activities (McElderry 2006; Course 2015). Compared to VMS it is obvious that EM offers much higher resolution information. VMS alone only enables the monitoring of geographical location, speed and direction. Compared to aerial and patrol vessel surveillance, the major benefit of EM with video is the potential coverage (i.e. amount of monitoring) that can be achieved. While surveillance through aerial and water-borne vehicles can only cover a small percentage of the fishing fleet and activity, EM has the potential for 100% surveillance of fishing activities, including catch monitoring, and at much lower cost.

Compared to onboard observers, while these can also offer full coverage, EM represents only a fraction of the cost (see below). Another major benefit of EM to onboard observers is its potential to offer 24/7 coverage, as it is not affected by differences in working times or by weather and is also less intrusive than accommodating an extra person onboard. On the other hand, EM cannot collect certain types of data otherwise provided by onboard observers, including tissue samples, weight measurements and otoliths. Onboard observers will therefore always be necessary if such data is required (Kennelly and Borges 2018). EM can neither provide a bridge between science and industry, improving communication and understanding.

Compared to self-sampling, a major advantage of data collected through EM is that data is anticipated to not be biased. Though research has found that information from self-sampling often reflects that from the EM videos, there is a lack of confidence in data collected when no surveillance and auditing is present (Ulrich et al. 2015). EM allows for the quality of self-reported data to be checked, and quality assured.

18.4.2 EM Costs

A number of studies have compared the costs of EM with observers. In the early days of development, McElderry and Turris (2008) stated that EM could be provided at a quarter of the daily cost of observers, Ames et al. (2007) a third and Kindt-Larsen et al. (2011) a tenth. Start-up and installation costs have remained high because of the limited consumer market and the specific requirements for the technology. However, operating and amortized costs are low and it takes only short time for the cumulated investment to become comparatively cheaper than observers (Needle et al. 2015). Improved technology using 3G/4G networks rather than hard disks and ensuring better connectivity between boat and shore (and reverse) has already contributed to reducing transmission costs (Mortensen et al. 2017).

A cost that has remained important in EM concerns data analysis. Video footage still needs to be manually reviewed and this may appear as a tedious and often expensive procedure. However, trials conducted over several years have contributed to the development of efficient analysis software and streamlined procedures that have significantly reduced review time. Bergsson et al. (2017) estimated that the catches of five gadoid species in a standard demersal trawl haul could be viewed and analysed in about 20 min. In the near future, it can be expected that technical advances involving computer learning and automatic image analysis will further reduce analysis costs.

Ultimately, the most important element in estimating analysis costs remains thus the number of hauls to be viewed and the amount of data to be collected. These depend on the design of the MCS program, its objectives and the required accuracy and precision of estimates. To reduce costs, EM can be used in combination with self-reporting. Self-reported catch data from fishers can be in broad agreement with EM analysis, provided that protocols are clear and that there is regular quality control and follow-up with fishers. EM can thus be used not as the main source of catch data but only to audit self-reported data, like black boxes used in trucks and airplanes. In doing so, a smaller amount of footage would be analysed, reducing costs of onshore viewers. For example, Needle et al. (2015) estimated that to obtain accurate estimates of all discarded species in a Scottish mixed demersal fishery from video footage alone, around 40% of footage must be reviewed. But other studies have found that in order to audit self-reported data, reviewing only 5–10% of hauls was sufficient (Roberts et al. 2015; Stanley et al. 2015).

18.4.3 Combination of Tools for Successful MCS Programs Design

Successful MCS programs have in reality involved a combination of tools. For example, in the Canadian Ground Fish Hook and Line Catch Monitoring Programme, dockside monitoring is used in conjunction with self-reported data and EM or onboard observers. This program is unique as fishers are offered a choice between EM and onboard observers, though observers are rarely used. Both EM and observers' data are used to audit self-reported data, with full dockside monitoring providing further validation of data regarding catch (Stanley et al. 2015). This combination of tools results in the increased reliability of self-reported data, and gives the fishers some buy-in because they are collecting the data (both logbook and audit data) themselves, which gives them more ownership.

Iceland provides a different example. For a long time, compliance with the discard ban has been performed using patrol vessel surveillance combined with logbooks and catch comparisons (European Union 2011). Fish are monitored throughout the whole supply chain. Data from the electronic logbooks, official weighings at harbour scales, purchasing receipts /receipts from fish auction, reweighing by processors, processing arrangement slips/production reports, sales- / export reports are all sent to the Directorate of Fisheries, which is then able to monitor for consistency in the mass balance (Óskarsdóttir and Gunnlaugsson 2015). Similar regulations are in place in some other countries where electronic data sharing and transparency is well advanced e.g. Faroe Islands and Norway. This type of monitoring is efficient in combination with other MCS tools and gives the authorities an indication of where they need to focus extra attention. The success of the discard ban in Iceland is also attributed to changes in social perception, with fishers themselves having the opinion that discarding is unacceptable and even reporting others if they are seen discarding (Karp et al., this volume). Nevertheless, none of the countries referred above have an independent large scale at-sea monitoring program, where discards quantities can be audited and verified. It is therefore noticeable that Iceland is currently moving towards introducing EM in its MCS programme. At the time of writing, the Directorate of Fisheries is considering a regulation which will require all commercial fishing vessels to be equipped with EM with video to remotely and electronically monitor potential discarding (Karp et al., this volume). Drones may also be introduced.

18.5 Conclusions

In conclusion, there are existing options to appropriately monitor and control the Landing Obligation, and the increased experience with their use together with technological developments will contribute to enhancing their capacity and reducing their costs. Nevertheless, MCS technology is only a tool and will not solve the discard issue alone. The crucial elements for the successful implementation of the Landing Obligation remain the MCS coverage level and compliance from the fishing industry. If the industry support remains low, there will always be ways to render MCS programs ineffective, especially if their coverage is low. Moving forward, this means that MCS is a necessary but insufficient tool for the successful reduction of discards, and MCS programs must thus be integrated into a broad mind shift within the fisheries and seafood sectors towards better accountancy, transparency and sustainability, and/or implemented with a high level of coverage.

Acknowledgments This chapter is largely based on work carried out in the DiscardLess project and the Life iSEAS project. The DiscardLess project has received funding from the European Union's Horizon 2020 research and innovation programme under grant agreement DiscardLess No 633680. The Life iSEAS project has been co-funded under the LIFE+Environment Program of the European Union (LIFE13 ENV/ES/000131). This support is gratefully acknowledged. Authors also want to thank Ms. Esther Abad, Mr. Xesús Morales, Ms. Marta Quinzán and Mr. Julio Valeiras for their contribution to the development of the iObserver.

References

Aanes, S., Nedreaas, K., Ulvatn, S. (2011). Estimation of total retained catch based on the frequency of fishing trips, inspections at sea, transhipment, and VMS data. *ICES Journal of Marine Science, 68*(8), 1598–1605.

Ames, R., Leaman, B., Ames, K., (2007). Evaluation of video technology for monitoring of multispecies longline catches. *Northern American Journal of Fisheries Management, 27,* 955–964.

Bailey, N., Campbell, N., Holmes, S., Needle, C., Wright, P. (2010). *Real time closure of fisheries*. European Parliament Directorate-General for Internal Policies. Brussels. 56 pp.

Bastardie, F., Nielsen, J.R., Ulrich, C., Egekvist, J., Degel, H. (2010). Detailed mapping of fishing effort and landings by coupling fishing logbooks with satellite-recorded vessel geo-location. *Fisheries Research, 106*(1), 41–53.

Benoit, H.P., & Allard, J. (2009). Can the data from at-sea observer surveys be used to make general inferences about catch composition and discards? *Canadian Journal of Fisheries and Aquatic Sciences, 66*(12), 2025–2039.

Bergsson, H., Plet-Hansen, K.S., Jessen, L.N., Jensen, P., Bahlke, S.Ø. (2017). *Final report on development and usage of REM systems along with electronic data transfer as a measure to monitor compliance with the Landing Obligation – 2016.* Danish AgriFish Agency, Ministry of Food, Agriculture and Fisheries. 61 pp. https://doi.org/10.13140/RG.2.2.23628.00645.

Borges, L. & Dalskov, J. (2018). Workshop 1 – European Union landing obligation. In Kennelly, S. J., & Borges, L. (Eds.). *Proceedings of the 9th International Fisheries Observer and Monitoring Conference, Vigo, Spain.* ISBN: 978-0-9924930-7-3, 397 pages.

Catchpole, T., Ribeiro-Santos, A., Mangi, S.C., Hedley, C., Gray, T.S. (2017). The challenges of the landing obligation in EU fisheries. *Marine Policy, 82,* 76–86.

Cotter, A.J.R., & Pilling, G.M. (2007). Landing, logbooks and observer surveys: Improving the protocols for sampling commercial fisheries. *Fish and Fisheries, 8*(2), 123–152.

Course, G. (2015). *Electronic monitoring in fisheries management*. WWF, UK. 38 pp.

Davis, J.M. (2000). *Monitoring control surveillance and vessel monitoring system requirements to combat IUU fishing*. FAO, Rome. 13 pp.

Diamond, B., & Baukers-Stewart, B. (2011). Fisheries discards in the North Sea: Waste of resources or a necessary evil? *Reviews in Fisheries Science 19*(3), 231–245.

Ducharme-Barth, N.D., Shertzer, K.W., Ahrens, R.N. (2018). Indices of abundance in the Gulf of Mexico reef fish complex: A comparative approach using spatial data from vessel monitoring systems. *Fisheries Research, 198,* 1–13.

European Union. (2011). *Studies in the field of the common fisheries policy and maritime affairs. Lot4: Impact assessment studies related to the CFP. Impact assessment of discard reducing policies: Case study annex*. European Commission. Brussels.

European Union. (2016). Commission Implementing Decision (EU) 2016/1251 of 12 July 2016 adopting a multiannual Union programme for the collection, management and use of data in the fisheries and aquaculture sectors for the period 2017-2019 (notified under document C(2016) 4329). C/2016/4329. *Official Journal of the European Union*, L207/113.

Faunce, C.H., & Barbeaux, S.J. (2011). The frequency and quantity of Alaskan groundfish catch-vessel landings made with and without an observer. *ICES Journal of Marine Science, 68*(8), 1757–1763.

Fernandes, P., Coull, K., Davis, C., Clark, P., Catarino, R., Bailey, N., et al. (2011). Observations of discards in the Scottish mixed demersal trawl fishery. *ICES Journal of Marine Science, 68*(8), 1734–1742.

French, G., Fisher, M.H., Mackiewicz, M., Needle, C. (2015). Convolutional neural networks for counting fish in fisheries surveillance video. In Amaral, T., Matthews, S., Plötz, T. et al. (Eds.). *Proceedings of the machine vision of animals and their behaviour (MVAB)*, (pp. 7.1–7.10). BMVA Press, Swansea. ISBN 1-901725-57-CX.

Gerritsen, H., Lordan, C., Minto, C., Kraak, S. (2012). Spatial patterns in the retained catch composition of Irish demersal otter trawlers: High-resolution fisheries data as a management tool. *Fisheries Research, 129–130,* 127–136.

Graham, N., Grainger, R., Karp, W.A., MacLennan, D.N, MacMullen, P., Nedreaas, K. (2011). An introduction to the proceedings and a synthesis of the 2010 ICES Symposium on Fishery-Dependent Information. *ICES Journal of Marine Science, 68*(8), 1593–1597.

Gray, C.A., Kennelly, S.J. (2017). Evaluation of observer- and industry-based catch data in a recreational charter fishery. *Fisheries Managment and Ecology, 24*(2), 126–138.

Hintzen, N.T., Bastardie, F., Beare D., Piet, G.J., Ulrich, C., Deporte, N., et al. (2012). VMStools: Open-source software for the processing, analysis and visualization of fisheries logbook and VMS data. *Fisheries Research, 115–116,* 31–43.

ICES. (2017). *Interim Report of the Working Group on Spatial Fisheries Data (WGSFD), 29 May – 2 June 2017, Hamburg, Germany*. ICES CM 2017/SSGEPI:16.

Karp, W.A., & McElderry, H. (1999). Catch monitoring by fisheries observers in the United States and Canada. In: *Proceedings of the International Conference on Integrated Fisheries Monitoring*. Rome: FAO.

Karp, W.A., Breen, M., Borges, L., Fitzpatrick, M., Kennelly, S. J., Kolding, J., et al. (this volume). Strategies used throughout the world to manage fisheries discards – Lessons for implementation of the EU Landing Obligation. In S.S. Uhlmann, C. Ulrich, S.J. Kennelly (Eds.), *The European Landing Obligation – Reducing discards in complex, multi-species and multi-jurisdictional fisheries*. Cham: Springer.

Kelleher, K. (2005). *Discards in the world's marine fisheries – An update* (FAO Fisheries Technical Paper. No. 470, p. 131). Rome: FAO. ISBN:92-5-105289-1.

Kennelly, S.J. (Ed.) (2016). *Proceedings of the 8th international fisheries observer and monitoring conference, San Diego, USA* (p. 349). ISBN:978-0-9924930-3-5.

Kennelly, S.J., & Borges, L. (Eds.) (2018). *Proceedings of the 9th international fisheries observer and monitoring conference, Vigo, Spain* (397 pp). ISBN:978-0-9924930-7-3.

Kindt-Larsen, L., Kirkegaard, E., Dalskov, J. (2011). Fully documented fishery: a tool to support a catch quota management system. *ICES Journal of Marine Science, 68*(8), 1606–1610.

Kindt-Larsen, L., Berg, C.W., Tougaard, J., Sørensen, T.K., Geitner, K., Northridge, S., et al. (2016). Identification of high-risk areas for harbour porpoise Phocoena phocoena bycatch using remote electronic monitoring and satellite telemetry data. *Marine Ecology Progress Series, 555*, 261–271.

Kraan, M., Uhlmann, S., Steenbergen, J., Van Helmond, A.T., Van Hoof, L. (2013), The optimal process of self-sampling in fisheries: Lessons learned in the Netherlandsa. *Journal of Fish Biology, 83*, 963–973. https://doi.org/10.1111/jfb.12192.

Lee, J., South, A.B., Jennings, S. (2010). Developing reliable, repeatable, and accessible methods to provide high-resolution estimates of fishing-effort distributions from vessel monitoring system (VMS) data. *ICES Journal of Marine Science, 67*(6), 1260–1271.

Liggins, G.W., Bradley, M.J., Kennelly, S.J. (1997). Detection of bias in observer-based estimates of retained and discarded catches from a multispecies trawl fishery. *Fisheries Research, 32*(2), 133–147.

Linchant, J., Lisein, J., Semeki, J., Lejeune, P., Vermeulen, C. (2015). Are unmanned aircraft systems (UASs) the future of wildlife monitoring? A review of accomplishments and challenges. *Mammal Review, 45*(4), 239–252.

Lordan, C., Cuaig, M., Graham, N., Rihan, D. (2011). The ups and downs of working with industry to collect fishery-dependent data: the Irish experience. *ICES Journal of Marine Science, 68*(8), 1670–1678.

Mangi, S.C., Smith, S., Armstrong, S., Catchpole, T.L. (2014). *Self-sampling in the inshore sector (SESAMI).* Final report, CEFAS, UK, p. 53.

Mangi, S.C., Dolder, P.J., Catchpole, T.L., Rodmell, D., Rozarieux, N. (2015). Approaches to fully documented fisheries: practical issues and stakeholder perceptions. *Fish Fish, 16*, 426–452. https://doi.org/10.1111/faf.12065

Mangi, S.C., Smith, S., Catchpole, T.L. (2016). Assessing the capability and willingness of skippers towards fishing industry-led data collection. *Ocean & Coastal Management, 134,* 11–19.

McElderry, H. (2006). *At-Sea observing using video-based electronic monitoring* (p. 25). ICES CM 2006/N:14.

McElderry, H., & Turris, B. (2008). *Evaluation of monitoring and reporting needs for groundfish sectors in New England.* Pacific fisheries management incorporated and Archipelago Marine Research Ltd., p. 68.

Miller, D.G.M., Slicer, N.M., Hanich, Q. (2013). Monitoring, control and surveillance of protected areas and specially managed areas in the marine domain. *Marine Policy, 39,* 64–71.

Mortensen, L.O., Ulrich, C., Olesen, H.J., Bergsson, H., Berg, C.W., Tzamouranis, N. et al. (2017). Effectiveness of fully documented fisheries to estimate discards in a participatory research scheme. *Fisheries Research, 187*:150–157.

Murawski, S.A., Wigley, S.E., Fogarty, M.J., Rago, P.J., Mountain, D.G. (2005). Effort distribution and catch patterns adjacent to temperate MPAs. *ICES Journal of Marine Science, 62*(6), 1150–1167.

Natale, F., Gibin, M., Alessandrini, A., Vespe, M., Paulrud, A. (2015). Mapping fishing effort through AIS data. *PLoS One, 10*(6), e0139746.

Needle, C., & Catarino, R. (2011) Evaluating the effect of real-time closures on cod targeting. *ICES Journal of Marine Science, 68,* 1647–1655.

Needle, C.L., Dinsdale, R., Buch, T.B., Catarino, R.M., Drewery, J., Butler, N. (2015). Scottish science applications of remote electronic monitoring. *ICES Journal of Marine Science, 72*(4), 1214–1229. https://doi.org/10.1093/icesjms/fsu225.

Nuevo, M., Morgado, C., Sala, A. (this volume). Monitoring the implementation of the Landing Obligation: Last Haul programme. In S.S. Uhlmann, C. Ulrich, S.J. Kennelly (Eds.), *The European Landing Obligation – Reducing discards in complex, multi-species and multi-jurisdictional fisheries*. Cham: Springer.

Óskarsdóttir, K., & Gunnlaugsson, V. (2015). Flæði gagna milli aðila í sjávarútveginum. Reykjavík: Matís.

Pennington, M., & Helle, K. (2011). Evaluation of the design and efficiency of the Norwegian self-sampling purse-seine reference fleet. *ICES Journal of Marine Science, 68*(8), 1764–1768.

Piovano, S. & Gilman, E. (2017). Elasmobranch captures in the Fijian pelagic longline fishery. *Aquatic Conservation: Marine and Freshwater Ecosystems, 27*(2), 381–393.

Plet-Hansen, K.S., Eliasen, S.Q., Mortensen, L.O., Bergsson, H., Olesen, H.J., Ulrich, C. (2018). Remote electronic monitoring and the landing obligation – some insights into fishers' and fishery inspectors' opinions. *Marine Policy, 76,* 98–106.

Porter, R.D. (2010). Fisheries observers as enforcement assets: Lessons from the North Pacific. *Marine Policy 34*(3), 583–589.

Roberts, J., Course, G., Pasco, G., Sandeman, L. (2015). *Catch quota trials – South West Beam Trawl*. Marine Management Organisation, London. 22 pp.

Rodríguez-Gutierrez, J., Castro, J., Salinas, I., Araujo, H., Marin, M. (2018). Improving protocol of selection of vessels to reduce bias. In Kennelly, S.J. & Borges, L. (Eds.) *Proceedings of the 9th International Fisheries Observer and Monitoring Conference, Vigo, Spain* (p. 397). ISBN:978-0-9924930-7-3.

Ruiz, J. (2013). *Pilot study of an electronic monitoring system on a tropical tuna purse seine vessel in the Atlantic Ocean*. Norwegian College of Fisheries Science: Master's Degree Thesis in International Fisheries Management. Tromso: UiT, p. 59.

Ruiz, J., Krug, I., Gonzalez, O., Gomez, G., Urrutia, X. (2014). Electronic eye: Electronic monitoring trial on a tropical tuna purse seiner in the Atlantic Ocean. SCRS/2014/138. *Collect Vol Science Papers ICCAT 71*(1), 476–488, (2015).

Ruiz, J., Krug, I., Justel-Rubio, A., Restrepo, V., Hammann, G., Gonzalez, O., Legorburu, G., Alayon, P.J.P., Bach, P., Bannerman, P., Galán, T. (2016). Minimum standard for the implementation of electronic monitoring systems for the tropical tuna purse seine fleet. SCRS/2016/180. *Collect Vol Science Papers ICCAT, 73*(2), 818–828 (2017).

Russo, T., D'Andrea, L., Parisi, A., Martinelli, M., Belardinelli, A., Broccoli, F., et al. (2016). Assessing the fishing footprint using data integrated from different tracking devices: Issues and opportunities. *Ecological Indicators, 69,* 818–827.

Russo, T., Morello, E.B., Parisi, A., Scarcella, G., Angelini, S., Labanchi, L., et al. (2018). A model combining landings and VMS data to estimate landings by fishing ground and harbour. *Fisheries Research, 199*, 218–230.

Sandeman, L., Royston, A., Roberts, J. (2016). *North Sea Cod catch quota trials: Final report 2015* (p. 38). London: Marine Management Organisation.

Selbe, S. (2014). *Monitoring and Surveillance Technologies for Fisheries* (p. 132). La Jolla: Waitt Institute.

Skaar, K.L., Jorgensen, T., Ulvestad, B.K.H., Engås, A. (2011). Accuracy of VMS data from Norwegian demersal stern trawlers for estimating trawled areas in the Barents Sea. *ICES Journal of Marine Science, 68*(8), 1615–1620.

Stanley, R.D., Karim, T., Koolman, J., McElderry, H. (2015). Design and implementation of electronic monitoring in the British Columbia groundfish hook and line fishery: a retrospective view of the ingredients of success. *ICES Journal of Marine Science, 72*(4), 1230–1236.

Starr, P. (2000). *Fishery management innovations in New Zeal and New Zealand Seafood Industry Council* (p. 6).

Ticheler, H., Kolding, J., Chanda, B. (1998). Participation of local fishermen in scientific fisheries data collection: A case study from the Bangweulu Swamps, Zambia. *Fisheries Management and Ecology, 5*(1), 81–92.

Ulrich, C., Olesen, H.J., Bergsson, H., Egekvist, J., Håkansson, K. B, Dalskov, J., et al. (2015). Discarding of cod in the Danish fully documented fisheries trials. *ICES Journal of Marine Science*, *72*(6),1848–1860.

van Helmond, A.T.M., Chen, C., Poos, J.J. (2015). How effective is electronic monitoring in mixed bottom-trawl fisheries? *ICES Journal of Marine Science, 72*(4), 1192–1200.

van Helmond, A.T.M., Chen, C., Trapman, B.K., Kraan, M., Poos J.J. (2016). Changes in fishing behaviour of two fleets under fully documented catch quota management: Same rules, different outcomes. *Marine Policy , 67,* 118–129.

van Helmond, A.T.M., Chen, C., Poos, J.J. (2017). Using electronic monitoring to record catches of sole *(Solea solea)* in a bottom trawl fishery. *ICES Journal of Marine Science, 74*(5), 1421–1427. https://doi.org/10.1093/icesjms/fsw241.

Vilas, C., Antelo, L.T., Ordoñez, T., Pérez-Martín, R.I., Alonso, A.A., Valeiras, J. et al. (2018a). An Innovative Technology for On Board Automatic Identification and Quantification of the Catch.

In: *Proceedings of the international conference on advances in marine technologies applied to discard mitigation and management (MARTEC 18), Vigo, Spain.*

Vilas, C., Antelo, L.T., Ordoñez, T., Pérez-Martín, R.I., Alonso, A.A., Valeiras, J. et al. (2018b). On board automatic identification and quantification of the total catch: The iObserver. In: Kennelly, S.J., & Borges, L. (Eds.), *Proceedings of the 9th Fisheries Observer & Monitoring Conference (IFOMC), Vigo, Spain.* ISBN:978-0-9924930-7-3. 395 pp.

Wallace, F., Faunce, C., Loefflad, M. (2013). *Pressing rewind: A cause for a pause on electronic monitoring in the North Pacific.* ICES CM 2013/J:11.

Zeller, D., Cashion, T., Palomares, M., Pauly, D. (2018). Global marine fisheries discards: A synthesis of reconstructed data. *Fish Fish, 19,* 1–10. https://doi.org/10.1111/faf.12233

Chapter 19
Monitoring the Implementation of the Landing Obligation: The Last Haul Programme

Miguel Nuevo, Cristina Morgado, and Antonello Sala

Abstract The collection of catch composition data during inspections at sea by EU Member States occurs under the framework of joint deployment plans (JDP). It is known as "the last haul" (LH) programme and has been a fundamental tool in allowing the estimation of discards and the derivation of indicators of compliance with the landing obligation (LO). During sea inspections, measures of quantities of fish below and above the minimum conservation reference size and grade categories of the legal-size catch are used to derive estimates of discards. The methods to estimate discards assume that the relative catch composition (discard ratios) obtained with the data collected during LH inspections reflects the true catch composition of the fleet segment operating with the same gear and mesh size and in that area. The comparison between these discard ratios and with what is reported in fishers' logbook is then used to estimate the discard component. The background of the LH programme, the methodologies for deriving discard ratios using LH data and the statistical analysis of the data are explained in this chapter.

Keywords Catch composition · Compliance · Inspections · Joint deployment plans · Risk assessment

19.1 Introduction

The landing obligation (LO) introduced under Article 15 of the latest reform of the EU's Common Fisheries Policy, adopted in 2013 (EU No 1380/2013), constituted a significant policy change and presented a number of challenges for control authorities working towards ensuring its uniform and effective implementation across all

M. Nuevo (✉) · C. Morgado
EFCA – European Fisheries Control Agency, Vigo, Spain
e-mail: Miguel.Nuevo@efca.europa.eu

A. Sala
Italian National Research Council, Institute of Marine Biological Resources and Biotechnologies (CNR-IRBIM), Ancona, Italy

S. S. Uhlmann et al. (eds.), *The European Landing Obligation*,
https://doi.org/10.1007/978-3-030-03308-8_19

EU Member States (MS). With the obligation to land all catches of quota species being progressively implemented from 2015 to 2019, the whole approach of monitoring what was landed, for the purpose of fisheries control, had to change to monitoring catches at sea to detect possible illegal discards.

The European Fisheries Control Agency (EFCA) considered that a coordinated implementation of the landing obligation using common methodologies was a prerequisite to ensure efficiency, effectiveness and a level playing field for the fishing industry. EFCA, in accordance with its multiannual work programme, and in its role supporting MS and the European Commission in the implementation of the Common Fisheries Policy (CFP) requirements, carried out a technical reflection on the definition of procedures and systems for monitoring the implementation of discard plans. This reflection was driven by the goal of assisting MS and the European Commission to develop simple and cost-efficient methods for monitoring implementation and evaluating compliance with the obligation to land all catches.

Joint deployment plans (JDPs) are one of the EFCA main instruments to ensure effective enforcement and equal treatment for all those involved in a particular fishery. They are the vehicle through which the Agency organises the deployment of MS's human and material means of control and inspection pooled together. Two criteria have to be met before a JDP can be devised: the fish stock(s) concerned must be subject to a long-term recovery plan or a multiannual management plan, and a specific control and inspection programme, adopted by the European Commission, must be in place.

EFCA has no mandate to formulate fisheries policy, which is the responsibility of the European Commission. Nevertheless, it is within the Agency's mandate to make technical recommendations in the context of providing assistance to MS regarding the range of compliance tools which could be employed to help meet their obligations vis-a-vis both Article 15 of the CFP and existing control provisions.

Some key objectives of these recommendations were:

- To ensure compliance with the requirements for accurate recording of discards
- To assist MS in the development of practical control and monitoring tools for the enforcement of the landing obligation through the detection of discarding practices
- To support the development of specific discard plans with suggested guidelines to facilitate the controllability of the landing obligation

The controllability of the landing obligation is also complicated by exemptions built into the various regional discard plans (see Borges and Penas Lado, this volume; Rihan et al., this volume). These complexities give rise to a high level of risk of a non-uniform implementation of the LO, both within and across regions. It is for this reason that efforts to ensure a level playing field in terms of the implementation of the LO became imperative for EFCA. Compliance with the LO is likely to be improved if fishers observe a common approach to inspection in all the areas in which they operate. Indeed, a common point raised by the fishing industry, through fora such as the EU Advisory Councils, is the need for a level playing field in terms of control and enforcement. Experience with the phased implementation of the LO has already highlighted the importance of this issue.

Considering the EFCA recommendations and conclusions agreed during a stakeholder seminar in January 2014[1], the launching of a so-called "last haul" (LH) project was endorsed, an inspection programme coordinated by EFCA in the framework of the JDPs with the main aim to obtain estimates of discards for control and compliance purposes. It focused initially in the Baltic Sea, Western Waters (Pelagic) and Mediterranean Sea (Adriatic) JDP areas, as the species covered by these JDP areas were the first to be subject to the LO from 1 January 2015. This dedicated project was put in place in close cooperation with MS. The underlying driver for this initiative is the need to maintain a level playing field, which should be achieved by developing a harmonised and standardised approach to inspections focused on the LO. The LH programme is now being implemented throughout the main JDPs and integrated in routine sea inspection procedures.

The LH inspection programme as a specific monitoring scheme on the implementation of the LO was introduced following a timeline according to the active JDPs in each area and to the phased introduction of the LO:

- June 2014: Baltic Sea (cod, salmon, herring and sprat) and Adriatic Sea (anchovy and sardine)
- August 2014: Pelagic fisheries in Western Waters
- May 2015: Demersal fisheries for cod, sole and plaice in the North Sea
- May 2016: Demersal fisheries for other demersal species in the North Sea

To implement the LH programme in areas without a JDP, such as for other demersal species in the North Sea in addition to cod, sole and plaice, cooperation with regional bodies occurred through the EFCA PACT (Partnership, Accountability (compliance), Cooperation and Transparency) concept. This allowed assistance to be given by EFCA to the MS. EFCA cooperated with the control expert groups (CEGs) of the main regional bodies created in the framework of regionalisation, such as BALTFISH, Scheveningen, SWW and NWW CEGs, and enlarged its assistance in areas and for species where there is no legal mandate via the specific control and inspection programmes (SCIP) in place and thus not covered by the JDP framework (i.e. demersal fisheries in Western Waters).

The LH inspection work has now been encompassed within the JDPs routine control and inspection effort and target sea inspections by amending the respective JDPs and introducing this specific objective in some campaigns. These specific actions are planned according to the results of the regional risk assessment performed by EFCA in cooperation with relevant MS. A methodology was developed by EFCA to derive discard ratios from the LH data using different methods according to the discarding characteristics (see Sect. 19.4).

[1]https://www.efca.europa.eu/en/content/pressroom/efca-coordination-new-cfp-provisions

19.2 Conducting LH Inspections and Processing the Data

The LH inspection programme consists of inspections at sea with catch data being collected by MS inspectors. The catch composition of the last observed haul of the inspected vessel is recorded in terms of live weight per species and quantities above or below the minimum conservation reference size (MCRS). These data are recorded on a template form, which is then submitted to EFCA for compilation and analysis. The LH concept considers the differences between quantities of catches observed during the sea inspections and quantities of fish reported in the logbook. These differences are used to derive the true discard ratio. With the entry into force of the LO, the difference between the quantities of fish by species subject to the LO below the minimum size (BMS) observed during the LH inspections and the quantities reported in the logbook can be used as an indicator of illegal discarding practices. The LH data are considered as reference data and provide knowledge on the catch composition (number of fish above and below MCRS defined at EU level and ratio of different species) in the sampled hauls.

Additional data are also recorded during LH inspections by some MS on more detailed size compositions of legal-size catches (LSC), i.e. fish above the MCRS. In these cases, additional to the species catches quantities above and below the MCRS, the quantities of fish above the MCRS are recorded by commercial size grade (mainly for cod). The differences between the grades declared in the sales and the grades recorded during the LH inspections provide an indication of the discarding of LSC, also designated as high-grading (see Method D, described in Sect. 19.4.4).

To facilitate LH classification and the data analyses, the fisheries within a JDP were categorised in several fleet segments according to gear, mesh size, area and species caught (see example in Annex 19.2). This categorisation by fleet segment was developed by EFCA in close cooperation with MS, within the JDP steering groups (SGs) and the regional control expert groups (CEGs) constituted in the framework of the CFP regionalisation. The analysis of the LH data and subsequent estimation of discards are conducted at fleet segment level.

Categorising the fisheries within a JDP area into fleet segments assumes implicitly that the catch profile of the fishing trips using the same gear and mesh size and operating in the same areas, i.e. belonging to the same fleet segment, is similar. However, there are variations in the proportion of undersized fish depending on the areas, the type of gear and time of year. In order to have a qualified knowledge of these variations and their interdependencies, it is necessary to obtain a large number of samples to have representative reference data.

19.3 Data and Analysis

The data are collected through the LH inspections together with the declared catches by category (i.e. discards, landed catches below and above the minimum conservation reference size, BMS and LSC, respectively). EFCA compiles these data provided by the MS concerned, which allows completing and updating a matrix of discard ratios by segment and by period, which serves as a baseline for future analysis.

In combination with the LH data collected during inspections, the datasets from each fleet segment are compiled based on information on annual quantities reported in logbooks provided by MS and aggregated by:

(a) Month: from January to December.
(b) Fleet segment: fleet defined based on the gear, mesh size and area of fishing activity. In some case the target species is an additional component to define the fleet segment.
(c) Areas: area of fishing activity.
(d) Species: the species caught.

19.3.1 Weighted Mean and Standard Deviation of the BMS Ratios

In the LH inspections, the following quantities are collected by species:

BMS_{RET}^{LH}: quantity (in kg) of the fish retained below MCRS

LSC_{RET}^{LH}: quantity (in kg) of the fish retained above MCRS

Considering the BMS ratio derived from the LH, $rBMS_{RET}^{LH}$, the percentage of fish below MCRS in relation to the total catch (*BMS* + *LSC*), for each LH, the ratio *rBMS* is:

$$rBMS^{LH} = \frac{BMS_{RET}^{LH}}{BMS_{RET}^{LH} + LSC_{RET}^{LH}} \tag{19.1}$$

The LH *rBMS mean, weighted by the catch of the respective haul*, is then calculated. Whereas weighted means generally behave in a similar way to arithmetic means, they do have a few counter instinctive properties.

Considering the weight (w_i) for each *ith* observation (or LH) the total catch (*BMS* + *LSC*), the LHs with higher catches contribute more than the LHs with lower catch quantities:

$$\overline{rBMS_{RET}^{LH}f,s} = \frac{\sum_{i=1}^{n}\left(w_{f,s,i} \times rBMS_{RET}^{LH}f,s,i\right)}{\sum_{i=1}^{n} w_{f,s,i}} \tag{19.2}$$

$$\forall f,s \;\; i = \{1,\ldots,n\}; \quad n \in \mathrm{N}_0$$

where *f* and *s* define a specific *fleet segment* and *species*, respectively, and *n* is the number of LHs for that specific segment.

The weighted standard deviation (sd_w) of the mean $\overline{rBMS}$ is therefore:

$$sd_{w}{}^{LH}_{RET}f,s = \sqrt{\frac{\sum_{i=1}^{n}\left[w_{f,s,i} \times \left(rBMS^{LH}_{RET}f,s,i - \overline{rBMS}^{LH}_{RET}f,s\right)^2\right]}{\frac{(n' - 1) \times \sum_{i=1}^{n} w_{f,s,i}}{n'}}} \tag{19.3}$$

$$\forall f,s \quad i = \{1,\ldots,n\}; \qquad n \in \mathrm{N}_0$$

where n' is the number of non-zero weights and $\overline{rBMS}$ is the weighted mean of the LH observations for a specific *fleet segment (f)* and *species (s)*.

19.3.2 Standard Error and Confidence Limits of the Mean

One of the best ways to assess the reliability of the precision of a measurement is to repeat the measurement several times and examine the different values obtained. Without variation, all the repeating measurements should give the same value, but in reality the results deviate from each other. Statistics treats each result of a measurement as an item or individual (i.e. each LH) and all the measurements as the sample. All possible measurements, including those which were not done, are called the population. The basic parameters that characterise a population are the **mean**, μ, and the **standard deviation**, σ. The latter indicates the variation or dispersion of the values around the mean. In order to determine the true μ and σ, the entire population should be measured, which is usually impossible to do. In practice, measurement of several items is done, which constitutes a sample. Estimates of the mean and the standard deviation are calculated based on data from sampling and are denoted by $\bar{x}$ and s, respectively. The values of $\bar{x}$ and s are used to calculate the **confidence interval** (*CI*), which is a range of values which is likely to contain the population parameter of interest. The formula for a confidence interval for a population with unknown standard deviation is therefore given by the formula:

$$CI = \bar{x} \pm t^*_{n-1} \times s/\sqrt{n} \tag{19.4}$$

where t^*_{n-1} is the critical t^*-value from the t-distribution with *n-1* degrees of freedom (where n is the sample size). The plus-or-minus figure usually is called **margin of error** and expresses the *statistical uncertainty* or the maximum expected difference between the true population parameter and a sample estimate of that parameter. To be meaningful, the margin of error should be qualified by a probability statement (often expressed in the form of a **confidence level**). The confidence level informs how sure the value is. It is expressed as a percentage and represents how often the true percentage of the population lies within the confidence interval. For example, the 95% confidence level (which is usually used) means the population parameter will be within that range with 95% certainty. The 95% confidence interval for the mean is calculated as:

$$Lower\ limit:\ \ LL = \bar{x} - t_{.95} \times s_M \quad Upper\ limit:\ \ UL = \bar{x} + t_{.95} \times s_M$$

The standard error of the mean is designated as σ_M. It is the standard deviation of the sampling distribution of the mean, which is $\sigma_M = \sigma/\sqrt{N}$ where σ is the standard deviation of the original distribution and N is the sample size (the number of scores each mean is based upon). When s is used as an estimate of σ, the estimated standard error of the mean is $s_M = s/\sqrt{N}$. The larger the sample size, the smaller the standard error of the mean. More specifically, the size of the standard error of the mean is inversely proportional to the square root of the sample size (Fig. 19.1).

19.3.3 Evaluating the Effect of the Number of LH Samples

The mean of a sample of measurements, $\bar{x}$, provides an estimate of the true value, μ, of the quantity we are trying to measure. However, it is quite unlikely that $\bar{x}$ is exactly equal to μ, and an important question is to find a range of values in which we are certain that the value lies. This range depends on the number of measurements done and on the question of *how certain we want to be*. The more certain we want to be, the larger the range we have to take.

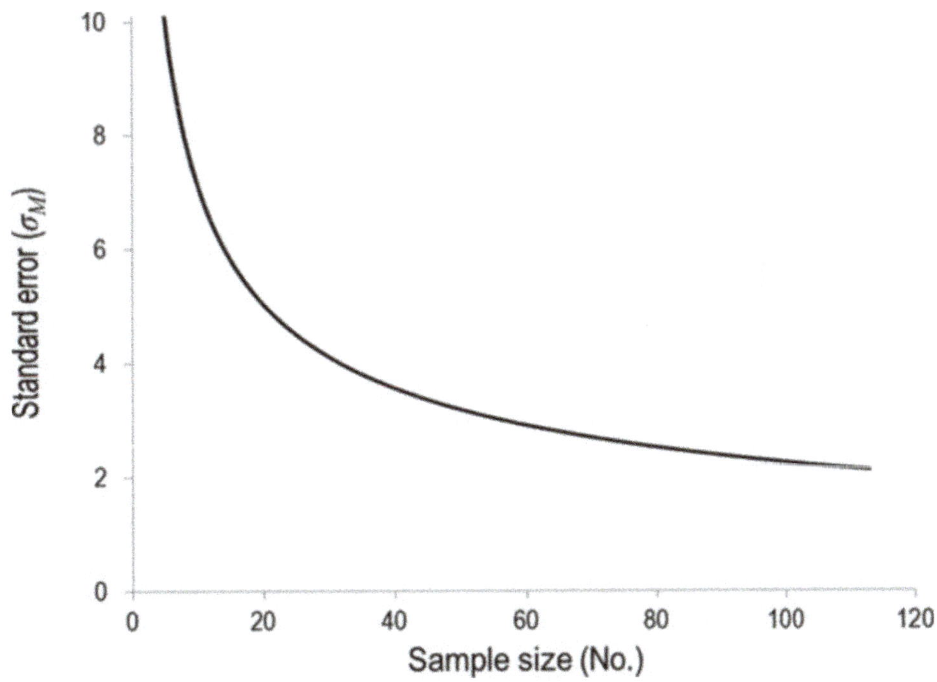

Fig. 19.1 Example of an effect of the sample size on the standard error for a standard deviation

The larger the number of experiments done, the closer $\bar{x}$ is to μ, and a smaller range has to be taken for the same percentage of certainty. The error in the estimate of the mean is proportional to the standard deviation of the sample, s, and the sample size, n. It can be visualised by plotting the mean and its 95% confidence interval (*CI95*). To calculate the sample mean $(\bar{x})$ is likely to be anywhere in the shaded region of the graph in Figure 19.2a, which shows the variation of the confidence interval around the mean $\overline{rBMS}$ or $(\bar{x})$ for different sample size (n).

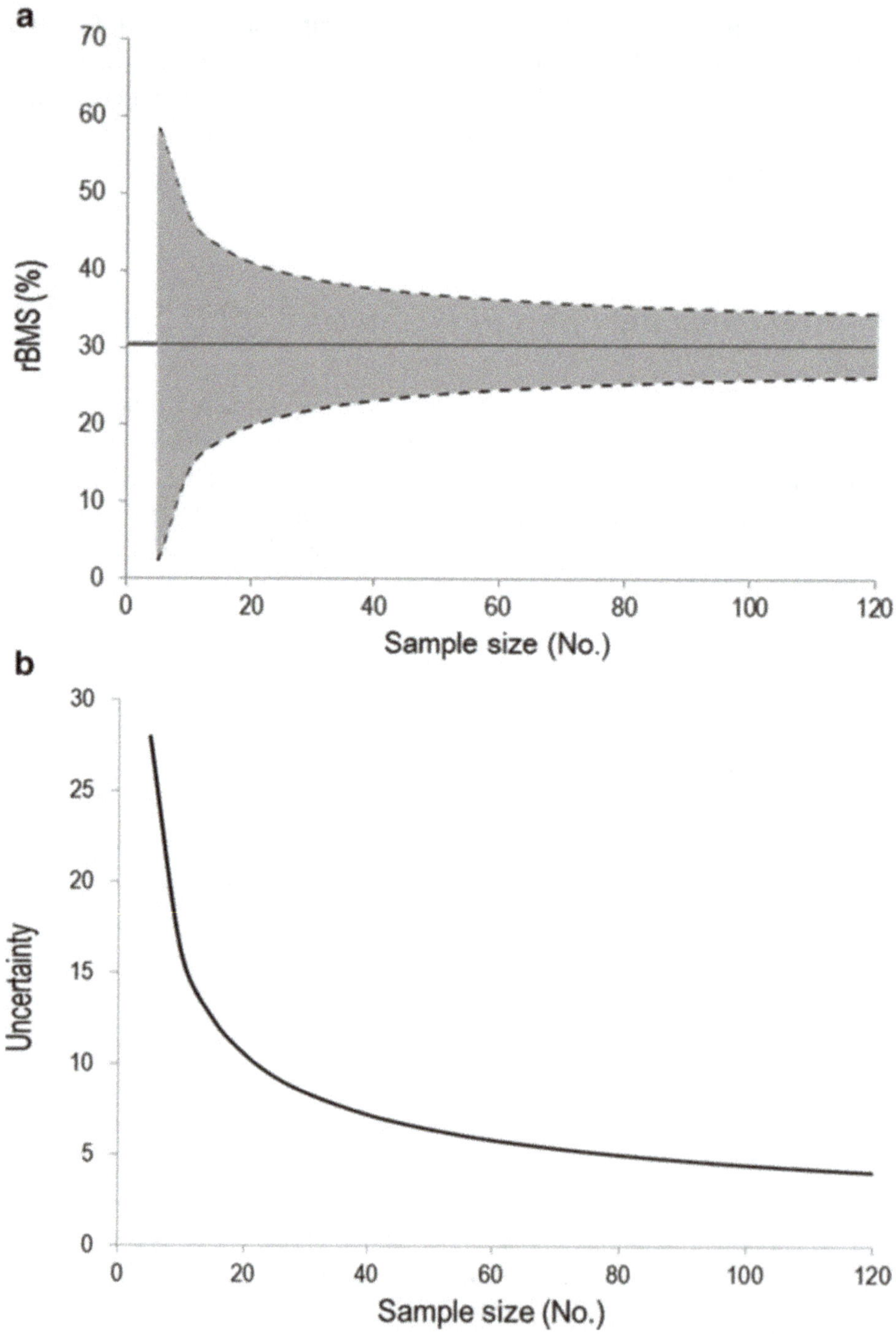

Fig. 19.2 Representation of the 95% confidence region (**a**) around a mean rBMS of 30.4% (bold line) contrasted with sample size (number of observations). Effect of the sample size (**b**) on the margin of error, e.g. statistical uncertainty of the mean

For sufficiently large number of samples, n, both the curve of the Uncertainty (Fig. 19.2b) and of the Standard Error (see Fig. 19.1) level off. To know the mean with a required certainty, there is a need to choose the right size of sample, i.e. the right number of repeated experiments, n.

After a critical analysis of the LH data, it was decided to look at the rate of variation of the Statistical Uncertainty (e.g. variation of the Margin of Error), hereby named as dMoE and set an arbitrary reference or threshold of $dMoE = 2.5\ \%$ (see the green line in Fig. 19.3) to identify the most appropriate number of samples (or LH), n, limit beyond which it would not be convenient to increase the sample size (e.g. number of LH).

Mathematically, the tax of variation of a certain variable is called **derivative,** and the official definition is:

$$f'(x) - \lim_{h \to 0} \frac{f(x+h) - f(x)}{h} \tag{19.5}$$

Thus, for the example reported in Fig. 19.3, the experiment would have had at least 92 samples to achieve the threshold of $dMoE = 2.5\ \%$, noteworthy additionally LHs would not have brought further "Certainty" (or less uncertainty) to the mean estimation of rBMS.

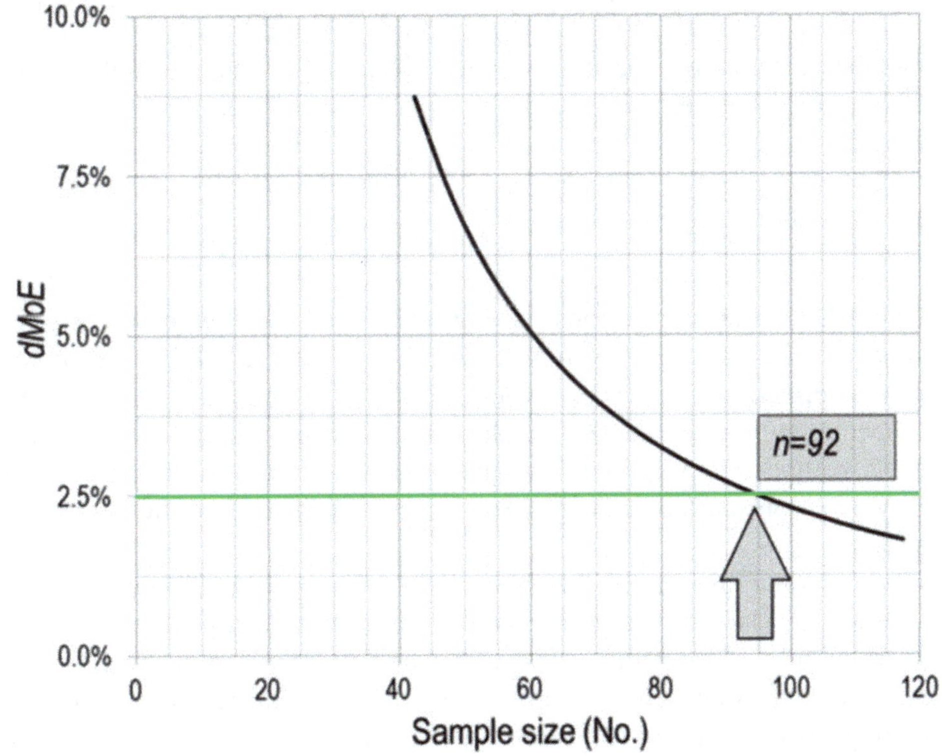

Fig. 19.3 **Example** of the variation of the margin of error (dMoE) contrasted with the sample size (number of observations). The arrow shows the appropriate number of samples (e.g. n = 92 LH) based on the reference threshold of dMoE = 2.5%

19.4 Discarding and Methods to Derive Discard Ratios

There are several reasons for discarding, and usually discarding is a combination of several factors, either legal or economic (Pascoe 1997; Hall et al. 2000; Catchpole et al. 2015; Damalas et al. 2015). To estimate discards there is a need to understand the reasons why discarding takes place, so that it becomes possible to identify more appropriate data and methodologies to be used. The decision tree below identifies the methodology to be used according to the main reason for discarding (Fig. 19.4).

Method A: Discards of fish below minimum conservation reference size (MCRS) not subject to the landing obligation (LO).
Method B: Discards of fish below MCRS subject to the LO.
Method C: Discard of fish species subject to the LO below and above the MCRS, but without reference data on fish size structure. The proportion of the species composition in the catch is used as reference information.
Method D: Discard of fish species subject to the LO below and above the MCRS, with reference data on fish size structure. This method is a combination of Method B, to estimate the discard of the catches below the MCRS, and the estimation of high-grading, i.e. discarding of fish of legal size (legal-size catch, LSC).
Method E: Discard of species subject to the LO with quota limitation (choke species). This method is similar to Method C but should take into account a component of temporal variability, since discarding could be higher when close to quota exhaustion.
Method F: Discard of species subject to the LO with exemption cases. This method provides discard estimates based on either Method B or C (depending on the main discarding motive).

In Sect. 19.4.1, the calculations used in each method to estimate the discard ratio are described (see also Annex 19.1 for an overview of these methods). Method A is applied for species not subject to the LO, while all the other methods concern species subject to the LO. Further details are provided here on when the use of each method is appropriate. The quality of the discard estimates depends on the quality and representativeness of the reference data used and the validity of the assumptions.

Note that in some cases in the literature, the term "discard rate" is used instead of "discard ratio". The latter is the appropriate terminology because a "ratio" represents the proportion of two quantities measured using the same units (e.g. weight in tonnes), while "rate" represents the proportion of two quantities measured in different units (e.g. speed in km/hr is a rate). Nevertheless, the term "rate" is very often used to denote the proportion of discarded fish in relation to the total catch.

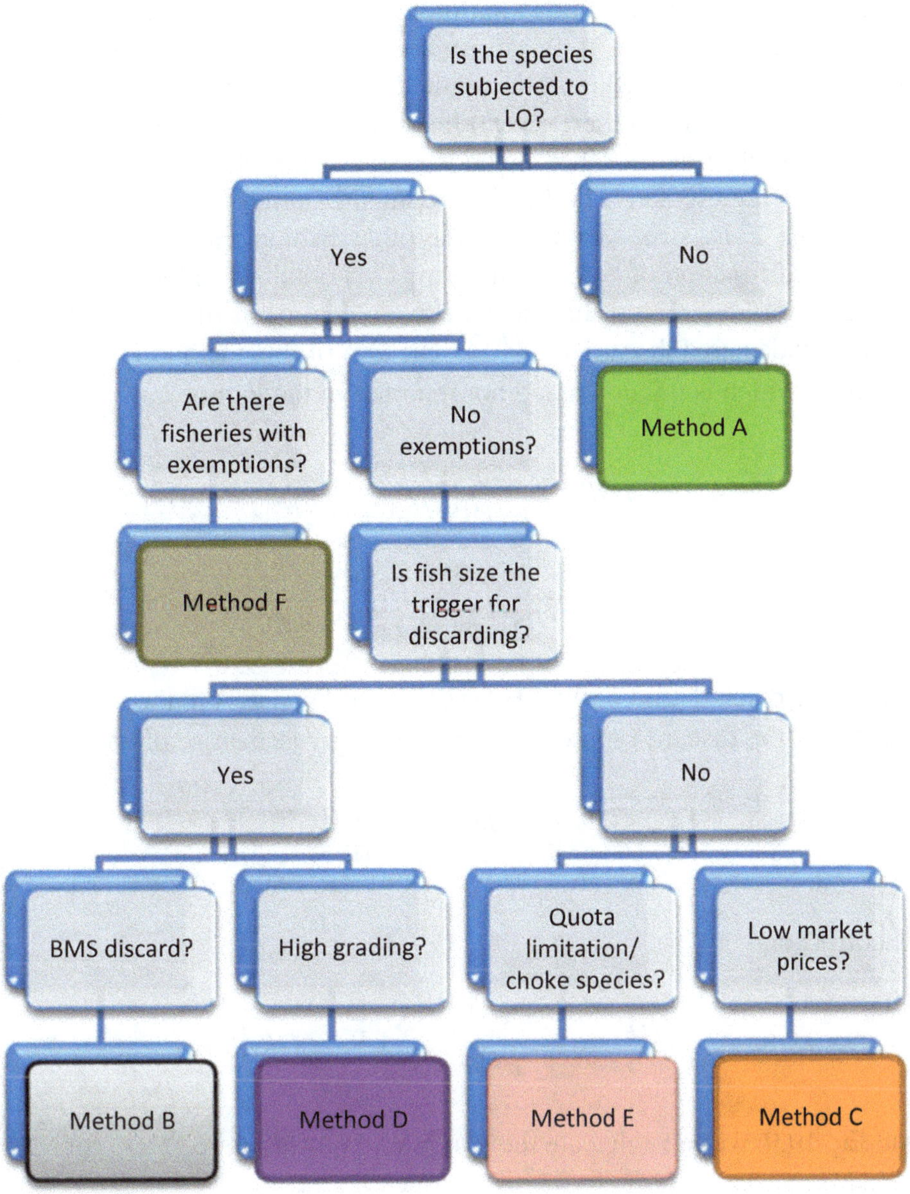

Fig. 19.4 Decision tree of the several methodologies used to estimate the discard ratio using reference data

19.4.1 Method A: Discard of BMS Catches for Species Not Subjected to the Landing Obligation

This method is applicable to estimate the discard ratio for species not subjected to the LO. To apply this method, reference data are needed on the catch size composition of the species being analysed. These data are obtained during detailed inspections on board (LH).

During LH inspections, catch data on unwanted and wanted catch quantities are collected. If a given *Species A* is not subjected to the LO, the unwanted catch component is discarded (DIS), and the wanted component is retained (LSC).

The method assumes that there are no high-grading practices, i.e. all the LSC are landed. The method, as all the other methods, also assumes that the LH data of a given fleet segment is a representative of that fleet segment. Therefore, the catch composition *Species A* of the LHs is the same as all the other fishing trips within a given fleet segment. The information of "other fishing trips" is obtained in the declared discards and landings in logbooks. Note that there are two components of the discards: declared $\left(DIS_A^{LB}\right)$ and undeclared discard$\left(DIS_A^{LB*}\right)$. The aim of this method is to calculate the latter quantity, which is unknown and not reported in the logbook.

$$rDIS_A^{LH} = \frac{DIS_A^{LH}}{DIS_A^{LH} + LSC_A^{LH}} \tag{19.6}$$

$$rDIS_A^{LB} = \frac{DIS_A^{LB} + DIS_A^{LB*}}{DIS_A^{LB} + DIS_A^{LB*} + LSC_A^{LB}} \tag{19.7}$$

By equating the discard ratio of the LH $\left(rDIS_A^{LH}\right)$ of *Species A* to the logbook discard ratio $\left(rDIS_A^{LB}\right)$, the discard ratio of *Species A* in a given fleet segment is:

$$rDIS_A^{LH} = \frac{DIS_A^{LB} + DIS_A^{LB*}}{DIS_A^{LB} + DIS_A^{LB*} + LSC_A^{LB}} \tag{19.8}$$

This corresponds to:

$$DIS_A^{LB*} = \frac{rDIS_A^{LH} \times LSC_A^{LB}}{1 - rDIS_A^{LH}} - DIS_A^{LB} \tag{19.9}$$

Using Eq. 19.9, we can calculate the unreported discard ratio $rDIS_A$ for *Species A*:

$$rDIS_A = \frac{DIS_A^{LB*}}{DIS_A^{LB} + DIS_A^{LB*} + LSC_A^{LB}} \tag{19.10}$$

Replacing $\left(DIS_A^{LB*}\right)$ of Eq. 19.9 in Eq. 19.10, we end with:

$$rDIS_A = \frac{rDIS_A^{LH} \times \left(LSC_A^{LB} + DIS_A^{LB}\right) - DIS_A^{LB}}{LSC_A^{LB}} \tag{19.11}$$

If high-grading (discarding of legal size catch) occurs, the estimation resulting from Method A does not take that into account and should be considered an underestimate. This is because the high-grading component is not recorded in the LH data (e.g. it is not expected that fishermen are high-grading during a LH inspection).

19.4.2 Method B: Discard of BMS Catches for Species Subjected to the Landing Obligation

This method is very similar to Method A but should be used when the species of interest is subjected to the LO. The non-reference data continues to consist of two catch components: the wanted and unwanted catch, as in Method A. However, in Method B there is an additional catch component to be considered in the non-reference data – the unwanted catch component that is now landed. This method is appropriate when the discarded catch consists only of fish below the minimum conservation reference size (MCRS), designated as below minimum size (BMS).

Similar to Method A, this method requires reference data of the size catch composition of the species being analysed, such as data collected during detailed inspections on board (LH). If the species is subjected to the LO, when conducting a LH inspection, the unwanted (BMS) and wanted catch (legal-size catch, LSC) quantities are recorded separately. The estimation of the discards of non-reference data is based on the proportion of unwanted catch of the species in the reference data, which is the BMS ratio $\left(rBMS_B^{LH}\right)$. For a given fleet segment, the BMS catch ratio of ***Species B*** in the LH data $\left(rBMS_B^{LH}\right)$ is compared to the BMS catch ratio in the logbook $rBMS_B^{LB}$.

From the LH data, the available information is the quantity of the fish retained below MCRS $\left(BMS_B^{LH}\right)$ and above MCRS $\left(LSC_B^{LH}\right)$. Therefore, the BMS ratio from LH of *Species B*$\left(rBMS_B^{LH}\right)$ is:

$$rBMS_B^{LH} = \frac{BMS_B^{LH}}{BMS_B^{LH} + LSC_B^{LH}} \tag{19.12}$$

However, the non- reference data $\left(rBMS_B^{LB}\right)$ might have an additional illegal and unreported component of fish below MCRS that is discarded, DIS_B^{LB*}. This component is not recorded in the LH data because it is not expected that fishers discard a species subjected to the LO in the presence of an inspector.

$$rBMS_B^{LB} = \frac{BMS_B^{LB} + DIS_B^{LB*}}{BMS_B^{LB} + DIS_B^{LB*} + LSC_B^{LB}} \tag{19.13}$$

Method B calculates the quantity of illegal discards, DIS_B^{LB*}, which is unknown and not reported in the logbook.

Assuming that $rBMS_B^{LH} = rBMS_B^{LB}$, the total BMS ratio of non-reference data is:

$$rBMS_B^{LH} = \frac{BMS_B^{LB} + DIS_B^{LB*}}{BMS_B^{LB} + DIS_B^{LB*} + LSC_B^{LB}} \tag{19.14}$$

which corresponds to:

$$DIS_B^{LB*} = \frac{rBMS_B^{LH} \times LSC_B^{LB}}{1 - rBMS_B^{LH}} - BMS_B^{LB} \quad (19.15)$$

Using the Eq. 19.15, the discard ratio $rDIS_B$ for a given *Species B* subjected to the LO is:

$$rDIS_B = \frac{DIS_B^{LB*}}{BMS_B^{LB} + DIS_B^{LB*} + LSC_B^{LB}} \quad (19.16)$$

Replacing $\left(DIS_A^{LB*}\right)$ of Eq. 19.15 in Eq. 19.16:

$$rDIS_B = \frac{rBMS_B^{LH} \times \left(LSC_B^{LB} + BMS_B^{LB}\right) - BMS_B^{LB}}{LSC_B^{LB}} \quad (19.17)$$

19.4.3 Method C: Discards of Low-Market Species Subjected to the LO Regardless of Size

In this method, discards of a certain *Species C* might occur because of its low market value, and the part DIS_C^{LB} is made of both the BMS and the LSC component. For such *Species C*, subjected to the LO, the recorded BMS from the reference data (LH) is not indicative of the usual discarding fishers' behaviour, as it is not expected that fishers discard either BMS or LSC components during an inspection.

Since the discards also occur in the LSC component, Method B is not appropriate. Method C to estimate of the discard ratio ($rDIS_C$), for a given *Species C* subjected to LO, is based on the catch profile similarity between the two species. As in the other methods, LH data are used as references and compared with the data recorded in the logbooks. The model assumes that within each fleet segment, the catch proportion of *Species C*, named r_C in the following equations, in relation to the total catch, is similar. Considering two species both subjected to the landing obligation, *Species B* and *Species C*, but with *Species C* discarded regardless of fish size, it is assumed that $r_C^{LH} = r_C^{LB}$. Therefore,

$$r_C^{LH} = \frac{t_C^{LH}}{t_C^{LH} + t_B^{LH}} \quad (19.18)$$

$$r_C^{LB} = \frac{t_C^{LB}}{t_C^{LB} + t_B^{LB}} \quad (19.19)$$

where t_B and t_C correspond to the total catch of *Species B* and *Species C*, respectively, in the LH and logbook (LB). The total catches for each species in the logbook are:

$$t_C^{LB} = DIS_C^{LB*} + BMS_C^{LB} + LSC_C^{LB} \tag{19.20}$$

$$t_B^{LB} = DIS_B^{LB*} + BMS_B^{LB} + LSC_B^{LB} \tag{19.21}$$

where DIS_C^{LB*} is the unknown discarded quantity of *Species C*. The discard component of *Species B*,DIS_B^{LB*}, also unknown, is estimated by applying Method B by Eq. 19.15.

Based on the above-mentioned assumption $r_C^{LH} = r_C^{LB}$, the ratio of *Species C* is:

$$r_C^{LH} = \frac{t_C^{LB}}{t_C^{LB} + t_B^{LB}} \Rightarrow t_C^{LB}\left(1 - r_C^{LH}\right) = r_C^{LH} \times t_B^{LB} \tag{19.22}$$

Replacing t_B^{LB} in Eq. 19.23 with $DIS_B^{LB*} + BMS_B^{LB} + LSC_B^{LB}$ from Eq. 19.20, the total catch of *Species C* is:

$$t_C^{LB} = \frac{r_C^{LH}}{1 - r_C^{LH}} \times \left(DIS_B^{LB*} + BMS_B^{LB} + LSC_B^{LB}\right) \tag{19.23}$$

Substituting the resulting t_C^{LB} from Eq. 19.23 into Eq. 19.20, the DIS_C^{LB} can be calculated, and discard ratio, $rDIS_C$, of *Species C* is (Eq. 19.28):

$$DIS_C^{LB*} = t_C^{LB} - \left(BMS_C^{LB} + LSC_C^{LB}\right) \tag{19.24}$$

$$rDIS_C = \frac{DIS_C^{LB*}}{t_C^{LB}} \tag{19.25}$$

$$rDIS_C = \frac{t_C^{LB} - \left(BMS_C^{LB} + LSC_C^{LB}\right)}{t_C^{LB}} \tag{19.26}$$

$$rDIS_C = 1 - \frac{BMS_C^{LB} + LSC_C^{LB}}{\frac{r_C^{LH}}{1 - r_C^{LH}} \times \left(DIS_B^{LB*} + BMS_B^{LB} + LSC_B^{LB}\right)} \tag{19.27}$$

$$rDIS_C = 1 - \frac{\left(1 - r_C^{LH}\right) \times \left(BMS_C^{LB} + LSC_C^{LB}\right)}{r_C^{LH} \times \left(DIS_B^{LB*} + BMS_B^{LB} + LSC_B^{LB}\right)} \tag{19.28}$$

19.4.4 Method D: High-Grading Discards

Method D should be used in similar cases as Method C, i.e. when discarding also involves the legal-size component of the catch. However, Method D should be applied when there are data available on the size structure of the catch for both reference and non-reference data sets. The size structure considered here is the grade size information, information which is easily collected. Other size structure information, such as the length frequency distribution of the catches, could be used but are more difficult to collect and therefore are not addressed here.

To estimate total discards, there is a need to consider two components: one to estimate the high-grading (discarding of LSC) and another to estimate the discarding of BMS, using Method 1 or 2, depending if the species is subjected to the LO or not, which should only be applied if the LSC are corrected with the estimates of LSC discards.

Method D assumes that high-grading practices are similar within a fleet segment or within the unit used for the analysis. Also, it assumes that high-grading does not occur for larger fish.

Considering the catch proportion of each grade size x, denoted as P_x, as:

$$P_x = \frac{C_x}{C_{TOTAL}} \tag{19.29}$$

Being C_x, the catch of grade x and the total catch of all grades, as:

$$C_{TOTAL} = \sum_{x=1}^{5} C_x = C_1 + C_2 + C_3 + C_4 + C_5 \tag{19.30}$$

In Eq. 19.30, five grade sizes are considered, but a different grade size number could be adopted as a different stratification, such as considering a group of combined grades (e.g. grade size equal or smaller than grade 3).

If the catch proportion of the non-reference data to be assessed ($P_{n,x}$) of grade size x is lower than the catch proportion of the same grade size x of the reference data ($P_{f,x}$), then there is an indication of discarding, because the method assumes similar size structures between reference (f) and non-reference data (n).

Based on Eq.19.29, the catch proportion of reference data could be calculated as:

$$P_{f,x} = \frac{C_{f,x}}{C_{f,TOTAL}} \tag{19.31}$$

To estimate the high-grading of non-reference data, it is necessary to identify the grade sizes for which high-grading is not perceived to be an issue, i.e. grades with similar catch proportion in reference and non-reference data.

$$P_{n,x} \sim P_{f,x} \tag{19.32}$$

Assuming that the grade size with no discarding (high-grading) are grade size 1 and 2, then Eq. 19.31 for reference data would be:

$$P_{f,1-2} = \frac{C_{f,1+}C_{f,2}}{C_{f,TOTAL}} \tag{19.33}$$

There is a need to verify that the catch proportion of reference and non-reference data is similar for those grade sizes (Eq. 19.32). After identifying the grades with no discards (100% retained), it is then possible to estimate the total catch of non-reference data, $C_{n,TOTAL}$, using Eq. 19.29. Considering that the grades with no discarding are grades 1 and 2, the catch proportion of reference data of non-discarded grades defined in Eq. 19.33 can be written as:

$$C_{n,TOTAL} = \frac{rC_{n,1} + rC_{n,2}}{P_{f,1-2}} \tag{19.34}$$

where $rC_{n,1}$ and $rC_{n,2}$ are the declared landings of grade sizes 1 and 2 of non-reference data, respectively, which are assumed to be grades with no high-grading.

The total weight of high-grading, H, is:

$$H_{TOTAL} = C_{n,TOTAL} - \sum_{x=1}^{5} C_{n,x} \tag{19.35}$$

ETo calculate the high-grading weight of a given grade size, x, the proportion of the reference data of that grade size is applied to the total catch of non-reference data (from Eq. 19.34), to obtain the total catch of each grade, which is then subtracted from the reported catch of that grade, as:

$$H_x = (P_{f,x} \cdot C_{n,TOTAL}) - rC_x \tag{19.36}$$

The estimation of the total high-grading ratio, HR, is:

$$HR = \frac{H_{TOTAL}}{C_{n,TOTAL}} \tag{19.37}$$

Note that the high-grading ratio is a discard ratio only for the legal-size component of the catch and does not consider the discards of fish below the MCRS.

19.4.5 Method E: Discard of Species with Quota Limitations (Choke Species) and Subjected to the Landing Obligation

Method E should not be considered as a method but a combination of different methods based on available information of discard practices. Discards in this case originates due to quota limitations. The more appropriate method to estimate the discard ratios is the application of Method C, but data analysis should be conducted regularly throughout the year as the discard pattern might change depending on the quota availability throughout the year.

19.4.6 *Method F: Discard of Species Subjected to the Landing Obligation with Exemption Cases*

Method F is a combination of methods according to the main discard reasons and should only be applied to the catch proportion that is not exempted. Therefore, there is a need to estimate the catch proportion that is subjected to the LO and the exempted component.

19.5 Outcome and Way Forward

The results from the LH programme are an important input into the annual risk assessment workshops, for the evaluation of the likelihood of non-compliance with the landing obligation. The outcome of the risk assessment is then used as a key input for the planning of the JDP for the upcoming year control activities. It should be noted that the objective of the collection of LH data is not to obtain precise discard ratios but to identify where problems are present, and in which magnitude, to focus monitoring and control efforts. The information collected by MS under their scientific discard observers' programmes is also taken into account when evaluating compliance. This information is compiled in a public database available at https://stecf.jrc.ec.europa.eu/web/stecf/reports. The Joint Research Centre is assembling fishery data collected under the EU Data Collection Framework (DCF)[2,3] and acting as secretariat of the Scientific, Technical and Economic Committee for Fisheries (STECF).

During the EFCA JDP risk assessment workshops, the main threats, including non-compliance with the LO and associated misreporting, are identified at fleet segment level, and the spatial and temporal distributions of fisheries are assessed, allowing for the planning of future JDPs. A set of possible risk treatment measures is then developed, and, on this basis, a series of "specific actions" addressing the main threats are implemented in the JDPs. In addition to the risk assessment workshops,

[2]Council Regulation (EC) No 199/2008 of 25 February 2008 concerning the establishment of a community framework for the collection, management and use of data in the fisheries sector and support for scientific advice regarding the Common Fisheries Policy.

[3]Regulation (EU) 2017/1004 of the European Parliament and of the Council of 17 May 2017 on the establishment of a union framework for the collection, management and use of data in the fisheries sector and support for scientific advice regarding the Common Fisheries Policy and repealing Council Regulation (EC) No 199/2008.

the CEGs formally requested the assistance of EFCA to facilitate a joint compliance evaluation with the provisions of the LO. In this context, EFCA is using the methods described in this chapter to derive indicators that can be used in the evaluation of compliance with the LO, one of which being based on the comparison of reference fishing activity information (inspected activity using LH, with information coming from electronic monitoring (EM) with video or observers) versus non-inspected activity. Differences could be measured and utilised in the form of a compliance indicator. This approach has been followed in the context of the LO and assisted in deriving the characterisation of the nature of possible non-compliance behaviours.

Regarding standardisation of how LH inspections are conducted, EFCA developed basic guidelines, produced in cooperation with MS in 2015. The standardised sampling procedure differs by region and type of fishery (segment) and includes observing the whole catch or taking a determined minimum sample size for large hauls (i.e. 300 kg). For these larger hauls, it is recommended that three sets are taking at the beginning, middle and end of the haul. A good approach to standardisation is also the exchange of experiences and best practices among MS in dedicated workshops, exchanges of inspectors and participation of EFCA coordinators. The issue of last haul standardisation was also raised in other meetings, specific LH and discard workshops organised by EFCA, in which relevant scientists, MS authorities and other stakeholders' feedback on the LH programme was acknowledged. The current guidelines provide a common basis for promoting standardisation and harmonisation, and more detailed guidelines to further improve LH data quality are being developed between EFCA and MS.

Additionally, reference data on catch composition could be observed through fully documented fisheries (FDF), including EM with video. These observed data could then be systematically compared with catch composition data from the reported landings of vessels of the same fleet segment that have operated in the same area at the same time.

Incentives may continue to exist for discarding, for example, of specimens below MCRS, smaller market categories [high-grading], species that threaten to choke the fishery, species of low market value, etc. If such specimens are being discarded due to non-compliance with the LO, it is expected that these will be found in smaller proportions of the reported landed catch than in the observed catch. One limitation of the LH programme could be that discrepancy between the observed and the reported catch composition cannot be used as evidence of discarding in any individual case (because catch compositions can vary by chance or due to differences in the skills of skippers), but trends in the magnitude of these discrepancies at aggregate level are being considered by EFCA and the MS as an indicator to evaluate compliance with the LO (see, e.g. Valentinsson et al., this volume). At the same time, declining trends in the proportions of unwanted catch (below MCRS or of the choke stock) in the inspected catch could be an indicator of progression in avoidance behaviour. These trends could be interpreted by looking at changes in selective gear uptake or changes in spatiotemporal effort allocation.

Annexes

Table Annex.19.1 Overview of different methodologies used by EFCA to estimate discards and the discard ratio

	Subject to LO	Discarded component	Type of reference data needed	Assumptions	Discard ratio
Method A	No	BMS	Reference catch data discriminated as retained and discarded	No high-grading in the retained component Uniform catch size composition within the fleet segment	$uDR_n = \frac{uDIS_n}{uDIS_n + rDIS_n + RET_n}$
Method B	Yes	BMS	Reference catch data discriminated as BMS and LSC	No high-grading in the LSC component Uniform catch size composition within the fleet segment	$uDR_n = \left(\frac{DR_f \cdot LSC_n}{1 - DR_f} - rBMS\right) \cdot \left(\frac{1 - DR_f}{LSC_n}\right)$
Method C	Yes	BMS and LSC	Reference catch data of two species discriminated as BMS and LSC	Uniform catch size and species composition within the fleet segment	$uDR_{A,n} = \frac{uA_n}{A_{n,TOTAL}}$
Method D	Yes	BMS and LSC	Reference data with size structure information (i.e. grade size)	Uniform catch size composition within the fleet segment No high-grading of the grade size classes of larger fish and similar retained proportion of reference and non-reference data	$DR_n = \frac{uBMS_n + H_{TOTAL}}{uBMS_n + rBMS_n + C_{n,TOTAL}}$
Method E	Yes	BMS and LSC	Similar to Method C	Similar to Method C	Similar to Method C
Method F	Yes, partially	See other methods[a]	See other methods[a]	See other methods[a]	See other methods[a]

[a]Dependent on the discarding practices in the nonexempted component

Annex 19.2 Baltic Sea JDP fleet segments

Fishery	Gear	Segment	Area	Segment code
Demersal fisheries, active gears all vessels lengths	Demersal active	OT[a] (≥105)	22–24	BS01
		SX[b] (≥105)	22–24	BS02
		OT[a] (≥105)	25–27	BS03
Pelagic fishery for sprat and herring, active gears and all lengths	Pelagic active	OT[a], PT[c] (≥16 and <32)	22–27	BS04
		OT[a], PT[c] (≥32and <90)	22–27	BS05
		OT[a], PT[c] (≥16 and <105)	28–32	BS06
Salmon South	Pelagic passive	GN[d] (≥157)	22–29	BS07[f]
		LL[e]	22–29	BS08[f]
Salmon North	Pelagic passive	FIX (national rules)	22–32	BS09[f]
Passive gear fishery	Demersal passive	GN[d] (≥110), LL[e]	22–24	BS10
		GN[d] (≥110), LL[e]	25–27	BS11
	Pelagic passive	GN[e] (≥32 and <110), FIX (national rules)	22–32	BS12
Other	Other	Other non-reported in segments 1–12	22–32	BS13

Notes:
[a]*OT* includes the following gear codes according to Annex XI of Regulation (EU) No 404/2011: OTB, TBN, TBS, TB, OTT, OTM
[b]*SX* includes the following gear codes according to Annex XI of Regulation (EU) No 404/2011: SDN, SSC, SPR, SX, SV
[c]*PT* includes the following gear codes according to Annex XI of Regulation (EU) No 404/2011: PTB, PTM
[d]*GN* includes the following gear codes according to Annex XI of Regulation (EU) No 404/2011: GN, GNS, GNC, GTN, GTR
[e]*LL* includes the following gear codes according to Annex XI of Regulation (EU) No 404/2011: LHP, LHM, LLS, LLD, LL, LTL, LX
[f]Direct fishing for salmon (i.e., when salmon catches are > 50% of total catches per fishing trip)

Glossary of Terms and Symbols Used in the Equations

Glossary of Terms

Below minimum size (BMS) Marine organism with size below the minimum conservation reference size (MCRS).

Discard(s) Catch component that is not retained (discarded) and not landed. Abbreviated in this document as DIS. Discarding is the practice of returning unwanted catches to the sea, either dead or alive, because they are undersized, due to market demand or due to specific regulations such as quota exhaustion or catch composition rules.

Discard rate See discard ratio definition.

Discard ratio Proportion of discard quantities in relation to the total catch. Usually expressed in percentage (%) and often also designated as "discard rate".

Grade size Size category determined according to weight or number of fishes in 1 kg. The EU grade definition is laid down in Council Regulation (EC) No 2406/96.

High-grading Discarding of legal-size catch (see LSC definition below).

Landing Obligation EU term for the obligation to land all catches of regulated commercial species on board and to count those catches against quota. As defined in Article 15 of Regulation (EU) No 1380/2013, the LO applies to all catches of species subject to catch limits or, in the Mediterranean, species which are subject to minimum sizes. The LO includes some exemptions: de minimis (that allows the discarding of a small percentage of catches), high survivability (defined in the discard plans and that allows the discarding of specimens with a high chance of surviving) or for catch damaged by predators, disease or contaminated and therefore unfit for human consumption.

Last haul (LH) Hauls in which fisheries inspectors have recorded the amounts of the different components of the catch (i.e. BMS, LSC, etc.) per species. Data collected at those hauls are considered reference data to estimate discards.

Legal-size catch (LSC) The catch of marine organisms of size above the minimum conservation reference size (MCRS).

Non-reference data In this document, it refers to data recorded in logbooks or landing declarations, where the discarding component has not been fully verified.

PACT Partnership, Accountability (compliance), Cooperation and Transparency. This concept allowed assistance to be given by EFCA to the MS in accordance with provisions of Articles 7 and 15 of EFCA's founding regulation. EFCA cooperated with control expert groups from the main regional bodies created in the framework of regionalisation and enlarged the assistance in areas and for species where there is no SCIP in place and thus not covered by the JDP framework.

Reference data Data assumed to be representative of the true catch composition. In opposition to non-reference data, in reference data, the discarding component is likely to have been fully verified.

REM Remote electronic monitoring. Vessels which are equipped with video cameras and a system of sensors and vessel monitoring tools, to record fishing and fish sorting operations. Very often fishing data from these vessels are considered reference data.

Retained catch Catch component that is retained (not discarded) and landed. Abbreviated in this document as RET.

Unwanted catch Designates the catch that was, usually, discarded prior to the coming into force of the LO. Very often, associated with undersized or low market demand fish or cases of catches for which the quota had been exhausted or in contradiction with catch composition rules.

Wanted catch Is used to designate the quantity of fish that would be landed in the absence of the LO.

Symbols Used in the Equations

BMS Catch below minimum size

rBMS Ratio of catch below minimum size

t Total catch

DIS Discards

rDIS Discard ratio, also designated as discard rate

H High-grading weight

HR High-grading ratio

LSC Legal-size catch

r Catch proportion of a certain species in relation to the total catch

P Proportion (e.g. of grade sizes in relation to the total catch)

References

Borges, L., & Penas Lado, E. (this volume). Discards in the common fisheries policy: The evolution of the policy. In S.S. Uhlmann, C. Ulrich, S.J. Kennelly (Eds.), *The European Landing Obligation – Reducing discards in complex multi-species and multi-jurisdictional fisheries*. Cham: Springer.

Catchpole, T.L., Frid, C.L.J., Gray, T.S. (2015). Discards in North Sea fisheries: Causes, consequences and solutions. *Marine Policy, 29*, 421–430.

Damalas, D., Maravelias, C.D., Osio, G.C., Maynou, F., Sbrana, M., Sartor, P., Casey, J. (2015). Historical discarding in Mediterranean fisheries: A fishers' perception. *ICES Journal of Marine Science, 72*, 2600–2608.

European Fisheries Control Agency. (2017). EFCA Annual Report 2017. Legal Base: Articles 14 and 23(2)(b) of Council Regulation (EC) No 768/2005 as last amended by Regulation (EU) 2016/1626, Article 47 of the Financial Regulation of the European Fisheries Control Agency (EFCA). Available in EU Bookshop (http://bookshop.europa.eu).

Hall, M.A., Alverson, D.L., Metuzals, K.I. (2000). By-catch: Problems and solutions. *Marine Pollution Bulletin, 41*, 204–219.

Pascoe, S. (1997). Bycatch management and the economics of discarding. FAO Fisheries Technical Paper. No. 370; FAO: Rome, 137p.

Rihan, D., Uhlmann, S., Ulrich, C., Breen, M., Catchpole, T. (this volume). Requirements for documentation, data collection and scientific evaluations. In S.S. Uhlmann, C. Ulrich, S.J. Kennelly (Eds.), *The European Landing Obligation – Reducing discards in complex multi-species and multi-jurisdictional fisheries*. Cham: Springer.

Valentinsson, D., Ringdahl, K., Storr-Paulsen, M., Madsen, N. (this volume). The Baltic cod trawl fishery: The perfect fishery for a successful implementation of the Landing Obligation? In S.S. Uhlmann, C. Ulrich, S.J. Kennelly (Eds.), *The European Landing Obligation – Reducing discards in complex multi-species and multi-jurisdictional fisheries*. Cham: Springer.

Chapter 20
Possible Uses of Genetic Methods in Fisheries Under the EU Landing Obligation

Magnus Wulff Jacobsen, Brian Klitgaard Hansen, and Einar Eg Nielsen

Abstract While genetics has assisted fisheries management for over 50 years, genetic applications aiming to alleviate or eliminate discards have received little attention. In this chapter, we focus on how genetics can be applied under the EU Landing Obligation, to identify and prevent unwanted catches and to estimate the composition of products made from such catches. Three themes are covered: (i) the genetic identification of bycatch; (ii) the genetic analysis of species composition in nutritional products made from unwanted fish; (iii) the potential of using so-called environmental DNA (DNA shedded from aquatic organisms into the water) to reduce bycatch. For all themes, we introduce and explain the relevant genetic techniques, including data formats and analyses. We present the most significant limitations of the methodologies for their implementation in fisheries and provide examples of their use through relevant case studies. Finally, we discuss the potential future perspectives, with emphasis on the rapid progress in portable and automatic DNA devices, which may revolutionize the use of real-time onsite genetic analyses.

Keywords Discard · Environmental DNA · Fisheries bycatch · Genetics · Species composition · Species identification

20.1 Introduction

For more than 50 years, genetics has been assisting fisheries management and marine conservation efforts (Ovenden et al. 2015). During this period, the molecular techniques for studying genetic variation and associated analytical methods have changed tremendously (e.g., Goodwin et al. 2016; Herbert et al. 2003), which in turn have prompted new possibilities in relation to fisheries (Ovenden et al. 2015). Examples of the application of genetics are diverse and cover issues like species

M. W. Jacobsen (✉) · B. K. Hansen · E. E. Nielsen
Section for Marine Living Resources, Technical University of Denmark (DTU), Silkeborg, Denmark
e-mail: lmwj@aqua.dtu.dk

S. S. Uhlmann et al. (eds.), *The European Landing Obligation*,
https://doi.org/10.1007/978-3-030-03308-8_20

identification, analysis of stock structure and food traceability, which are all relevant for conservation, fisheries managers and consumers (reviewed by Ovenden et al. 2015). So far, most uses have focussed on ecosystem management and sustainable harvest of fish resources, while genetic applications aiming to alleviate or eliminate discards have received little attention.

In this chapter, we focus on how genetics can be applied under the EU Landing Obligation, in order to assess and mitigate unwanted catches (e.g., undersized juvenile and low-profit species that will have to be retained and landed under the Landing Obligation) and identify the composition of processed products made from these catches. For all themes, we introduce the relevant genetic techniques, the associated data format and downstream analyses. Moreover, we present the most significant limitations of the techniques, provide examples of their use through relevant case studies and finally discuss potential future perspectives.

20.2 Genomes, Genes and Genetic Markers

All genetic approaches described in this chapter rely on the analysis of Deoxyribonucleic acid (DNA), which constitutes the genetic code. When populations of organisms are isolated from each other, they follow separate evolutionary trajectories and accumulate genetic differences over time due to the evolutionary forces of mutation, selection and genetic drift (stochastic changes in the frequencies of genetic variants accumulating from generation to generation) (Coyne and Orr 2004; Mayr 1942, 1963). Accordingly, when two sympathetic species have been geographically separated for long enough, their DNA will contain genetic differences, which can be used to unambiguously identify these species using various techniques.

Within a normal animal cell, two different genomes can be found, i.e. the nuclear genome and the mitochondrial genome also referred to as the "mitogenome" (Fig. 20.1). The nuclear genome is located within the cell nucleus in one copy per cell. This genome contains most of the organisms' genes and is by far the largest, consisting of millions or billions of base pairs. The mitochondrial genome is a much smaller circular genome of around 15–17,000 base pairs in length, and normally includes 37 genes (Ballard and Whitlock 2004). One mitogenome is located in each mitochondrion, i.e. the small organelles responsible for energy production (Saraste 1999), which is also found in the cell. The number of mitochondria varies between cell types, with numbers ranging from ~80 to ~2000 in mammals (Cole 2016). Hence, there are many copies of the mitogenome, but only one nuclear genome within each cell. The high number of mitogenomes per cell is an advantage for many genetic applications, as it is easier to obtain a sufficient amount of intact DNA suitable for downstream applications (Ballard and Whitlock 2004; Galtier et al. 2009). Combined with other attributes like a highly conserved gene content and architecture, which promote easy inter-species comparisons, this feature has made the mitochondrial DNA the most widely used marker for many genetic applications.

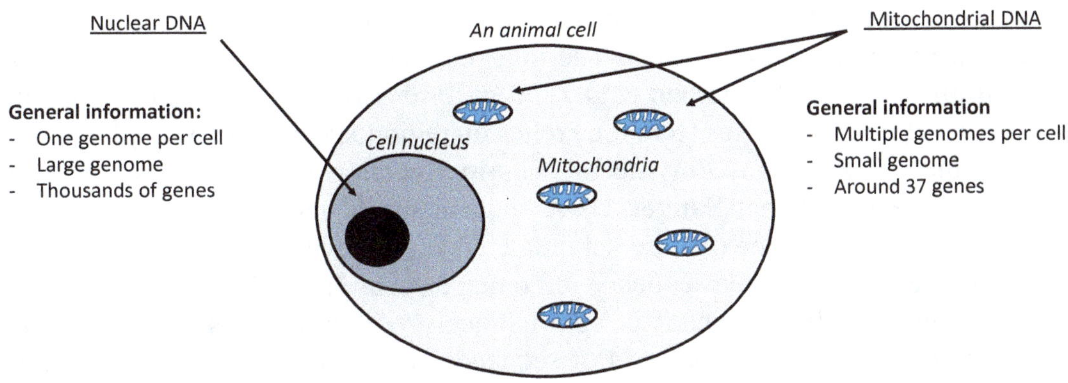

Fig. 20.1 Key differences between the nuclear and mitochondrial genomes. The drawing depicts an animal cell and the arrows show the location of the nuclear and mitochondrial DNA. For simplicity, not all cellular organelles are shown

20.3 Species Detection and Bycatch Assignment

20.3.1 The Technology and Its Applications in Fisheries

With the implementation of the Landing Obligation, many species that previously would have been discarded, e.g. undersized juvenile and low-profit species will now have to be retained and landed. This will challenge the fishing industry, requiring authorities and managers to ensure the correct identification of all landed fish. This information will allow for a more accurate assessment of fisheries' impacts on species and populations, improving possibilities for evaluating the effect of fishing on the viability of the species or populations of interest (Ovenden et al. 2015). Thus, if fishing has a significant impact on certain species, legal actions can be put in place to reduce bycatch.

Species identification is often based on morphological characters alone; however, in many cases proper species assignment can be problematic. For example, diagnostic characters may not be present across juvenile life stages of some species (McEachran and Musick 1973). Moreover, species may lack morphological characters for easy taxonomic identification because they are badly preserved (or damaged by the fishing gear), are poorly defined (due to limited taxonomic knowledge), or do not exist (cryptic species, i.e. different species that are morphological identical) (Ovenden et al. 2015). Even when good diagnostic morphological characters exist, taxonomic analysis can be difficult, labour intensive and expensive, as it relies on the expertise of highly specialised taxonomists, the number of which is declining worldwide (Hopkins and Freckleton 2002; Wheeler et al. 2004). Hence, alternative ways of conducting species assignment are warranted.

DNA-based identification methods offer fast, standardized and accurate tools for species assignment. The only prerequisites are that the focal species and its sister-species have been described taxonomically, that regions of their DNA have been sequenced and that there is sufficient variation within these sequences to discriminate between target and other taxa. The most commonly used technique is "**DNA barcoding**" described by Herbert et al. (2003) (for description of technical terms in bold, see text box 20.1). This technique relies on region-specific DNA amplification using a method known as **Polymerase Chain Reaction (PCR)** and is typically followed by conventional **Sanger** DNA sequencing. Typically, for animals, the mitochondrial cytochrome oxidase subunit 1 gene (abbreviated COI) is sequenced and compared to a validated sequence reference database (Fig. 20.2a).

The amplified gene segment is approximately 650 base pairs long and was originally chosen based on the level of sequence variation, for most taxa showing only minute genetic variation within the species, while still enabling the discrimination of even closely related species (Hebert et al. 2003). Moreover, several universal **PCR assays** have been developed rendering it possible to amplify and sequence COI across large taxonomic groups, e.g. fish, vertebrates and invertebrates (Ivanova et al. 2007). However, other gene regions of either mitochondrial or nuclear origin are also used, but to a lesser degree, provided they are diagnostic for the species under investigation. For example, in fish the mitochondrial cytochrome b gene (Cytb) is extensively used as a DNA marker for species identification (Teletchea 2009).

Textbox 20.1 Descriptions of the Molecular Methods Mentioned in This Chapter

PCR

Polymerase Chain Reaction or PCR is a method to amplify a specific region of the DNA, allowing the generation of billions of DNA copies of specific genomic regions. The technique relies on two short DNA sequences, so-called primers, and an enzyme, a polymerase, which copies DNA during repeated cycles of heating and cooling (so-called PCR cycles). Primers only bind to a specific region of the target DNA and serve as starting point for the DNA polymerase. PCR can be conducted with species-specific primers, targeting a single species, or universal primers that amplifies a specific region across taxa. The latter is normally used for DNA barcoding when the identity of the specimen is not known *a priori*.

PCR assay

The specific combination of primers used in PCR. For qPCR and ddPCR the assay include primers and reporter molecule.

DNA barcoding

DNA barcoding relies on PCR using universal primers that amplifies a specific region, normally the mitochondrial COI gene, across taxa. Following amplification, the region can be purified, sequenced and finally compared to a database in order to establish species identity.

(continued)

TextBox 20.1 (continued)

Sanger sequencing

Sanger sequencing is a technique that allows for DNA sequencing of one sequence at the time of up to around 1000 base pairs.

Next generation sequencing

Next generation sequencing or NGS refers to DNA sequencing, which allows individual sequencing of million of sequences in parallel. Many platforms only generate short sequences up to few hundred base pairs.

Meta-barcoding

Meta-barcoding is a method that combines DNA based identification and Next generation Sequencing DNA sequencing. It relies on the use of universal primers, which allows amplification of a specific target region across species groups (e.g. fish) in environmental or other mixed species samples, which subsequently can be sequenced. Output sequences can following be matched to a reference database in order to analyse the number of species present in the sample. Moreover, a measure of relative species composition can be made by counting the number of sequences from each observed species. However, in some cases, the primers may work better for some species than other, which will bias such estimates.

qPCR

Real-time PCR, also called quantitative PCR or qPCR, is a method that allows for analysis of a PCR reaction in real-time, which enables enumeration of DNA copies. qPCR is based on the same approach as PCR but additionally contains a fluorescence region-specific reporter molecule, which allows for visualization of the increase in DNA copies during PCR. A positive detection is based on the accumulation of the fluorescent signal and the cycle threshold (Ct), which is defined as the number of cycles required for the signal to be significantly higher than the background fluorescence level. If a qPCR reaction is analysed together with a dilution series (a standard curve) of a known number of targets, the number of DNA molecules in the reaction can be estimated.

ddPCR

Like in qPCR, droplet digital PCR or ddPCR relies on PCR using region specific primers and a DNA specific reporter. However, differently from qPCR the total PCR reaction is fractionated into thousands of small oil droplets, and PCR is subsequently carried out separately within each droplet. Following PCR each droplet is analysed for positive amplification and number of initial DNA targets. Compared to qPCR, ddPCR provide an absolute count of target DNA copies per sample without the need for running a standard curve and enables more accurate measurements of small fold changes in target DNA amongst samples.

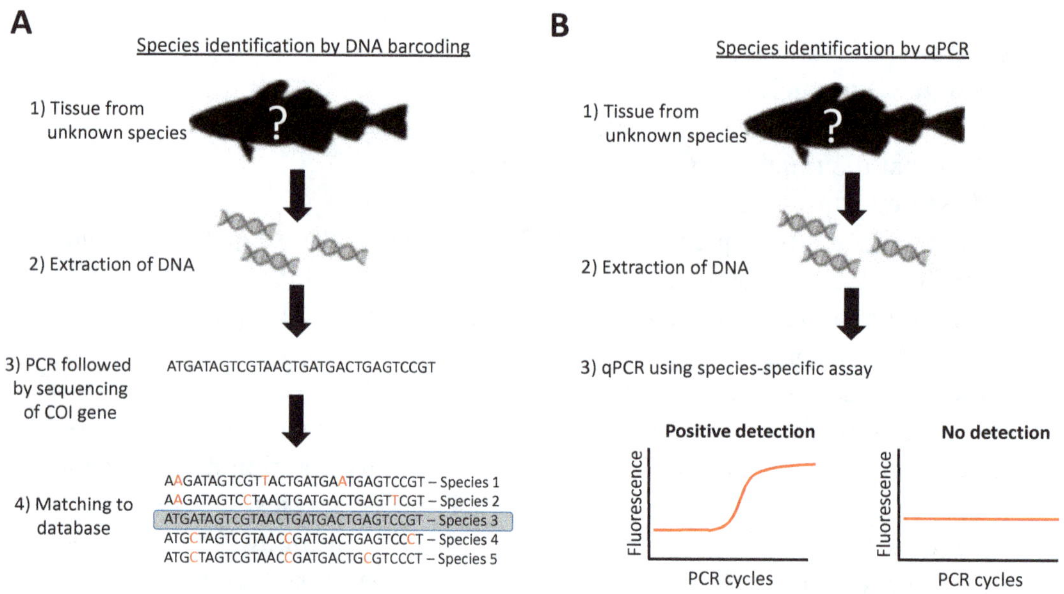

Fig. 20.2 Overview of the analytical steps in genetic species-identification either by (**a**) DNA barcoding or (**b**) Quantitative PCR (qPCR). In DNA barcoding, the barcoding gene (e.g. COI) is amplified by PCR and the product subsequently sequenced and finally compared to a database for species-match. In qPCR species-specific assays are used. Such assays are initially tested for possible cross-amplification to other species in order to control for false-positive detections. Hence, a positive amplification signal denotes a positive identification for the tested species

When comparing a sequence from a specimen with unknown taxonomic identity to the database, it is important to make sure that the obtained sequence is of high quality and that the database includes all closely related species. If this is the case, the identity of the specimen can be inferred with very high precision, except in cases where diagnostic variants do not exist. A lack of diagnostic variants can be found e.g. between extremely closely related species and in groups of organisms with low levels of variation in the chosen barcoding region or due to introgression events where the DNA is shared between species because of hybridization. When no close matches of the DNA sequence can be found in the database, the sample likely belongs to an unknown species or one that is not included in the database (Ovenden et al. 2015). Genetics can also be used to assign individuals to populations within species (Nielsen 2016). In these cases, several nuclear genetic markers are normally used in a probabilistic framework, combining the statistical power of individual genetic markers into an overall likelihood of origin. This is necessary, as populations are rarely fixed for specific (diagnostic) gene variants of either mitochondrial or nuclear genes.

20.3.2 Current Limitations

The accuracy of DNA barcoding is generally high with most species showing distinctive COI sequences (Hebert et al. 2003). For example, among 207 Australian fish species, the identification success rate was 93–98% across groups

(Ward et al. 2005), and a 90% accuracy has been documented for North American freshwater fish (April et al. 2011). Nonetheless, the overall precision of the methods critically depends on the reference database used (Ovenden et al. 2015). To assure transparency, sequences should optimally be matched against DNA sequence data from voucher specimens or other curated museum specimens to ensure the reliability of the data (Ratnasingham and Hebert 2007). Currently several public DNA databases exist. GenBank (https://www.ncbi.nlm.nih.gov/genbank/) and the Barcode of Life (http://www.boldsystems.org/index.php/databases) are probably the best known and contain sequences from thousands of species. While the Barcode of Life database consists of several recognized DNA barcoding genes with high quality assurance assessment, Genbank acts as a general repository for genetic data and has therefore a lower quality standard. Due to a general increase in the use of genetics for species identification and taxonomic purposes, as well as the establishment of databases for commercial species, the databases already contain sequences from many species and are constantly expanding. For example, the Barcode of Life database (on the 16th of June 2018) contained 18,843 species of ray-finned fishes. However, the databases remain incomplete, especially for some cryptic, rare and non-commercial species and for geographical regions with high species diversity.

Another limiting factor is the labour expenditure needed for sequencing and analysing of the data. However, other genetic methods can be used for performing species-level assignment that do not involve direct visualization of the genetic code, thereby potentially allowing faster and easier species identification. One such technique is real-time PCR, also known as **quantitative PCR (qPCR)** (Fig. 20.2b). Quantitative PCR relies on species-specific detection and utilizes a chemistry, which emit light when in contact with DNA from the target organism. The technique allows for simultaneous analysis of one or multiple samples in parallel thereby permitting a fast, sensitive and a potentially cost-efficient alternative to conventional DNA barcoding for routine presence/absence applications. While the risk of false-positives in DNA barcoding both depends on the accuracy of the reference databases and the target region's ability to discriminate between species, false-positives in qPCR analysis depend on potential cross-amplification to non-target species (the method provides a signal of presence of the species, while it is not there). Hence, for qPCR analysis thorough testing for cross-amplification is needed before an assay can be validated. This can be particularly laborious and expensive for marine species due to the often-large number of related and co-occurring species. Moreover, since a single qPCR assay is specific to a particular species or related group of organisms, qPCR is most useful to identify samples where the number of candidate species can be restricted to a few (for example by using morphological characteristics, see Helyar et al. 2014). In cases where the species identity is completely unknown (e.g. highly damaged or processed individuals) many independent qPCR analyses might be needed to provide a positive detection, which increases the cost significantly. In these cases, DNA barcoding may be more cost effective.

20.3.3 Case Studies

Skates and rays are particularly vulnerable to overharvesting due to their life history patterns involving slow growth rates, high age of maturity and low fecundity and slow reproductive rates (Holden 1973). Currently, several species are considered threatened and are therefore protected, while others are still targeted or caught as bycatch and sold commercially (STECF 2017). These include several species that can be difficult to distinguish from protected species based on morphology alone (McEachran and Musick 1973): e.g., the thorny skate (*Amblyraja radiata*) is protected in the North Sea fisheries (STECF 2017) but can be difficult to distinguish from the thornback ray (*Raja clavata*), which may be landed (Larsen et al. 2013). Moreover, identification is commonly based on the number of dorsal spines, which in cases where only the ray "wings" are retained, as practiced in some countries (e.g., Denmark), renders species assessment nearly impossible - even by trained specialists. Hence, to monitor rays and skates, genetic tools should be applied. This may especially become relevant with the introduction of the Landing Obligation, as there is a risk that protected species will be landed. Here a standardized genetic approach may be of vital importance for accurate species monitoring of the landed catches, which in turn should eventually lead to better management of the species. Examples of this already exist for other species. For example, albatrosses have suffered high mortality as bycatch in long line fisheries (Burg 2007; Walsh and Edwards 2005). These species are highly mobile, migrating thousands of kilometres between breeding and feeding grounds. As feeding habitat often overlaps between species or populations, and because bycatch carcasses may be highly degraded, it has been difficult to assign individuals to species or populations. Analyses of genetic markers either mitochondrial DNA (Walsh and Edwards 2005) alone or in combination with nuclear markers (Burg 2007) have demonstrated that species, sub-species and even breeding populations could be assigned with high accuracy. This has permitting the impacts of bycatch on specific populations or species to be more accurately estimated.

20.4 Monitoring Species Composition in Processed Fish Products

20.4.1 The Technology and Its Applications in Fisheries

Millions of tonnes of fish are thrown back to the sea as bycatch every year (Kelleher 2005). With the introduction of the EU Landing Obligation, all catches of regulated commercial species will have to be landed and counted against quota. This procedure is expected to lead to large amounts of bycatch being landed, which can be utilized in the feed or food industry (Iñarra et al., this volume). Given a low price for unwanted catch (e.g. due to low valued species or individuals), methods for minimizing handling time and costs are likely to be implemented, which will probably lead to storing of catch as mixed or bulk products. One promising approach is the application of fish silage (Viðarsson et al., this volume). Fish silage is a liquid product made from grounded fish in the presence of added acids or lactic-acid-producing bacteria. This product can be pre-processed on board at sea, requires low investments and can be stored at room temperature for an extended period (Arruda et al. 2007). Due the expanding aquaculture industry, especially salmon farming, the demand for fish feed has grown. Accordingly, the price for prime fish silage has increased and is now in a price range where it is a valid alternative to other storage methods like freezing (personal communication, Erling P. Larsen, Senior Executive Consultant, DTU Aqua, Technical University of Denmark). Frozen fish can be used for production of feed (fish meal) but also for food like surimi. However, freezing of low cost fish will reduce the storage capacity for more high-priced catch rendering fish silage an economical and practical alternative. Fish silage can also be produced from the discarded fraction of the primary catch, such as guts, liver and other organs (Arruda et al. 2007), which increases the economic motivation for its use. However, before such products can be introduced to the relevant industry, accurate labelling of species composition according to the food safety standard regulation of the European Union, needs to be ensured (EU Commission Directive 2002/86/EC). As this task cannot be done by visual inspections of bulk products like fish silage, other approaches are needed to analyse and quantify species composition.

Extensive research has been conducted into molecular methods enabling assessment of the species-identity of processed meats. Immunological methods have been used for species detection in meat (Ballin et al. 2009). These methods typically rely on the detection of a species-specific protein, but have now largely been replaced by genetic methods. Targeting DNA compared to proteins has several advantages. First of all, DNA has a higher thermal stability, allowing detection in both processed and raw products. DNA is present in all living cells, regardless of tissue origin and hence easier to detect than proteins that may be located in specific tissues (Lockley and Bardsley 2000). Secondly, the variation in the genetic code between species makes DNA more suitable for species discrimination, especially between closely related species. Finally, PCR-based techniques have lower limits of detection (defined as the lowest concentration of target that can be detected on 95% of occasions) with estimates of 0.001%–1% of the total weight in mixed meat samples (Ballin et al. 2009).

Overall, the methods used for species identification can also detect species presence in processed products. For example, DNA Barcoding, has been used in investigations of adulterated labelling of fish products like fresh or frozen fish fillets, fish fingers, fish cakes and surimi (Helyar et al. 2014; Huxley-Jones et al. 2012; Pepe et al. 2007). However, as conventional Sanger sequencing techniques do not allow for analysis of multiple DNA sequences, this technique only qualifies when used on products originating from a single species. Moreover, in processed samples, shorter and more variable DNA segments may be preferred, as the DNA normally is degraded into short fragments often only few hundred of base pairs long (e.g. Pepe et al. 2007). Quantitative PCR can be used for detection of short species-specific DNA sequences (often < 200 bases) in mixed species products and has been used for example for testing the presence of haddock (*Melanogrammus aeglefinus*) and cod (*Gadus morhua*) in products labelled as whitefish and sold from supermarkets in the UK (Helyar et al. 2014). This technique also allows for quantitative estimation of DNA copy numbers when analysed in combination with a standard dilution series of known DNA concentration. It therefore combines fast identification with quantitative applications (Fig. 20.3a). Another well-established PCR-based method is the **droplet digital PCR (abbreviated ddPCR)** (Fig. 20.3b). This technique builds on the same principles as qPCR but is more precise when used for quantification and does not require a standard curve for quantification of DNA copies. Both qPCR and ddPCR have been used with success to analyse species composition in mixed meat products (e.g., Floren et al. 2015), but so far few studies exists for fish (Bojolly et al. 2017). However, with the increased focus on species identification in mixed fish products and a continuous declining cost of the chemistry used, it is likely to become a more common practice within a few years.

20.4.2 Current Limitations

Quantitative DNA based methods can be used to accurately determine the number of DNA molecules present in a mixed-species product. However, translating this into percentages by weight, which is the normal standard within the food industry, is difficult. For example, the number of mitochondria and hence mitochondrial DNA copies per cell varies across cell types, as well as species (Cole 2016). For this reason, several researchers have advocated for the use of nuclear markers, as they exist in a fixed number of copies per cell (Ballin et al. 2009). However, for products like fish silage generated from whole un-gutted fish, complications may still arise. For example, liver and intestines contain more cells and hence nuclear DNA copies per weight than white muscle. As organ size is not fixed relative to the size of the fish, but varies according to size, age, maturity and species, inaccuracies using nuclear DNA is also expected. A solution may be to use a small fraction of the total catch to estimate species weight composition, which then can be used to estimate a correction factor (Thomas et al. 2016) or to label mixed products in a genome to genome equivalent, instead of weight (Ballin et al. 2009). A final issue with using whole fish, or fish guts, is that DNA from prey items also can be detected. As part of the prey may be other commercial species, which in theory would then be deducted from the quota. However, since prey items normally constitute a minute fraction of the entire product, this risk may be avoided by using a minimum threshold for quota deductions.

Another limitation relates to the fact that both qPCR and ddPCR rely on species-specific detections, which make these techniques expensive for screening products of unknown composition. In such cases, **meta-barcoding** may be a better

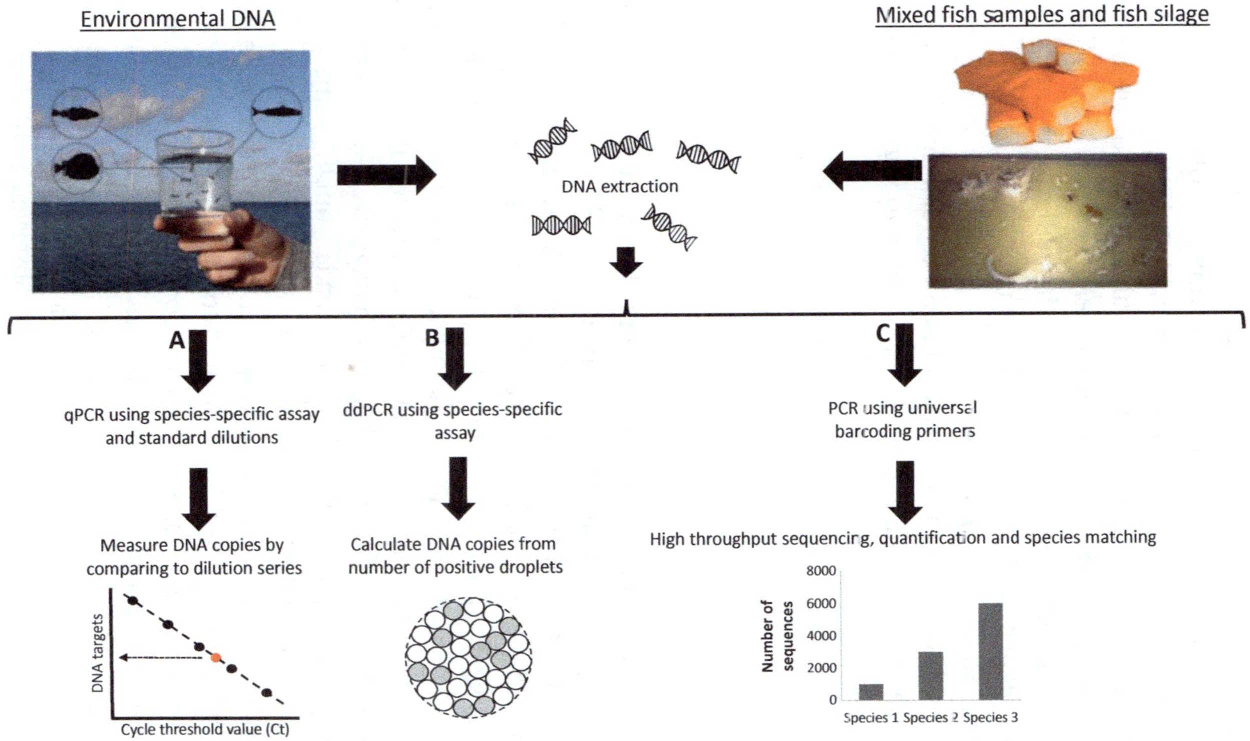

Fig. 20.3 Overview of the methods used for DNA quantification using either mixed species samples or eDNA. Initial DNA extraction is universal for all applications. In qPCR studies, (**a**) a dilution series (standard curve) with a known number of DNA copies (depicted in black) can be used to estimate the number of copies in the unknown sample (red dot). Here cycle threshold value (Ct) is the PCR cycle where the fluorescence signal becomes significantly higher than the background and can be used to estimate the amount of DNA in the unknown sample. Compared to qPCR, (**b**) ddPCR provides an absolute count of target DNA copies per sample without the need for running a standard curve enabling more accurate measurements of target DNA content among samples. In metabarcoding, (**c**) the gene in focus (e.g. COI) is first amplified in order to up-concentrate the number of DNA copies from a specific group of taxa, e.g. fish. Following this, DNA is sequenced using a Next-generation (NGS) platform generating millions of sequences that subsequently can be sorted, quantified and compared to a data base

alternative. In meta-barcoding, a gene, like COI, is sequenced across one taxon (e.g. fish) using a high-throughput **next generation sequencing (NGS) platform**. This allows for the sequencing of millions of individual gene sequences, which can be matched to a reference database and counted, and hence combines DNA barcoding with possible quantification (Fig. 20.3c). This method has been used to assess species identity and composition for different kinds of mixed species samples. Meta-barcoding has been used for assessing animal diet by analysing the gut content (e.g., Murray et al. 2011) and has been one of the favoured tools for analysing environmental DNA samples (see next section) (Hansen et al. 2018). However, often only a subset of species can be targeted simultaneously due to variation in the DNA code used for PCR amplification, which moreover can lead to quantification bias if some species amplify better than others during PCR. In addition, as most NGS platforms generate short sequences, the sequenced DNA region will be shorter than for conventional Sanger sequencing. This represents a potential problem, as the shorter sequences contain less variation to discriminate between species. Hence, often sequences cannot be resolved to a species level but only to higher taxonomic levels (Pompanon et al. 2012; Stat et al. 2017).

20.4.3 Case Study

In order for fish silage to be a valid alternative to frozen fish, several questions need to be addressed. Earlier unpublished studies conducted by DTU Aqua, Technical University of Denmark, have demonstrated that DNA is still present in fish silage after 21 days at both 5 and 20 degrees, as confirmed by amplification and sequencing of the mitochondrial COI gene. However, whether the product also allows accurate quantification of species composition remains unknown. This was investigated in a preliminary study, conducted as part of the Discardless project funded by the European Union. Here specific mitochondrial qPCR assays were developed for cod, haddock and whiting. Subsequently fish silage was made from the same species and DNA samples were extracted over a period of 21 days. Finally, qPCR analysis was undertaken in order to estimate the distribution of DNA copy number among species in the mixture, which then could be compared to the relative difference in weight. The results show that, overall, there is a good correspondence between the input (in weight) of the three species and the number of DNA copies present. The largest difference is found at the first day of the experiment, probably indicating insufficient liquefaction and mixing (Fig. 20.4). After that, the difference decreased significantly, although smaller differences could still be observed until day 21. In general, whiting, which represented the smallest biomass, was overrepresented in the DNA quantification. Similar results have been documented in other studies, but can be improved by calculating a correction factor from a subsample (Thomas et al. 2016). Hence, although more empirical studies are needed, the results hold promise for the application of DNA-based methods for determining the composition of species in silage and other mixed products from bycatch.

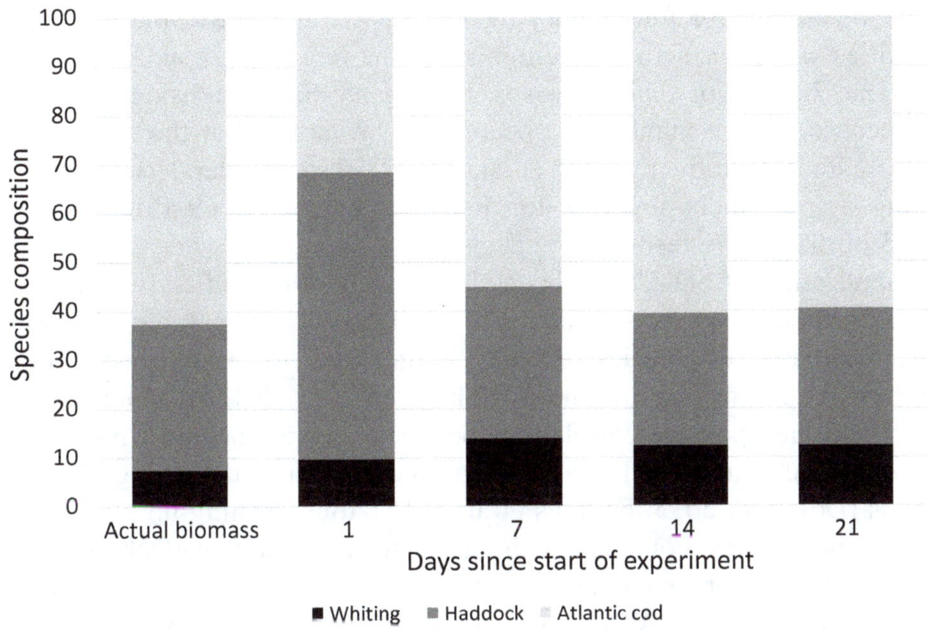

Fig. 20.4 Analysed species composition in fish silage made from whiting, haddock and Atlantic cod. The estimates are based on qPCR analysis of the number of species-specific DNA copies. The first column represents the actual biomass while the latter show the qPCR estimates following 1, 7, 14 and 21 days from when the silage was produced

20.5 Environmental DNA and Pre-assessment of Appropriate Fishing Grounds

20.5.1 The Technology and Its Applications in Fisheries

Traditional DNA-based monitoring of fisheries entails direct DNA sampling of the catch. However, non-invasive monitoring is possible through analysis of so-called environmental DNA (abbreviated eDNA). Environmental DNA simply refers to genetic material, obtained directly from an environmental sample (water, air or soil), without any obvious parts of macro-organisms (Hansen et al. 2018). DNA is continuously shed by all organisms to the surrounding environment through skin (scales and mucus from fish), faeces, gametes, etc. (Hansen et al. 2018). Since marine organisms shed DNA "particles" directly into the sea, eDNA can be obtained from a water sample. The procedure is relatively simple: following water collection (normally ranging from typically 250 mL–5 L (Goldberg et al. 2016)), the water is filtered through a filter that retains the minute particles containing DNA, which can be extracted and analysed by either qPCR, ddPCR or meta-barcoding methods (Fig. 20.3). Quantitative PCR has been the preferred method to analyse the presence (and abundance) of specific species of interest (e.g., Thomsen et al. 2012b). For example, qPCR has been used extensively as a tool for the detection of rare and invasive species, which otherwise can be difficult to detect by conventional methods (e.g. Thomsen et al. 2012b). On the other hand, meta-barcoding has been the favoured tool for identifying and describing entire marine biological communities within or across taxonomic groups (e.g. Stat et al. 2017; Thomsen et al. 2012a, 2016). For such purposes, meta-barcoding is considered superior to qPCR because a large number of species can be targeted at the same time. For example, a recent study by Stat et al. (2017) tested several 'universal' PCR barcoding assays, including one for fish that recovered 69 different taxa, of which 33 of them were resolved to species.

The application of eDNA has been shown to be a cost-effective and sensitive sampling method (Sigsgaard et al. 2015) and shows comparable or even more accurate estimates of species diversity compared to established monitoring approaches like net fishing and diving (Thomsen et al. 2012a). Moreover, several studies have found positive correlations between species-specific DNA concentrations and biomass and abundance in controlled experimental settings like aquaria and ponds (Doi et al. 2015; Klymus et al. 2015), thus opening up possibilities for interesting applications. For example, it has been suggested that eDNA may be used as a future substitute or supplement to established stock assessments of commercial fish (Thomsen et al. 2016). Although we are still quite far from this scenario (Hansen et al. 2018), eDNA may be used in the future to reduce potential bycatch by focusing fishing efforts in areas with low numbers of potential bycatch species. In fact, Thomsen et al. (2016) have already tested such an application indirectly. They used a meta-barcoding approach to analyse seawater samples collected in South-western Greenland and compared it to trawling catch data. They generally observed weak but significant correlations between numbers of sequences belonging to specific groups of fish and biomass/abundance data, indicating a potential application for using eDNA as a tool for targeted fishing.

20.5.2 Current Limitations

Several issues exist which needs to be properly addressed before eDNA can be implemented to assist in more selective fishing efforts. First, faster ways of processing and analysing samples are needed to allow analysis directly at the fishing grounds. Such techniques, however, are already under development (see below section on future perspectives). Other more fundamental issues relate to (i) retention time of DNA (i.e. how long can it be detected in the environment) and (ii) its correlation with biomass. The first issue depends on the effects of abiotic and biotic factors that influence the production, degradation and transport of eDNA (see Hansen et al. (2018)). Here, abiotic factors like water temperature, pH and UV light determine the rate of DNA decay in the water, while biotic factors like size, age, species, health, sex and food condition influence eDNA production. Moreover, ocean currents will determine the distribution of eDNA, which may be carried hundreds of kilometres away from the source within days (Thomsen et al. 2012a). Still, marine eDNA studies report good concordance between eDNA analyses and visual surveys of local vertebrate communities (Port et al. 2016; Thomsen *et al.*

2012a). Nonetheless, environmental conditions play a large role in the production and turnover of eDNA and will likely vary significantly due to the widely different geospatial environmental conditions in our oceans (Klymus et al. 2015; Strickler et al. 2014).

An improved understanding of DNA production is crucial for assessing the relationship between the number of observed eDNA copies and the biomass in natural environments. While experimental studies in general have shown solid positive correlations between DNA and biomass (Doi et al. 2015; Klymus et al. 2015), relationships from natural systems are less pronounced, or in some cases even lacking (Spear et al. 2015; Thomsen et al. 2016). This is likely due to the complexity of such systems. For example, compared to a controlled experiment, natural habitats comprise many different size and age groups of fish, which will shed different amounts of DNA per mass, e.g. due to differences in metabolism, life stage or condition (Hansen et al. 2018). Moreover, different species of fish may shed more DNA than others, rendering it difficult to directly compare eDNA levels across species with widely different biology. Hence, before eDNA can be used in targeted fisheries, more empirical research is needed. This will improve our understanding about how these issues affect eDNA quantification and pave the way for future research into fisheries application.

20.5.3 Case-Study

In a current study, conducted by DTU Aqua, Denmark, the potential for using eDNA to avoid bycatch is being directly investigated. Specifically, this study aims at testing the potential for applying qPCR as a tool for monitoring absolute and relative biomass of European sprat (*Sprattus sprattus*), Atlantic herring (*Clupea harengus*) and sand eel (*Ammodytidae*). While all species are harvested commercially, the bycatch of herring is a well-known problem, especially in sprat fisheries, and to some extent, in sand eel fisheries. Currently, bycatch of herring are discarded when the bycatch quota is exceeded. However, with the introduction of the Landing Obligation, all bycatch of herring should, in theory, be deducted from this quota. This may lead to a scenario where herring is a “choke species” for sprat and sand eel fisheries leading to their closure. Thus, approaches facilitating more selective fisheries are warranted by the industry.

In the study, water samples were collected on board the fishing vessels while fishing and estimates of catch and bycatch were conducted. Subsequently the water samples were analysed for the quantity of species-specific DNA to assess potential correlations. The data are still being analysed, thus only a few preliminary results are available. However, they demonstrate large differences in the relative proportion of DNA from Atlantic herring in the European sprat fisheries across fishing localities (Fig. 20.5). Whether this is also well correlated with the actual biomass caught is currently being investigated.

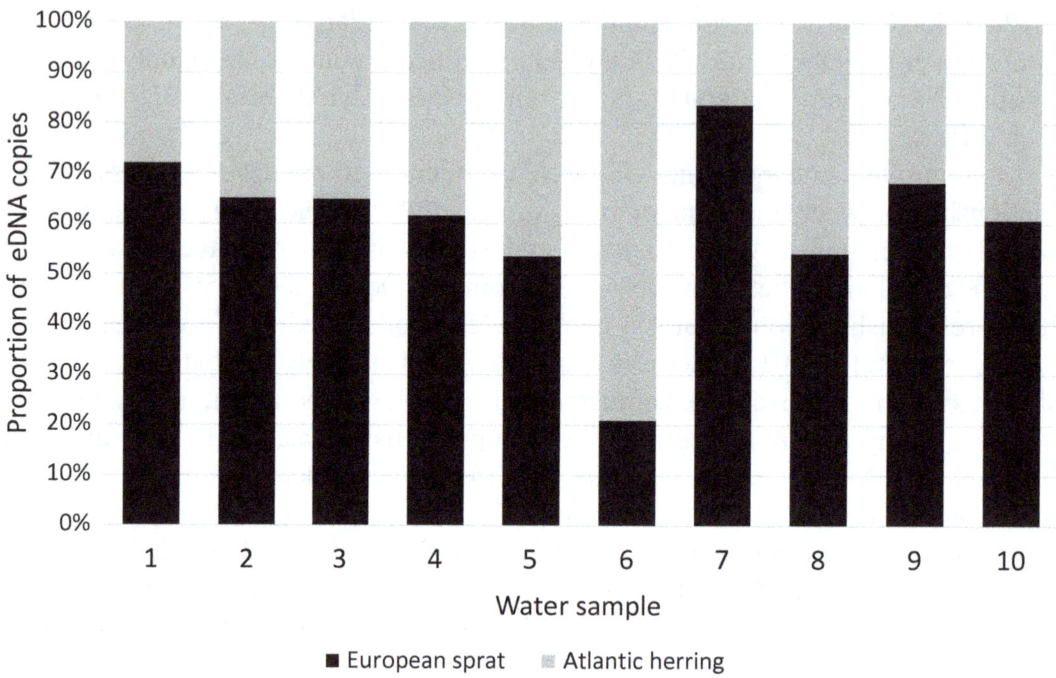

Fig. 20.5 Quantitative PCR (qPCR) estimates of the difference in species-specific DNA from European sprat and Atlantic herring observed in the North Sea fisheries for European sprat. The samples were collected before the fishery was initiated at 10 different locations

20.6 Future Perspectives

Some of the major limitations of the methods described in this chapter are their accessibility, time from sampling to results, the required manual labour and associated cost. Most of the techniques are currently only performed by highly trained personnel in well-equipped molecular laboratories. However, in the near future, it is possible that people with little molecular training can use many of the techniques *in situ*, e.g. on board fishing vessels or at fishing ports. This should lower the costs and facilitate faster communication between fishers and managers regarding bycatch and catch composition, potentially leading to more efficient fisheries management and conservation. On board analyses of eDNA in water samples prior to fishing would, moreover, enhance the potential of using eDNA for targeted fishing efforts, as it would allow near real-time analysis at the fishing grounds.

Already today, novel portable devices and kits exist that allow for DNA extraction and sequencing outside the laboratory. For example, Oxford Nanopore Technologies has designed a device for automated sample preparation and even PCR (https://nanoporetech.com/products/voltrax). Combined with small portable sequencing machines like, for example, the MinION or the smartphone compatible SmidgION (Oxford Nanopore Technologies), this allows for fast onsite DNA barcoding (Menegon et al. 2017). The MinION device has the same size as a normal smart phone but can sequence up to 20 Gigabases of DNA (20 billion A, C, T and

G's) per run. The sequences are directly uploaded to a laptop computer and can therefore be analysed in close-to-real-time. The major disadvantage of the MinION is a high sequencing error rate, compared to laboratory DNA sequencing. However, this is to some degree neutralized by the large sequencing output and recent development in analytical methods. This poses another challenge, as analysis of genomic data is computationally intensive and hence some applications may need internet access to upload the data to an online server for fast on site analysis (Srivathsan et al. 2018) and subsequent storing of results. Moreover, automated methods and software, that can transfer gigabytes of DNA data from these devices into a format that can be interpreted by the non-expert (fishers), is a field, which needs more research and development. Further, the cost per run is still high compared to conventional DNA barcoding, however; sequencing methods exist for tagging, pooling and subsequent sequencing of up to hundreds of samples per MinION run, thereby lowering the cost per sample significantly (Srivathsan et al. 2018). Another option is stopping the sequencing run when just enough sequences have been obtained for producing proper species identification. Subsequently, the device can be washed and used again until reaching the overarching sequencing capacity.

Onsite analysis of species composition and quantification in mixed species products or eDNA may also be conducted using qPCR. For example, the US-based company Biomeme (https://biomeme.com) has developed an ultra-small qPCR device that attaches to a smartphone and can run up to nine samples with up to two different target species within an hour (e.g. sprat and herring). The same company has also developed fast DNA extraction kits that allows for DNA preparation within a few minutes without need of any lab equipment. This kit uses disposable, single-use consumables, which decreases the possibility for DNA contamination among samples. Lastly, future eDNA analysis may also be conducted by so called "ecogenomic sensors", which are submersible instruments for autonomous *in situ* automatic DNA analysis of water samples (Scholin 2009). The 2nd generation environmental sample processor (2G ESP) is such a device and is essentially a DNA lab in a waterproof can. The 2G ESP has already shown practical applications for monitoring microorganisms (Ottensen et al. 2014), zooplankton (Harvey ct al. 2012) and eDNA from fish (personal communication, Brian K. Hansen)). The 2G ESP can be moored to a specific location (e.g. a dockside or buoy), be mobile (free drifting) or carried on board fishing vessels where it can conduct automatized water sampling, DNA extraction and qPCR analysis. The ESP provides two-way communications, which allows the end-user to remotely assess the results in near real-time. This has potential interesting applications for fisheries, as this may allow for eDNA analysis directly at the fishing grounds, which then can be communicated to the fishing fleet and act as a supplement in order to avoid bycatch.

In conclusion, DNA-based methods can already provide important information that can be used under the Landing Obligation to identify and prevent bycatch as well as in control, management and consumer protection. The rapid and continuous development of portable and automatic real-time devices for DNA analysis are expected to become more powerful, user-friendly and much cheaper in the near future. This will likely increase the incentives to use genetics in fisheries management, including subjects related to discards.

Acknowledgments This work has received funding from the Horizon 2020 Programme under grant agreement DiscardLess number 633680. This support is gratefully acknowledged.

References

April, J., Mayden, R.L., Hanner, R.H., Bernatchez, L. (2011). Genetic calibration of species diversity among North America's freshwater fishes. *Proceedings of the National Academy of Sciences of the United States of America, 108*, 10602–10607. https://doi.org/10.1073/pnas.1016437108.

Arruda, L.F., Borghesi, R., Oetterer, M. (2007). Use of fish waste as silage: A review. *Brazilian Archives of Biology and Technology, 50*(5), 879–886. https://doi.org/10.1590/S1516-89132007000500016.

Ballard, J.W.O., & Whitlock M.C. (2004). The incomplete natural history of mitochondria. *Molecular Ecology, 13*, 729–744. https://doi.org/10.1046/j.1365-294X.2003.02063.x.

Ballin, N.Z., Vogensen, F.K., Karlsson A.H. (2009). Species determination – Can we detect and quantify meat adulteration? *Meat Science, 83*, 165–174. https://doi.org/10.1016/j.meatsci.2009.06.003.

Bojolly, D., Doyen, P., Le Fur, B., Christaki, U., Verrez-Bagnis, V., Gard, T. (2017). Development of a qPCR method for the identification and quantification of two closely related tuna species, bigeye tuna (*Thunnus obesus*) and yellowfin tuna (*Thunnus albacares*), in canned tuna. *Journal of Agricultural and Food Chemistry, 65*, 913–920. https://doi.org/10.1021/acs.jafc.6b04713.

Burg, T.M. (2007). Genetic analysis of wandering albatrosses killed in longline fisheries off the east coast of New Zealand. *Aquatic Conservation-Marine and Freshwater Ecosystems, 17*, 93–101. https://doi.org/10.1002/aqc.907.

Cole, L.W. (2016). The evolution of per-cell organelle number. *Frontiers in Cell and Development, 4*, 85. https://doi.org/10.3389/fcell.2016.00085.

Coyne, J., & Orr, H. (2004). *Speciation*. Sunderland: Sinnauer Associates.

Doi, H., Uchii, K., Takahara, T., Matcuhashi, S., Yamanaka, H., Minamoto, T. (2015). Use of droplet digital PCR for estimation of fish abundance and biomass in environmental DNA surveys. *PLoS One, 10*. https://doi.org/10.1371/journal.pone.0122763.

Floren, C., Wiedemann, I., Brenig, B., Schutz, E., & Beck, J. (2015). Species identification and quantification in meat and meat products using droplet digital PCR (ddPCR). *Food Chemistry, 173*, 1054–1058. https://doi.org/10.1016/j.foodchem.2014.10.138.

Galtier, N., Nabholz, B., Glemin, S., Hurst, G.D.D. (2009). Mitochondrial DNA as a marker of molecular diversity: A reappraisal. *Molecular Ecology, 18*, 4541–4550. https://doi.org/10.1111/j.1365-294X.2009.04380.x.

Goldberg, C.S., Turner, C.R., Deiner, K., Klymus, K.E., Thomsen, P.F., Murphy, M.A., et al. (2016). Critical considerations for the application of environmental DNA methods to detect aquatic species. *Methods in Ecology and Evolution, 7*, 1299–1307. https://doi.org/10.1111/2041-210X.12595.

Goodwin, S., McPherson, J.D., McCombie, W.R. (2016). Coming of age: Ten years of next-generation sequencing technologies. *Nature Review Genetics, 17*, 333–351. https://doi.org/10.1038/nrg.2016.49.

Hansen, B.H., Bekkevold, D., Clausen, L.W., & Nielsen, E.E. (2018). The sceptical optimist: Challenges and perspectives for the application of environmental DNA in marine fishes. *Fish and Fisheries, 1*, 1–18. https://doi.org/10.1111/faf.12286.

Harvey, J.B.J., Ryan, J.P., Marin, R., Preston, C.M., Alvarado, N., Scholin, C.A., et al. (2012). Robotic sampling, in situ monitoring and molecular detection of marine zooplankton. *Journal of Esperimental Marine Biology and Ecology, 413*, 60–70. https://doi.org/10.1016/j.jembe.2011.11.022.

Hebert, P.D.N., Ratnasingham, S., deWaard, J.R. (2003). Barcoding animal life: Cytochrome c oxidase subunit 1 divergences among closely related species. *Proceedings of the Royal Society B-Biological Sciences, 270*, S96–S99. https://doi.org/10.1098/rsbl.2003.0025.

Helyar, S.J., Lloyd, H.A., de Bruyn, M., Leake, J., Bennett, N., Carcalho, G.R. (2014). Fish product mislabelling: Failings of traceability in the production chain and implications for illegal, unreported and unregulated (IUU) fishing. *PLoS One, 9*, https://doi.org/10.1371/journal.pone.0098691.

Holden, M.J. (1973). Are long-term sustainable fisheries for elasmobranchs possible? Rapports et Procés Verbaux des *Rèunionsdu Conseil International pour l'Exploration de la Mer, 164*, 360–367.

Hopkins, G.W., & Freckleton, R.P. (2002). Declines in the numbers of amateur and professional taxonomists: Implications for conservation. *Animal Conservation, 5*, 245–249. https://doi.org/10.1017/S1367943002002299.

Huxley-Jones, E., Shaw, J.L.A., Fletcher, C., Parnell, J., Watts, P.C. (2012). Use of DNA barcoding to reveal species composition of convenience seafood. *Conservation Biology, 26*, 367–371. https://doi.org/10.1111/j.1523-1739.2011.01813.x.

Iñarra, B., Bald, C., Cebrián, M., Antelo, L.T., Franco-Uría, A., Vázquez, J.A., et al. (this volume). What to do with unwanted catches: Valorisation options and selection strategies. In S.S. Uhlmann, C. Ulrich, S.J. Kennelly (Eds.), *The European Landing Obligation – Reducing discards in complex, multi-species multi-jurisdictional fisheries*. Cham: Springer.

Ivanova, N.V., Zemlak, T.S., Hanner, R.H., Hebert, P.D.N. (2007). Universal primer cocktails for fish DNA barcoding. *Molecular Ecology Notes, 7*, 544–548. https://doi.org/10.1111/j.1471-8286.2007.01748.x.

Kelleher, K. (2005). Discards in the world's marine fisheries. An update. FAO Fisheries Technical Paper. No. 470. Rome, FAO. Available at: http://www.fao.org/3/a-y5936e.pdf

Klymus, K.E., Richter, C.A., Chapman, D.C., Paukert, C. (2015). Quantification of eDNA shedding rates from invasive bighead carp Hypophthalmichthys nobilis and silver carp *Hypophthalmichthys molitrix. Biological Conservation, 183*, 77–84. https://doi.org/10.1016/j.biocon.2014.11.020.

Larsen, E., Dalskov, J., Nielsen, E.E., Kirkegaard, E., Nielsen, J.W., Tørring, P., et al. (2013). Dansk fiskeris udnyttelse af discardforbuddet: En udredning. Charlottenlund: DTU Aqua. Institut for Akvatiske Ressourcer. *DTU Aqua-rapport; No. 275-2013*, (p. 106).

Lockley, A.K., & Bardsley, R.G. (2000). DNA-based methods for food authentication. *Trends in Food Science & Technology, 11*, 67–77. https://doi.org/10.1016/S0924-2244(00)00049-2.

Mayr, E. (1942). *Systematics and the origin of species*. New York: Columbia University Press.

Mayr, E. (1963). *Animal species and evolution*. Cambridge, MA: Harvard University Press.

McEachran, J.D., & Musick, J.A. (1973). Characters for distinguishing between immature specimens of the sibling species, *Raja erinacea* and *Raja ocellata* (Pisces: Rajidae). *Copeia*, 238–250. https://doi.org/10.2307/1442962.

Menegon, M., Cantaloni, C., Rodriguez-Prieto, A., Centomo, C., Abdelfattah, A., Rossato, M., et al. (2017). On site DNA barcoding by nanopore sequencing. *PLoS One, 12*, https://doi.org/10.1371/journal.pone.0184741.

Murray, D.C., Bunce, M., Cannell, B.L., Oliver, R., Houston, J., White, N.E., et al. (2011). DNA-based faecal dietary analysis: A comparison of qPCR and high throughput sequencing approaches. *PLoS One, 6*, https://doi.org/10.1371/journal.pone.0025776.

Nielsen, E.E. (2016) Population or point-of-origin identification. In A.M. Naarum & R.H. Hanner (Eds.), *Seafood authenticity and traceability* (pp. 149–169). London: Academic.

Ottensen, E.A., Young, C.R., Gifford, S.M. Eppley, J.M., Marin, R., Schuster, S.C., et al. (2014). Multispecies diel transcriptional oscilliations in open oceab heterotrophic bacterial assemblages. *Science, 345*, 207–212. https://doi.org/10.1126/science.1252476.

Ovenden, J.R., Berry, O., Welch, D.J., Buckworth, R.C., Dichmont, C.M. (2015). Ocean's eleven: A critical evaluation of the role of population, evolutionary and molecular genetics in the management of wild fisheries. *Fish and Fisheries, 16*, 125–159. https://doi.org/10.1111/faf.12052.

Pepe, T., Trotta, M., Di Marco, I., Anastasio, A., Bautista, J.M., Cortesi, M.L. (2007). Species identification in surimi-based products. *Journal of Agricultural and Food Chemistry, 55*, 3681−3685. https://doi.org/10.1021/jf063321o.

Pompanon, F., Deagle, B.E., Symondson, W.O.C., Brown, D.S., Jarman, S.N., Tarberlet, P. (2012). Who is eating what: Diet assessment using next generation sequencing. *Molecular Ecology, 21*, 1931–1950. https://doi.org/10.1111/j.1365-294X.2011.05403.x.

Port, J.A., O'Donnell, J.L., Romero-Maraccini, O.C., Leary, P.R., Litvin, S.Y., Nickols, K.H., et al. (2016). Assessing vertebrate biodiversity in a kelp forest ecosystem using environmental DNA. *Molecular Ecology, 25*, 527–541. https://doi.org/10.1111/mec.13481.

Ratnasingham, S., & Hebert, P.D.N. (2007). BOLD: The barcode of life data system (http://www.barcodinglife.org). *Molecular Ecology Notes, 7*, 355–364. https://doi.org/10.1111/j.1471-8286.2007.01678.x.

Saraste, M. (1999). Oxidative phosphorylation at the fin de siecle. *Science, 283*, 1488−1493. https://doi.org/10.1126/science.283.5407.1488.

Scholin, C.A. (2009). What are "ecogenomic sensors?"—A review and thoughts for the future. *Ocean Science Discussions, 6*, 191–213. https://doi.org/10.5194/os-6-51-2010.

Scientific, Technical and Economic Committee for Fisheries (STECF). (2017). Long-term management of skates and rays (STECF-17-21). Publications Office of the European Union, Luxembourg, 2017, ISBN 978-92-79-67493-8. https://doi.org/10.2760/44133, JRC109366.

Sigsgaard, E.E., Carl, H., Møller, P.R., & Thomsen, P.F. (2015). Monitoring the near-extinct European weather loach in Denmark based on environmental DNA from water samples. *Biological Conservation, 183*, 46–52. https://doi.org/10.1016/j.biocon.2014.11.023.

Spear, S.F., Groves, J.D., Williams, LA., Waits, L.P. (2015). Using environmental DNA methods to improve detectability in a hellbender (*Cryptobranchus alleganiensis*) monitoring program. *Biological Conservation, 183*, 38–45. https://doi.org/10.1016/j.biocon.2014.11.016.

Srivathsan, A., Baloğlu, B., Wang, W., Tan, W.Z., Bertrand, D., Ng, A.H.G., et al. (2018). A MinION™-based pipeline for fast and cost-effective DNA barcoding. *Molecular Ecology Resources*. https://doi.org/10.1111/1755-0998.12890.

Stat, M., Huggett, M.J., Bernasconi, R., DiBattista, J.D. Berry, T.E., Newman S.J., et al. (2017). Ecosystem biomonitoring with eDNA: Metabarcoding across the tree of life in a tropical marine environment. *Scientific Reports, 7*. https://doi.org/10.1038/s41598-017-12501-5.

Strickler, K.M., Fremier, A.K., Goldberg, C.S. (2014). Quantifying effects of UV- B, temperature, and pH on eDNA degradation in aquatic microcosms. *Biological Conservation, 183*, 85–92. https://doi.org/10.1016/j.biocon.2014.11.038.

Teletchea, F. (2009). Molecular identification methods of fish species: Reassessment and possible applications. *Reviews in Fish Biology and Fisheries, 19*, 265–293. https://doi.org/10.1007/s11160-009-9107-4.

Thomas, A.C., Deagle, B.E., Eveson, J.P., Harsch, C.H., & Trites, A.W. (2016). Quantitative DNA metabarcoding: Improved estimates of species proportional biomass using correction factors derived from control material. *Molecular Ecology Resources, 16*, 714–726. https://doi.org/10.1111/1755-0998.12490.

Thomsen, P.F., Kielgast, J., Iversen, L.L., Møller, P.R., Rasmussen, M., Willerslev, E. (2012a). Detection of a diverse marine fish fauna using environmental DNA from seawater samples. *PLoS One, 7*, https://doi.org/10.1371/journal.pone.0041732.

Thomsen, P.F., Kielgast, J., Iversen, L.L., Wiuf, C., Rasmussen, M., Gilbert, M.T., et al. (2012b). Monitoring endangered freshwater biodiversity using environmental DNA. *Molecular Ecology, 21*, 2565–2573. https://doi.org/10.1111/j.1365-294X.2011.05418.x.

Thomsen, P.F., Møller, P.R., Sigsgaard, E.E., Knudsen, S.W., Jørgensen, O.A., Willerslev, E. (2016). Environmental DNA from seawater samples correlate with trawl catches of subarctic, deepwater fishes. *PLoS One, 11*. https://doi.org/10.1371/journal.pone.0165252.

Viðarsson, J.R., Larsen, E.P., Valeiras, J., & Ragnarsson, S.Ö. (this volume). Onboard and vessel layout modifications. In S.S. Uhlmann, C. Ulrich, S.J. Kennelly (Eds.), *The European Landing Obligation – Reducing discards in complex, multi-species and multi-jurisdictional fisheries*. Cham: Springer.

Walsh, H.E., & Edwards, S.V. (2005). Conservation genetics and Pacific fisheries by-catch: Mitochondrial differentiation and population assignment in black-footed albatrosses (*Phoebastria nigripes*). *Conservation Genetics, 6*, 289–295. https://doi.org/10.1007/s10592-004-7824-8.

Ward, R.D., Zemlak, T.S., Innes, B.H., Last, P.R., Hebert, P.D.N. (2005). DNA barcoding Australia's fish species. *Philosophical Transactions of the Royal Society B-Biological Sciences, 360*, 1847–1857. https://doi.org/10.1098/rstb.2005.1716.

Wheeler, Q.D., Raven, P.H., Wilson, E.O. (2004). Taxonomy: Impediment or expedient? *Science, 303*, 285–285. https://doi.org/10.1126/science.303.5656.285.

Index

A

Analytic hierarchy process (AHP), 341, 356
Automated classification, 323
Avoidance, 4, 20, 21, 37, 53, 64, 98, 172, 214, 230, 243, 262, 265, 272, 280–292, 314, 368, 401

B

Baltic Sea, 34, 35, 53, 62, 170, 198, 200, 202, 204, 206, 208, 209, 259, 260, 385, 403
Behavioural economics, 221–231
Being watched, 221, 224, 229, 230, 232
Biomolecules, 334, 336
Biorefineries, 341, 348, 352, 356
Bulk landing, 323
Bycatches, 42, 73, 102, 112, 183, 184, 191, 198, 298, 334, 409–414, 418, 420–422
Bycatch and Slipping Mitigation, 297

C

Carrion, 130, 132–134, 136, 138–144, 151–153
"Carrots" or "sticks", 230, 231
Catch avoidance, 10, 37
Catch composition, 28, 29, 32, 73, 91, 98, 99, 120, 125, 166, 170, 180, 211, 244, 249, 250, 274, 311, 370, 371, 386, 394, 395, 401, 422
Challenge trials, 258–261, 263, 264, 276
Common Fisheries Policy (CFP), 5, 27–43, 50, 91, 165, 171, 180, 221, 320, 383, 400
Compliance, 7, 11, 19, 21, 22, 40, 41, 62, 63, 78, 82, 83, 105, 110, 126, 180, 187, 192, 207, 211, 220–232, 252, 314, 320, 335, 364, 365, 368, 369, 371, 372, 377, 384, 385, 400, 401
Costs and earnings, 110

D

Data collection, 10, 13, 19, 21, 50–65, 90, 136, 171, 174–175, 200–202, 221, 366, 369, 370, 400
Decision support tools (DST), 265, 268–272, 274, 275
Delayed mortality, 51, 52
Delegated acts, 34, 83, 170, 171, 189, 206
de minimis, 31, 33, 54–59, 110, 118, 121, 122, 124, 125, 168, 170, 173–175, 185, 191, 321
de minimis and high survival exemptions, 191
Discard
- avoidance, 280–292
- ban, 5–8, 11, 13, 15, 16, 18, 20, 21, 30, 73, 82, 151, 153, 166, 180, 183, 184, 190, 215, 328, 334, 367, 377
- consumption, 152
- GIS platform, 241, 243–248
- management, 72, 170, 241
- plans, 22, 31, 34, 53, 55, 56, 73, 76, 77, 83, 165, 166, 170–174, 180, 185, 189, 206, 208, 384
- reduction, 18, 19, 21, 73, 84, 183, 240, 259, 368, 369

S. S. Uhlmann et al. (eds.), *The European Landing Obligation*,
https://doi.org/10.1007/978-3-030-03308-8

E
Economic repercussions, 110, 111, 124, 125
Electronic monitoring (EM), 8, 11, 41, 65, 78, 214, 221, 229, 252, 258, 259, 270, 314, 364, 366, 371–372, 401
Environmental DNA, 418–421
ε-Constraint approach, 349
EU landing obligation, 4–22, 72–84, 131, 183–186, 198, 280, 298, 299, 304, 312, 407–424
European Commission (EC), 7, 28–30, 36, 42, 51, 72, 73, 78, 101, 165, 166, 169–174, 180–185, 189, 334, 384
European Union (EU), 5, 16, 28, 34, 90–105, 167, 169, 170, 182, 220, 240, 254, 280, 320, 366, 367, 369, 377, 418

F
Fine-scale mapping, 368
Fish distributions, 271
Fisheries bycatch, 409
Fisheries Control, 78, 169, 211, 384
Fisheries governance, 82
Fisheries management, 13, 17, 21, 38, 53, 64, 76, 83, 105, 145, 166, 174, 181–183, 192, 227, 228, 230, 240, 243, 250, 251, 367, 369, 407, 422, 424
Fishers, 4, 28, 72, 91, 111, 166, 181, 202, 220, 240, 258, 280, 299, 320, 364, 384, 422
Fishing strategies, 243, 244, 320, 334, 343
Fish protein, 20, 326, 328, 336, 341, 349, 352, 355
Fish reactions to fishing gear, 286
Fish silage, 325, 415, 416, 418, 419
Fleet adjustment, 110
Flexibility, 9, 22, 30–32, 34, 36, 39–43, 54, 58, 64, 79, 82, 118, 121, 122, 124, 142, 185, 208, 214, 230, 289
Food subsidies, 152
Food web models, 130, 145
Full retention, 4

G
Genetics, 198, 200, 370, 407–424

H
Hidden cost of control, 226
High survival, 5, 14, 31–33, 37, 50–53, 64, 78, 79, 100, 169–171, 185, 191, 192, 240, 304, 309, 313, 315
Honest reporting, 220–232
Hot spot maps, 270
Hydrolysates, 337, 340, 341, 345, 349, 352, 357

I
Impacts, 7, 11, 15, 20, 28–30, 38, 50, 55, 58–65, 75–78, 82, 83, 91–105, 120, 125, 130–154, 166, 169, 187, 240, 250, 264, 307, 324, 349, 350, 368, 409, 414
Implementation, 5, 31, 50, 72, 90, 110, 146, 165, 180, 198, 220, 240, 263, 280, 299, 320, 334, 372, 383, 409
Industry participation, 74
Industry-science collaborations, 78, 288
Inland management, 348–350
Inspections, 12, 211, 213, 253, 368, 372, 384–386, 395, 396, 415
International Council for the Exploration of the Sea (ICES), 31, 35, 37, 43, 50, 51, 53, 58–60, 62, 64, 133, 151, 174, 199–201, 204, 207, 208, 211–214, 271, 282, 304, 309, 313, 348, 368

J
Joint deployment plans (JDPs), 384–386, 400
Joint recommendations (JR), 34, 51, 55, 61, 76, 83, 170–174, 189, 205, 206, 208, 214

L
Lack of trust, 226, 227, 231
Landing obligation (LO), 4–22, 29–35, 38–42, 50–53, 58–63, 65, 72–84, 90–105, 110, 151, 153, 165–175, 180–192, 198–215, 220–232, 240–255, 258, 280, 291, 311, 320–322, 328, 364, 370, 371, 377, 383–401, 409
Loss aversion, 224, 230, 231

M
Maps of discard probability, 241
Marine Spatial Planning (MSP), 240–255
Minimum conservation reference size (MCRS), 5, 35, 64, 110, 111, 113, 114, 116–119, 125, 126, 184, 201, 202, 206, 209, 211, 212, 220, 262, 269, 304, 311, 321–323, 343, 351, 386, 387, 392, 399, 401
Moral priming, 225
Multi-criteria decision analysis (MCDA), 341, 342, 356
Multi-level governance, 180–192

N
Nonmonetary incentives, 232, 288

O
Observers, 6, 59, 78, 102, 153, 175, 198, 221, 247, 258, 304, 369, 400
Omnibus regulation, 34, 40, 62, 166–169, 185, 191
Onboard handling, 319–322, 326, 329

P
Pelagic fish, 31, 34, 95, 101, 113, 150, 248–251, 287, 298–315, 385, 403
Purse seine, 7, 52, 95, 102, 241–243, 248–251, 255, 281, 298–315, 322, 370, 374

R
Regionalisation, 171, 180, 182, 190, 191, 204, 206, 385, 386
Risk assessment, 385, 400

S
Scavengers, 98, 104, 130–154
Scientific, Technical and Economic Committee for Fisheries (STECF), 28, 32, 34, 36, 38, 40, 50–56, 58, 60, 62–65, 79, 99, 168–170, 172–175, 211, 212, 309, 324, 400, 414
Selective fishing gears, 18, 185
Selectivity, 5, 35, 54, 75, 91, 110, 153, 166, 184, 198, 240, 258, 280, 320, 334, 364
Self-decision, 231
Self-selected groups, 231
Simulating fishing closures, 241, 243, 249–251
Small-scale fisheries (SSF), 15, 72, 90–105, 190
Southern Europe, 90–105, 312
Species composition, 133, 248, 269, 311, 314, 343, 372, 392, 414–419, 423
Species identification, 287, 366, 374, 407, 409, 410, 412, 413, 415, 416, 423
Stakeholder engagement, 72
Subsidiarity, 181–183, 191, 192

T
TACs and quotas, 125
Top-down policy, 22, 29, 30
Trawls, 12, 35, 52, 75, 100, 110, 132, 198, 222, 255, 258, 280, 307, 322, 348, 376, 420

U
Unreported discard, 394
Unwanted catches (UWC), 5, 6, 9, 17–21, 40, 42, 50, 53–56, 59, 64, 75, 76, 79, 84, 97, 98, 100, 105, 110, 112, 116, 118, 124, 166, 168, 172, 175, 180, 184, 185, 187, 192, 198, 202, 205, 208, 209, 211–214, 240–243, 254, 258, 264–272, 275, 276, 298, 311, 312, 320, 322, 324–326, 334–357, 364–378, 394, 395, 401, 408, 415
Utilisation, 4, 7, 9, 15, 17–22, 77, 114, 115, 118, 140, 322

V
Valorisation, 188, 211, 334–357
Vessel modifications, 328, 329
Video, 8, 11, 41, 78, 131, 229, 252, 270, 271, 366, 371–372, 374–376, 401
Voluntary compliance, 224–229, 231

www.ingramcontent.com/pod-product-compliance
Lightning Source LLC
LaVergne TN
LVHW082109200926
848484LV00008B/1931
* 9 7 8 1 0 1 3 2 7 3 6 9 8 *